T4-AQK-756

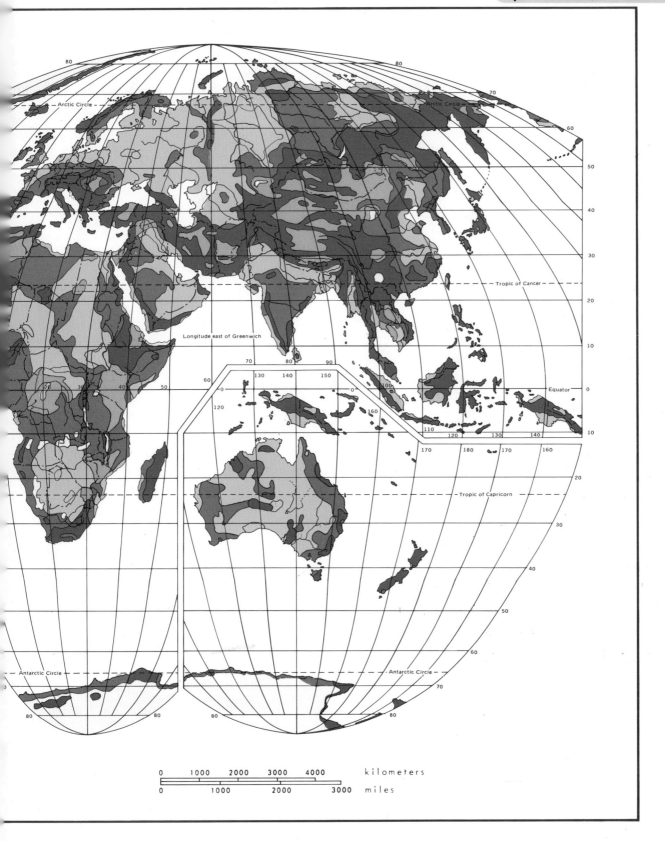

Arctic Circle

Arctic Circle

70

80   80   60

50

40

30

Tropic of Cancer

20

10

Longitude east of Greenwich

70   80   90

60   130   140   150   Equator   0

120   0   0   100

160

20   30   40   50   110

120   130   140   10

170   180   170   160

20

Tropic of Capricorn

30

40

50

60

Antarctic Circle   Antarctic Circle   70

80   80   80   80   80

0   1000   2000   3000   4000   kilometers

0   1000   2000   3000   miles

CONTEMPORARY PHYSICAL GEOGRAPHY

JOHN GABRIEL NAVARRA

Jersey City State College

SAUNDERS COLLEGE PUBLISHING

Philadelphia New York Chicago
San Francisco Montreal Toronto
London Sydney Tokyo Mexico City
Rio de Janeiro Madrid

Address orders to:
383 Madison Avenue
New York, NY 10017
Address editorial correspondence to:
West Washington Square
Philadelphia, PA   19105

This book was set in Caledonia by Hampton Graphics.
The editors were Don Jackson, Lloyd Black, Sally Kusch and Carolyn Newbergh.
The art director and cover designer was Nancy E. J. Grossman.
The text was designed by Betty Binns Graphics.
The production manager was Tom O'Connor.
New artwork was drawn by John Hackmaster, Dimitri Karetnikov, Mario Neves,
    Linda Savalli, Linda Clark, Tony Crocetto and Suanne Riehm.
The printer was Hampshire Press.

Front cover photo: Tidbinilla Range in Australia with eucalyptus in left foreground. (John Navarra)

Back cover photo: Edisto River in South Carolina. (South Carolina Industrial Development Commission)

LIBRARY OF CONGRESS

CATALOG CARD NO.:   80-53922

Navarra, John G.
    Contemporary physical geography.

Philadelphia, Pa.:   Saunders College
160 p.
8101   801010

CONTEMPORARY PHYSICAL GEOGRAPHY   ISBN   0-03-057859-0

© 1981 by CBS College Publishing.   Copyright under the International Copyright Union.   All rights reserved.   This book is protected by copyright.   No part of it may be reproduced, stored in a retrieval system, or transmitted in any form or by any means, electronic, mechanical, photocopying, recording, or otherwise, without written permission from the publisher.   Made in the United States of America.   Library of Congress catalog card number 80-53922.

1234 146 987654321

CBS COLLEGE PUBLISHING
Saunders College Publishing
Holt, Rinehart and Winston
The Dryden Press

Preface

This book is an introduction to the study of physical geography. It is written for the future teachers, lawyers, journalists, artists, political leaders, and philosophers who recognize the importance of such information in their development as articulate literate citizens. For a person who wishes to continue in the field of physical geography, this book and the course it serves are an initial step toward more advanced studies.

Geography is a comprehensive subject. It deals with the whole Earth in general and with many aspects of the Earth and its regions in particular. But, the essence of geography is the study of spatial relations and the development of theory regarding the space relations on the face of the Earth.

Physical geography—as a branch of geography—is a very special kind of study; it deals with our curiosity as to how the natural world is organized. As such, physical geography shares with other sciences a concern for a common arena—the Earth's surface. But physical geography looks at that arena from the viewpoint of humanity; that is, physical geographers are concerned with the Earth as the environment for humans—an environment that influences how humans live and organize themselves. Physical geographers also pay attention to the fact that humans, themselves, have helped to modify and to build the environment.

Physical geography, then, as developed within the pages of this book, involves the investigation of (1) surface features produced by the form, structure, and dynamics of the Earth, (2) the climates of the Earth, and (3) the interaction of these two to modify each other and to influence the soils and vegetation of the Earth. The underlying theme in this book is "dynamics through time"; my objective is to guide you through the study of a time-space continuum of climate-landforms-vegetation-soils. Thus, we will concern ourselves with the features that characterize the Earth's surface, with the soils that are derived from that surface, with the climate that regulates the life cycle of plants, with freshwater and saltwater on the land and in the seas, and with the carpet of vegetation that covers large areas of this Earth.

Maps and charts, which are two-dimensional graphic representations of the spatial distribution of selected phenomena, are the best means we have of recording, tracking, interpreting, and analyzing the time–space continuum of physical geography. Thus, I have selected and used maps to display data in such a way that you can readily assimilate and understand the information. A special presentation on mapping and charting data can be found in Appendix VI. This information is presented so you can familiarize yourself with the design and language of various map projections used throughout the book.

Important features—designed to be helpful to you—are built into this book: The outline at the end of each chapter gives you a capsule view of the text's organization and the salient principles being developed. Within the body of the text, when significant concepts are introduced, they are set in boldface type; these terms are then gathered together and further identified in a section on the language of physical geography at the end of the chapter. A complete glossary and nine appendices are available for ready reference. The self-testing review questions for each chapter can be used to assess whether you understand the important concepts, facts, and explanations. The master list of reading suggestions was developed from current magazines and periodicals that are readily accessible in most libraries; you may select from among these articles to pursue your individual interest.

Many colleagues, friends, and associates have contributed to the development of this book. All of the chapters have been fully and meticulously reviewed, and I am grateful to James Bingham (Western Kentucky University), Orville Gab (South Dakota University), Don L. Gary (Nicholls State University), H. D. Hays (University of Alabama), Richard Jarvis (State University of New York at Buffalo), Arthur Limbird (University of Calgary), Alexander G. McLellan (University of Waterloo), Albert Melvin (Trenton State College), Richard Riess (Salem State College), Stephen Sandlin (San Bernardino State College), and Marshall Stevenson (Towson State University). I am also indebted for research assistance to Gloria Navarra (First Boston Corporation), John G. Navarra, Jr. (Seton Hall University), and L. Michael Treadwell (South Carolina Industrial Development Commission). My friends at Saunders College Publishing are among the best in the profession, and I am indebted particularly to Don Jackson (publisher), Lloyd Black (developmental editor), Nancy Grossman (art director), Tom O'Connor (production manager), Sally Kusch (project editor) and Randall Grossman and Linda Peirce (make-up and illustration processing). Finally, John Hackmaster (senior illustrator) and his group—Dimitri Karetnikov, Mario Neves, Linda Savalli, Linda Clark, Tony Crocetto, and Suanne Riehm—and George Laurie (manager), and Pat Morrison (assistant) rendered all details of the illustration program with a special creative flair; and, I am deeply appreciative.

JOHN GABRIEL NAVARRA

Contents

CONTEMPORARY PHYSICAL GEOGRAPHY

PART ONE

THE EARTH IN SPACE AND TIME

Physical geographers have never been bound by the somewhat arbitrary nature of most subject matter divisions. They undertake the study of all phenomena and events associated with the development of spatial patterns at the Earth's surface. Our fundamental objective as we embark on this study is to deal with locational concepts and to sensitize you to the richness and variety of patterns found on the Earth's surface. Spatial relations will be examined through description and classification of places as well as through the examination of theory that gets at underlying causes for the arrangement of these spatial patterns.

A proper account and examination of the spatial variation at the Earth's surface must of necessity be based on the interrelationship of the Earth with the larger space of the Universe. Thus, our initial steps in this first part will place the Earth in its cosmic setting and clarify some essential facts about its evolution and motions as a planet. Then, an examination of three spatial patterns—the world ocean, the gaseous envelope of the atmosphere, and the distribution of water above and below the land surface—establishes some basic facts that are useful for the study of the time-space continuum of climate-landforms-vegetation-soils set out in parts 2, 3, and 4.

A. The cavern called *Lost World* in West Virginia was sculptured below ground by the relentless flow and work of subterranean water. The giant stalagmite formations were built over millions of years by persistent drops of water. (Department of Tourism, West Virginia)

B. The levees along the Mississippi River are worked on constantly to make sure the mightly flow is contained within the river's banks. (Army Corps of Engineers)

C. The water that drops from the clouds onto the distant mountains drains into the Humboldt Sink of Pershing County, Nevada, from which it has no outlet to the sea. (Nevada Magazine)

D. The water of the North Atlantic surrounds Prince Edward Island, Canada, and works its will on the land. (Department of Tourism, Prince Edward Island)

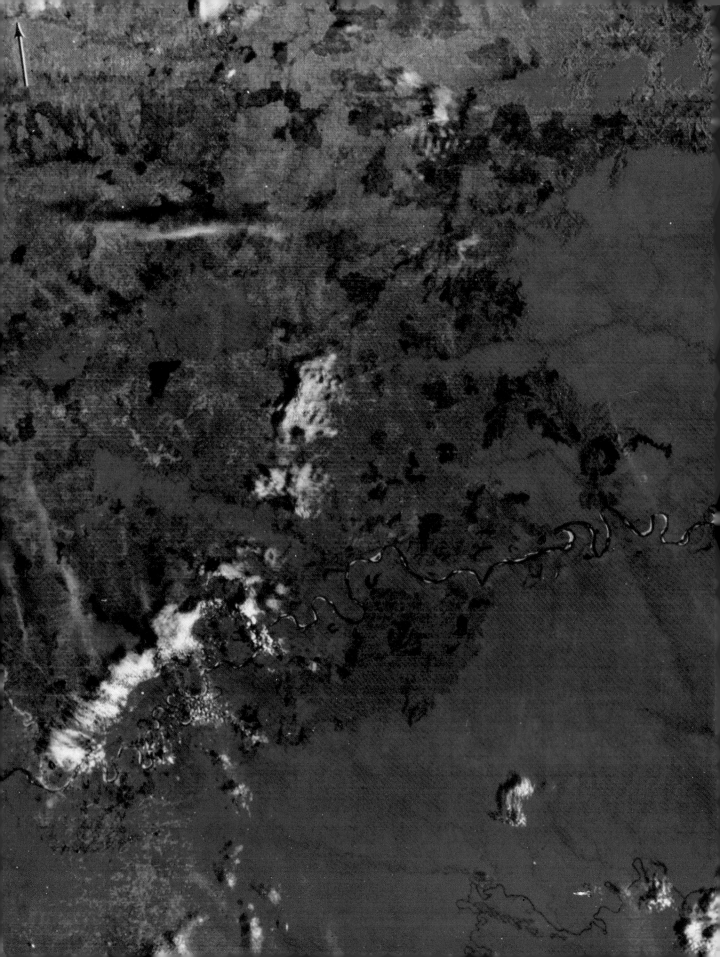

1 The Earth: a spherical platform in space

Even a casual observer recognizes the Earth is not alone and isolated in space. We are surrounded by a multitude of objects including, quite obviously, stars and planets. The Sun, which dominates our daytime sky, and the Moon, which dominates our nighttime sky, are daily reminders that the Earth is an interrelated part of the Universe.

Prior to 1957, before the first satellite was launched, our information about the space around and beyond our planet was quite limited; our only observational platform was the spherical surface of the Earth. The development of artificial orbiting-satellites and remote-sensing deep-space probes has provided us with new platforms from which to make observations and gather data. Most of us, however, are still confined to the surface of our planet; our personal view of events is of necessity from the vantage point of the Earth's spherical platform.

An untutored observer on this platform of ours sees the heavens move across the sky and believes the Earth is stationary. Most of us recognize that this is not the case; it is, however, our understanding of the Earth's rotation that helps us to put our observation of the heavens in perspective. What we "see" is conditioned by what we understand!

This platform of ours is undergoing some very complicated movements; many of our concepts—time and the seasons, for example—are related to these motions. Proper interpretation of the many and varied observations we make depends on an understanding of the essential characteristics of our sphere, the nature of its evolution, and the nature of its motions. We will begin this orientation to our spherical platform by placing the Earth in its cosmic setting and tracing the chronology of its evolution.

Data pertaining to the Earth as a planet are of relatively recent vintage, which is difficult for most of us to comprehend because the concepts seem so "obvious" and widely understood. But the fact of the Earth's **rotation**, which is taught in the primary school of today, was published

Today we use satellites to gather information about the space beyond our Earth, and we also use satellites to get a better perspective and view of the Earth's surface. Sunlight consists of electromagnetic radiation in the visible region and in the near-visible infrared region. Different objects reflect different relative proportions of each wavelength. Vegetation, for example, reflects proportionately more shortwave infrared radiation than does bare soil. This Landsat-satellite photo of the broad basin of the Colombian interior was taken at an altitude of 568 miles (914 kilometers). Bogota lies about 100 miles (160 kilometers) northwest of the upper left corner of this image (see arrow). The blue portion of the photo is a grassland with widely scattered trees, called a *savanna;* the grasses of the savanna were burnt by the natives to remove weeds, destroy insects, and add ash to the soil. The burnt-over savanna gives way to a rain-forest region, which stands out in red. (NASA)

as a highly speculative theory by a Polish scientist, Nicholas Copernicus, in 1543; and it was not until the nineteenth century that any proofs of the theory independent of the motions of the heavenly bodies were available. Copernicus's sixteenth-century theory broke with the conventional wisdom in another way, too; he suggested that the Earth is revolving around the Sun. Again, the concept of **revolution** is not very revolutionary to us; but there was quite a time span between this theoretical proposal and the proof—the proof of this motion of revolution was not available until 1725. In Table 1–1 you will find a summary of commonly accepted data—distance from the Sun, size, motion, mass, and satellites—for the Earth and its companion planets.

EVOLUTION OF THE PLANETARY SYSTEM

A **nebula** is an immense, diffuse body, or cloud, of interstellar gas and dust. About 4.6 billion years ago a great nebula that occupied space along one of the curved arms of the Milky Way Galaxy—that is, the cluster of stars in which we find ourselves today—began to contract (Fig. 1–1a). The great cloud of gas and dust collapsed and spun rapidly; it eventually took the shape of a disk, with a massive body accreting at the center of the disk (Fig. 1–1b). The central mass became so dense and hot that its nuclear fuel,

hydrogen, ignited and it became a star—our Sun (Fig. 1–1c).

As the **proto-star**, which we call our Sun, began to form, some particles of gas and dust were left behind at the edge of the contracting cloud. Thus, the proto-star was surrounded by large numbers of dust grains and gas atoms that were in orbit around the central contracting cloud. When nuclear reactions began in the core of the proto-star, the orbiting dust grains and gas atoms condensed to form planets. This event, which occurred about 4.6 billion years ago, established the Sun and its family of planets—Mercury, Venus, Earth, Mars, Jupiter, Saturn, Uranus, Neptune, Pluto, and possibly a tenth planet. All satellites—more than 40 moons in the case of our planetary system—formed at the same time and in the same way, that is, as small condensations around their parent planets.

EVOLUTION OF THE EARTH

The current most popular hypothesis of the Earth's origin infers a gradual condensation and accretion to a solid planet as the Earth swept up enormous quantities of small particles from the nebular disk. Thus, the developing **proto-planet,** which we know as Earth, was an aggregate composed largely of hydrogen and helium gases and dust, with some solid bodies resembling the various kinds of meteorites. The surface of the young planet was pockmarked

TABLE 1–1   PLANETARY-DATA

PLANETS	MEAN DISTANCE FROM SUN Miles (In Millions)	Km	SIDEREAL REVOLUTION PERIOD (In Earth-Years*)	SIDEREAL ROTATION PERIOD (In Earth Hours Or Earth-Days)	MASS (Earth's Mass = 1)	EQUATORIAL DIAMETER Miles	Km	AVERAGE DENSITY (Water = 1)	NUMBER OF SATELLITES
Mercury	36.0	57.9	0.241	58^d16^h	0.055	3,025	4,865	5.46	0
Venus	67.2	108.1	0.615	243^d	0.815	7,525	12,110	5.23	0
Earth	92.9	149.5	1.000	$23^h56^m04^s$	1.000	7,927	12,756	5.52	1
Mars	141.5	227.8	1.880	$24^h37^m23^s$	0.107	4,218	6,788	3.93	2
Jupiter	483.4	778.0	11.860	9^h50^m-9^h55^m	318.000	88,700	143,000	1.33	15
Saturn	886.7	1427.0	29.460	10^h14^m-10^h38^m	95.200	75,100	121,000	0.69	15
Uranus	1782.0	2869.0	84.010	10^h49^m	14.600	29,200	47,000	1.56	5
Neptune	2794.0	4497.0	164.800	16^h	17.300	31,650	50,900	1.54	2
Pluto	3664.0	5900.0	247.700	$6^d9^h17^m$	0.200	1,600	2,700	1.80	1

*One sidereal earth-year is equal to 365.26 days. Multiple 365.26 by 0.241 to get 88 days, which is the sidereal period of revolution for Mercury expressed in earth-days.

FIGURE 1–1   The development of the planetary system according to the nebular hypothesis: (a) a fine cloud of gas and dust is in the process of contracting, (b) most of the cloud's mass is in the process of condensing to form a central body—a proto-star with dust grains and gas in orbit around it, (c) consolidated accumulations of debris, called *planets*, are revolving around a true star.

by the infall of millions of rocks from surrounding space. Its accretion atmosphere was rich in hydrogen and the noble gases—helium, neon, and argon.

As the mass of the Earth increased during some tens of millions of years, it began to heat up. The increase in the proto-planet's temperature was the result of the combined effect of the gravitational infall of its mass, the impact of meteorites on its surface, and the heat generated by the decay of radioactive elements such as uranium and thorium. Eventually the interior of the Earth became molten and melted.

The melting of its interior caused a vast reorganization of the entire mass of the Earth. For example, molten drops of iron and nickel sank to the center of the planet. Today the Earth has a molten iron-nickel **core** that remains largely liquid in the outer regions. As vast quantities of iron and nickel sank, other elements floated upward to form what we now call the **mantle,** which is a molten layer of rock-forming material, and the **crust,** which is the solidified zone of rock. Thus, the Earth was differentiated into a core, a mantle, and a crust (Fig. 1–2).

Two important effects of the heating of the Earth's interior were the inception of volcanic eruptions and mountain building. Both of these activities contributed significantly to the shaping of the Earth's surface.

The Earth's accretion atmosphere was swept away by the solar wind, that is, an outpouring of particles from the Sun. The heating of the Earth's interior, however, caused gases to begin leaking to its surface in a process called **outgassing.** The gases had been locked in the materials of the Earth when those materials originally accreted. During the period of reorganization and differentiation of the interior, gases leaked to the surface in tremendous volumes. The outgassing included carbon dioxide, methane, water, and gases containing sulfur. The Earth's gravity was strong enough to prevent all but the lightest gases from escaping into space.

Between 3.7 and 2.2 billion years ago the surface of the Earth cooled. Gaseous water condensed from the dense atmosphere and fell to the surface as rain. During this period the world ocean developed. The process of erosion by wind and water began to operate in much the same way that it does today. Gases dissolved in rainwater combined chemically with elements such as calcium and magnesium that were leached from surface rocks. Liquid water became the medium for transporting and redistributing the debris of erosion. River systems developed on the surface of the Earth and carried eroded material to the ocean.

GEOCHRONOLOGY

To maintain chronological order in analyzing the development of the Earth's landscape and environment, scientists use a standard time scale, which serves as a framework for discussion. This scale is referred to as the geologic **timetable.** It is divided into major divisions, called **eras;** subdivisions of the eras are known as **periods;** and divisions of the periods are called **epochs.** (Table 1–2).

Various dating procedures make it possible to establish the age of fossils and rocks with a high degree of certainty. One of these procedures is based on the radioactive decay of an isotope of uranium that has an atomic weight of 238—referred to as **uranium-238.** The radio-

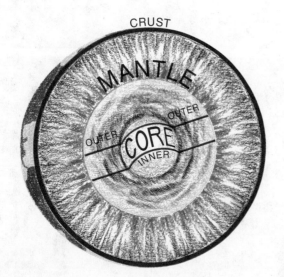

FIGURE 1–2   The geophysical evidence indicates the Earth is differentiated into core, mantle, and crustal zones.

Volcanic eruptions have been occurring for more
than 4 billion years of the Earth's history.

The Mayon volcano *(upper photo)*, with its peak
at 8077 feet (2462 meters) is a perfect cone-
shaped volcano that rises from the Bicol region on
the Philippine island of Luzon. Its 1861 eruption
buried an entire town; only the church belfry
in the left foreground survived that event.

The Taal volcano *(lower photo)* is a volcanic
island surrounded by a lake, which is within the
crater of a larger volcano. This volcano, which
was photographed emitting a large jet of steam, is
located south of Manila on Luzon. (Republic of
the Philippines)

TABLE 1–2   GEOLOGIC TIMETABLE[1]

ERAS	PERIODS	EPOCHS	PERIOD BEGAN—ERA BEGAN YEARS BEFORE PRESENT	
	Quaternary[2]	Holocene Pleistocene Pliocene Miocene Oligocene Eocene	2,000,000	
Cenozoic (Recent life)	Tertiary[2]	Paleocene	65,000,000	65,000,000
	Cretaceous		136,000,000	
	Jurassic		190,000,000	
Mesozoic (Medieval life)	Triassic		225,000,000	225,000,000
	Permian		280,000,000	
	Carboniferous[3]		345,000,000	
	Devonian		395,000,000	
	Silurian		435,000,000	
	Ordovician		500,000,000	
Paleozoic (Ancient life)	Cambrian		570,000,000	570,000,000
Proterozoic (Primitive life)				1,500,000,000
Archeozoic (Beginning life)				3,000,000,000
Azoic (Before life)				4,700,000,000

[1]Sources: Based on the time scale published by the Geological Society of London in 1964 and the Geological Timetable compiled by F. Van Eysinga in 1975.

[2]In one scheme, the Cenozoic era is divided into two periods: Tertiary and Quaternary. In another scheme, the Cenozoic is divided into two periods referred to as the Paleogene period and the Neogene period. The Paleogene started 65 million years ago and is said to span the Paleocene, Eocene, and Oligocene epochs. The Neogene period—in this second scheme of dating—is placed as starting 26 million years ago and spans the Miocene, Pliocene, Pleistocene, and Holocene epochs.

[3]The Carboniferous is regarded as a single system by scientists outside of North America. Geologists in the United States generally regard the Carboniferous as consisting of two periods: the Mississippian (beginning 345 million years before the present) and the Pennsylvanian (beginning 325 million years ago).

active decay of uranium-238 occurs at a steady, measureable rate; for example, one-half of a given amount of uranium-238 will disintegrate to lead-206 in 4.5 billion years. It is this fact that makes this uranium isotope usable as a way to establish the age of a rock. The figure 4.5 billion years is referred to as the *half-life of uranium-238*.

Uranium is contained in more than 100 minerals. Thus, many rocks can be accurately dated by determining the relative amounts of uranium-238 and lead-206 in a sample. The older the rock, the less uranium-238 and the more lead-206 it will contain. The age of the sample is fixed by getting the ratio of lead-206 to uranium-238 and relating this ratio to the half-life of the uranium isotope.

Other chemical procedures are also used to date fossils and rocks. Among these are the potassium-argon method, lead-alpha method using zircon, a carbon-14 method, and a rubi-dium-strontium method. These procedures are too technical for us to consider the details at this point. But remember this: Scientists have made great strides in determining the absolute ages of rocks and fossils and the dates when various events occurred.

Refer to Table 1–2. The first three eras—**Azoic, Archeozoic,** and **Proterozoic**—are usually grouped together and referred to as the *Pre-cambrian era.* The first period of the **Paleozoic** era is, as you will note, the Cambrian. The **Mesozoic** and **Cenozoic** complete the list of eras.

The names of the periods have been derived in a variety of ways: The Latin name for Wales, for example, is *Cambria.* The strata of this period were originally studied in Wales and, thus, the name *Cambrian* was assigned. The word *Ordovician* is taken from the name of an ancient tribe that resided in Wales, the Ordo-vices. *Jurassic* is derived from the Jura Moun-

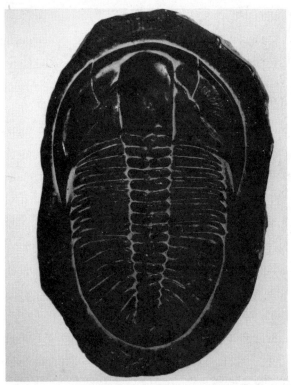

The imprint in this piece of shale rock from British Columbia is called a *fossil.* **The outline that you see is that of a trilobite—the name given to a group of sea animals that lived during the Paleozoic era. The trilobite that became the fossil imprint in the sea-floor mud died during the Cambrian period of the Paleozoic.** (Larry D'Andrea)

tains of France and Switzerland while Cretaceous comes from the Latin word for chalk, *creta.*

In the chapters that follow, I will indicate the dating of an event by simply referring to an era or period. By referring to Table 1–2 you will be able to fix the event in the chronology of the Earth's history and to get an idea of how long ago the event occurred.

WANDERING CONTINENTS

It was not too long ago that most people thought of the Earth as a rigid object with fixed continents and ocean basins. During the 1960s, most scientists began to think of the Earth's outer shell, or **lithosphere,** as a mosaic of perhaps 20 great segments, known as **plates.** The outlines of the 20 plates are sketched in Figure 1–3. The continents—raftlike inclusions embedded in these plates—travel slowly and ponderously across the face of the Earth.

A single, all-encompassing theory, called *plate tectonics,* dominates modern scientific thought and is used to explain the origin and position of our present continents as well as the positions and characteristics of mountain systems, deep-sea trenches, and ocean basins. According to this theory the continents originated from one large supercontinent by a process of fragmentation and drifting. The hypothetical **supercontinent,** referred to as **Pangaea,** may have had an outline similar to that in Figure 1–4. Pangaea is postulated to have existed 225 million years ago at the beginning of the Mesozoic era, also known as the Age of Reptiles.

THE GATHERING EVIDENCE
When you glance at a globe of the Earth, the bulging coastlines of South America and Africa seem to be stretching toward one another across the Atlantic. It appears as though these continents were once part of some gigantic jigsaw puzzle that has come apart. In fact, the scientific evidence that has been accumulating supports this conclusion; in 1965, a British group at Cambridge University used a computer to demonstrate the fit of South America and Africa. Computer matching of the contours revealed that there is a nearly precise fit along the Atlantic coastlines of these continents at depths of 500 fathoms (3000 ft, 914 m). The fit of South America against Africa (Fig. 1–5) occurs along the continental slopes of these two continents, where they fall away into the deep ocean basins.

Two years after the Cambridge group detailed the location of the South American-African fit, investigators at the Massachusetts Institute of Technology and the University of São Paulo in Brazil went a step further. In 1967, they used radioactive-isotope dating to compare the rock strata on the east coast of South America with those on the west coast of Africa along the "contour fit" made by the Cambridge group. The data show the ages of the rock strata match with a precision that makes the "fit" very convincing.

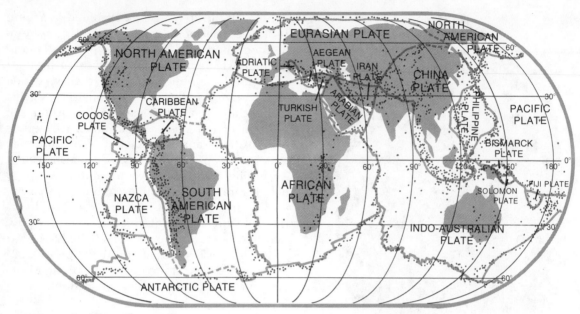

FIGURE 1–3   The outline of 20 principal tectonic plates are superimposed on this map of the world.

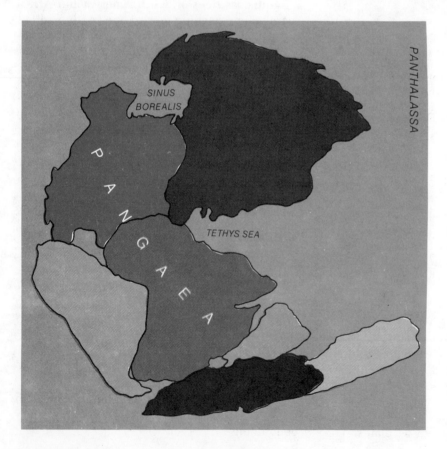

FIGURE 1–4   The Tethys and Sinus Borealis were seas in the world ocean, Panthalassa, which surrounded the supercontinent of Pangaea.

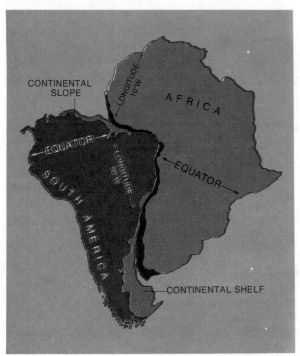

FIGURE 1–5   The bulging Brazilian coastline of South America fits snugly into the indentation of Africa's Ivory Coast. The contour fit is very precise along the continental slopes of these continents. Two reference points—the Equator and a line of longitude—marked on each landmass will give you some idea of the change in orientation and the drift that has occurred since the separation, which took place about 150 million years ago.

SEA-FLOOR SPREADING

Evidence supporting the existence of a supercontinent had been accumulating for a long time prior to the discovery of the contour and age fit of South America and Africa. Nineteenth-century scientists, for example, had noted that fossils found in European coal fields were closely related to fossils found in the eastern United States; this, of course, is an interesting relationship that points in the direction of a single continent. But the clearest and most direct evidence to support the concept of a fragmented supercontinent came from the extensive exploration of the sea floor, which began in the late 1940s.

A mid-ocean ridge was known to exist in the Atlantic Ocean since the time the first trans-atlantic cable was laid about a century ago. The extent of the Mid-Atlantic Ridge was not suspected, however, until scientists began using the new and more sophisticated instruments developed during World War II to sound the sea floor. By the early 1950s the charts showed the Mid-Atlantic Ridge to be a part of the longest mountain range on Earth—about 40,000 miles (64,400 km) long.

A large crack, called a **rift,** runs the length of the ridge in its center. Measurements made during the 1950s indicate that hot, molten rock is rising from the mantle and forcing its way to the surface of the sea floor through the rift in the Mid-Atlantic Ridge (Fig. 1–6). To accommodate the new material, the older lithosphere on either side moves away from the rift. The fact that the plates are growing and spreading as new rock material is added to them at the rift was confirmed by assessing the ages of the rocks on the sea floor.

The rocks making up the Atlantic's floor are very young in the neighborhood of the Mid-Atlantic Ridge; they become progressively older as the distance from the ridge increases. The oldest rocks are found close to the continental boundaries. Furthermore, no rocks older than 150 million years are found anywhere in the

A variety of sophisticated techniques are used to probe, explore, and photograph the sea floor. (U.S. Navy)

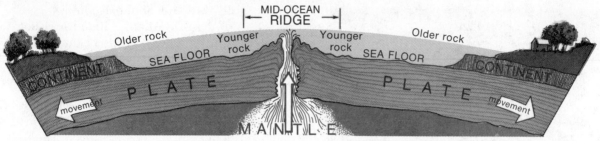

FIGURE 1–6   The plates on either side of a mid-ocean ridge are forced to move in opposite directions as a result of a spreading sea floor that grows by additions of molten material rising through the rift of the ridge. The sea floor rocks on either side of a ridge become progressively older as the distance from the ridge increases.

Atlantic between South America and Africa, which indicates this part of the Atlantic Ocean did not exist prior to that time. The information on sea-floor spreading proves conclusively that Africa and South America were a single landmass with no ocean between them until 150 million years ago.

The Mid-Atlantic Ridge (Fig. 1–7) is almost precisely in the middle of the Atlantic Ocean, which is precisely why the prefix *mid* is used in its name. The ridge extends southward part way around the Cape of Good Hope at the southern tip of Africa and then veers northeastward into the Indian Ocean. At about the

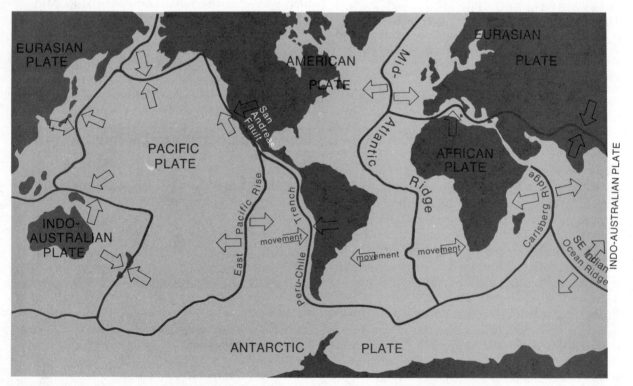

FIGURE 1–7   Six principal plates were proposed by Le Pichon in 1968. The plates are growing along spreading ridges—indicated by arrows moving in opposite directions, that is, diverging. Plate material is being consumed in trenches, where the paired arrows are converging.

middle of the Indian Ocean the ridge splits. One branch of the ridge, the Carlsberg Ridge, extends northward to the mouth of the Gulf of Aden. The other branch, the Southeast Indian Ocean Ridge, extends southeastward. All along these ridges, as the arrows in Figure 1–7 indicate, the plates are being forced apart by the formation of new sea floor.

PLATE TECTONICS
This theory about the structure and behavior, or *tectonics,* of the Earth's plates is the foundation for today's thinking and serves as a unifying theme for observation and study of the Earth. In broad outline, the modern view of **plate tectonics** considers the Earth's lithosphere to be divided into blocks or segments that are drifting in various directions with respect to one another. Mountains are raised along plate boundaries that are in collision. New crust that forms along an ocean ridge drifts away on either side of the ridge and sinks into the mantle of an ocean deep, or trench.

As you read more about the subject in scientific as well as popular periodicals, you may find that various authors refer to different numbers of plates. In a previous paragraph in this chapter, for example, the lithosphere's 20 plates were referred to, and they are: Pacific, Antarctic, Nazca, Cocos, Caribbean, North American, South American, Adriatic, Aegean, Turkish, Arabian, Iran, Eurasian, African, China, Philippine, Bismarck, Solomon, Fiji, and Indo-Australian plates. Review for a moment the outlines of these plates as sketched in Figure 1–3.

The scale at which a person considers the data will condition the number of plates identified. In 1968, for example, Xavier Le Pichon suggested that the strong brittle lithosphere consists of six large blocks or plates. The six major plates identified by Le Pichon are outlined in Figure 1–7 as the Pacific, American, African, Eurasian, Indo-Australian, and Antarctic plates.

These large blocks can include both oceanic and continental crust. Compare Le Pichon's six plates with the 20 plates identified in Figure 1–3. Le Pichon's American Plate, for example, includes the continental crust of both North America and South America as well as the

oceanic crust of the western Atlantic extending eastward to the Mid-Atlantic Ridge. In the sketch that outlines 20 plates you will find the North American, Caribbean, and South American plates occupying the same area as Le Pichon's American Plate. It is apparent that some of the large plates identified by Le Pichon include smaller plates. The smaller plates were probably created by a breakup of the large plates during collisions.

PLATE BOUNDARIES
Three kinds of plate boundaries have been identified: (1) mid-ocean ridges, (2) trenches, that is, zones where crust is being carried downward into the Earth's interior, and (3) faults, that is, fractures which separate crustal blocks as they slide past one another. These boundaries are characterized as being either divergent, convergent, or shear.

The Mid-Atlantic Ridge is a typical example of a divergent junction (Fig. 1–6). This divergent boundary is the point of contact between the American plates, on the one side, and the Eurasian and African plates, on the other. With this type of junction the continental zones on either side are forced to drift apart. The term sea-floor spreading describes the divergence of plates away from mid-ocean ridges as new crust is added along the common boundary at the ridge.

A convergent junction develops when two plates move toward each other and collide. As the plates collide, one slips and plunges below the other, and the boundary is marked by a trench. The lithosphere of the plunging plate is subducted, or pushed down, into the Earth's deep mantle. The plane in which the collision occurs is called a *subduction zone.* The sediments and rocks within the plunging plate are squeezed, subjected to great pressure, and heated. Over a period of millions of years these rocks are driven into the deep mantle where they are absorbed and reincorporated into the mantle's general circulation.

The Peru-Chile deep-sea trench is produced by a subduction zone that runs along most of the west coast of South America (Fig. 1–8). This subduction zone along with its accompanying trench is the longest in the world—running

Mountains and plateaus are produced on the South American Plate as a result of convergence. A plateau of Bolivia stretches before us; the plateau region lies between two great parallel ranges of the Andes: the Andes of Chile to the west and the Andes of Bolivia to the east. This relatively flat-floored depression starts at Lake Titicaca in the north and runs 500 miles (800 kilometers) to the south; it is 80 miles (130 kilometers) wide, with an elevation of about 12,000 feet (3700 meters). Lake Poopó (latitude 18°45′S, longitude 67°W) occupies a very shallow depression in the plateau and is nowhere more than 15 feet (4.5 meters) deep; this salt lake covers about 1000 square miles (2600 square kilometers) at low stage. The only outlet from Lake Poopó moves under the sand and empties in Salar de Coipasa, which is a salt-encrusted marsh with a small permanent body of water in its lowest part. The Salar de Uyuni is a totally arid windswept salt flat that covers 3500 square miles (9000 square kilometers). (NASA)

about 10,800 miles (17,380 km). The subduction rate—that is, the rate at which the plunging plate is being consumed—is about 3.7 inches (9.4 cm) per year. The South American Plate along its western boundary is referred to as the *overriding plate* because the Nazca Plate plunges under it. Continental crust has a lower density than oceanic crust; thus, in a collision of plates the "lighter" crust tends to become the overriding plate while the denser crust becomes

the plunging plate. In this type of convergence the overriding plate's edge may crumple into mountains; and, in fact, this is happening along the western boundary of the South American Plate where we find the Andes Mountains (Fig. 1–8).

The third kind of plate boundary—a shear junction—exists where two plates slide past one another. The sliding takes place along a deep, linear, essentially vertical fracture, which is referred to as a *fault*. Where a shear boundary exists the plates are neither added to nor consumed. The San Andreas Fault in California is a shear boundary between two plates (Fig. 1–7). The Pacific Plate is on the west side of the fault; the North American Plate is on the east side. The fault zone, or rift, cuts obliquely across the California Coast Ranges from the ocean at Point Arena on the north through the south-

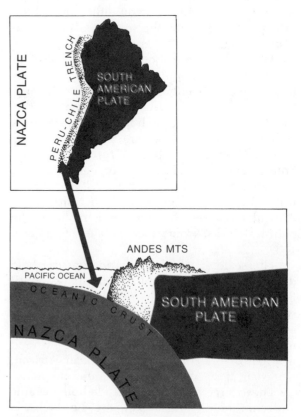

FIGURE 1–8   This is an example of a convergent junction, or boundary, between two plates. The Nazca Plate is plunging under the South American Plate.

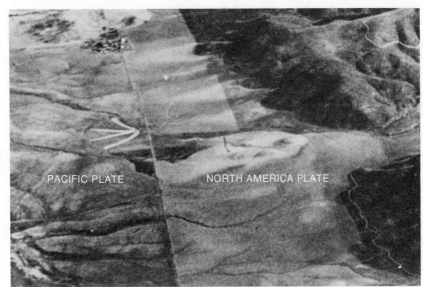

PACIFIC PLATE   NORTH AMERICA PLATE

The arrow on the ground at the left in this aerial photo points to the San Andreas Fault. The fault appears as a line, which runs in a south-easterly direction. The land to the left of the fault is on the Pacific Plate; the flat land and mountains to the right are on the North American Plate. (U.S. Geological Survey)

western corner of San Francisco and on to the southeast of San Bernardino (Fig. 1–9). The main strand of the fault is lost in the deep deposits of the Colorado Desert of California. The fault is traceable for some 650 miles (almost 1050 km) across California. Obviously, then, California's coastal edge is in the Pacific Plate; the rest of the state rides the North American Plate.

The Pacific Plate is grinding and sliding past the North American Plate in a north-westerly direction. Los Angeles rides the Pacific Plate; most of San Francisco is on the North American Plate, which is moving to the west. Recent satellite measurements indicate that Los Angeles and San Francisco will be side by side in 10 million years!

DEFINING POSITION ON THE EARTH

The Earth is round; but it is not perfectly round. The Earth is actually somewhat flattened at its poles, which makes its equatorial diameter about 27 miles (43 km) greater than its polar diameter. Because of this polar flattening, the Earth is referred to as a **spheroid,** which simply means a less-than-perfect sphere.

Since spatial relationships are fundamental to geography, geographers use a well-established system for locating points on the Earth's surface. The position of a point on the Earth is defined by latitude, longitude, and elevation—or at times by depth below sea level. Latitude and longitude are expressed by angular coordinates; elevation and depth are usually given as vertical distances.

LATITUDE

The **latitude** of a point on the surface of the Earth is its angular distance from the Equator, which is expressed in degrees, minutes, and seconds (Fig. 1–10). It is, in reality, an arc of the Earth's surface that runs from the Equator toward either pole; the angle swept out by the arc is measured at the center of the Earth. Latitude ranges from 0° at the Equator to 90°N at the North Pole, or to 90°S at the South Pole.

The small arc, Arc 1, depicted on the Earth's surface in Figure 1–10, sweeps out an angle of 20°. The latitude of point A is reported as 20°N. The large arc, Arc 2, extends from the Equator to the North Pole along the surface of the Earth

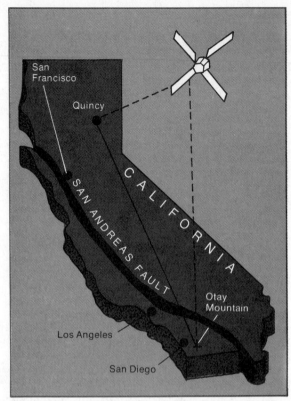

FIGURE 1–9 The San Andreas Fault is an example of a shear boundary between two plates. Los Angeles and San Diego are riding on the Pacific Plate; they are being carried in a northwesterly direction. The actual movement along the fault is currently being measured by satellite.

(Fig. 1–10). It sweeps out an angle of 90°. The latitude of the North Pole is reported as 90°N.

The length of a degree of latitude, measured along a meridian,* is everywhere the same on a perfect sphere. For navigational purposes the Earth is treated as a perfect sphere. One degree of latitude, from the Equator to the poles, is said to be equal to 60 nautical miles, and 1 minute of latitude is equal to 1 nautical mile.†

*The geographic grid is a network of intersecting lines inscribed on a globe. The north–south lines connecting the geographic poles are called **meridians**. The east–west lines running parallel to the Equator are called **parallels**.
†The **nautical mile** is defined as the length of one minute of arc along the Earth's Equator; 1 nautical mile = 1.1508 statute miles (1.852 kilometers). The statute mile, commonly used on land, is equal to 5280 feet; the international nautical mile, always used in navigation, is equal to 6076.1 feet. The nautical mile is 15 percent longer than the statute mile.

If you wish to be very precise, however, you must take into account the polar flattening, or oblateness, of the Earth. And, as a precise fact, the length of a degree of latitude does change as you move north or south from the Equator. The length of 1 degree of latitude at the Equator is 59.7 nautical miles; at the poles it is 60.3 nautical miles. Compare these figures; 1 degree of latitude at the poles is approximately 1 percent longer than at the Equator.

The most common and simplest way to determine latitude is by means of a sextant, which is used to measure the altitude of the Sun at noon. When this information is determined, a simple formula is used to convert the observed altitude of the Sun into the actual latitude of the location at which the observation is made.

LONGITUDE

The **longitude** of a place is its angular distance east or west from a standard reference point on

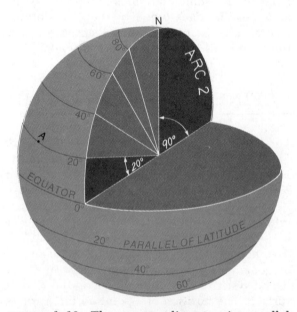

FIGURE 1–10 The east-west lines running parallel to the Equator are known as *parallels of latitude*; they are circles that girdle the Earth. In principle, parallels can be drawn on a globe. The angular distance of any parallel of latitude from the Equator is measured at the center of the Earth.

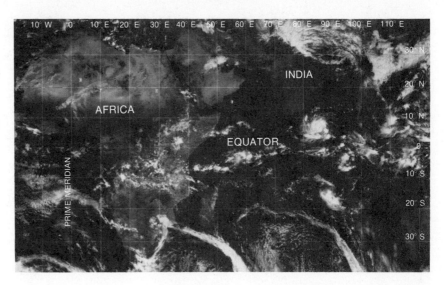

Africa, the Middle East, India, Southeast Asia, and the western portion of Australia are visible in this satellite photo. The latitude-longitude grid that is printed on the photo helps to locate particular areas of interest. For example, there is a well developed typhoon off the southeast tip of India between the Equator and 10°N latitude. The west-to-east stretch of this storm is found between 80°E and 90°E longitude. (NASA)

the Earth's surface. Internationally the most generally accepted reference point for longitude is the Royal Observatory at Greenwich, England. The angular distance from Greenwich to any point on the Earth's surface is measured from

0° to 180° east or west at either end of the axis of rotation and anywhere in between (Fig. 1–11).

The concept of longitude is based on imaginary planes that run from the North Pole to the South Pole. Since these planes run from pole to pole, they pass through the Earth's axis of rotation. The lines that result from the intersection of the Earth's surface by these imaginary planes are referred to as *meridians*. When you determine longitude, you are actually measuring the angle between meridian planes at their common junction along the axis of rotation. In Figure 1–11 the angle between the Greenwich meridian and the meridian of point C is 20°, which is reported as 20°E.

Longitude is usually found with the aid of a chronometer, a very accurate clock, which is set to Greenwich mean time (GMT). Longitude is computed by a formula: Local longitude is equal to 15 times the difference between GMT and the local solar time. When the Sun is at its highest point in the sky and casts its shortest shadow, it is noon. If the chronometer reads 5:00 P.M. GMT when local solar time is noon, then local longitude computed according to the formula is 75°W.

In Table 1–3 the nautical mile equivalent of 1° of longitude at various locations on the Earth's surface is catalogued. The length of a degree of longitude, measured along a parallel of the geographic grid, decreases as you move from the Equator toward the poles (Fig. 1–12).

FIGURE 1–11 The Equator is a great circle girdling the Earth exactly midway between the geographic poles. The meridians cut the Equator at right angles. The meridian that passes through Greenwich, England, is called the *prime meridian*. In principle, meridians can be drawn at any desired angular distance from the prime meridian.

TABLE 1–3 LENGTH OF DEGREES OF LONGITUDE

LOCATION OF POSITION (In Latitude Degrees)	LENGTH OF 1° OF LONGITUDE (In Nautical Miles)
0 ←———————— Equator ————————→	60.1
5	59.9
10	59.2
15	58.1
20	56.5
23.5 ←———————— Tropic of { Cancer / Capricorn } ————————→	55.1
25	54.5
30	52.1
35	49.3
40	46.1
45	42.6
50	38.7
55	34.6
60	30.1
65	25.5
66.5 ←———————— { Arctic / Antarctic } Circle ————————→	24.0
70	20.6
75	15.6
80	10.5
85	5.3
90 ←———————— { North / South } Pole ————————→	0.0

On the Equator, for example, 1 degree of longitude is equal to 60 nautical miles; at latitude 60°, 1 degree of longitude is equal to 30 nautical miles. At either of the poles, 1 degree of longitude has a linear distance of zero.

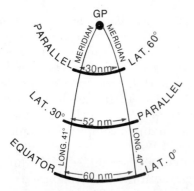

FIGURE 1–12 The length of 1 degree of longitude measured along a parallel in nautical miles (nm) decreases as you move from the Equator toward a geographic pole (GP).

USING SATELLITES

In Figure 1–13 the geographic grid has been placed on the globe; the parallel and meridian that run through St. John, New Brunswick, Canada, are identified. St John is located at latitude 45°16′N, longitude 66°3′W. Ordinary methods for determining latitude and longitude require the observation of the Sun, or stars. A number of sightings must be made and the sky must be sufficiently clear to allow unobscured observation. In many ways the ordinary methods are cumbersome, and inaccuracies can creep into the calculations.

In January 1964, a new system for determining latitude and longitude became operational. The U.S. Navy Navigation Satellite System, called NAVSAT, is a highly accurate, all-weather, worldwide navigational system that can be used by surface ships, submarines, or aircraft in flight.

In NAVSAT the navigational satellite is placed in a circular polar orbit about 600 nautical miles above the Earth. The satellite orbits the Earth in 105 minutes and broadcasts data every 2 minutes. The radio signals transmitted by the

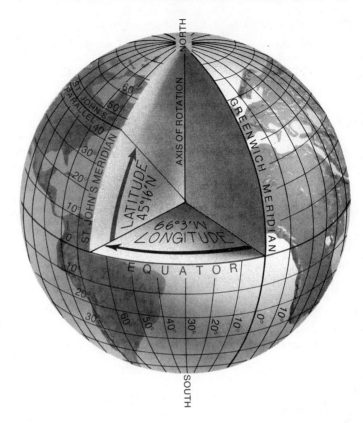

FIGURE 1–13 A network of intersecting lines, called the *geographic grid,* is inscribed on this globe. The north-south lines connecting the geographic poles are known as *meridians.* The east-west lines running parallel to the Equator are called *parallels.* St. John's parallel is 66°3′W of the Greenwich meridian.

satellite are interpreted by an onboard computer to give an immediate read-out of the latitude and longitude of the receiving station.

THE SEA-LEVEL CONCEPT

Landmasses are elevations on the spheroidal figure of the Earth. Sea bottoms are depressions in this same geometrical figure. Measurements of elevations and depressions are made with reference to the level of water in the ocean basins, called ideal **sea level.**

Of course, near shore the actual sea level is far from ideal since it varies with every high tide and low tide. In and around shore areas the difference in level from high tide to low tide can be significant. Over any 24-hour period, however, the variations in the level of the sea surface in the open ocean are very small—generally less than 3 feet (1 m). Thus, soundings taken far from shore can be referred to the actual sea surface with minimal error because inaccuracies in the sounding equipment will be greater than the difference between the ideal and actual level of the sea.

The depth of water bordering a coastal zone must be ascertained quite accurately because a shoal, or shallow area, is a significant hazard to navigation. In coastal waters soundings are made with respect to a single reference point, which is usually indicated on the navigational chart for the area. Along the coasts of the United States the reference point is mean low water, that is, average low tide; in fact, worldwide, most charts have low water as the reference point for the documented depths.

AN EARTH IN MOTION

This spherical platform of ours is not stationary; it travels through space and moves in a number of directions—all at the same time. Two commonly known motions are the Earth's rotation on its axis and its revolution around the Sun; but, in addition to rotation and revolution there are other more complicated and less commonly understood motions:

■ In this expanding Universe of ours, the galaxies are flying apart. The Milky Way is moving in the direction of the Andromeda galaxy at a velocity of 180 miles (290 km) per second. The Earth, as a part of the Milky Way, is being carried along toward Andromeda.

■ The Sun is revolving around the center of the Milky Way and the Earth is making the trip, too.

■ The Sun is moving in a random way toward a bright star called *Vega*; the Earth is also being swept along this random path. In fact, the Sun's motion toward Vega actually causes the Earth's path of revolution around the Sun to take the shape of a spiral (Fig. 1–14).

A ROTATING EARTH

When time-exposure photographs of the heavens are made from different points on the Earth, the star trails recorded on film do not have the same pattern. A photo taken at the Equator has the stars moving in a series of straight lines across the film. Photographs made at either pole show the stars forming circular paths directly above the camera. It is more than obvious that stars cannot move in straight lines and in circular paths at the same time. Today, we recognize that the spinning of the Earth around its **polar axis**—that is, rotation—produces these visual effects.

A TILTED AXIS
The Earth's **axis** of rotation maintains a rather constant orientation in relation to surrounding space—that is, the axis points almost directly

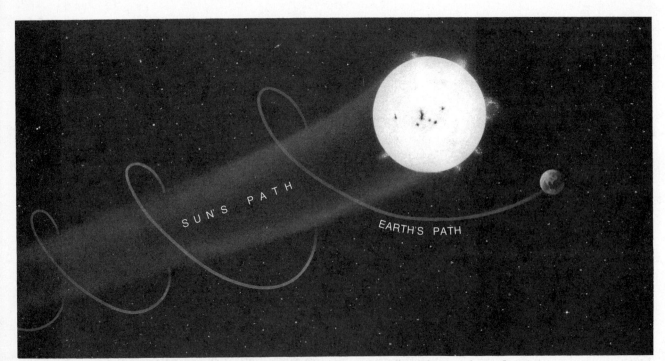

FIGURE 1–14 The Earth's orbit around the Sun takes the shape of an elongated spiral as a result of the Sun's random motion toward Vega.

(Left) Polar star trails. (Lick Observatory)
(Right) Equatorial star trails. (Lick Observatory)

toward Polaris, a star in the constellation Ursa Minor. Polaris is, in fact, called the **polestar.**

The Earth, as you know, revolves around the Sun; the plane in which the Earth orbits around the Sun is called the plane of the **ecliptic.** At the present time,* with respect to the plane of the ecliptic, the Earth's axis has an inclination from the vertical of 23°27′, generally reported as 23.5° (Fig. 1–15). The tilt of the Earth's axis produces a number of phenomena—different elevations of the Sun at high noon on June 22 and on December 22, for example.

DIRECTION OF ROTATION

The direction in which the Earth rotates can be described in a number of ways, depending on the reference point selected. For example, a person in space looking down on the Earth's North Pole sees the Earth rotating in a counter-clockwise direction. On the other hand, the rotation is observed as being clockwise when the observer in space is peering down on the Earth's South Pole. If a compass orientation is used as the point of reference, then the Earth's rotation is described as being from west to east, that is, having an eastward motion.

Viewed from our spherical platform the Sun, Moon, and stars seem to rise above the eastern horizon, travel westward across our sky, and then set below the Earth's western horizon. These apparent motions are simply the

---

*The tilt of the Earth's axis changes with time. For a discussion of this and related phenomena, see pages 135–137.

result of the rotation of the Earth. It is the eastward rotation of the Earth that makes the stars and other heavenly objects appear to rise in the east, wander westward, and then sink out of sight in the west.

PERIOD OF ROTATION

The Earth completes one rotation on its axis in a period of 24 hours, which we refer to as a **day.** In order to define one complete rotation of the Earth on its axis with precision, however, a point of reference is needed. The Sun or some distant star can be used as the reference point.

The best reference point is a star that is at a great distance from the Earth and appears as a true point—that is, an object with position but no appreciable dimension. One complete Earth-rotation of 360°, defined by using a distant star as the reference point, is referred to as a **sidereal day** (Fig. 1–16).

During a 24-hour sidereal day the Earth moves through approximately 1 degree of its orbit around the Sun. In Figure 1–16 this orbital motion is depicted as a movement from position 1 to position 2—the angles in this diagram are, of course, greatly exaggerated. Since the star is far away, the movement of the Earth in its orbit does not affect the sighting of the star on the next day. When the Sun is used as the point of reference, this is not the case; a distortion is

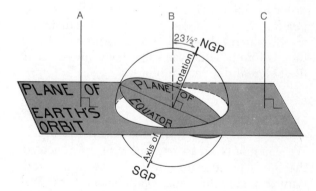

FIGURE 1–15 The geographic poles (NGP and SGP) are at either end of the Earth's axis of rotation. The axis of rotation is perpendicular to the equatorial plane. Lines A, B, and C have been erected so they are perpendicular to the Earth's orbital plane around the Sun; the Earth's axis has an inclination of 23.5° with respect to these lines.

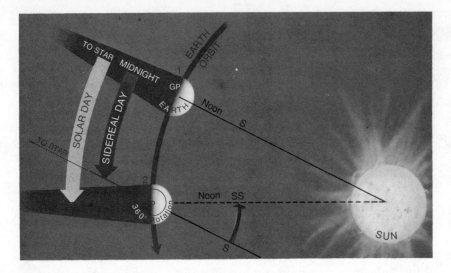

FIGURE 1–16 One complete rotation of 360°, defined by using a distant star as a reference point, is referred to as a *sidereal day.* Time measured with respect to the Sun's position gives us the solar day. The solar day is slightly longer than the sidereal day.

produced because the Sun is much closer to the Earth than is the star. Thus, a complete rotation of the Earth with reference to the Sun is not the same as a complete rotation with respect to the distant star.

Refer to Figure 1–16 and examine the situation in some detail: At position 1, line ZS passes through the Earth's axis of rotation and extends from the center of the distant star (off to the left) to the center of the Sun. By the time the Earth arrives at position 2 in its orbit around the Sun, line ZS has rotated throught 360° and is again pointing to the center of the distant star; but in order for ZS to point to the center of the Sun in position 2, ZS must rotate the additional distance to SS, which is equivalent to 1 degree or 4 minutes of time. Thus, the solar day is longer than the sidereal day by 4 minutes.

DIVIDING THE DAY

We commonly measure time with reference to the Sun's position in our sky. Time measured with respect to the Sun is called **solar time.** We divide the solar day into 24 segments, called **hours;** this is not a very old way of marking time, however. In fact, the day was not divided into 24 hours until A.D. 1600.

Sunrise and sunset were the only times of day used by the Romans of 500 B.C. This two-part division of the day worked very well for them because their activities were simple and they did not need more exact ways of keeping time. Eventually more sophisticated Romans gave the name **noon** to that moment of the day when the Sun is highest in the sky, and they used noon as a third time of day. As the business activity of the Roman Empire grew, a need for more time periods developed; by the time Christ was born, the Roman day had been divided into five periods.

The five-period day of the Romans served the world until A.D. 605 when Pope Sabinianus of the Roman Catholic Church added two more hours, which divided the day into seven periods—known as the seven canonical hours. This seven-part division of the day was the only form of time used for centuries in Europe. The canonical hours were also used aboard ships; and, in 1492, the ships of Columbus used an hourglass filled with sand to mark the seven canonical hours.

Our modern concept of a 24-hour day is built with two meridians in mind: a noon meridian and a midnight meridian. The noon meridian, a north-south line extending from pole to pole, is located where the Sun's rays strike the Earth's surface at the greatest possible angle; of course, there is a continuous westward shift of the noon meridian as the Earth rotates eastward. The midnight meridian is positioned on the other side of the Earth—directly opposite the noon meridian; it, too, is continuously shifting westward. The two meridians are separated by 180° of longitude.

The unit called the *day* is defined as the time it takes the Earth to complete one rotation of 360°. If we divide this sweep of 360° into 24 segments of equal size, we end up with units called *hours*, which are equivalent to 15° of longitude. A simple calculation indicates the Earth rotates at the rate of 15° of longitude per hour.

Our day begins at midnight. The noon meridian separates the forenoon (A.M.) and afternoon (P.M.) of the same day.* The dividing line between one day and the next is at the midnight meridian.

STANDARD TIME

About 100 years ago every town in the United States kept its own local time. A central point in the town was selected; and all the clocks were set to stay in step with the Sun, that is, to read 12:00M.* when the Sun was directly over the selected point.

Local time was an adequate concept for people within a community who were essentially isolated from the world beyond the town's borders. Travelers, of course, found that their watches were out of step when they arrived in a new town. If you traveled west, your watch was fast. If you traveled east, your watch was slow.

There were more than 300 local times used in the United States during the 1860s. The rapid building of railroads during this period brought the problem of time to a focus because a railroad must run on a schedule so people know when the trains will arrive and depart. Since it was impossible to make a schedule for each local time, the railroads set their own time, which became known as *railroad time*. The system of railroad time was used from 1860 to 1883. The Pennsylvania Railroad, for example, used Philadelphia time, which was 5 minutes slower than New York's local time and 5 minutes

Meridies (M.) is the Latin word for noon. The letters A.M. stand for ante meridiem, which means before noon. The letters P.M. stand for post meridiem, which means afternoon.

faster than Baltimore's local time, to schedule all the trains in its system.

England was one of the first countries to face the problem of each town keeping its own local time. In 1828, Sir John Herschel, a famous British scientist, proposed that England establish one time for the whole country. The British put Herschel's plan into operation during 1850; they used the local Sun time of Greenwich for all of England, Scotland, and Wales. Greenwich time became known as **standard time.**

The United Kingdom is a relatively small country compared to the United States. Local clocks were not that much out of step with the Sun when Greenwich time was used throughout Britain. Special planning was needed, however, to make standard time work in the United States; a plan was devised and put into operation in 1883. The main idea of the plan was to establish standard time zones and have all places within a zone keep the same time.

The seven standard time zones that span the 50 states of the United States are identified in Figure 1–17. The Sun is at its noon position over 105°W longitude in the Mountain Time Zone. The longitude of the standard meridians for each zone are: Eastern, 75°W; Central, 90°W; Mountain, 105°W; Pacific, 120°W; Yukon, 135°W; Alaska-Hawaii, 150°W; Bering, 165°W. The local time on each of these meridians is used as the standard time for the whole zone.

The boundaries of the four zones that span our 48-contiguous states and most of southern Canada are shown in Figure 1–18. Natural boundaries as well as state and county boundaries were followed in establishing the zones. Generally but not always, entire states were kept in one time zone; for example, all of Texas was placed in the Central Time Zone; but South Dakota, Kansas, and Nebraska have a time-zone boundary moving through their central sections.

WORLD TIME ZONES

An international meeting, called the Prime Meridian Conference, was held at Washington, D.C., in 1884. Twenty-six nations sent repre-

We are at the Royal Observatory Building, Greenwich, England. The Prime Meridian is marked as a line between the cobblestone walk on either side. We are facing south so the area of east longitude is to our right. As the Earth rotates the rays of the Sun will, at one moment in time, pass through the open roof area and fall on the line of the Prime Meridian; at that moment it is noon at Greenwich. (British Tourist Authority)

sentatives to consider establishing a system of standard time zones throughout the world. They were successful and, today, the whole world is divided into 24 time zones (Fig. 1–19).

The meridian running through Greenwich, England, is the reference meridian; it is assigned 0° and is known as the *prime meridian.* The standard meridians are 15 longitude degrees apart, starting from Greenwich. The time zones in Figure 1–19 are numbered 1 through 12. To distinguish whether the zone lies east or west of Greenwich a plus (+) or minus (−) sign is used. Time is designated as being fast— that is, plus—for all places east of Greenwich and slow—that is, minus—for all places west of Greenwich.

The Eastern Standard Time Zone of the United States is zone number−5 in Figure 1–19. This means that there is a 5-hour difference in time between the standard meridian, 75°W, of this zone and the Greenwich meridian. The minus sign means the time is slow; it is slow by 5 hours. If you add the number of the zone to the actual standard time in the zone, you will have Greenwich time at that moment. For example at 7:00 A.M. eastern standard time (EST), it is 12:00 M. Greenwich mean time (GMT).

In 1968, England, Scotland, Wales, and Northern Ireland went off Greenwich time and joined Western Europe by adopting the time of Europe's standard meridian, which is 15°E. Thus, although the prime meridian runs through Greenwich, England, the United Kingdom was shown as belonging to zone +1 on maps during the late 1960s and early 1970s. This procedure was dropped during the 1970s for a variety of

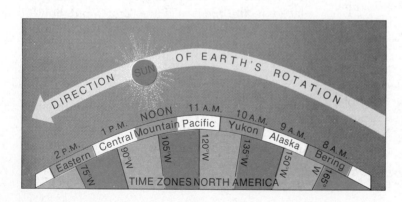

FIGURE 1–17 We are at the North Pole facing south—east is to your left and west is to your right. The highest angle rays of the Sun are falling at longitude 105°W. As the Earth rotates, 15 degrees of longitude pass beneath the Sun every hour. Each unit of 15 degrees marks off one time zone.

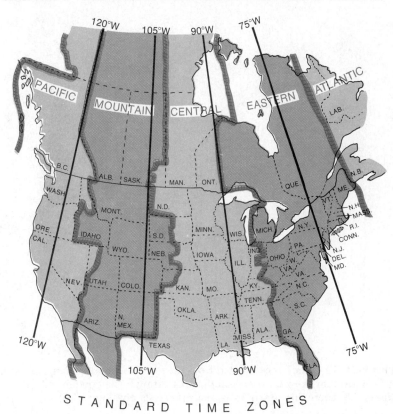

FIGURE 1–18 The clocks throughout a time zone are set to the local solar time of the standard meridian for the zone. The clocks in western Texas, for example, are set to the solar time of the meridian at 90°W.

STANDARD TIME ZONES

reasons and today the United Kingdom is on Greenwich time.

Some countries—Iran, Pakistan, India, and Australia, for example—have their standard time fast or slow by some multiple of one-half hour (Fig. 1–19). The countries have adopted, as an internal policy within their own borders, a zoning every 7.5°, giving time intervals to one-half hour. Saudi Arabia keeps Arabic time, with watches set to midnight at every sunset.

DAYLIGHT SAVING TIME

Benjamin Franklin had an opinion on almost every topic and a suggestion for the solution of almost every human concern. He was, for example, among the first to suggest that we set our activities according to the Sun—our natural timepiece. Franklin noted that human activities, especially in urban areas, start well after sunrise and continue long after sunset. He proposed that we set clocks ahead during the summer months so people would get up closer to sunrise and enjoy the greatest span of daylight.

Dr. Franklin's suggestion was ignored for more than 100 years. Then during World War I the idea was revived. Franklin's system, called **daylight saving time,** was used effectively from 1914 to 1918 to save some of the fuel that would have ordinarily been used to produce light for after-dark activities.

Daylight saving time takes advantage of the sunlight hours to the greatest extent possible. Today most communities in the United States keep daylight saving time—that is, standard time plus 1 hour—during spring and summer. The clock is set ahead 1 hour at the beginning of spring and set back 1 hour at the beginning of autumn. In fact, many countries use daylight saving time throughout the year. The Sun rises and sets 1 hour later by the clock when daylight saving time is used.

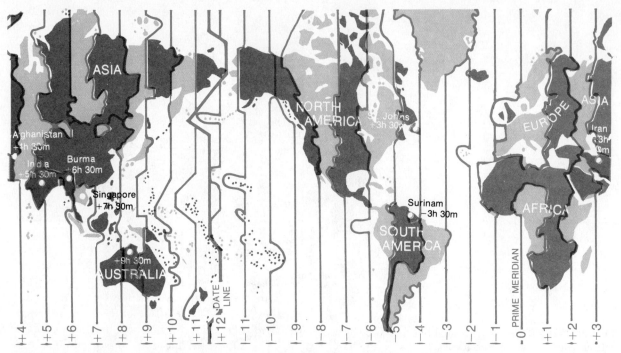

FIGURE 1–19 Time is calculated from the Greenwich meridian, called the *prime meridian.* The middle of the zero time zone passes through Greenwich with its east and west limits 7.5° each side. Each 15° zone represents one hour of time.

THE INTERNATIONAL DATE LINE

Ferdinand Magellan led one of the most daring expeditions of the sixteenth century. He did not complete the intended journey; but one of his five ships and some members of his crew did. These survivors of the Magellan expedition were the first to circumnavigate the Earth and the first ever to lose a day!

Magellan's journey started on August 10, 1519. He sailed west, south through the Atlantic Ocean, around the tip of South America, and across the Pacific Ocean to the Philippines where he was killed on the island of Mactan. One of Magellan's ships finally crossed the Indian Ocean, rounded the Cape of Good Hope, and arrived at the Cape Verde Islands in the Atlantic. The crew had kept careful records throughout the trip and the log showed the day to be Wednesday; much to their surprise, they learned the day was Thursday! They finally arrived in the harbor of Seville, Spain, on Monday, September 8, 1522; but by their records it was only Sunday, September 7.

East and west longitude meet at the 180th meridian, called the **international date line** (Fig. 1–19). If you move westward from Greenwich, the time at 180°W longitude is 12 hours slow. If you move eastward from Greenwich, the time at 180°E longitude is 12 hours fast. The difference in time between 12 hours slow and 12 hours fast is 24 hours—which is a full day. The calendar day on the west side of the international date line is one day ahead of that on the east.

The international date line falls in a very ideal location in the Pacific Ocean. (Fig. 1–19). The date line by agreement has been made somewhat irregular so as not to cut through groups of islands and through Alaska. At this date line the calendar is changed by one day. For example, as you move from west to east across the international date line, you are advancing your time—you gain a day. Move from east to west across the line and you lose a day. If you board a plane in Hong Kong on Thursday and fly eastward, it will be Wednesday when you land in Hawaii a few hours later. You can live through Thursday again.

A REVOLVING EARTH

Although the true motion of the Earth around the Sun is a spiral (page 22), it is much too difficult to visualize and analyze the Earth's relative position with respect to the Sun using this concept. In order to simplify the picture, the motion of the Sun toward Vega is "wiped-out" and the path of the Earth around the Sun is plotted in two dimensions, which causes it to take the shape of an ellipse. The elliptical shape of the Earth's orbit in Figure 1–20 has been greatly exaggerated in order to illustrate the various relationships and positions within the very restricted limits of the page.

Figure 1–20 shows the Earth at perihelion, that is, closest to the Sun, on January 3. The distance to the Sun at perihelion is 91.5 million miles (147.3 million km). At aphelion, the most distant position, on July 4, the Earth is separated from the Sun by 94.5 million miles (152 million km). The average distance over the circumference of the orbit is 93 million miles (149.7 million km).

Although the distance from the December 5 to the January 3 position along the circumference of the ellipse in Figure 1–20 is greater than the distance from the June 6 to the July 4 position, the Earth moves through each distance in approximately the same length of time. The explanation, of course, is that the Earth travels at a greater average velocity during December than during June. In fact, the velocity of the Earth in its orbit around the Sun is not constant; it changes in a natural rhythmic cycle. For example, at perihelion, the Earth's velocity is at a maximum; then, orbital velocity decreases as the Earth moves toward aphelion, where it is at a minimum; the latter half of the cycle begins when the Earth's orbital velocity increases as it moves from aphelion toward perihelion. The average orbital velocity of the Earth is 18.5 miles (29.8 km) per second.

THE CIRCLE OF ILLUMINATION
The surface intersection of a plane passed through a sphere's center is called a **great circle**.

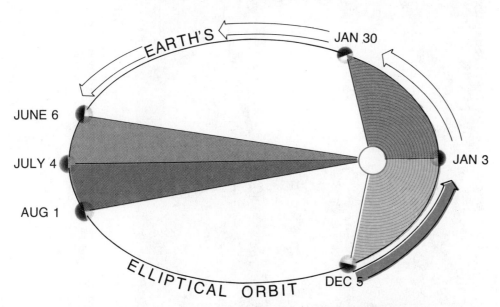

FIGURE 1–20 The velocity of Earth in its orbit around the Sun changes in a natural rhythmic cycle. The Earth's velocity is at a maximum when it is closest to the Sun.

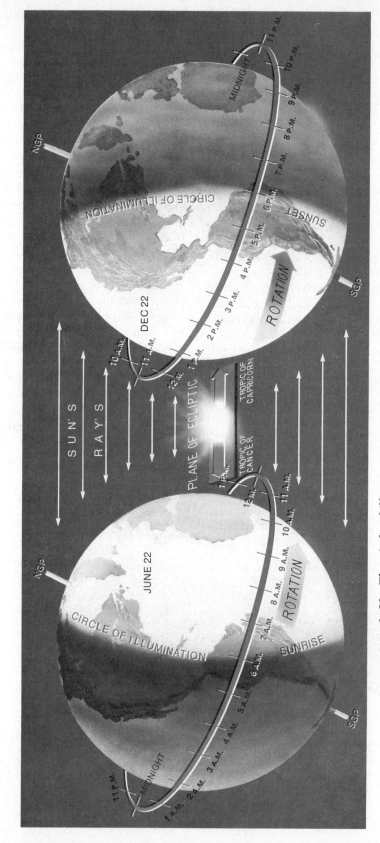

FIGURE 1-22 The circle of illumination is a great circle. The plane that creates this great circle is, at all times, perpendicular to the plane of the ecliptic. Note the contrasting positions assumed by the circle of illumination during June and December.

A plane passed through the center of a sphere cuts the sphere exactly in half. In Figure 1–21 the plane used to create the great circle was passed through the center of the sphere from pole to pole. A plane can be passed through the center of a sphere in a number of ways; for example, a plane passed through the Earth's center perpendicular to its axis of rotation creates the great circle that is known as the *Equator.*

As the Earth revolves around the Sun, half the Earth's surface receives solar radiation, called **insolation,** and half is in shadow. The **circle of illumination,** which marks the division between the half receiving insolation and the half in shadow, is a great circle (Fig. 1–22). The plane of this great circle of illumination passes through the Earth's center and is, at all times, perpendicular to the plane of the ecliptic and the Sun's rays.

ENERGY INPUT

The fact that the axis of the Earth has an inclination of 23.5° and maintains a constant orientation toward Polaris causes the circle of illumination to assume different positions as the Earth revolves around the Sun. At two times during the

FIGURE 1–21 The great circle is a line on the surface of a sphere that divides the sphere into two equal parts; it is formed by the intersection of the sphere's surface with any plane passing through the sphere's center. All meridians are great circles.

year, that is, the autumnal and vernal equinoxes,* the circle of illumination passes through the poles. At summer and winter solstices,† the circle of illumination is tangent to the parallels of latitude at 66.5°N and S—the Arctic and the Antarctic circles, respectively (Fig. 1–22).

At autumnal and vernal equinoxes, there are 12 hours of daylight and 12 hours of darkness at all points on the Earth's surface; the noon altitude of the Sun is 90° at the Equator, 66.5° at the Tropic of Cancer and Tropic of Capricorn, 23.5° at the Arctic Circle and the Antarctic Circle, and 0° at the poles. The greatest input of energy during these periods is at the Equator because the noon altitude of the Sun decreases as you move north and south of the Equator until it diminishes to 0° at the poles.

The summer solstice occurs on or about June 22 (Fig. 1–22). At the summer solstice, there are approximately 24 hours of daylight at the Arctic Circle, 14 hours on the Tropic of Cancer, 12 hours on the Equator, 10 hours of daylight on the Tropic of Capricorn, and 0 hours on the Antarctic Circle. The noon altitude of the Sun is 90° at the Tropic of Cancer (23.5°N), 47° at the Arctic Circle, 23.5° at the North Pole, 66.5° at the Equator, 43° at the Tropic of Capricorn, and exactly on the horizon at the Antarctic Circle. The greatest input of solar energy per unit of time is at 23.5°N; the input of energy in any given unit of time decreases as you move north and south of this latitude.

The winter solstice occurs on or about December 22 (Fig. 1–22). At the winter solstice the noon altitude of the Sun is 90° at the Tropic of Capricorn (23.5°S), 47° at the Antarctic Circle, 23.5° at the South Pole, 66.5° at the Equator, 43° at the Tropic of Cancer, and exactly on the horizon at the Arctic Circle. The greatest input of energy during a given unit of time occurs at the Tropic of Capricorn. The approximate length of daylight hours is 24 on the Antarctic Circle, 14

*Equinox refers to either of the two times each year when the Sun crosses the Equator and day and night are everywhere of equal length, being about March 21 (vernal equinox) and September 23 (autumnal equinox).
†Solstice refers to either of two times each year when the Sun is at its greatest distance from the celestial equator; about June 22 when the Northern Hemisphere is tipped toward the Sun, and December 22 when the Northern Hemisphere is tipped away from the Sun.

on the Tropic of Capricorn, 12 on the Equator, 10 on the Tropic of Cancer, and 0 on the Arctic Circle.

THE SEASONS

Summer, fall, winter, and spring are the orderly progression of the seasons in each hemisphere. The amount of solar energy received over the whole Earth is essentially constant from day to day. The heating effectiveness of the insolation at any moment, however, depends upon latitude and the length of the daylight hours.

OUR CALENDAR

The **calendar** we use today has a long, long history. Its origin goes back to ancient Rome of 738 B.C. when a calendar was begun by Romulus, the lengendary founder of the city. Martius, named for Mars, the god of war, was the first month of the Romulus calendar. The first day of the Romulus year, called the *vernal equinox,* was the beginning of spring. The ancient Romans could easily identify this period because of the many changes taking place in nature; it is, as you know, the warm-up and green-up time in the Northern Hemisphere.

The second month of the Romulus year was *Aprilis,* which means to open. Aprilis was the month in which the Roman world bloomed as it came out of the cold time of winter. The third month of the Romulus calendar was named for the goddess of growth, *Maius.* The fourth month, *Junius,* was named for youth, a period when we think of things in full bloom. There were six more months in the Romulus calendar; each, beginning with the fifth, was given a numerical name: the fifth, *Quintilus;* the sixth, *Sextilis;* the seventh, *September;* the eighth, *October;* the ninth, *November;* and the tenth, *December.*

The Romulus calendar had only 304 days for a 10-month period, which did not bring the Romans full cycle to the beginning of spring. They had an additional 61 days to wait until the vernal equinox arrived again. The days-of-waiting, which were during the deepest part of winter in the Roman provinces, were not counted and were called *useless.*

Numa Pompilius, the ruler who followed Romulus, gave the Romans a new calendar based on the Moon. In 700 B.C. he added two new months, Januarius and Februarius, to make a year of 12 months. Each month of this new calendar began with the new Moon and had an average of 29.5 days, which means the 12 months only accounted for 355 days and the calendar was still out of step with the year of the seasons.

In 46 B.C. Julius Caesar ordered a change in the calendar. The work of changing the Roman calendar was placed in the hands of Sosigenes, an Egyptian. The calendar Sosigenes created for Julius Caesar, called the *Julian calendar,* kept the 12 months of the Roman calendar but it was based on a tropical year—that is, a solar year*

*A *solar year* is the time it takes the Sun to return to the vernal equinox.

Throughout Egypt's long history the country has had a dry climate. The Egyptians depended on the annual flooding of the Nile River, which left behind a black soil that gave them rich farm lands. Ancient Egyptians discovered that the annual flood of the Nile came soon after Sirius, the Dog Star, appeared in the eastern sky after several months of invisibility. The Egyptians used this event to fix their calendar and recognize a year of 365 days. (Egyptian Tourist Bureau)

of 365.25 days. The odd-numbered months were each given 31 days; each even-numbered month was given 30 days. February was treated as a special case and was given 29 days.

The fifth month of the original Roman calendar, Quintilus, had become the seventh month when Numa Pompilius added January and February to the beginning of the Romulus calendar. After Caesar's death, Quintilus was named *Julius* in his honor.

The solar year of the Julian calendar is 365.25 days. Since it is difficult to add one-fourth of a day to the end of a calendar year, the year was assigned 365 days in the original Julian calendar. The one-fourth day was allowed to accumulate and was added to February each fourth year. Each fourth year of 366 days was called *leap year*.

After Julius Caesar's death, leap year was made to occur every third year instead of every fourth. This caused difficulties; and by 8 B.C. the calendar was out of step and needed to be changed again. Augustus Caesar ordered corrections to be made; but in addition, he changed Sextilis to Augustus and then added an extra day to his month, which he took from February. Thus, February was reduced to 28 days; of course, every fourth year, leap year, February continued to have a day added.

The Julian calendar was based on a solar year of 365.25 days. Sosigenes' observation of the solar year was not as precise as it should have been. The solar year is, in fact, 11 minutes and 14 seconds shorter than 365.25 days, which does not seem to be a great amount of time; but it causes an error of 1 day to accumulate over a period of 128 years. This error eventually put the Julian calendar out of step with the seasons. For example, during A.D. 325 the vernal equinox, that is, the first day of spring, occurred on March 21 but in A.D. 1093 the vernal equinox arrived on March 15. During 1582 the first day of spring corresponded to March 11 of the Julian calendar; spring was coming on the scene during the winter!

As a result of the confusion, Pope Gregory XIII ordered a calendar change that dropped 10 days from 1582 and, in addition, drops 3 leap years every 4 centuries. The Gregorian calendar is the one we use today. It will take more than 30 centuries before this calendar is 1 day out of step with the seasons.

The Catholic countries carried out the changes and adopted the Gregorian calendar immediately; but countries under the Eastern Church and most Protestant countries did not. When we purchased Alaska from the Russians in 1867, for example, its calendar had to be changed over to the Gregorian calendar. England and the American colonies waited until 1752 to make the change—almost 200 years! In fact, although George Washington's birthday is noted in present-day calendars on February 22, when he was born the Julian calendar was in use in the American colonies and the calendar date was February 11, 1732.

THE EARTH-MOON SYSTEM

The Moon, which is one-fourth the diameter of the Earth, controls the tidal movement of our ocean waters and is, in reality, the Earth's partner in space. Many scientists describe the Earth-Moon combination as a double planet, which is a departure from the usual description of the Moon as the Earth's satellite.

THE BARYCENTER

The common description of the Earth–Moon relationship indicates that the Moon orbits around the Earth. According to Sir Isaac Newton's law of gravitation, however, two objects attract each other with a force proportional to the product of their masses and inversely proportional to the square of the distance between them. In the case of the Earth and Moon, both masses are relatively large and they have a common center of gravity known as the **barycenter**. In actual fact, both the Earth and the Moon revolve around the barycenter.

The barycenter, which is located at a point on a line connecting the central areas of both

Surface of the Moon is covered by lightweight, fragmented material containing rocks of all sizes. Thousands of craters are visible in this photograph; some were formed when huge meteors struck the Moon's surface, and others were formed by volcanic eruptions. The dark areas are plains, called seas. (Lick Observatory)

bodies, is about 1074 miles (1728 km) below the Earth's surface on the side facing the Moon (Fig. 1–23). This means that as the Earth rotates on its own axis, the barycenter is constantly changing its position with respect to the Earth's continental and oceanic plates. It is, in fact, the barycenter that describes an elliptical orbit around the Sun—not the center of the Earth (Fig. 1–23).

REVOLUTION AROUND THE BARYCENTER

From the center of the Earth to the barycenter is a distance of 2886 miles (4645 km). The distance from the center of the Moon to the barycenter averages 236,000 miles (379,800 km). The Moon moves at a rate of 13.19° per day in its orbit around the barycenter. This means that the Moon, M1 in Figure 1–24, completes a star-determined 360° orbit around the barycenter and arrives at position M2 after 27.3 days. This period of time is referred to as the *sidereal*

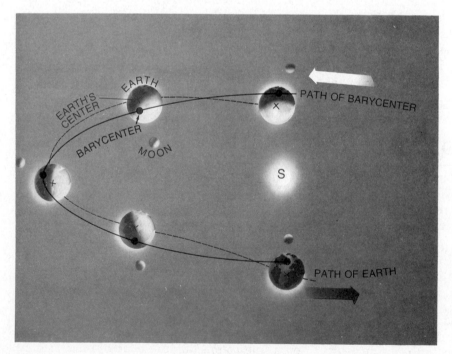

FIGURE 1–23 The barycenter is located below the Earth's surface on a line connecting the central areas of the Moon and the Earth. The barycenter takes an elliptical path around the Sun. The center of the Earth revolves around the barycenter and "wobbles" from one side of the barycenter's elliptical orbit to the other.

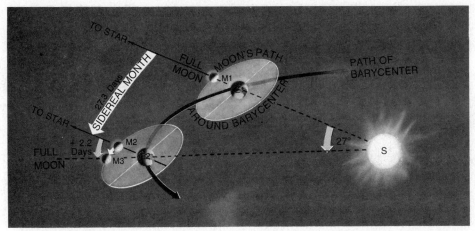

FIGURE 1–24   The synodic month—that is, the month of the phases—is 2.2 days longer than the sidereal month.

month. The center of the Earth also completes an orbit around the barycenter in one sidereal month of 27.3 days.

The barycenter itself is moving around the Sun. It takes 365.25 days for the barycenter to complete one orbit of 360° around the Sun. This means that the barycenter moves at a rate of 0.99° per day in its orbit around the Sun.

Let's assume that we begin a count at the time of full Moon—M1 in Figure 1–24—when the Moon, Earth, and Sun are along the same line in space. During the 27.3 days of the sidereal month, the barycenter revolves through 27° of the 360° it must complete to make its way around the Sun. For the return of the full Moon, that is, for the Moon to come into line with the Earth and Sun again at position M3, the Moon must travel an extra distance beyond the end of the sidereal month, which amounts to 2.2 days at the Moon's rate of travel. Thus, the **synodic month**—from full Moon to full Moon—is 29.5 days.

The same side of the Moon is always facing the Earth. At times it is mistakenly thought that this means the Moon does not rotate on its axis; if this were the case, the fixed orientation of one side of the Moon with respect to the stars would, in fact, allow us to see all sides as the Moon revolves around the barycenter. The Moon rotates at the same rate that it revolves around the barycenter; it is this synchronous rotation period that keeps turning one face of the Moon in our direction.

PHASES OF THE MOON

The changes in the appearance of the Moon that we see throughout the synodic month are called *phases*. Radiation from the Sun falls on the Moon's surface and illuminates it. Half the Moon is illuminated at all times. The phases of the Moon are simply the varying proportions of the illuminated surface that are turned toward us as the Moon revolves around the barycenter. The sequence of phases is from new Moon to waxing crescent, to first quarter, to gibbous, to full Moon, to gibbous, to last quarter, to waning crescent, to new Moon again (Fig. 1–25).

The synodic month can be thought of as beginning at the new Moon when the illuminated surface of the Moon is not visible to us. The shadow side faces us at new Moon. Each day the Moon moves 13.19° in its orbit around the barycenter (Fig. 1–24). About three days after the onset of the new Moon, we begin to

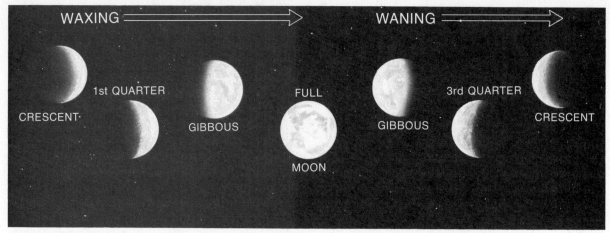

FIGURE 1–25   The Moon is called a waxing Moon as it progresses from the first crescent to the full-moon phase. The waxing crescent rises around 9:00 A.M. and sets about 9:00 P.M. Each succeeding night the Moon rises 50 minutes later; thus, the waning crescent rises at about 3:00 A.M.

see the waxing crescent. The waxing crescent rises around 9:00 A.M. and sets about 9:00 P.M. Each succeeding night the Moon rises 50 minutes later than the night before. There is a span of 12 hours from moonrise to moonset

each night. The Moon is called a waxing Moon as it progresses toward the full Moon phase.

The Moon is "full" about 14.7 days after the onset of the new Moon. The full Moon rises generally at 6:00 P.M. and sets at 6:00 A.M. In the

When the Moon is almost on a line between the Earth and the Sun, its dark side is turned toward the Earth and it is invisible to us. Then within 24 hours a slim crescent begins to appear. As the Moon revolves around the Earth we see more of its surface as sunlight floods on to the portion facing the Earth. This photo was made four days after the new Moon. (Lick Observatory)

second half of the cycle—from full Moon to new Moon—it is called a *waning* Moon. The waning crescent generally rises at 3:00 A.M. and sets at 3:00 P.M.

THE WEEK

Although the unit of time known as the *week* is not exactly and precisely consistent with any of the periodicities in the celestial bodies, there is a pattern in the cycle of the Moon's phases that might have stimulated the ancients to devise the 7-day week. For example, it takes about 7 days for the first quarter Moon to fill out to the full Moon; and it takes about the same period of time for the full Moon to wane to the last quarter.

The Babylonians in 2000 B.C. and the ancient Hebrews observed a 7-day week; one day out of seven was devoted to rest. In fact, reference to the 7-day week of the Hebrews is found in the Old Testament; the Sabbath, a day for rest and happiness, was the seventh day of the Hebrew week.

The Romans named the 7 days of the week after the Sun, Moon, and the five planets that were known to them—Mars, Mercury, Jupiter, Venus, and Saturn. Each day was considered sacred to the Roman god who was associated with that planet. The days were known as Sun's day, Moon's day, Mars' day, Mercury's day, Jupiter's day, Venus' day, and Saturn's day. This system was used by the Romans at the beginning of the Christian era.

Generally, however, the 7-day week was not observed in northern Europe until A.D. 321. Constantine the Great, the first emperor of Rome to become a Christian, established the 7-day week in the European provinces. Before A.D. 321, market days were the only divisions of time that approximated the week. In many cities, every fourth or fifth day was market day. No regular "week" was followed. The Saxon names for the days of the week—Sun's day, Moon's day, Tiw's day, Woden's day, Thor's day, Frigg's day, Saturn's day—are very much like the names we use today.

LUNAR CALENDARS

The religious events in the Jewish world are set by a calendar that is more than 5000 years old. The Hebrew calendar is based on lunar cycles so it ordinarily has a year of 12 months. Six of the Hebrew months have 29 days and the others have 30 days. At times, because lunar calendars are not very exact, an extra month of 29 days is added to the year to keep the Jewish religious festivals in order. This means Hebraic years do not always have the same number of days—some years have as few as 354, others have as many as 383 days.

The beginning of the Hebrew calendar is set at autumn in the year 3761 B.C. The Gregorian date A.D. 1979 was the Hebraic date 5739 A.M. After Hebraic dates, A.M. is used. These letters stand for the Latin anno mundi, which means *in the year of creation.* For the Hebrews, each day begins at sunset; each month begins at sunset on the day of the new Moon. The new Moon in Hebrew tradition, however, is really the crescent Moon.

Devout Muslims use the Islamic calendar, which is based on lunar cycles rather than the solar year. An Islamic month is the time between two successive new Moons. The months of the Islamic year normally contain 29 and 30 days alternately, resulting in a total of 354 days, or 11 days less than the Gregorian year. This difference means that the beginning of any given Muslim month comes earlier in the season every year. The months of the Islamic year have no relation to the seasons and make a complete circuit of the seasons in approximately 33 Gregorian years; 100 years of the Islamic calendar are approximately equal to 97 years of the Gregorian calendar.

The first year of the Islamic calendar was the year of the Prophet Muhammad's emigration from Mecca to Medina (A.D. 622). The common European method for designating Islamic dates is by the letters A.H. These letters stand for anno hegirae, which means "in the time of the emigration" or the year of the Prophet's move. Our year A.D. 1980 was the Islamic year 1400 A.H. The Islamic calendar is the official calendar of the Saudi Arabian government.

THE LANGUAGE OF PHYSICAL GEOGRAPHY

A physical geographer uses a technical vocabulary to make explanations. Review your understanding of the vocabulary used to develop the concepts in this chapter. Among the important words and terms used are:

Archeozoic	great circle	noon	revolution
axis	hour	outgassing	rift
Azoic	hydrologic cycle	Paleozoic	rotation
barycenter	insolation	Pangaea	sea level
calendar	international date line	parallels	sidereal day
circle of illumination	latitude	periods	solar time
core	lithosphere	plates	solstice
crust	longitude	plate tectonics	spheroid
day	mantle	polar axis	standard time
daylight saving time	meridians	polestar	supercontinent
ecliptic	Mesozoic	Proterozoic	synodic month
epochs	nautical mile	proto-planet	timetable
equinox	nebula	proto-star	uranium-238
eras			

SELF-TESTING QUIZ

1. Compare the size of the Earth to that of the other planets in the solar system.
2. Trace the evolution of the planetary system.
3. Describe the most popular hypothesis of the Earth's origin.
4. What effects developed as a result of the heating of the Earth's interior?
5. Explain the dating procedures that make it possible to establish the age of fossils and rocks.
6. List the important concepts for the theory referred to as plate tectonics.
7. How did computer matching of the coastlines of South America and Africa contribute to our understanding of plate tectonics?
8. Describe the various kinds of boundaries found between plates.
9. What is meant by latitude?
10. Explain the significance of the standard reference point in developing the concept of longitude.
11. How is the sea-level concept used?
12. Why is Polaris called the *polestar*?
13. List the various phenomena produced as a result of the tilt of the Earth's axis.
14. In what ways do we define a day?
15. Explain the concept referred to as an hour.
16. Why is it necessary to have an international date line?
17. Describe the evolution of the Julian calendar.
18. Why is the Moon considered to be the Earth's partner in space?

IN REVIEW
THE EARTH: A SPHERICAL PLATFORM IN SPACE

I. OUR PLANET: THE EARTH
   A. Evolution of the Planetary System
   B. Evolution of the Earth
   C. Geochronology
   D. Wandering Continents
      1. The Gathering Evidence
      2. Sea-Floor Spreading
      3. Plate Tectonics
      4. Plate Boundaries

II. DEFINING POSITION ON THE EARTH
   A. Latitude
   B. Longitude
   C. Using Satellites
   D. The Sea-Level Concept

III. AN EARTH IN MOTION
   A. A Rotating Earth
      1. A Tilted Axis
      2. Direction of Rotation
      3. Period of Rotation

   B. Dividing the Day
   C. Standard Time
   D. World Time Zones
   E. Daylight Saving Time
   F. The International Date Line
   G. A Revolving Earth
      1. The Circle of Illumination
      2. Energy Input
      3. The Seasons
   H. Our Calendar

IV. THE EARTH-MOON SYSTEM
   A. The Barycenter
   B. Revolution Around the Barycenter
   C. Phases of the Moon
   D. The Week
   E. Lunar Calendars

2 *The ocean and the atmosphere*

The atmosphere drives the great ocean circulations and strongly affects the properties of seawater. To a significant extent the atmosphere in turn owes its nature to and derives its energy from the ocean. For example, evaporation from the sea's surface converts ocean water into water vapor. Once in the atmosphere, water vapor is carried by the wind until it condenses to form clouds; it is from the clouds that precipitation falls to the Earth's surface. Most precipitation falls directly into the sea; some falls on land, but it, too, eventually works its way back to the ocean. This cycle of evaporation and condensation, known as the **hydrologic cycle**, is one of the significant ways in which the ocean and atmosphere are linked. Our first concern in this chapter is with the regional characterization and description of the world ocean and the atmosphere. Then we will examine a few ocean-atmosphere interactions.

EXPLORING THE OCEAN'S BASINS

The **ocean's basins** are depressions that hold huge quantities of water. Before equipment to probe and to make measurements became available in the nineteenth century, very little was known about the basins. At first, mechanical soundings of the sea-bottom profile were made;

The ocean liner in the left foreground is approaching a fiord—a long narrow inlet of the sea—that cuts into the southwestern coast of South Island, New Zealand. This part of the South Pacific Ocean between New Zealand and Australia is called the *Tasman Sea*. Rugged rocks and cliffs rise from the Tasman Sea to form this portion of the New Zealand coastline. Bands of clouds formed by water evaporated from the sea surface hang above the ship and in the fiord. A dense cloud deck above the cliffs of the fiord hides the rugged, mountainous terrain beyond the fiord. (New Zealand Consulate)

later, sonar was used to develop more sophisticated profiles; and today, navigational satellites give a survey ship a continuous "fix" on its latitude and longitude, which allows very accurate charts of the ocean's floor to be made.

At the present time the ocean's basins are filled to the brim with water; in fact, the basins are so brimful that a considerable amount of water spills over on to large areas of the continental margins and produces shallow seas. The volume of the world ocean is estimated to be about 328 million cubic miles (1367 million km³), which is almost 97 percent of the world's free water.

The obvious, most remarkable fact about the ocean's basins is that they are interconnected and, thus, are part of one world ocean. This means that the waters of the ocean spread from the Arctic to the Antarctic as one huge interconnected flow (Fig. 2–1).

The ocean's basins are three in number: Atlantic, Pacific, and Indian. Each basin extends northward from the area surrounding the Antarctic continent (Fig. 2–1). By agreement the east-

ern border of the Atlantic Ocean in the area of Antarctica is placed at the 20°E meridian, which extends from the Antarctic continent to Cape Agulhas (34°52′S, 20°E) at the tip of South Africa (Fig. 2–2). The western border of the Atlantic is fixed by a line that connects the Palmer Peninsula on Antarctica to Cape Horn (56°S, 67°16′W) at the tip of South America. Within its eastern and western boundaries, the Atlantic extends northward to include the North Polar Sea, which is commonly called the *Arctic Ocean*. The Pacific Ocean extends from the Cape Horn line westward to longitude 147°E, which is the meridian of the South Cape of Tasmania, and then northward to Bering Strait. With these parameters in mind, the Indian Ocean simply lies between the Atlantic and the Pacific oceans.

SHAPES AND DEPTHS

The boundary between the Pacific and the Atlantic basins is, in reality, a sill which manifests itself as a narrow, curved shoal—bounded

FIGURE 2–1 The ocean basins are interconnected. They extend northward from the area surrounding the Antarctic continent. There is twice as much land standing above sea level in the Northern Hemisphere as in the Southern Hemisphere.

FIGURE 2–2   In the Southern Hemisphere, the eastern and western borders of the three oceans have been set by international agreement.

on the north by the Scotia Ridge and on the east by the South Sandwich Trench (Fig. 2–3). The sill runs from Cape Horn eastward in a gentle curve that ends at the South Sandwich Trench. The sill—located generally at latitude 56°S, lon-gitude 40°W—is a part of the Antarctic Plate. The water covering this sill is called the Scotia Sea.

Each of the three major ocean basins is unique in general outline and configuration. The Atlantic Basin is the longest; it is relatively nar-

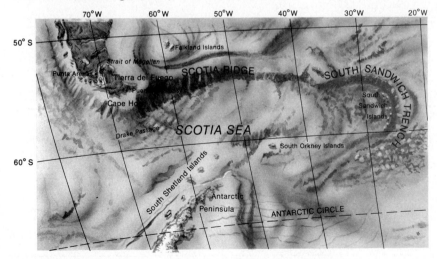

FIGURE 2–3   The Scotia Ridge and the South Sandwich Trench form a boundary between the South America Plate on the north and the Antarctica Plate on the south.

row and extends northward with an irregular twisting pattern that is roughly S-shaped (Fig. 2–1). The average depth of the Atlantic Basin, without considering the adjacent seas, is 12,881 feet (3926 m). The Pacific's basin is roughly symmetrical and it is the largest of the three; its east-west dimension as depicted in Figure 2–1 is about one-half the circumference of the Earth. There are many oceanic islands and island arcs in the Pacific Ocean. Excluding the adjacent seas, the average depth in the Pacific Basin is 14,049 feet (4282 m). The Indian Ocean (Fig. 2–1) is also roughly symmetrical; the average depth within this basin, excluding the adjacent seas, is 13,002 feet (3963 m).

The average depth of the world ocean, that is, considering all three basins, is 13,311 feet (4057 m). Let's compare the average depth within each basin to the world average: The Pacific's depth is within 5.5 percent of the world average, the Atlantic's is within 3.2 percent, and the Indian Ocean has an average depth that only varies by 2.3 percent from the world average. The fact that depths below sea level in the three basins are not so different is an indication that the sea floor is a uniform global feature as the theory of plate tectonics suggests.

LAND-WATER COMPARISON

Sea-surface water covers an area of 139.78 million square miles (362 million km²), which is 70.7 percent of the Earth's total surface. A mere 57.88 million square miles (149.69 million km²), or 29.3 percent of the Earth's surface, is exposed land. Use Figure 2–1 to compare the amount of the exposed-land surface found north of the Equator with the amount found south of the Equator. Most of the land not covered by water is in the Northern Hemisphere; in fact, there is twice as much exposed surface in the Northern Hemisphere as in the Southern. This, of course, was not always the case; at the end of the Paleozoic, for example, the landmass of **Pangaea** was distributed somewhat evenly over the two hemispheres.

The present-day percentages of the surface covered by water and land in each 10° latitude band is documented in Table 2–1. Between 80° and 90°N latitude, 90.2 percent of the Earth's surface is covered by water; 9.8 percent is covered by land. In the Southern Hemisphere for this same latitude band—that is, 80° to 90°S—100 percent of the surface is covered by land.

When we compute the percentage of water and land found within each hemisphere, we find that 60.7 percent of the Earth's surface in the Northern Hemisphere is covered by water and 39.3 percent is covered by land. In contrast, in the Southern Hemisphere, 80.9 percent of the surface is covered by water and only 19.1 percent by land. Water is the predominant surface cover in the Southern Hemisphere.

TABLE 2–1   DISTRIBUTION OF WATER AND LAND IN DIFFERENT LATITUDE BANDS

	SOUTHERN HEMISPHERE			NORTHERN HEMISPHERE	
Latitude °S	Percent of Water	Percent of Land	Latitude °N	Percent of Water	Percent of Land
90–80	0	100	90–80	90.2	9.8
80–70	27.0	73.0	80–70	70.3	29.7
70–60	90.5	9.5	70–60	29.5	70.5
60–50	99.2	0.8	60–50	42.7	57.3
50–40	97.0	3.0	50–40	47.7	52.3
40–30	88.6	11.4	40–30	57.2	42.8
30–20	76.8	23.2	30–20	62.4	37.6
20–10	78.0	22.0	20–10	73.6	26.3
10–0	76.4	23.6	10–0	77.2	22.8

ON THE OCEAN FLOOR

The first feature encountered, as you leave land and move into an ocean basin, is the **continental shelf**, which is a broad area from the **intertidal zone** to the shelf break (Fig. 2–4). The worldwide-average width of continental shelves is 42 miles (67.5 km). An individual shelf's width can vary quite a bit from this average, however. The shelf off the California coast, for example, is less than 1 mile (1.6 km) wide; along the northern coast of Siberia, on the other hand, the continental shelf is 800 miles (1290 km) wide; the shelf off Atlantic City, New Jersey, is shown as 100 miles (160 km) wide in Figure 2–4. A continental shelf slopes gently seaward with an average inclination of 10 feet per mile (2 m per km). A shelf drops off more sharply at the region known as the **shelf break**.

The shelf break marks the beginning of the region called the *continental slope* (Fig. 2–4).

The slope is the outer margin of a continent where it descends to the deep-sea floor. Continental slopes have inclinations ranging from 1.5° to 30°, with an average near 5°. The section of the slope that makes the final plunge into the depths is called the *continental rise*. Where it exists, the continental rise is a thick deposit of sediment that has been carried from the continent and deposited at the base of the continental slope. The water overlying a continental shelf at the shelf break may be as shallow as 160 feet (50 m) or as deep as 1500 feet (460 m). The depth of water at the shelf break in Figure 2–4 is about 600 feet (almost 200 m).

At the edge of many continents, the shelf and slope are scarred by submarine canyons. Dense, sediment-laden water flowing as a strong current along the shelf is probably responsible for cutting these canyons. The Hudson Submarine Canyon (Fig. 2–4) is cut into the shelf off our eastern coast; the canyon begins at the mouth of the

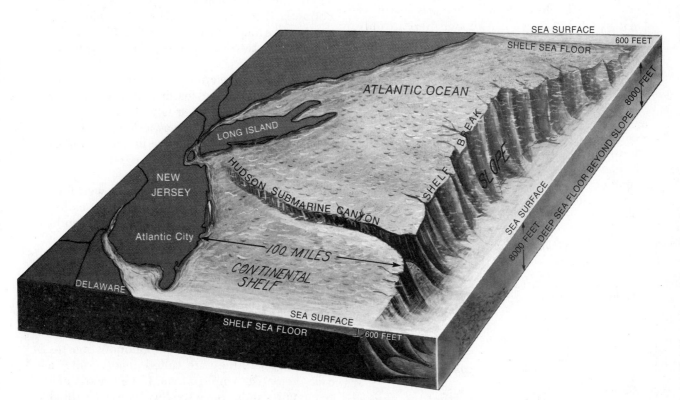

FIGURE 2–4 The Continental Shelf is a broad area running from the intertidal zone to the shelf break. The slope is the outer margin of a continent where it descends to the deep sea floor.

A variety of submersibles are used to explore the ocean's floor. This underwater photo shows the Ben Franklin beginning one of its exploratory trips. Two divers are shown working outside the submersible. (U.S. Navy)

Hudson River, where the river flows from New York Harbor.

The marginal **trench** is another feature of some continental margins (Table 2–2). In some locations, trenches are found in close proximity to the shelf-slope region. Although there are only two major trenches in the Atlantic, the Pacific Ocean contains more than a dozen; the Indian Ocean has one major trench, the Java Trench. The Pacific Ocean trench system is discussed on pages 54–55.

The deep-sea floor contains almost flat **abyssal plains,** that is, featureless sediment-covered areas. The plains are common in the Atlantic and Indian oceans. Some estimates indicate that 42 percent of the ocean's floor is covered by abyssal plains, which were formed by the deposition of sediment on recently formed sea floor. Although

abyssal plains exist in the Pacific, most of its floor has a hilly topography with small abyssal hills dotting the landscape and, at times, rising 1970 feet (600 m) above the plains.

The floor in all three basins contains gentle arches, called **rises,** which are elongated broad elevations. For example, the island of Great Bermuda, which is about 14 miles (22 km) long, is the top of a pedestal of rock that extends upward from the Bermuda Rise—a submerged and beveled volcanic region that sits on the Atlantic's floor (Fig. 2–5). The Bermuda Rise runs about 600 miles (970 km) in a north-south direction and 300 miles (480 km) from west to east.

Bermuda, in fact, consists of a group of 300 islands that are clustered in a chain shaped like a fishhook. The Bermuda-island chain, with an overall length of about 22 miles (35 km), is in the western North Atlantic, about 570 miles (920 km) east-southeast of Cape Hatteras, North Carolina. The islands consist of reef limestone that "sits on" windblown calcareous deposits capping the Bermuda Rise (Fig. 2–5). The windblown deposits were laid down during the Pleistocene epoch when the sea level was quite a bit lower than it is today.

Island arcs (described more fully on pages 54–56) are prominent features in the basin of the Pacific. An arc, which is generally in close proximity to a trench, is a seaward-curved island chain that is a region of active mountain building. Volcanic and earthquake activity are associated with an island arc.

All three basins are marked by east-west fracture zones that form steep cliff faces, called **scarps.** The scarps, facing northward and southward, stand as much as 1640 feet (500 m) above the sea floor. Another elevated feature of the sea floor, called a **seamount,** has the appearance of a mountainous mass. Seamounts are comparatively isolated mountains standing 3000 feet (910 m) or more above the deep-sea floor. A seamount has a rather steep slope and a relatively small summit area; when these mountainous masses are flat-topped, they are called *guyots*—after a Swiss scientist, Arnold Guyot. If the summit of a seamount stands above sea level, it is identified on maps as an island.

Another feature found in all three basins is

TABLE 2–2   MAJOR OCEAN TRENCHES

OCEAN	TRENCH	DEPTH Feet	Meters
Pacific	Aleutian	28,500	8,690
	Japan	32,680	9,810
	Kermadec	32,960	10,050
	Kuril	34,600	10,550
	Marianas	36,160	11,020
	Philippine	37,780	11,515
	Tonga	35,700	10,880
	Peru-Chile	26,410	8,050
Atlantic	Puerto Rico	30,180	9,200
	South Sandwich	27,600	8,410
Indian	Java	25,340	7,720

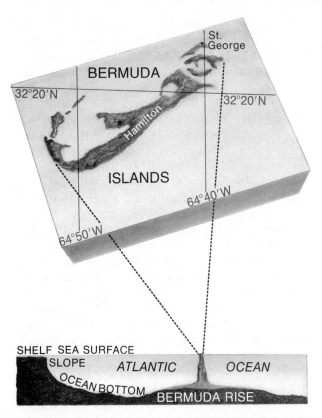

FIGURE 2–5   The Bermuda-Island chain sits on a pedestal of rock that extends upward from the Bermuda Rise—a submerged and leveled volcanic region on the Atlantic's floor.

the coral reef. A reef consists of calcium deposits that are built as a housing network by colonies of anthozoan coelenterates. In warm equatorial waters these animals attach their shelflike abode, called a **fringing reef**, directly to an island. If the island subsides or there is a rise in sea level and the colony thrives and continues to grow, new coral is emplaced close to the sea's surface, and develops into a structure known as a **barrier reef**, which is characterized by a lagoon between the reef and the original island. If the island is totally submerged, that is, the top is beneath the sea's surface, the reef that encircles it is referred to as an **atoll**. An atoll—from the point of view of an observer at sea level—consists of a lagoon surrounded by a coral reef.

MOUNTAIN RANGES

The Mid-Atlantic Ridge extends from the Arctic Basin southward to Bouvet Island at latitude 54°26′S (Fig. 2–6). At certain places individual peaks of this system rise above sea level to form islands. Iceland, the Azores, Ascension, Saint Helena, Tristan da Cunha, and Bouvet islands are identified in Figure 2–6 as peaks of the Mid-Atlantic Ridge system. The nine islands that make up the Azores group rise about 3 miles (5 km) from the ocean floor; on the average, however, the Mid-Atlantic Ridge rises between 1.2

Two islands make up the Midway Islands at latitude 28°13′N, longitude 177°23′W. Midway is a circular atoll in the Central Pacific Ocean at the upper end of the Hawaiian chain. This atoll, which is 6 miles (10 kilometers) in diameter, is 1300 miles (2090 kilometers) northwest of Honolulu. We are east of the atoll, looking toward the west; Eastern Island is in the foreground, and Sand Island is in the background. (U.S. Navy)

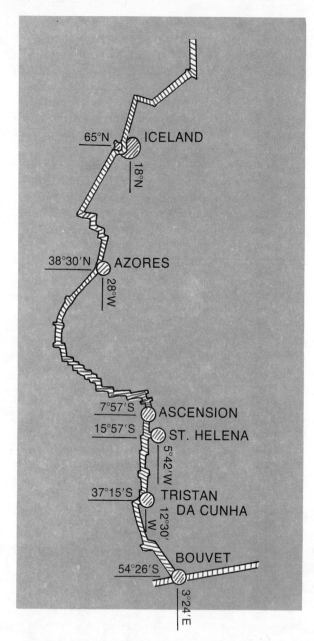

FIGURE 2–6 Each of the islands identified in this figure is a peak of the Mid-Atlantic Ridge system that rises above sea level.

and Pacific oceans; thus, there are well-developed ridges in all major ocean basins. However, the ridges of the Pacific Ocean are not as conspicuous or as numerous as those in the Atlantic and Indian oceans. The East Pacific Rise—one of the few ridges in the Pacific—is found from 1500 to 3000 miles (2400 to 4800 km) off the western coast of South America. It does not follow a mid-ocean course; rather, it extends in a general northerly direction toward the west coast of Mexico where it becomes obscure just short of the Gulf of California. In addition, the Pacific Basin has several short ridges in the southwest Pacific and in the northeast off Vancouver Island.

Along spreading ridges there is a continuous injection of **magma** from below. The heating effect caused by the hot magma expands the central zone of the ridge so that it rises high above the sea floor. Ridges are elevated on the average about 1.5 miles (2.4 km) above the normal level of the floor. The magma that works its way to the crustal surface and becomes part of the spreading ridge is called **lava.** The newly formed crust cools as it drifts away from the ridge; in addition, as the outward drifting hot rock cools, it contracts and gradually subsides until it sinks to the normal level of the ocean floor.

A giant underwater volcano erupted along the Mid-Atlantic Ridge and by November 14, 1963 it had built a lava mountain above the sea level. The 1963 above-sea-level eruption is recorded in this photo. The eruption occurred a few miles from the Westman Islands off the southern coast of Iceland. The volcanic island is called *Surtsey.* (Icelandic Airlines)

and 1.9 miles (2 and 3 km) above the floor, which is well-below sea level.

As noted on page 13 of Chapter 1, the Mid-Atlantic Ridge is part of a world-encircling ridge system that passes through the Atlantic, Indian,

The portion of the mid-ocean ridge system that has received the greatest attention lies in the Atlantic Ocean. The ridge crest as shown in Figure 2–7 is bisected by a deep rift valley, called a **cleft** or **graben.** The rift valley is about 15 to 30 miles (25 to 50 km) wide; it lies about 6560 feet (2000 m) below the adjacent peaks. The rift valley is a continuous north-south fracture zone. There is good evidence to believe that a world-wide rift exists within the ridge system, which means there is a double mountain chain with a rift valley between the ridges.

The lower portion of Figure 2–7 is a profile of the Atlantic's floor. The section runs from Massachusetts on the North American Plate to Gibraltar on the Eurasian Plate. The portion of the profile identified as the Mid-Atlantic Ridge spreads across almost one-third of the cross section in Figure 2–7. The ridge system is, in fact, about 1000 miles (1600 km) wide; it occupies about 30 percent of the entire Atlantic Ocean Area.

SEDIMENTS

The sediments covering the ocean floor are quite varied. They are presently being studied to gather information about the Earth's climatic and geologic history. In addition, many of these sediments—manganese nodules, for example—have commercial value in the markets of the world. One of the best systems for categorizing the ocean sediments classifies them by origin. Based on origin there are four kinds of sediment on the ocean's floor: **terrigenous** (from the continents), **biogenous** (an organic beginning), **halymyrogenous** (chemical precipitation), and **cosmogenous** (from extraterrestrial space—meteorite remains, for example).

SEA LEVEL

The surface area of the ocean and, thus, sea level is dependent on the volume of seawater as well as the configuration and extent of the basins. The total volume of the Earth's water throughout the eras of geologic time seems to have been rather constant. Water is, however, occasionally removed from the hydrologic cycle in various ways; and the level of the sea, of course, has varied from time to time. The glaciers of the Pleistocene epoch of the Cenozoic era, for example, "tied up" large amounts of water for long periods of time and significantly reduced the sea's surface below its present level. Today, glaciers cover approximately 6 million square miles (15.5 million km²), which is 3 percent of the Earth's surface. If the Earth's glaciers should suddenly melt and return their water to the sea, sea level would rise more than 200 feet (60 m).

FIGURE 2–7   This profile of the Atlantic Ocean's floor runs from Massachusetts on the North American Plate to Gibraltar on the Eurasian Plate.

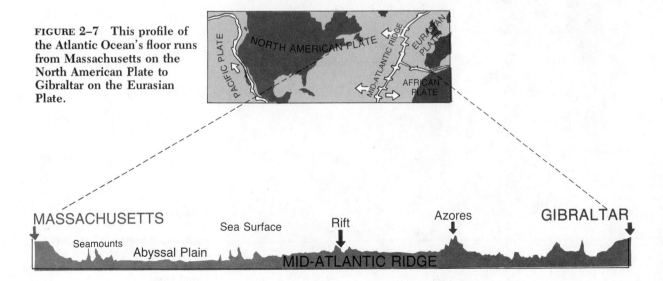

EVOLUTION OF THE OCEAN'S BASINS

The present basins of the world ocean have developed as a result of the breakup of Pangaea at the end of the Paleozoic. Figure 2–8 depicts the situation after 20 million years of drift during the Triassic of the Mesozoic era. The northern group of continents, called **Laurasia**, has split away from the southern group, referred to as **Gondwana**. In fact, Figure 2–8 shows that Gondwana is not intact at this time; for example, India as well as the Antarctic-Australia landmasses have been set free and a rift has begun to separate South American from Africa. The positions of the continents changed radically throughout the Mesozoic and Cenozoic eras. The ocean basins were shaped and sculptured during the process of plates diverging, converging, colliding, and subducting.

THE ARCTIC BASIN

The ancestral Arctic Basin, **Sinus Borealis**, existed at the end of the Paleozoic as an indenta-tion in the outline of Pangaea (See Fig. 1–4). The Sinus Borealis spread from 45°N to 60°N latitude prior to the breakup of Pangaea. During the Triassic the rotation of Laurasia repositioned Sinus Borealis farther to the north so that it spread from 60°N to almost 80°N latitude (Fig. 2–8). Toward the end of the Cretaceous, a rift in the North Atlantic worked its way to the east side of Greenland; then, during the Cenozoic the Mid-Atlantic Ridge moved into the Arctic Basin and finally detached Greenland from Europe. Thus, today, the Arctic Basin, which stretches from the Bering Strait across the North Pole to Spitzbergen (Fig. 2–9), is physically part of the Atlantic.

By tradition, however, the water filling the Arctic Basin is referred to as the Arctic Ocean. Figure 2–9 shows that these waters include the Greenland, Norwegian, Barents, Kara, Laptev, and Beaufort seas, as well as Baffin Bay and the waters of the Canadian arctic archipelago. The total area of the system is estimated at 4.73 mil-

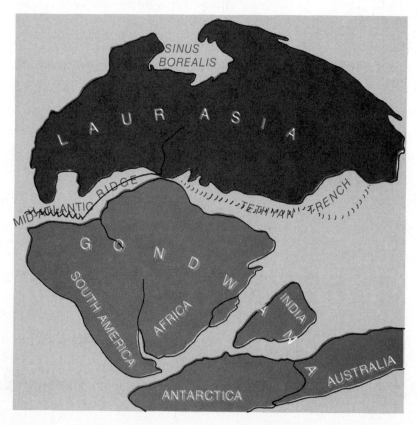

FIGURE 2–8   The universal landmass known as *Pangaea* was split into a northern group of continents, Laurasia, and a southern group of continents, Gondwana.

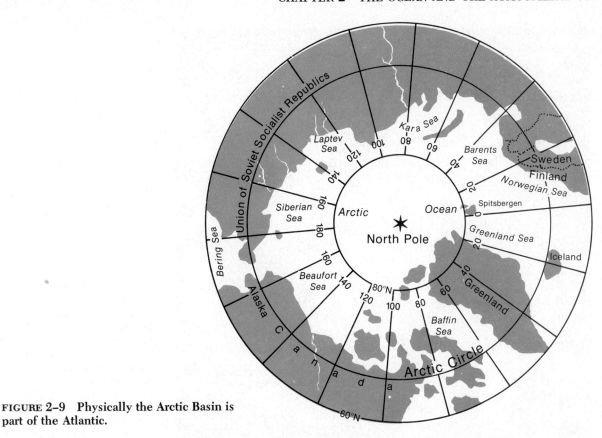

FIGURE 2-9   Physically the Arctic Basin is part of the Atlantic.

lion square miles (12.25 million km²). Enormous areas of the Arctic Ocean are covered with ice floes. The floes drift with the prevailing winds and currents, moving generally from the northern coast of Siberia to the northeastern coast of Greenland, crossing the North Pole on the way. When the water filling the Arctic Basin is considered as part of the Atlantic Ocean, it is referred to as the *North Polar Sea*.

THE ANTARCTIC BASIN

The Antarctica-Australia landmass was severed from the rest of Gondwana as a result of the rifting that took place during the Triassic (Fig. 2–8). Australia, however, was not separated from Antarctica until the Cenozoic. Today the waters of the Atlantic, Indian, and Pacific oceans meet to produce a water belt that encircles the landmass of Antarctica.

It was quite reasonable to apply the name *Antarctic Ocean* to this great expanse of water since the ocean currents, winds, and weather are arranged in a circumpolar pattern around Antarctica. The northern limit of this ocean has been set quite arbitrarily, however. The British Admiralty defines the Antarctic Ocean as all the waters south of latitude 55°S. The total area of water in the region from 55° to the edge of Antarctica's ice shelf is 12.45 million square miles (32.25 million km²).

Anarctica, with an area of 5.39 million square miles (13.96 million km²), is covered by glacial ice, some of which protrudes from the continent into the surrounding ocean. A rather significant portion of the sea butting Antarctica, is, in fact, covered by floating ice shelves that are commonly included in the area assigned to the landmass. The Ross Ice Shelf, for example, covers about 160,000 square miles (414,400 km²) of ocean.

The average thickness of Antarctica's glacial ice is 5300 feet (1600 m). One-quarter of the area, however, is overlain by ice that has a thickness

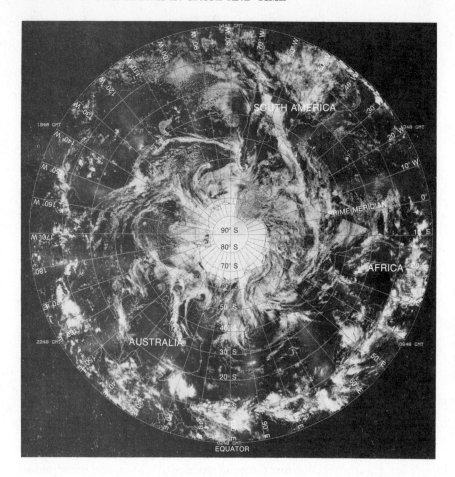

This satellite photo gives us a view of the complete Southern Hemisphere on January 19, 1973, which was the summer period in this hemisphere. The outline of the glacier-covered Antarctic continent is clearly visible in the center of the photo. There is an extensive cloud cover over the Antarctic Ocean, which is defined as the water south of latitude 55°S. (NOAA)

greater than its height above sea level. This probably means the continental area is down-warped or depressed as a result of the mass of the overlying glacier. It may mean the "continent" is, in reality, an island archipelago with the bottom of the glacier resting on a former shallow-sea floor.

In any event, Antarctica's continental ice thins toward the coasts and discharges huge, flat-topped icebergs into the ocean from the glaciers and the shelf-ice system. Some of these icebergs reach heights of more than 300 feet (90 m) and extend more than 2500 feet (760 m) below the surface of the water. The largest iceberg ever reported was 60 miles (96.5 km) wide and 208 miles (335 km) long, which is equivalent to an area of 12,480 square miles (3230 km²).

THE ATLANTIC BASIN

The modern North Atlantic Ocean had its start in a rift that split Pangaea from east to west along a line slightly to the north of the Equator. Figure 2–8 shows the North Atlantic as an opening between the western portion of Laurasia (North America) and the South American–African landmass of Gondwana. The rifting that split South America away from Africa to open the South Atlantic Ocean began during the Triassic (Fig. 2–8); during the Jurassic the rift worked northward to the position of present-day Nigeria. At the beginning of the Cretaceous, a portion of South America was still joined to Africa; rifting throughout the Cretaceous, however, finally separated the two continents and made the South Atlantic and the North Atlantic one basin.

The Atlantic Ocean is, as noted earlier, roughly S-shaped. It spreads in a north-south direction for 12,810 miles (20,615 km) from the Bering Strait toward and across the North Pole and then southward to the edge of Antarctica. The Atlantic covers an area of 36.41 million square miles (94.31 million km²).

The North and South Atlantic together get about half of all the Earth's rain. The greatest portion of this rainwater comes indirectly by means of rivers because most of the larger rivers of the world drain into the Atlantic. The Amazon River, for example, with an overall length of 4080 miles (6565 km), discharges 6.18 million cubic feet (175,000 m³) of water per second into the Atlantic.

The Romanche Fracture Zone, which is located near the Equator, is an important break in

the Mid-Atlantic Ridge system and is often used as a dividing line between the North and South Atlantic basins. When the prevailing wind patterns, ocean currents, and sea-surface temperatures are considered, however, the boundary between the North Atlantic and the South Atlantic is set at 5°N latitude.

THE PACIFIC BASIN

Panthalassa, the ancestral Pacific, was the universal ocean that surrounded Pangaea (See Fig. 1–4). When the rifting that opened the Atlantic and Indian oceans began, Panthalassa got smaller. The floor of Panthalassa was consumed in the zones of subduction, commonly called *trenches*, that developed around its rim. During the Triassic, for example, the Tethyan Trench extended from Gibraltar to Borneo (Fig. 2–8). A vast amount of Panthalassa floor was consumed in this subduction zone. Today as a result of subduction and new sea-floor rock solidifying along oceanic rifts, the Pacific is floored with rock and sediments far younger than the continents that rim it. In fact, the oldest sediments

There are many very productive fishing grounds in the Atlantic Ocean. Menhaden—an ocean fish of the herring family—which is used primarily as a high-protein supplement for animal feeds, represents more than half the tonnage of commercial fish caught in U.S. waters each year. The *Grand Batture* is a 180-foot (55-meter) long menhaden fishing vessel. It carries two 36-foot (11-meter) long "purse boats" on its stern. The purse boats are released when a menhaden school is sighted; each purse boat carries one end of a 1200-foot (366-meter) net that is used to encircle the school. The fish are pumped from the net into the refrigerated holds of the *Grand Batture*. (Zapata Corp.)

found in this ocean were retrieved in the Northwest Pacific; they were deposited only about 160 million years ago during the Jurassic.

The Pacific Ocean and its adjacent seas cover an area of 70 million square miles (181 million km²). The north-south line of the Pacific extends from Bering Strait to Antarctica. The distance from Bering Strait to Cape Colbeck on Antarctica is 9060 miles (14,580 km). The Pacific is widest between Panama and the Philippine Islands—a distance of 10,600 miles (17,060 km).

The ring of trenches around the central basin of the Pacific is shown in Figure 2–10. Note that the trenches in the northern and western reaches of the basin are found in association with island arcs. Most of the Aleutian Trench (52°N, 176°W), for example, is on the convex side of an island arc. The Kuril Trench (46°N, 152°E) is also on the convex side of an island arc—the Kuril Islands that stretch from the Kamchatka Peninsula to the northern Japanese island of Hokkaido. The Izu, Bonin, Mariana, and Yap trenches are associated with island arcs, too.

Oceanic lithosphere is being subducted in all of the trenches identified in Figure 2–10. The Tonga and Kermadec trenches, which lie on a north-south line that stretches almost to North Island, New Zealand, are a classic example of the events taking place along and to the west of these narrow trenches in the Pacific. Figure 2–11

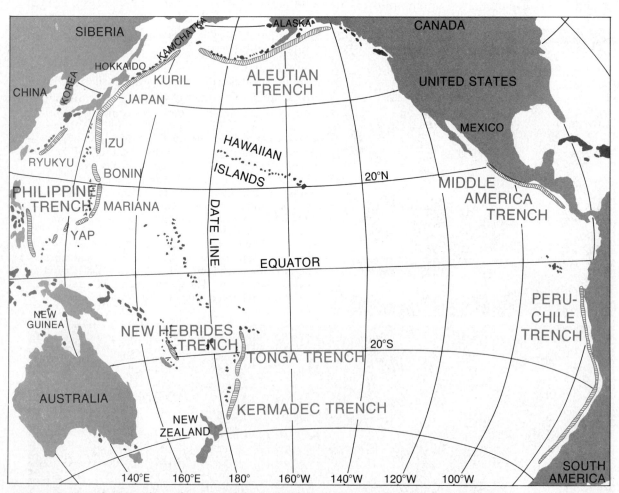

FIGURE 2–10 The ring of trenches around the central basin of the Pacific is shown in red. The Peru-Chile trench is the longest; it spreads from just above 10°S almost to the Equator.

shows that two plates—the Pacific and the Indo-Australian—are in collision in this section of the Pacific Basin. The subduction of the Pacific Plate produces the Tonga Trench.

The islands of Tonga (20°S, 175°W) are on the Indo-Australian Plate just to the west of the trench. About 160 islands with a total area of 270 square miles (700 km²) make up the group. The islands lie in two parallel lines from north to south; the line closest to the trench rests on broad shelves of drowned coral and consist of coral–limestone formation laid down in shallow water. The line of islands farthest from the trench was formed by volcanoes. The molten material that spewed onto the ocean floor to build this second line was from the melt of the subducting plate.

About 500 islands make up the Fiji group. The Fiji Islands, shown just to the west of the international date line in Figure 2–11, are about 300 miles (480 km) west of the zone of subduction. Viti Levu, the largest island, makes up more than half the Fiji group's area. The lava that formed the high volcanic peaks of Viti Levu (3709 ft, 1130 m) and Vaniura Levu (2420 ft, 738 m) was derived from the melt of the plunging plate. Thus, an island arc associated with a trench is, in fact, the product of the plunging plate that forms the trench.

THE INDIAN BASIN

Of the three major oceans, only the Indian Ocean does not extend very far into the Northern Hemisphere. The Indian Ocean has a number of adjacent seas—Red Sea, Arabian Sea, and Bay of Bengal, for example. The ocean and its adjacent seas cover an area of 28.6 million square miles (74.1 million km²).

The Y-shaped rift that began to open the Indian Ocean at the beginning of the Triassic was well-developed after 20-million years of divergence; the Y-shaped ocean area around India is clearly evident in Figure 2–8. Madagascar is shown attached to Africa in Figure 2–8; it was not carved away from the African landmass until the Cretaceous period of the Mesozoic.

Throughout the Mesozoic, India was moving toward the north. Early in the Cenozoic, in the vicinity of the Equator, the western section of the Indian landmass crossed a fixed source of magma, a **hot plume**, rising from the mantle (Fig. 2–12). Molten magma erupted through the crust and spread across the continent's surface. As the lava cooled, the Deccan Plateau began to take shape. India continued its journey northward and left the hot spot behind. The present-day position of the Deccan Plateau and the flood basalts is detailed by latitude and longitude in Figure 2–12. India has moved northward by more than 22 latitude degrees since its encounter with the hot plume early in the Cenozoic.

The 65 million years of the Cenozoic have been a busy time in the Indian Ocean: The Indian landmass completed its northward journey and collided with Asia. Rifting split Arabia away from Africa and opened the Gulf of Aden and the Red Sea. Today the major north-south rift in the Indian Ocean has largely stopped spreading and has become a shear zone; this conversion from divergence to shear is significant because it accommodates the present counterclockwise and northward rotation of the African Plate.

OTHER BASINS

The present Mediterranean Sea is a remnant of the Tethys Sea—a large triangular indentation, called a **bight,** on the east side of Pangaea (See Fig. 1–4). Throughout the breakup of Pangaea during the Triassic, the Laurasian landmass rotated clockwise around a point of rotation that is now Spain (Fig. 2–8). A subduction zone that formed in the Tethys, called the **Tethyan Trench,** served as an area of crustal uptake. The contined clockwise rotation of Laurasia during the Jurassic began to close the eastern end of the ancestral Mediterranean. The shear zone that developed between Africa and Spain forced Spain to rotate counterclockwise and, thus, produced an opening that we now call the Bay of Biscay. The Mediterranean Sea was clearly recognizable by the end of the Cretaceous. Today it is surrounded by land with a tiny connection, the Strait of Gibraltar (35°57'N, 5°36'W), to the Atlantic Ocean (Fig. 2–13). The Mediterranean, as shown in Figure 2–13, has a west-east line that

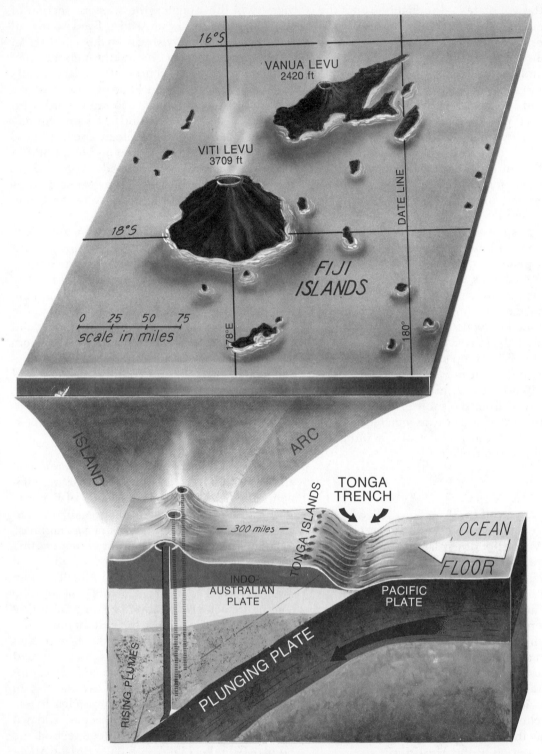

FIGURE 2-11 The high volcanic peaks in the Fiji Islands are built of lava derived from the melt of a subducting plate.

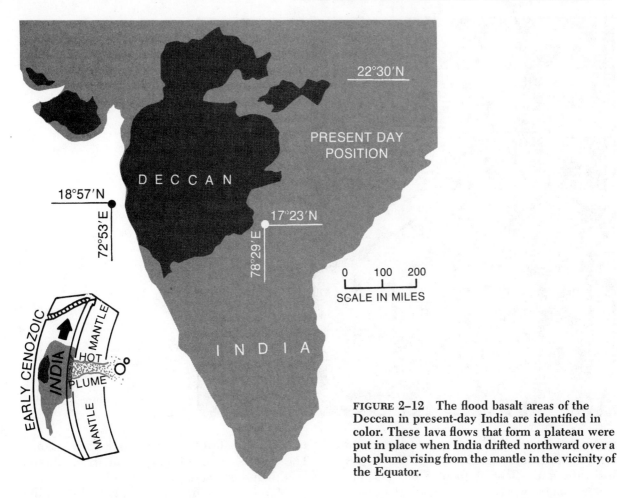

FIGURE 2–12 The flood basalt areas of the Deccan in present-day India are identified in color. These lava flows that form a plateau were put in place when India drifted northward over a hot plume rising from the mantle in the vicinity of the Equator.

stretches from Gibraltar (5°36′W) to Antioch 36°7′E); its north-south line runs form Trieste (45°30′N) to Al'Ugaylah (30°16′N).

The Mediterranean Sea is considered to be part of the Atlantic Ocean, as are the Gulf of Mexico and the Caribbean. These large seas are simply arms of the ocean that are located between landmasses. The term **mediterranean** is used to refer to seas between land areas; based on this definition, the Arctic Ocean is a mediterranean of the Atlantic Ocean. The Gulf of Mexico and the Caribbean Sea are sometimes called the *American mediterranean*. The Baltic Sea, Hudson Bay, Persian Gulf, and the Red Sea are considered intercontinental mediterraneans. The Red Sea was created when a branch of the rift in the Indian Ocean split Arabia away from Africa. The Gulf of California—similar in many ways to the Red Sea and the Gulf of Aden—is being

created by a spreading rift that is separating Baja California from the rest of Mexico. There is an Australian-Asiatic mediterranean in the Pacific that includes the Timor, Java, and South China Seas.

The Red Sea is a most extraordinary body of water and deserves some comment. Figure 2–14 shows that the Nile River stretches the full length of the Red Sea. The Nile, however, simply parallels the Red Sea on its run north to Cairo; none of the Nile's water finds its way into the Red Sea. In fact, along the Red Sea's total length, nearly 1120 miles (1800 km), not one river empties into it, which means there is no surface runoff.

The annual rainfall in the immediate area of the Red Sea averages less than 1 inch (2.54 cm). The searing heat from the Sun at these latitudes and the low humidity in the surrounding deserts induce a high rate of evaporation; the hot spots

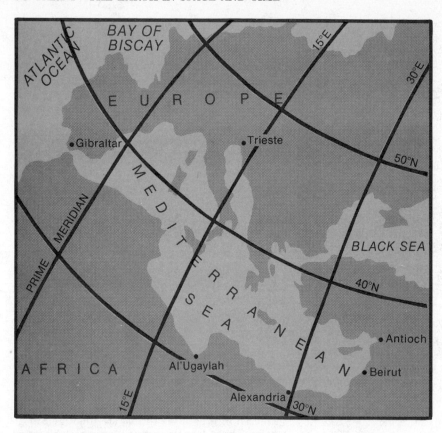

FIGURE 2-13   The Mediterranean Sea is part of the Atlantic Ocean. The term *mediterranean* is used to refer to seas between land areas.

in the sea's floor warm its deep waters and, thus, tend to increase evaporation, too. The Red Sea loses the equivalent of about 6 feet (1.8 m) of water from its surface each year, which causes its water to have a very high salt content compared to the three major oceans. To replace the loss brought about by evaporation, Indian Ocean water flows into the Red Sea from the Gulf of Aden through a shallow strait at its southern end. An insignificant amount of water also enters the sea from the northwest through the Suez Canal. The Red Sea holds billions of dollars in mineral wealth and an incredible array of animal life.

Marginal seas are merely indentations of the ocean into the continental coasts. Among the Atlantic's marginal seas are the Gulf of St. Lawrence, the North Sea, Bay of Biscay, and the English Channel. Other marginal seas of the world are the East China Sea, Arabian Sea, Bay of Bengal, and the Bering Sea.

THE ATMOSPHERE'S REGIONAL CHARACTER

A thin envelope of air surrounds the Earth. Humans and other living things depend on this envelope for survival. During respiration, for example, we extract oxygen from the air, use it in the process of metabolism, and then exhale carbon dioxide. Green plants, on the other hand, absorb carbon dioxide from the air, use the carbon for growth through the process of photosynthesis, and return oxygen to air.

Studies using sounding balloons, rockets, and satellites provide the best data from which to develop a description of the atmosphere. These studies, for example, indicate that the atmosphere changes in pressure, in temperature, and in composition with increasing altitude above the Earth's surface. In fact, these data allow the atmosphere to be divided into several sharply defined layers with reference to any one of a

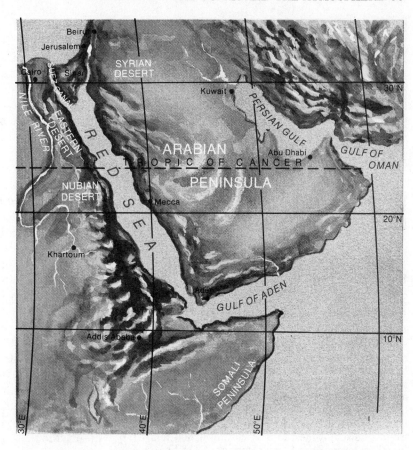

FIGURE 2–14   The Red Sea stretches from the Sinai to the Gulf of Aden. Not one river empties into this sea from the surrounding deserts.

number of factors. We can, for example, define regions in the atmosphere on the basis of their composition; or we might select thermal structure to identify layers of the atmosphere. Chemical processes, ionization, and magnetism can also be used to identify regions in the atmosphere.

BASED ON COMPOSITION

The atmosphere can be described as a "blanket of air" surrounding the Earth. This enveloping mixture of gases rotates with our planet, and it is truly a part of the Earth. The mass of the atmosphere is enormous. Expressed in units of weight, the total mass of the Earth's dry air is more than 5600 trillion tons. In similar units, the total amount of water vapor in the atmosphere is about 146 trillion tons. The atmosphere can be divided into two broad regions on the basis of its chemical composition: the **homosphere** and the

heterosphere. The relative extent of these regions is depicted in Figure 2–15.

HOMOSPHERE

In this lower region of the atmosphere, which extends from the surface of the Earth to an altitude of approximately 55 miles (88 km), a uniform mixture of gases is found (Fig. 2–15). The uniformity of the mixture is the result of turbulent mixing in this region of the atmosphere. Nitrogen (78.08%), oxygen (20.94%), argon (0.934%), and carbon dioxide (0.03%) are the major constituents in this region of uniform composition.

The homosphere furnishes the vital elements of life—carbon, oxygen, hydrogen, and nitrogen—in the form of gases: carbon dioxide, free oxygen, free nitrogen, and water vapor. Each of these is a truly important gas that makes the existence of life possible. Carbon dioxide, for example, is the source of the element carbon, which is an essential ingredient in the carbohydrate compounds found in all living things.

The area of the Earth shown in this photograph extends from the Mediterranean Sea in the north to Antarctica's ice cap in the south. The Red Sea between the Arabian Peninsula and Africa is clearly visible. The Malagasy Republic is the large island off the southeastern coast of Africa. There is a very heavy cloud cover over most of the Southern Hemisphere; all of this cloud cover is in the troposphere and most of it is within 20,000 feet (6100 meters) of the Earth's surface. (NASA)

HETEROSPHERE

In the region of the atmosphere above 55 miles (88 km)—referred to as the *heterosphere*—turbulent mixing is negligible and the composition of the gas is no longer a uniform mix. The molecules and atoms of elements found here tend to separate and arrange themselves in layers, each of distinctive composition (Fig. 2–15). The layer at the lowest level in the heterosphere is made up of the heaviest of these molecules. The uppermost layer consists of the lightest atoms.

There are four layers in the region identified as the heterosphere. A molecular nitrogen layer, mostly of molecules of nitrogen (N_2), girdles the Earth from 55 to 125 miles (88 to 200 km). From 125 to 700 miles (200 to 1125 km), there is a layer consisting primarily of atomic oxygen (O). A helium layer, composed largely of helium atoms (He), extends from 700 to 2200 miles (1125 to 3540 km). The fourth and highest band of the heterosphere, an atomic hydrogen layer consisting largely of hydrogen atoms (H), begins at about 2200 miles and stretches to an altitude of more than 6000 miles (9660 km).

The upper reaches of the heterosphere are often referred to as the *exosphere*. At extreme altitudes of 6000 miles (9660 km), atmospheric density is very low and a molecule with sufficient velocity can escape from the gravitational attraction of the Earth. At an altitude of 500 miles (800 km), an atom of oxygen, for example, will not encounter another atom for 100 miles (160 km).

The four layers of the heterosphere do not

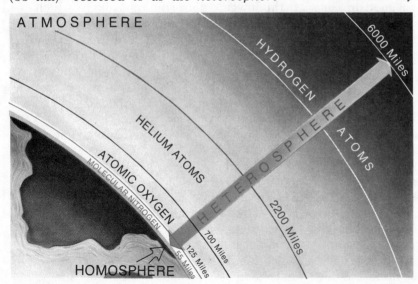

FIGURE 2–15   The homosphere is actually a very thin shell that surrounds the Earth's surface. The four shells of the heterosphere have a tremendous volume compared to the volume of the homosphere.

have sharply defined surfaces of separation. The boundaries between the layers are, in fact, transitional zones in which one kind of atom gradually becomes dominant. The atoms and molecules of the heterosphere are bound to the Earth by the force of gravity and, therefore, turn with the Earth's rotation.

BASED ON TEMPERATURE

A small package of sensors attached to a radio transmitter and referred to as a *radiosonde* is sent aloft to probe the vertical structure of the atmos-

The radiosonde hanging below the balloon is in flight. Radiosonde observation of the atmosphere is sometimes made to heights of 18 miles (30 kilometers). The instruments carried aloft measure temperature, pressure, and humidity. A transmitter sends the data to the ground where the signals are received and interpreted. (WMO)

phere. The instrument package is carried into the atmosphere by a large balloon, which can reach altitudes of 100,000 feet (30,480 m). On the basis of data collected by radiosondes and other types of probes, the atmosphere's thermal structure is fairly well known.

Four distinct regions of the atmosphere—based on temperature characteristics—have been identified: **troposphere** (0 to 7 miles,* 0 to 11 km), **stratosphere** (7 to 31 miles, 11 to 50 km), **mesosphere** (31 to 53 miles, 50 to 85 km), and **thermosphere** (above 53 miles, 85 km) (Fig. 2–16). The first three of these regions—troposphere, stratosphere, and mesosphere—occupy the same general range of altitudes as does the homosphere. The thermosphere occupies the same approximate altitude range as the heterosphere.

TROPOSPHERE

Based on temperature characteristics, the layer of the atmosphere nearest to the Earth is called the *troposphere*. The altitude range of this region varies with the latitude and the season. At the North Pole in July, for example, the upper range of the troposphere is at 5.5 miles (8.9 km) whereas in January the upper range is found at a lower altitude of 5.3 miles (8.5 km). In July at latitude 60°N, the top of the troposphere is at 6.2 miles (10 km); in the winter month of January, at the same latitude it stands lower at 5.6 miles (9 km). At latitude 45°N, the upper reach of the troposphere is at 9 miles (14.5 km) in July and at 8 miles (12.9 km) in January. The upper range of the troposphere over the Equator is found at 11 miles (17.7 km) in January and later in July at 10 miles (16 km). These data indicate that the upper level of the troposphere is lower at higher latitudes. The upper range of the polar troposphere is lowest in winter; the upper range of the equatorial troposphere is lowest during the Northern Hemisphere's summer period.

The troposphere is comparatively dense, and it contains almost all the water vapor of the atmosphere. This layer is also characterized by a great deal of turbulence. For all practical purposes, all our clouds, precipitation, and storms develop within the troposphere. As different air

*The ranges cited for these regions are for the middle latitudes.

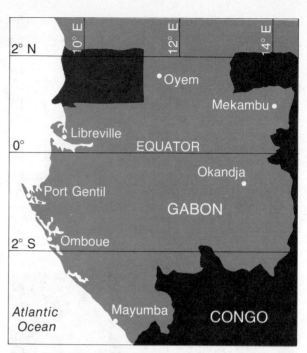

FIGURE 2–17   Gabon is a heavily forested country. It lies on the west coast of Africa where it straddles the Equator. Libreville sits at the mouth of an estuary that empties into the Atlantic Ocean.

FIGURE 2–16   The troposphere, stratosphere and mesosphere are in the range of the homosphere with its uniform mix of gases. The thermosphere occupies the region of the heterosphere.

masses circulate throughout the troposphere, they make contact with one another and begin to change as they mix.

Since air is a gas, its behavior can be predicted by the **universal gas law:** the temperature of a given mass of gas is proportional to its volume and pressure. This means that when the pressure of a given mass of gas increases, either its volume decreases or its temperature goes up. By the same token, when the temperature of a given volume of gas decreases, its pressure decreases, too. Everyday experience confirms these relationships. For example, an automobile tire filled with air on a warm day to a pressure of 26 pounds per square inch (1.8 kg per cm²) will have a lower pressure when the temperature drops on a colder day. Tire pressure will be higher after a trip because the temperature of the air in the tire has increased as a result of friction.

When air ascends, the pressure on it decreases and the gases expand according to the gas law. Expansion of the gases constitutes work in the physical sense and requires the use of energy. The energy used to expand a gas is heat energy, and the effect is to cool the air.

Examination of the troposphere reveals that temperatures decrease on the average at a rate of 3.5°F for every 1000 feet of height (6.4°C per km). This phenomenon is commonly known as the **normal lapse rate** but may also be referred to as the **environmental lapse rate.** It refers to the way temperature changes with height. The figure cited in this paragraph is the average normal decrease in the temperature of air with height in the troposphere.

In rising dry air,* the dry lapse rate, or temperature decrease, is 5.5°F for every 1000 feet of height (10°C per km). In rising moist air,** the wet lapse rate is about 3.2°F for every 1000 feet of height (5.8°C per km). In short, ascending air masses in which condensation of water vapor takes place cool more slowly than those in which no condensation takes place because the process of condensation releases heat.

*The term dry air is generally used in two different frames of reference: (1) Air without any moisture content, and (2) air that has not reached its total capacity to hold water vapor, which is the definition that applies in this discussion.
**Moist air, in this discussion, means air in which condensation has occurred.

These decreases in temperature with an increase in altitude result from the expansion of air; they are not related to any transfer of heat to or from air. The change in temperature is an internal change as a result of the change of pressure. When this same air descends, it warms up by compression at the same rate, that is, at 5.5°F per 1000 feet of descent if it is unsaturated air. The term **adiabatic,** which implies "without transfer of heat," is used to describe such temperature changes.

The normal temperature lapse rate continues as you rise through the troposphere; then the character of the lapse rate changes, and at certain altitudes minimum temperatures are reached with no further drop. The altitude at which a minimum temperature is reached is referred to as the **tropopause,** which is the boundary between the troposphere and stratosphere.

Libreville, the capital of Gabon, is located close to the Equator at latitude 0°23′N, longitude 9°26′E (Fig. 2–17). Helsinki, the capital of Finland, is located at 60°10′N, 24°57′E (Fig. 2–18). Both cities sit at the side of the sea and are in forested regions. Temperature data gathered over Libreville and Helsinki during July are plotted in Figure 2–19. The average July surface temperature at Libreville is 76°F (24°C); Helsinki's average air temperature at its surface during the same period is 64°F (18°C). The tropopause over Libreville occurred at almost 53,000 feet (16,150 m); the tropopause was encountered at about 32,500 feet (9900 m) over Helsinki. The actual lapse rate at Libreville was 3.3°F per 1000 feet of ascent; at Helsinki the rate was 3.7°F per 1000 feet. The difference in the rates is due to the greater amount of water in the air over Gabon. If we take the average of these rates, we get 3.5°F per 1000 feet of ascent—which happens to be the figure quoted previously as the normal lapse rate.

In January, over the North Pole, a minimum temperature of −76°F (−60°C) is reached at 5.3 miles (8.5 km); the tropopause temperature in July over the North Pole at an altitude of 5.5 miles (8.9 km) is −49°F (−45°C). At 60°N, a minimum of −67°F (−55°C) is reached in January at an altitude of 5.6 miles (9 km), while the July minimum of −58°F (−50°C) is reached

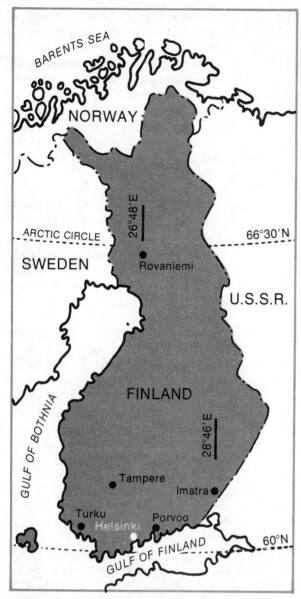

FIGURE 2–18 Helsinki sits at the edge of the Gulf of Finland. It is surrounded by woods and water. Finland, itself, is a country of lakes, swamps, and forests. Forests cover more than two-thirds of the country.

at 6.2 miles (10 km). Over the Equator in July a minimum of −100°F (−73°C) is reached at an altitude of 10 miles (16 km); and a January minimum of −112°F (−80°C) is reached at 11 miles (17.7 km). At heights of 10 miles (16 km), it

is colder over the Equator than over the North Pole.

A simplistic picture indicates the troposphere tends to bulge as you move from either the North Pole or the South Pole toward the Equator. But all recent data indicate the tropopause is not a continuous surface between the polar and equatorial regions. In fact, the polar tropopause extends to about 30° latitude in winter and 35° latitude in summer. The tropical tropopause can be followed as far as 45° latitude. Thus, there is a multiple tropopause at middle latitudes; for example, in January both the polar and tropical tropopauses extend to 30°N. Temperature probes in January at latitude 30°N show a tropopause of −67°F (−55°C) at an altitude of 7.5 miles (12 km), and then another tropopause of −94°F (−70°C) at 10.5 miles (17 km).

STRATOSPHERE

This is the second layer of the atmosphere; it is above the troposphere and below the mesosphere. The lower boundary of the stratosphere is the tropopause. As indicated in Figure 2–20, temperatures rise gradually in the stratosphere up to a level of about 32°F (0°C). The stratosphere is separated from the mesosphere by the **stratopause.**

The stratosphere is within the range of the homosphere, and it is a region whose chemical composition is essentially the same as that of the troposphere. The region of **ozone** production is centered in the stratosphere; maximum concentrations of ozone occur at altitudes between 12 and 19 miles (19 to 30 km). Ozone absorbs ultraviolet radiation; it is this process of energy absorption that causes the temperature rise throughout the stratosphere.

The relative proportion of molecular oxygen and nitrogen is the same in the stratosphere as in the troposphere, which indicates considerable vertical mixing takes place. However, the vertical mixing in the stratosphere is slow because there are positive temperature gradients, that is, temperature increases throughout the stratosphere with no colder air lying above warmer air.

The insignificant amount of water vapor in the stratosphere means the transfer of latent heat

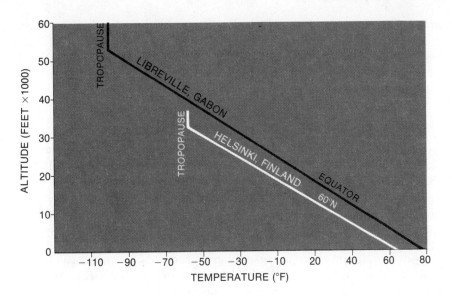

FIGURE 2–19 Temperature profiles through the troposphere over Libreville and Helsinki during July are detailed in this graph. The troposphere is characterized by a decrease in temperature with height.

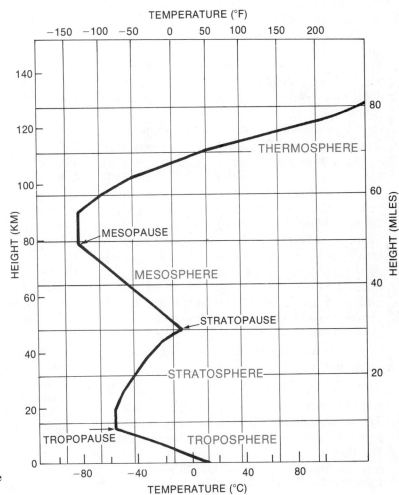

FIGURE 2–20 Each layer of the atmosphere has a unique lapse-rate pattern.

of vaporization and condensation is of no importance. Thus, clouds are rare in this layer, but occasionally clouds are observed between 12.5 and 18.5 miles (20 to 30 km). These stratospheric clouds are called mother-of-pearl, or **nacreous**, clouds, because of their coloring.

MESOSPHERE

The region above the stratosphere is called the *mesosphere*, which means middle sphere. The mesosphere extends from 31 to 53 miles (50 to 85 km); it is a region of diminishing temperature as depicted by the graph in Figure 2–20. At the **mesopause**, the temperature reaches a minimum value averaging about −120°F (−84°C). The minimum value, however, can be as much as 45 Fahrenheit degrees (25 C degrees) lower or higher than the average value. The mesopause is the region where the lowest temperatures in the atmosphere are found.

The relative proportions of molecular oxygen and nitrogen remain rather constant from the Earth's surface to the mesopause. For this reason, the troposphere, stratosphere, and mesosphere are referred to collectively as the *homosphere*.

An unusual aspect of the mesosphere is the appearance of clouds that are normally observed at high latitudes during the summer season. Since the clouds seem to glow after nightfall, they are called **noctilucent**, which means nightshining. The noctilucent clouds are roughly at altitudes of 50 miles (80 km), which allows them to direct reflected sunlight to a person on the Earth's surface long after the observer has experienced sunset. Rocket investigations of noctilucent clouds indicate they consist of cosmic dust with a high nickel content that is lightly covered with ice. With temperature in this region hovering on the average around −112°F (−80°C) to lows of −148°F (−100°C), it is surprising to find water in any state!

THERMOSPHERE

This layer starts at a height above 53 miles (85 km) and lies mostly in the range of the heterosphere. The thermosphere is a region of rapid temperature increases; in fact, values hover around 1300°F (700°C) at an altitude of 125 miles (200 km). Of course, the temperatures recorded in the thermosphere are a measure of the speeds of individual molecules, which are separated from neighbors by great volumes of space.

THE CHEMOSPHERE

Concepts other than composition and temperature can be used to regionalize and investigate the atmosphere. One productive approach is to study the chemical processes taking place in the atmosphere. The area of the atmosphere in which certain chemical effects are produced by solar radiation is called the **chemosphere**. Generally, this zone is considered to extend from the tropopause to an altitude of 120 miles (195 km). The chemosphere overlaps both the homosphere and the heterosphere.

THE OZONE LAYER

The region of the chemosphere in which ozone is produced is commonly called the **ozone layer;** a more descriptive term for this region is ozonosphere, since it really is a shell of ozone that surrounds the Earth.

It has already been noted that the zone of ozone production is centered in the stratosphere. Ultraviolet light with wavelengths in the vicinity of 1800 angstrom* units is absorbed by oxygen in the region of the stratosphere up to altitudes of 31 miles (50 km). The energy absorbed dissociates molecular oxygen (O_2) into atomic oxygen (O), which is continuously reformed into ozone (O_3). The process itself releases heat, and then the ozone formed absorbs ultraviolet radiation at 2550 Å, which tends to decompose the ozone and maintain a chemical balance in the region. The ozone layer thus extinguishes all wavelengths of ultraviolet radiation below 3000 Å. If these ultraviolet rays were not absorbed, they would bombard the surface of the Earth and make life as we know it impossible. The strong ultraviolet absorptions are largely responsible for the thermal state of the stratosphere.

Two widely used items, spray cans and nitrogen fertilizers, are believed to be threatening the

---

* Angstrom (Å) is a linear unit commonly used to measure very short wavelengths. An angstrom equals 0.00000001 or 10^{-8} centimeter.

Bacterial activity on the nitrogen fertilizer used to produce this potato crop near Harrah, Washington, in the Yakima Valley on the Columbia Plateau about 15 miles (24 kilometers) south of Yakima has an effect on the ozone layer. Therein lies a dilemma: How do we maintain productive farm areas without threatening the survival of the ozone layer? Fortunately, the ash from the May 1980 eruption of Mount St. Helens that fell south of Yakima was light and only of nuisance value (see page 171); the ash did not affect the productivity of this fertile area. (USDA)

survival of the ozone layer. Fluorocarbons used as propellants in some spray cans float up to the ozone layer, absorb ultraviolet light, and emit chlorine. The chlorine, scientists assert, sets off chemical reactions that destroy ozone. Oxides of nitrogen, notably nitrous oxide (N_2O), produced by bacterial activity on nitrogen fertilizer, are feared to have a similar effect. Air currents carry nitrous oxide aloft into the stratosphere where ultraviolet radiation converts it into nitric oxide (NO). Each nitric oxide molecule acts as an agent that destroys thousands of ozone molecules before it is carried back into the lower atmosphere.

A phenomenon called **airglow** occurs in the chemosphere. Its occurrence is fairly evenly distributed over all latitudes. Airglow is observed best on a moonless night; it is seen as a spread of diffuse light across the sky and accounts for several times as much radiant energy reaching the Earth in visible and ultraviolet wavelengths as all the stars combined. The luminous haze of airglow has been observed and reported by astronauts as they circled the Earth.

Reds, greens, and yellows are the predominant colors of the airglow phenomenon. Airglow is believed to be produced by the recombination of atoms and from reactions between atoms and molecules of oxygen, nitrogen, hydrogen, and

sodium that have been excited by solar radiation. The phenomenon has been observed generally but not exclusively at altitudes between 60 and 100 miles (95 to 160 km), which places it in the altitude range of the heterosphere at the lower reaches of the thermosphere.

THE IONOSPHERE

Another interesting effect produced by solar radiation in the upper atmosphere is ionization. Atoms of nitrogen and oxygen in the altitude range from 35 to 600 miles (55 to 965 km) absorb gamma rays, x-rays, and the shortest wavelengths of ultraviolet radiation. The input of energy causes the atoms to lose electrons and to become positively charged atoms, called *ions*. The electrons ejected from the nitrogen and oxygen atoms are set free to travel as electric currents, which gives the region a high electrical conductivity.

Four distinct regions of ionized particles have been identified as the **ionosphere**. By agreement, these layers of ionized particles are referred to as D, E, F_1 and F_2 regions, which are shown schematically in Figure 2–21. Both the ion concentration and the altitude of maximum electron density vary from day to day, with time

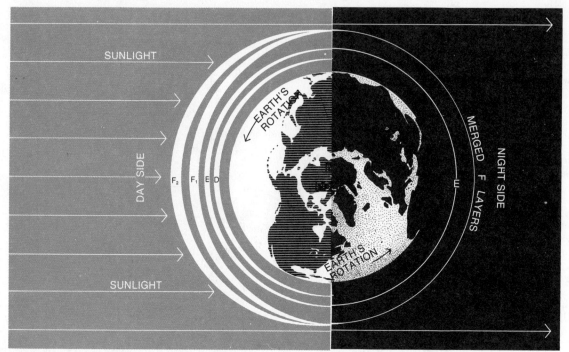

FIGURE 2–21 Radiation from the Sun produces a number of effects in the atmosphere. The four distinct regions which are identified in this sketch on the daytime side of the Earth, are known collectively as the *ionosphere*. The ionosphere is a Sun-produced effect.

of day, and with the season. For purposes of classification and simplification, however, we can say that the daytime expanse of each region is generally as follows: D, 35 to 55 miles (55 to 90 km); E, 55 to 90 miles (90 to 145 km); F_1, 90 to 150 miles (145 to 240 km); and F_2, 150 to 600 miles (240 to 965 km). At night, the D region disappears and the two F regions merge. These changes raise the base of the ionized layer during the nighttime. The D region of the ionosphere is just above the ozonosphere. In this area immediately above the ozonosphere, radiation at 1216 Å dissociates molecular oxygen (O_2) into atomic oxygen (O), giving rise to the D region. The E region is centered at an altitude of 62 miles (100 km) and owes its characteristics to the absorption of x-rays, which ionize molecules of nitrogen. In the F regions, absorption of radiation between 100 Å and 800 Å produces positive oxygen ions.

THE MAGNETOSPHERE

The Earth has a magnetic field, which is often referred to as a **dipole field** because it resembles the field of a bar magnet with its two magnetic poles, one designated north and the other south.

The Earth's dipole field is generated in the metallic liquid core of the Earth. The liquid core produces an effect similar to a gigantic bar magnet running not quite through the Earth's center and tilted about 11 degrees to the spin axis as depicted in Figure 2–22. Visualized in three dimensions, the idealized lines of force of a simple dipole magnet system within the Earth extend outward in great sweeping curves, which connect points in the Northern Hemisphere to corresponding points in the Southern Hemisphere. The bundles of forcelines create a figure shaped much like a doughnut, with the Earth at the center of the doughnut.

The lines of the Earth's magnetic field arrange themselves in a regular pattern as they emanate from the magnetic poles (Fig. 2–22). They therefore should be found in a symmetrical pattern around the Earth; but they are not! A force in space distorts the pattern. An outflow of high-speed protons and electrons from the Sun, the solar wind, exerts a pressure that compresses the boundary between the geomagnetic field and itself to a distance of 40,000 miles (64,370 km) on the day side. The boundary between the Earth's magnetic field and the solar wind de-

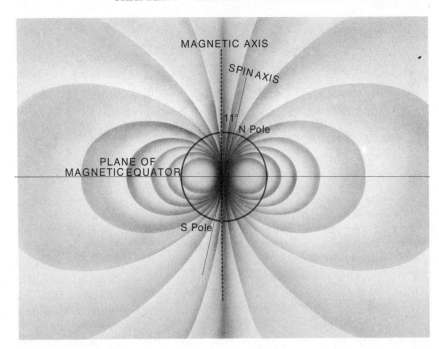

FIGURE 2–22 The Earth's magnetic field is generated by its metallic liquid core. The magnetic axis is tilted about 11 degrees with respect to the Earth's spin axis.

picted in Figure 2–23 is called the **magneto-pause.** Everything below the magnetopause is referred to as the **magnetosphere.**

VAN ALLEN BELTS
During periods of high solar activity, particles of radiation penetrate the magnetopause on the night side of the Earth where the boundary is not strong. The particles that penetrate arrange themselves under the influence of the Earth's magnetic field. These charged particles— protons and electrons—were discovered for the first time in 1958 by U. S. satellites, Explorer I and Explorer II. In 1959, James Van Allen, a

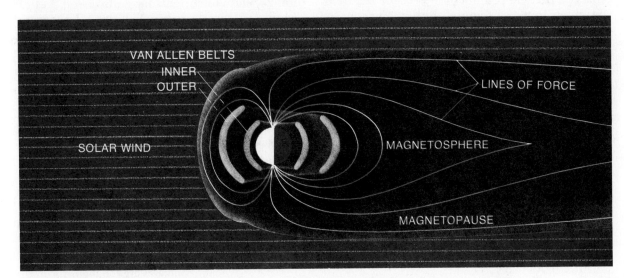

FIGURE 2–23   The distortion in the Earth's magnetic field is produced by the solar wind. The magnetosphere is the boundary between the geomagnetic field and the solar wind. Everything below the magnetopause is referred to as the *magnetosphere.*

FIGURE 2–24   The two doughnutlike rings that surround the Earth are known as the *Van Allen radiation belts.*

physicist at the State University of Iowa, described these belts which now bear his name.

The Van Allen radiation belts depicted in Figure 2–24 form two doughnutlike rings that surround the Earth. These rings are parallel to the plane of the geomagnetic equator. The inner belt lies at a distance of one to two earth-radii above the Earth. The outer belt encircles the Earth at a distance of three to four earth-radii.

Although traveling at high speeds individually, the charged particles of the **Van Allen belts** are clustered like a swarm of bees in flight. The charged particles cannot easily escape the belts in which they are trapped; but when they do, they are guided toward the Earth and have a pronounced effect on the ionosphere where they produce magnetic storms and disrupt radio communication.

THE AURORAE

The **aurora borealis** (northern lights) and the **aurora australis** (southern lights) are phenomena associated with the magnetosphere. These phenomena are seen, in a roughly circular band around the Earth at 65° to 70° magnetic latitude, as light bands that continually shift in pattern and intensity. The most familiar patterns— ribbon, rays, and arc—are sketched in Figure 2–25. Generally, the bottom of the aurora is never lower than 40 miles (65 km), whereas the top has been measured at altitudes up to 560 miles (900 km). The arcs are seldom more than 1 or 2 miles (2 or 3 km) thick, but a single aurora can stretch from east to west for more than 1000 miles (1600 km).

The best explanation of this light display indicates that it is produced by excitation of gaseous atoms in the heterosphere. The magnetic lines of force that bound the outer Van Allen radiation belt come down to the Earth's surface at the rings of maximum auroral intensity. Thus it appears that the electrons and protons of the outer Van Allen belt give rise to the auroral lights. For example, during the periods of great solar activity, charged particles enter the magnetosphere and are accelerated toward the poles. The charged particles oscillate along lines of the geomagnetic field and strike the gas molecules of the heterosphere. With each collision, the gas molecules emit the various glows of light. Most common are greenish lights, though red, blue, orange, and violet lights have also been seen. It is believed that the greenish lights are produced by the excitation of oxygen. The excitation of molecular nitrogen may account for the red to violet range of lights.

ATMOSPHERIC PRESSURE

The force of gravity draws all matter toward the Earth. The atmosphere is no exception; its

RIBBON   RAY   ARC

100 MILES

60 MILES

FIGURE 2–25   Ribbons, rays, and arcs are the typical patterns seen in aurorae displays. The lines of the aurora descend along the geomagnetic field.

gases are attracted earthward, too. The molecules of these gases tend to crowd together under the influence of gravity, and they become progressively more dense from the outer limits of the atmosphere to sea level.

MEANING OF PRESSURE

The force of gravity pulls the atmospheric gases toward the center of the Earth, and, thus, gives air its weight. Any one layer of the atmosphere is being compressed by the weight of the layers above it. Visualize, for a moment, a vertical column of air similar to the one depicted in Figure 2–26 that measures one inch (2.54 cm) on each side at its base. From this one square-inch (6.5 cm²) base, extend the column upward toward the outermost limits of the atmosphere. If it were possible to weigh this column of air that reaches from sea level to the top of the atmosphere, its weight would be 14.7 pounds (6.7 kg).

Pressure is defined as a ratio of force to area. The column of air described in the previous paragraph pushes on the Earth's surface with a force equivalent to its own weight of 14.7 pounds (6.7 kg). The force is distributed over an area of 1 square inch at the Earth's surface; it is reported as 14.7 pounds per square inch (1 kg per cm²), which is the ratio defined as pressure.

Atmospheric pressure is recorded on weather maps in a unit called the **millibar** (mb). A millibar represents a pressure of 1.0197 grams per square centimeter (1.0197 g/cm²). The millibar is derived from the bar, which is a conventional unit of pressure used by engineers. The bar is 1000 times larger than a millibar. Since 1000 grams is equal to 1 kilogram (1 kg), 1 bar is equivalent to a pressure of 1.0197 kilograms per square centimeter.

A pressure of 14.695 pounds per square inch is defined as one standard atmosphere by international agreement. The pressure equivalent in millibars of one standard atmosphere is 1013.2 millibars; or if you prefer to use metric units, the equivalent pressure is 1.033 kilograms per square centimeter.

VERTICAL DISTRIBUTION OF PRESSURE

As noted above, atmospheric pressure is simply the weight of the column of air above a unit-area at the point in question; the column, of course, extends to the top of the atmosphere. If a unit-area at an elevation above sea level is investigated, the pressure will be less than at sea level because there is a shorter column of air above the higher elevation. This means that the pressure found at a particular unit-area is a function of the unit-area's altitude.

The distribution of pressure in a vertical column of a representative middle-latitude atmosphere is given in Table 2–3. The pressure at any fixed level in the troposphere may vary by ±5 percent from its mean value as a result of travel-

FIGURE 2–26 The weight of a vertical column of air with a 1-inch square base that extends to the outermost limits of the atmosphere is 14.7 pounds.

TABLE 2–3 PRESSURE AS A FUNCTION OF ALTITUDE

ALTITUDE		PRESSURE
Km	Miles	Millibars
0	0	1013
2	1.2	795
4	2.5	616
6	3.7	472
8	5.0	356
10	6.2	264
15	9.3	120
20	12.4	55.21
30	18.6	11.52
40	24.8	2.78
50	31.0	0.93
60	37.3	0.35
70	43.5	0.12
80	49.7	0.03
90	55.9	0.008
100	62.1	0.003

ing pressure patterns known as *cyclones* and *anticyclones*, which are intimately related to wind and weather.

Since the atmosphere is very compressible, pressure falls very rapidly with each increase in elevation through the lower layers of the atmosphere and, then, much more slowly in the upper layers. In Table 2–3, for example, at an altitude of a mere 1.2 miles (2 km), the pressure reading of 795 millibars is only 78.5 percent of the sea-level reading. A pressure of 472 millibars at 3.7 miles (6 km) is a decrease of 53.4 percent from the reading at sea level. A mountain climber at an elevation of 5 miles (8 km) on the slopes of Mt. Everest is living and working in very rarefied air with pressure around 365 millibars, which is only 35 percent of the sea-level pressure. A simple computation shows that atmospheric pressure has fallen by almost 99 percent at 18.6 miles (30 km). Table 2–4 can be used as a ready reference to find the pressure equivalent in millibars for air pressure reported in inches of mercury.

OCEAN-ATMOSPHERE INTERACTION

It has often been suggested that the surface circulation of the ocean is wind-driven. Wind does play a role in the generation of the ocean's surface currents but the cause-and-effect relationship is not simple. The exchange of heat between the atmosphere and the ocean also makes a significant contribution to the development and maintenance of ocean currents. The important point at this juncture in our discussion, however, is that major currents—depicted in Figure 2–27—have been identified within the sea's surface, and these currents are the result of a complicated interaction between the ocean and the atmosphere. In addition, the salinity of seawater, sea-surface temperature, and air temperatures are also affected by ocean-atmosphere interaction.

SALINITY OF SEAWATER

The total amount of solid material dissolved in a kilogram of seawater is referred to as **salinity** and is expressed in grams per kilogram, or parts per thousand, written as %₀. Salinity is determined by chemical analysis of water samples that are collected from the surface and depths of the ocean.

The average salinity of seawater is about 35 parts per thousand. This means there are about 35 grams of salt in a kilogram of ocean water, which in more familiar terms amounts roughly to 1 ounce of salt in a quart of water. The fact that there is salt in seawater means that its freezing point is depressed and its boiling point is elevated.

Evaporation and precipitation affect salinity. When there is high evaporation from the sea's surface, the salt concentration increases. If high evaporation is accompanied by low precipitation—that is, very little rain or falling

TABLE 2–4   **PRESSURE EQUIVALENTS**

MERCURY IN COLUMN INCHES	MILLIBARS
29.00	982.00
29.05	983.70
29.10	985.39
29.15	987.08
29.20	988.78
29.25	990.50
29.30	992.16
29.35	993.86
29.40	995.55
29.45	997.24
29.50	998.94
29.55	1000.63
29.60	1002.32
29.65	1004.01
29.70	1005.70
29.75	1007.40
29.80	1009.09
29.85	1010.78
29.90	1012.48
29.9213 ———one atmosphere———	1013.20
30.00	1015.87
30.05	1017.56
30.10	1019.25
30.15	1020.95
30.20	1022.64
30.25	1024.33
30.30	1026.03
30.35	1027.72
30.40	1029.41
30.45	1031.10
30.50	1032.80

Giant waves develop under the influence of the wind in the roughest coastal waters of the United States outside Yaquina Bay, Oregon. A 52-foot Coast Guard motor lifeboat is covered with spume as it attempts to move through one of these giant waves. (U.S. Coast Guard)

snow are being carried into the sea—salinity increases at a rapid rate. A combination of low evaporation and heavy precipitation causes the salinity of seawater to decline because the seawater is diluted by the addition of freshwater.

The salinity of sea-surface water tends to be relatively low in equatorial regions because of heavy rainfall. In the subropical latitudes (25°N and S), there is a minimum of rainfall with a high rate of evaporation and, thus, this area of the world ocean tends to have the highest relative surface salinity. As you move from the subtropical areas to the poles, salinity tends to decrease; the decreasing salinity is the result of increasing rainfall, decreasing evaporation, and lower relative temperatures.

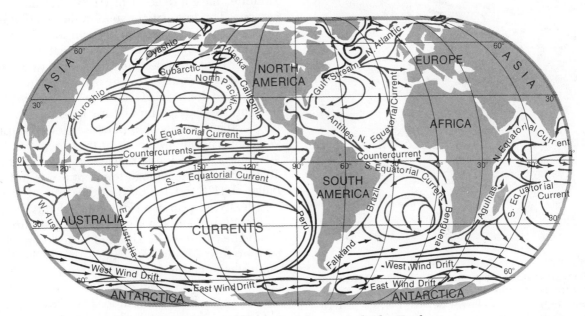

FIGURE 2–27   There is a pattern to the ocean currents that encircle the Earth.

The salt content at a depth of 110 fathoms (200 m) is plotted as **isohalines,** lines joining points of equal salinity, in Figure 2–28. The area of the Atlantic richest in salts is a zone that spreads from Gibraltar westward. The high salinity (36 to 36.5%) is probably due to the outflow of very salty water from the Mediterranean—an inland sea located in a relatively hot area subject to a lot of evaporation.

In enclosed basins—the Great Salt Lake (41°10′N, 112°30′W) and the Dead Sea (31°30′N, 35°30′E), for example—the water tends to have a high salinity because evaporation exceeds precipitation in these areas. The salinity in the Dead Sea is greater than 200 parts per thousand. The Red and Mediterranean Seas also have high salinities—around 40 parts per

thousand. This figure is understandable when you realize that the Mediterranean receives relatively little freshwater from rivers and the Red Sea receives none.

As salinity increases, the freezing point decreases. Water in a bay whose seawater is measurably diluted by freshwater may have a salinity of 25 parts per thousand and a freezing point of 29.7°F (−1.3°C). Ocean water of average salinity, 35 parts per thousand, has a freezing point of about 28.9°F (−1.7°C). The lowest water temperature found in the ocean is 28.9°F (−1.7°C).

SEA-SURFACE TEMPERATURES

The surface of the ocean is heated by the input of radiation from the Sun as well as by the transfer or conduction of heat to it from the atmosphere. On the other hand, the sea's surface is cooled when it radiates and conducts heat to the atmosphere and when evaporation carries water upward from its surface into the atmosphere. Below the sea surface there are horizontal currents that transfer heat from one area to another.

Since ocean water receives the preponderance of its heat from the Sun, sea-surface temperatures have a relationship to latitude and the seasons. The intensity of solar radiation, that is, the heat received per unit area on March 21 or September 22, for example, is greater at the Equator than at either the North Pole or South Pole. At the Equator in the Atlantic, the mid Pacific, and the Indian Ocean during September, the surface-water temperature is around 81°F (27°C). The average temperature decreases as you move north and south of these regions. At 40°N, the average temperature is generally about 59°F (15°C). The average temperature is in the range of 41°F (5°C) at 60°N at this time of year.

Ideally, sea-surface **isotherms**—that is, lines connecting points having equal temperature—should be parallel to the lines of latitude, and the temperature of surface waters should be the same at equal latitudes, but they are not! Surface currents distort the isotherms and cause them to cut across the lines of latitude. The cold California Current, for example, has a southward flow from 48°N to 23°N along the western coast of the

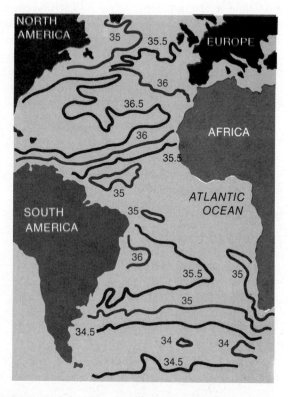

FIGURE 2–28 Isohalines, that is, lines joining points of equal salinity, are plotted for a depth of 110 fathoms (200 meters) in the Atlantic Ocean. There are different salinity trends at different depths; but the pattern represented here is not drastically different from the course of surface isohalines.

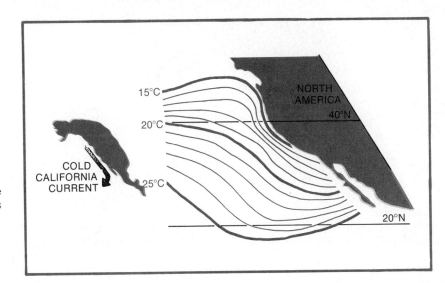

FIGURE 2–29   This plot of the average sea-surface isotherms for August documents the southward flow of the California Current off the western coast of the United States.

United States (Fig. 2–29); the sea-surface isotherms in Figure 2–29 show the colder water protruding southward. The isotherms in Figure 2–30 show cold water protruding northward; the distortion of these isotherms is caused by the Humboldt Current, which carries cold water from southern latitudes north along the coast of South America to the vicinity of the Equator.

The sea-surface isotherms off the East Coast of the United States during August are sketched in Figure 2–31. The area of the Atlantic Ocean depicted in Figure 2–31 is between the meridians 60°W and 80°W. Note that the 77°F (25°C) isotherm makes a gentle arc to the northeast just below the parallel of 40°N. From the more detailed view of this area (70°W to 75°W) in Figure

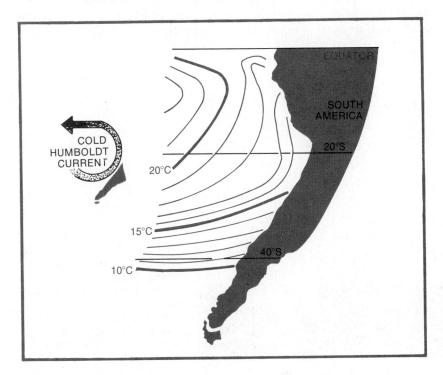

FIGURE 2–30   The distortion in these sea-surface isotherms during August in the Pacific Ocean shows that cold water is being carried northward along the South American coast.

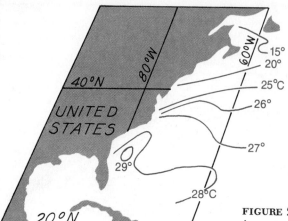

FIGURE 2–31 The sea-surface isotherms show that warm water is protruding north along the east coast of the United States.

2–32, we find the 77°F (25°C) isotherm actually represents the temperature of the water in the Gulf Stream—one of the most important of the marine currents which carries warm water from the Gulf of Mexico and the Caribbean Sea north and east across the Atlantic. The Gulf Stream starts its arc to the northeast off the coast of North Carolina around 34°N. At the mouth of the Chesapeake Bay (37°N, 76°W), the Gulf Stream is well beyond the 100 fathom line (600 ft, 180 m), which represents the shelf break. Marine cur-

rents resemble great rivers flowing through the sea; velocities in the Gulf Stream are identified by contour lines—a 4-knot contour is in the central region of the Gulf Stream in Figure 2–32.

If the yearly average sea-surface temperature is plotted, the maximum temperature is not found at the Equator. The average maximum temperature of seawater 81.5°F (27.5°C) maintains a general position between 5° and 10°N latitude. The location of the average maximum sea-surface temperature is called the *oceano-*

The ice-covered landmass of Antarctica is at the top of the photo. Three of the U.S. Navy's large icebreakers are pushing together to move a huge tabular iceberg from the channel of broken ice leading to McMurdo Station on Antarctica. (U.S. Navy)

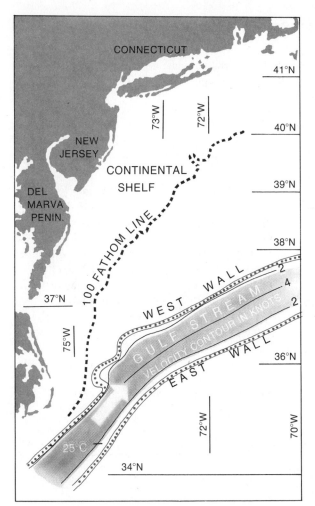

FIGURE 2–32 The Gulf Stream carries warm water north and east across the Atlantic. The greatest velocities within the stream are found in the central region.

graphic thermal equator. It is not a fixed location; the sea's thermal equator wanders but maintains a location that is in the Northern Hemisphere.

HEAT-ABSORBING CAPACITY OF WATER

The specific heat of any substance is defined as the quantity of heat required to raise 1 gram of the substance through 1° Celsius. The specific heat of water is 1 calorie per gram per 1° Celsius. This means that 10 calories of heat are required to raise 1 gram of water from 90°C to 100°C.

Water at a temperature of 100°C is at the boiling point. When water boils, liquid water is changed to gaseous water. To change 1 gram of liquid water at 100°C to 1 gram of gaseous water at 100°C, it takes a large input of energy—540 calories of heat energy to be exact. This quantity of heat, which is required to produce the change of state from liquid to gas without a change in temperature, is called **latent heat of vaporization**. When gaseous water condenses and changes to liquid water, an amount of heat equivalent to the latent heat stored is given up in the process of condensation.

About 61 percent of the Earth's surface is covered by water in the Northern Hemisphere, while 81 percent is covered by water in the Southern Hemisphere. The larger amount of water in the Southern Hemisphere—coupled with the high specific heat of water—gives the Southern Hemisphere a greater heat-absorbing capacity than the Northern Hemisphere. As a result, the average annual air temperatures for comparable latitudes tends to be lower in the Southern Hemisphere than in the Northern Hemisphere (Table 2–5). The continent of Antarctica, which has an area of 5.39 million square miles (13.97 million km²) or 9.3 percent of the Earth's surface, also helps to keep the Southern Hemisphere temperatures below those of the Northern Hemisphere. Antarctica is covered by a massive ice sheet, which functions like a huge icebox and tends to hold temperatures down in the Southern Hemisphere. Table 2–5 also shows that the belt of highest average yearly temperature, 80.1°F (26.7°C), is located at latitude 10°N; this belt is known as the **thermal equator**.

TABLE 2–5 **AVERAGE ANNUAL AIR TEMPERATURE**

	HEMISPHERE			
LATITUDE	*Northern*		*Southern*	
	°F	°C	°F	°C
80°	1.9	−16.7	− 3.6	−19.8
70	14.0	−10.0	11.3	−11.5
60	30.2	− 1.0	31.6	− 0.2
50	42.3	5.7	42.1	5.6
40	57.2	14.0	53.6	12.0
30	68.5	20.3	65.1	18.4
20	77.5	25.3	73.4	23.0
10	80.1	26.7	77.5	25.3
Equator	79.2	26.2	79.2	36.2

THE LANGUAGE OF PHYSICAL GEOGRAPHY

A physical geographer uses a technical vocabulary to make explanations. Review your understanding of the vocabulary used to develop the concepts in this chapter. Among the important words and terms used are:

abyssal plains	environmental lapse rate	magma	salinity
adiabatic	fringing reef	magnetopause	scarp
airglow	Gondwana	magnetosphere	seamount
atoll	graben	mediterranean	shelf break
aurora	halmyrogenous sediments	mesopause	Sinus Borealis
aurora australis	heterosphere	mesosphere	stratopause
aurora borealis	homosphere	millibar	stratosphere
barrier reef	hot plume	nacreous clouds	terrigenous sediments
bight	intertidal zone	noctilucent clouds	Tethyan Trench
biogenous sediments	ionosphere	normal lapse rate	thermal equator
chemosphere	island arc	ocean basin	thermosphere
cleft	isohalines	ozone	trench
continental rise	isotherms	ozone layer	tropopause
continental shelf	latent heat of	Pangaea	troposphere
continental slope	vaporization	pressure	universal gas law
cosmogenous sediments	lava	radiosonde	Van Allen belts
dipole field	Laurasia	rises	

SELF-TESTING QUIZ

1. In what ways are the ocean and atmosphere interlinked?
2. Describe the journey of a molecule of water as it moves through the hydrologic cycle.
3. What is the total volume of ocean water?
4. Trace the boundaries of the major ocean basins.
5. What are the depths in the three major ocean basins?
6. Compare the percentages of the surface covered by water and land in each hemisphere.
7. Describe the features encountered as you move eastward along the floor of the Atlantic Basin from the North American shore to the European shore.
8. What is meant by an island arc?
9. How do a fringing reef and barrier reef differ?
10. In what ways are oceanic sediments classified?
11. Trace the evolution of the Atlantic Basin.
12. Explain the significance of the trenches in the Pacific Basin.
13. How was the Deccan Plateau of India formed?
14. In what ways do the homosphere and heterosphere differ?
15. How is the region of the troposphere identified?

16. Why does temperature decrease with an increase in altitude through the troposphere?
17. What role does ozone play in the temperature pattern of the stratosphere?
18. Describe the chemical effects produced in the atmosphere by solar radiation.
19. How many regions are there in the ionosphere?
20. How are the Van Allen belts produced?
21. What is the pressure equivalent in millibars of one standard atmosphere?
22. Trace the change in pressure in a vertical column of a representative middle-latitude atmosphere.

IN REVIEW
THE OCEAN AND THE ATMOSPHERE

I. EXPLORING THE OCEAN'S BASINS

A. Shapes and Depths
B. Land-Water Comparison
C. On the Ocean Floor
   1. Mountain Ranges
   2. Sediments
D. Sea Level

II. EVOLUTION OF THE OCEAN'S BASINS

A. The Arctic Basin
B. The Antarctic Basin
C. The Atlantic Basin
D. The Pacific Basin
E. The Indian Basin
F. Other Basins

III. THE ATMOSPHERE'S REGIONAL CHARACTER

A. Based on Composition
   1. Homosphere
   2. Heterosphere

B. Based on Temperature
   1. Troposphere
   2. Stratosphere
   3. Mesosphere
   4. Thermosphere
C. The Chemosphere
   1. The Ozone Layer
D. The Ionosphere
E. The Magnetosphere
   1. Van Allen Belts
   2. The Aurorae
F. Atmospheric Pressure
   1. Meaning of Pressure
   2. Vertical Distribution of Pressure

IV. OCEAN-ATMOSPHERE INTERACTION

A. Salinity of Seawater
B. Sea-Surface Temperatures
C. Heat-Absorbing Capacity of Water

3 Water beneath the surface

In the circulatory system of the hydrologic cycle, water evaporates into the atmosphere and then falls back to the surface as precipitation. Some of the "fallen" water collects in bodies of surface water; the remainder penetrates the surface and moves underground. A portion of the Earth's subterranean water is absorbed by plant roots and reenters the atmosphere via leaf pores in a process called **transpiration;** another portion reaches the surface through streams and springs and reenters the hydrologic cycle by evaporation.

Surface water in the United States is only 4 percent of our potential freshwater supply; the remainder of our freshwater, that is, 96 percent, is below the surface. The study of these huge resources of groundwater, with particular emphasis given to its chemistry, mode of migration, and relation to the environment, is called **hydrology.** The supply of both surface and underground water is reported in a unit called an **acre-foot**—the volume of water that covers an acre to a depth of 1 foot.

Using today's technology, the total recoverable groundwater supply for the Unites States is estimated at 75 billion acre-feet. The aboveground reservoirs presently constructed and in use hold 1 billion acre-feet; our natural lakes have a capacity of 13 billion acre-feet. Even though our recoverable groundwater supply is significant, the Unites States currently gets only 20 percent of its water from this source.

Large quantities of groundwater are used in many countries of the world. Groundwater, for example, was the only source of water on Malta until the Maltese began to desalt ocean water for some of their supply. Through the 1970s, groundwater constituted more than 70 percent of the water Germans used and 54 per cent of that used in Israel.

The eruption of Yellowstone National Park's Old Faithful geyser is spectacular in summer; but the combination of winter's snow and cool temperatures makes Old Faithful's powerful eruption magnificent and awesome in the winter setting. The hot groundwater that is forced to the surface along joints and cracks in the bedrock derives its heat at depths between 8000 feet (2440 meters) and 3000 feet (910 meters). (Wyoming Travel Commission)

SUBTERRANEAN WATER

All the water beneath the land's surface is referred to as **subterranean water.** At varying depths below the surface, depending on whether it is a wet or dry season, there is a zone saturated with water, which is sometimes referred to as the **phreatic zone** (Fig. 3–1). The upper surface of the saturated zone is designated as the **water table.** Technically the term **groundwater** is only used when referring to the water in the zone of saturation. The region above the water table is called the **zone of aeration.**

FIGURE 3–1 Subterranean water can be divided into two general zones: the zone of aeration and the zone of saturation. The term *groundwater* should only be used when referring to the water in the saturated zone.

THE ZONE OF AERATION

Water in the zone of aeration, that is, above the water table, is also known as **vadose water.** The vadose water is differentiated into soil water and water of the capillary fringe (Fig. 3–1).

CAPILLARY WATER

Water in the **capillary fringe,** called **capillary water,** comes from and is connected to the zone of saturation. It is held above the zone of saturation by capillary forces, commonly referred to as **suction.** Suction, that is, the property of **capillarity,** is the ability of a liquid to "climb up" a surface against the pull of gravity. Water travels up from the water table for a few feet by capillarity to form the capillary fringe. The suction that forms the capillary fringe is similar to the force that operates when ink is drawn up to fill the pores between the fibers of an ink blotter.

The lower part of the capillary fringe may be saturated, but it is under less than atmospheric pressure. Thus, it is not part of the zone of saturation and its water will not flow into a well sunk into the fringe. On the other hand, groundwater in the zone of saturation, which is under atmospheric pressure, will enter a well sunk into the zone and will stand at the level of the water table.

SOIL WATER

A plant absorbs some water through its leaf openings, but most of the water used by a plant is absorbed by its roots from **soil water.** The roots of most plants are confined to the zone of aeration. If we are concerned with the optimum use of water in this zone, it is important to know how water moves into and through soil, how the soil stores water, and how to measure the moisture content of soil.

A raindrop that falls to the surface may remain above ground or it may enter small channels, called **interstices,** in the ground and infiltrate the soil. There are many interstices in fertile soil. They are formed by the successive moistening and drying of the soil, the boring of worms, tunnels that remain after roots decay, and the expansion of water when it freezes. Once

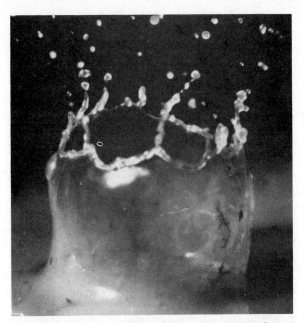

A raindrop shatters with explosive violence when it strikes the Earth's surface. The water that spreads from the point of impact may remain above ground or it may enter small channels in the ground and infiltrate the soil. (NOAA)

FIGURE 3–2   The relative positions of the hydrogen and oxygen atoms cause the water molecule to have a positive and negative end. The size of the molecule can be estimated from the relative size of the angstrom unit.

rainwater has infiltrated the soil, it is referred to as **soil moisture.** The water in a soil has an effect on soil formation, erosion, and the stability of the soil structure.

The soil acts as a reservoir for water; but it is, in fact, a leaky resevoir. When too much water is added to soil, the excess sinks into deeper soil layers. The downward movement of water through the pores and spaces in the soil under the influence of gravity is called **percolation.**

After a light rainfall all the water may be retained in the upper layers of the soil, and there is no excess to travel farther downward. The proportion held in the upper layers depends on a number of factors, including the dryness of the soil; for example, all of a short heavy downpour following a dry spell may be retained.

Forces of attraction—adhesive and cohesive—hold water in the soil against the pull of gravity. These forces are the result of the lop-sided shape of a water molecule—two hydrogen atoms bulge from one end of an oxygen atom (Fig. 3–2). The hydrogen end of the water molecule has a positive charge; the oxygen end

has a negative charge. Since opposite charges attract, a **cohesive force** develops as water molecules link together through positive-hydrogen ends attaching to negative-oxygen ends of other water molecules (Fig. 3–3). The attraction of the positive-hydrogen ends of a water molecule for negative-oxygen atoms in soil minerals is designated as an **adhesive force.**

The combination of large numbers of adhesive and cohesive forces causes water films of considerable thickness to be held on the surface of soil particles. Since both these forces are surface-attractive forces, the amount of adsorbed water is directly proportional to the soil-surface available—more surface means more adsorbed water. Soil consists of fine grains of mineral and organic matter, which vary in size from one soil type to another; clay soils, for example, have much smaller grains and, thus, more particle sur-

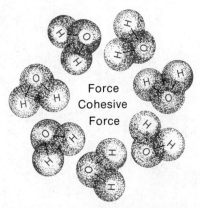

FIGURE 3–3   The positive hydrogen atom of one molecule of water attracts the negative oxygen atom of another molecule to produce a strong cohesive force.

face than sandy soils. It is natural, then, for a soil containing a lot of clay to hold more adsorbed water than a soil with less clay and more sand. The addition of organic matter to soil also increases its surface and, in turn, the quantity of adsorbed water it can hold.

A very useful number that indicates the force with which water is held in soil has been devised to make comparisons among soils. The number is derived by measuring the force required to push or to pull water off soil. In the United States, the unit of pressure called the **bar*** is used to measure these forces. The terms suction or tension are often used in conjunction with the pressure to designate the force required to pull the water from the soil. For example, when referring to the moisture content in a given soil sample, the measure "moisture at 2 bars suction" might be applied. This means that a force of at least 2 bars is needed to remove the most weakly held water from the soil.

Silts, clays, and other fine-textured soils with an adequate supply of decaying vegetation, or humus, have the greatest potential for holding on

*A bar is equal to 29.53 inches of mercury. One atmosphere, 29.92 inches of mercury, is equal to 1.0127 bars.

to moisture. They often take a long time to reach saturation, but they also resist heavy losses of their moisture. Sands and other coarse soils usually reach saturation quickly and they tend to lose their moisture just as rapidly through percolation as well as evaporation from the surface.

THE ZONE OF SATURATION

All pore spaces are filled with water in this zone and free-flowing water is present (Fig. 3–4). The pressure of water within the saturated zone is greater than atmospheric pressure; at the water table, however, water pressure is equal to air pressure. Groundwater that is not overlain by impermeable rock is said to be unconfined.

Gravity draws the groundwater to the lowest point it can reach; then, the groundwater moves along the path of least resistance, which is in the direction of a lower pressure. The movement toward a region of lower pressure can be in any direction—that is, laterally or vertically. The movement of groundwater is controlled by the nature and structure of the surrounding material.

In some locations, water percolating downward under the influence of gravity is impeded by an isolated lens-shaped layer of impermeable

FIGURE 3–4   A perched water table develops above the impervious bed of rock that is located in the zone of aeration. At a spring and along a stream bed the water table coincides with the ground surface.

material, which is above the water table of the region (Fig. 3–4). The collection of water above such an impervious bed of rock produces a zone of local saturation within the zone of aeration. This combination of circumstances is sketched in Figure 3–4 and is designated a **perched water table.** Perched water tables can serve as important sources of water when the regional water table is found at great depths—as in a desert, for example.

There is a lower limit to the saturated zone because the groundwater ultimately reaches bedrock that is squeezed so tightly that for all practical purposes no pores exist in it and no water-passage cracks are open. The maximum depth at which rocks making up the Earth's crust may still be porous is 10 miles (16 km). This depth, then, is the theoretical lower limit at which free groundwater may exist. Water has been obtained from a depth of 2 miles (3.2 km); but, in actual fact, very little groundwater is found at depths below 3000 feet (900 m).

THE WATER TABLE

In the real world of soil and rocks, the water table is not a well-defined, stable surface. Its depth under an area can fluctuate with the seasons and with the amount of precipitation that falls and infiltrates the ground. Ordinarily, the water table is no more than 10 feet (3 m) below the surface.

In some locations—a mountainous region, for example—it may lie more than 330 feet (100 m) below the surface. In an area in which the zone of aeration is not present, the water table may coincide with the surface of the land to produce a spring (Fig. 3–4). If the water table comes to the surface over an extensive area, a swamp or a marsh may develop. The water table is also at the surface along the shore of a lake or the bank of a stream.

ORIGIN OF GROUNDWATER

Groundwater is derived from three sources: (1) Chemical and physical processes deep within the Earth release water, which is called **juvenile water** to indicate it is moving toward the surface for the first time and it has never been part of the hydrologic cycle. Volcanic activity, for example, releases juvenile water. (2) Some water is stored in deep-lying sedimentary rocks and is a relic of the ancient seas in which the sediments were deposited. When this **connate water,** as it is called, is released from the sedimentary rocks in which it is trapped, it becomes part of the groundwater supply. (3) The third source of groundwater is **meteoric water,** that is, water derived from precipitation that infiltrates the ground.

The main way in which groundwater is replenished is through the downward percolation

The chief way in which groundwater is replenished is through the downward percolation of meteoric water. The water entering these waterspouts, which were photographed in the Bahamas, is moving from the sea surface to the dark cloud above. Eventually this water will fall from the cloud as precipitation; some of it will work its way below the surface to become groundwater. (NOAA)

of meteoric water. The total quantity of water from juvenile and connate sources is insignificant when compared to the quantities derived from meteoric water. The origin of this percolating water is either the direct infiltration of rainwater or snowmelt or infiltration from bodies of surface water that have, in turn, been supplied by rain or snowmelt.

GROUNDWATER RESERVOIRS

A layer below the surface through which groundwater moves is called an **aquifer.** Some of these underground formations are capable of holding and yielding significant volumes of water; they should be thought of as reservoirs. The value of an aquifer is based on the volume of water it holds and the rate at which the water flows. The areas depicted in Figure 3–5 are underlain by aquifers capable of yielding 50 gallons per minute (gpm) to individual wells tapped into them. The type of aquifer from which the water flows is also identified in Figure 3–5. Note, however, the 50-gpm yield is simply the lower limit of water flow from these aquifers; the scale

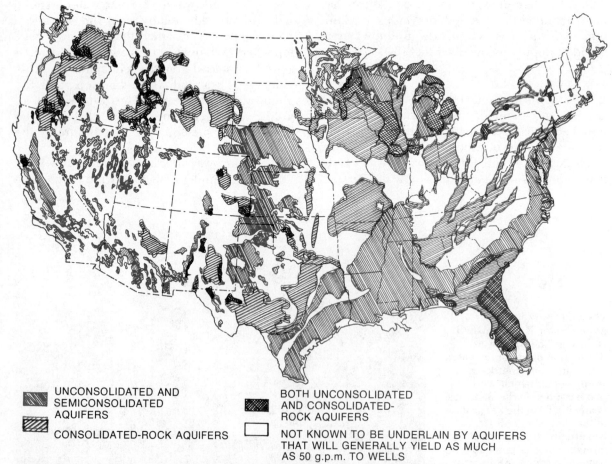

UNCONSOLIDATED AND SEMICONSOLIDATED AQUIFERS

CONSOLIDATED-ROCK AQUIFERS

BOTH UNCONSOLIDATED AND CONSOLIDATED-ROCK AQUIFERS

NOT KNOWN TO BE UNDERLAIN BY AQUIFERS THAT WILL GENERALLY YIELD AS MUCH AS 50 g.p.m. TO WELLS

FIGURE 3–5 With the use of the key, you can identify areas underlain by aquifers that yield a minimum flow of 50 gallons per minute to an individual well.

at which the data are presented allows no opportunity to define the range of the yield with any degree of accuracy.

Figure 3–5 shows the 50-gpm yields in South Carolina are primarily from unconsolidated and semiconsolidated aquifers. The consolidated rock aquifers with this minimum yield are in the southern tip of the state. A more detailed analysis of these South Carolina aquifers shows the yields actually range from a low of 100 gpm to a high of 1000 gpm. In fact, the South Carolina aquifer areas depicted in Figure 3–5 can be divided into four regions on the basis of yield (Fig. 3–6). The yield for wells in the zone labeled Region A ranges from 500 to 1000 gpm; in Region B, the yield per well is 200 to 600 gpm; Region C wells yield 200 to 400 gpm; and along the coastal area, Region D, a typical well yields 100 to 150 gpm. In the Piedmont of South Carolina, the minimum yield per well falls below 50 gpm; but even in this region the range is 25 to 200 gpm.

AQUIFER CAPACITY

The amount of water that can be stored in an aquifer depends on the size and number of spaces it contains. This property of an aquifer for containing voids, or interstices, is termed **porosity** and is expressed quantitatively as the percentage of the total volume of the aquifer that is occupied by openings. Porosity ranges from a high of 80 percent in newly deposited silt and clay down to a very small fraction of 1 percent in a compact rock.

Table 3–1 lists the average porosities for some selected sediments and rocks. Although a newly deposited clay may have a porosity of 80 percent, most clay deposits have porosities that run from 50 to 60 percent, with an average around 55. Loose, unconsolidated layers of well-assorted sand or gravel, which consist of grains of uniform size, are among the most porous sediments. The pore spaces in sand

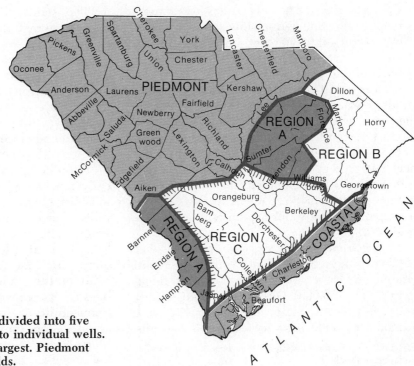

FIGURE 3–6   South Carolina is divided into five regions based on aquifer yields to individual wells. The yields in region A are the largest. Piedmont aquifers produce the lowest yields.

TABLE 3–1 AVERAGE POROSITIES

SEDIMENT	PERCENT	ROCK	PERCENT
Silt	30	Tuff	45
Clay	55	Sandstone	15
Sand	35	Limestone	5
Gravel	30	Slate	0.5

usually account for 30 to 40 percent of the volume of the sediment—in some instances it may run as high as 50 percent. Generally, 25 to 35 percent of the volume of gravel is void space. When unconsolidated sand or gravel deposits are poorly assorted with small grains of clay filling the space between the larger grains, the porosity drops significantly.

Shale is a sedimentary rock; it is formed when clay sediments are compacted and consolidated. The volume of pore spaces in shale can range from 15 percent to a low of 5 percent, depending on how much compaction has occurred in the consolidation of the clay. In some shales, the fine particles of clay are consolidated and compacted so well that the rock has a very low porosity.

Sandstone and limestone are also sedimentary rocks. The porosity of sandstone depends on compaction, too, and usually ranges from 5 to 25 percent. A very fine-grained limestone may have a low porosity of 0.1 percent; but a coarse limestone can have pore spaces that amount to 10 percent of the rock's volume.

Volcanic ash that has been transported by winds or water currents, deposited in layers, and then compacted is called **tuff** (Table 3–1). Since the rock is stratified it is classed as sedimentary. Volcanic tuff that has been tightly compacted may have a porosity of 10 percent. Some beds of tuff, however, have porosities of 80 percent. Alameda Square in the center of Mexico City sits on an underlying water-logged aquifer of volcanic ash.

The volume of pore spaces in igneous* and metamorphic* rock averages less than 1 percent. Granite, which is an igneous rock, can range from a low of 0.0001 percent to a porosity just under 1 percent. Slate, a metamorphic rock formed from shale, has very few pore spaces; its

*See pp. 163–171 for a detailed discussion of igneous and metamorphic rocks.

porosities range from lows of 0.001 to highs of 1 percent.

A porosity of less than 5 percent is regarded as small. When pore spaces in a sediment or rock range from 5 to 20 percent of the volume of the material, the porosity is classified as medium. For a porosity to be considered large it must range above 20 percent. A caution is in order: Porosity indicates only how much water a sediment or rock will hold, not how much it will yield!

FLOW IN AN AQUIFER

The rate at which groundwater flows through an aquifer is determined by the size of the pores and the extent to which they are interconnected. This capability of an aquifer to transmit water is called **permeability.**

The porosity of a sediment or rock has no direct relationship to an aquifer's permeability or water-yielding capacity. If the pores are small, the aquifer will transmit water very slowly. On the other hand if the pores are large and interconnected, the aquifer will transmit water readily.

The standard coefficient of permeability used in the hydrologic work of the U.S. Geological Survey is defined as the rate of flow of water at a standard temperature of 60°F (15.5°C) through a cross section of 1 square foot (0.3 m²). Under field conditions the adjustment to standard temperature is commonly ignored. Thus, permeability is often expressed at the prevailing water temperature in gallons per square foot per day or as liters per square meter per day.

In order for water to be withdrawn economically from an aquifer, the sediments or rocks must have high porosity as well as high permeability. For example, although clay has a high porosity, it is a poor aquifer because its permeability is negligible at less than 0.000008 gallon per square foot per day (0.00034 liter/m²/day). Sand, on the other hand, makes a good aquifer because its porosity is high (Table 3–1) and its permeability is also high, in some cases as much as 84 gallons per square foot per day (3400 liters/m²/day). Gravel also makes a good aquifer because it has high porosity and its permeability

Two-centuries-old water flows from a pipe tapped into a recently discovered underground reservoir beneath the Negev Desert. A new oasis to serve the Bedouins will develop as a result of this water. (Govt. of Israel)

can run as high as 8350 gallons per square foot per day (340,000 liters/m²/day).

Most limestones have medium to low porosities with permeabilities that range from 0.0000008 to 0.08 gallon per square foot per day (0.000034 to 3.4 liters/m²/day). Some limestones are good aquifers not because they are porous or permeable but because they are full of cracks, eroded joints, and solution cavities. The limestone aquifer of Malta, an island in the Mediteranean Sea, about 60 miles (100 km) from Sicily, is a good example of such a situation (Fig. 3–7). The island of Malta consists of a fractured and warped plateau of dense limestone. The highest point on Malta is 845 feet (257.5 m) above sea level. The water table as shown in Figure 3–7 lies only 10 feet (3 m) above sea level. Through a carefully planned system of wells the Maltese are able to extract 10 million gallons of water per day from their limestone aquifer.

AQUIFER TYPES

The area where water enters the ground to replenish an aquifer is called the *recharge area* (Fig. 3–8). The rate of recharge depends on the amount of precipitation that falls and the proportion that infiltrates the surface and percolates through the zone of aeration to the aquifer. Two types of aquifer are shown in Figure 3–8: an unconfined aquifer and a confined aquifer.

UNCONFINED AQUIFER
The groundwater in an **unconfined aquifer** is not overlain by an impermeable layer. Water falling on the recharge area in the vicinity of the pump well in Figure 3–8 percolates downward until it meets an impermeable layer. The impermeable layer of shale, in this case, is a barrier to the further downward movement of groundwater. An impermeable layer is referred to as an **aquiclude.**

Since the water table in an unconfined aquifer is at atmospheric pressure, it rises and falls according to the amount of inflow and outflow. A well is simply an opening that is dug deep enough to encounter the zone of saturation. When a well is sunk into an unconfined aquifer, the water must be brought to the surface by some kind of pumping action because the water table is at atmospheric pressure.

As water is removed from a pump well, a slight depression is produced in the water table immediately surrounding the well. Generally, these depressions are temporary and tend to disappear when the pumping action is stopped. When the pumping action is continuous and huge volumes of water are withdrawn to supply the needs of large irrigation projects, industrial processes, or cities, the depression can be extensive and can result in the lowering of the water table throughout a whole district.

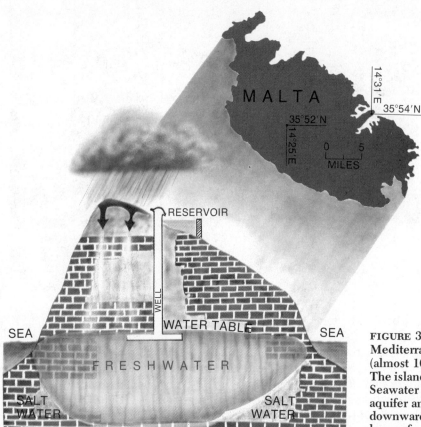

FIGURE 3–7 Malta is an island in the Mediterranean Sea, about 60 miles (almost 100 kilometers) south of Sicily. The island consists of fractured limestone. Seawater has intruded into the limestone aquifer and the freshwater that percolates downward from the surface sits on a layer of saltwater.

ARTESIAN AQUIFER

A **confined,** or **artesian, aquifer** is one that lies between two impermeable strata, or aquicludes (Fig. 3–8). Since much of the Earth's crust has a fairly well-defined, layered structure in which zones of high and low permeability alternate, situations are common in which groundwater moving laterally in a permeable sediment or rock passes between layers of relatively low permeability. An artesian aquifer often lies beneath an unconfined aquifer.

The recharge area for the confined aquifer is at the left in Figure 3–8. The bed of the confined aquifer is tilted; it moves down and below the surface. Its recharge area is at a greater elevation than the bed of the unconfined aquifer; the dif-

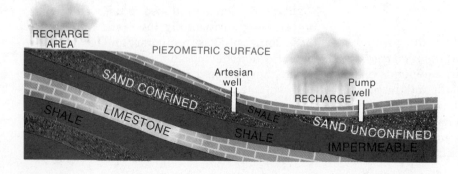

FIGURE 3–8 The confined aquifer of sand is wedged between two impermeable layers of shale. There is no impermeable layer above the unconfined sand aquifer.

ference in elevation establishes the **hydraulic "head,"** or pressure. Thus, groundwater in an artesian aquifer is pressing upward against the confining bed because its pressure is greater than atmospheric pressure.

When a well is drilled into a confined aquifer, the water rises under pressure to a level higher than the aquifer itself. Wells tapped into a confined aquifer are called **artesian wells.** The water pressure in an artesian well is roughly equal to the difference in elevation between the well and the recharge area. The pressure also depends to a small extent on the weight of the overlying aquiclude. The level to which the artesian water rises under pressure for every point in the aquifer is called the **piezometric** surface (Fig. 3–8).

Artesian wells are an important source of water in many localities. Artesian conditions, for example, exist in and around Artois, France; along much of the Atlantic Coastal Plain of the United States; in North and South Dakota; in Nebraska, Kansas, Illinois, Indiana, Missouri, and Arkansas; in the northern Sahara; on the Arabian Peninsula; and in six large basins in Australia.

One of the largest artesian systems in the world—the Great Artesian Basin—is in the Central Lowlands of Australia (Fig. 3–9). The recharge area for the artesian aquifer that runs

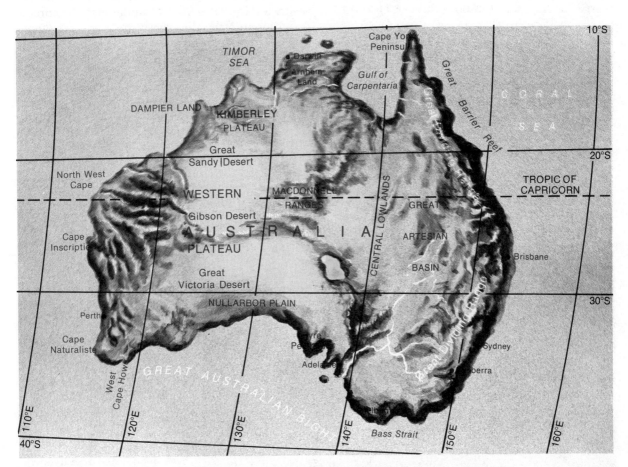

FIGURE 3–9 **The Great Artesian Basin is one of the largest artesian systems in the world. A gravel-bed aquifer, which is sandwiched between impermeable layers, is the source of the artesian water. The aquifer runs from the western slopes of the Great Dividing Range to Lake Eyre region.**

under the Great Artesian Basin is in the Eastern Highlands. The Eastern Highlands, with an average width of 150 miles (240 km), stretch along the entire eastern coast; a very narrow coastal plain separates this region from the sea. The Central Lowlands extend from the Gulf of Carpentaria in the north to the eastern shore of the Great Australian Bight in the south. The Great Artesian Basin covers 678,000 square miles (1,756,000 km²) of the Central Lowlands and supplies 492 million gallons of artesian water per day to wells in the area.

A gravel-bed aquifer of Jurassic age, which is sandwiched between impermeable rock, dips downward from the Great Dividing Range toward the Central Lowlands. The aquifer, which is overlain by impermeable shale, stretches westward for almost 1000 miles (1600 km) toward the Western Plateau. It comes to the surface there in the Lake Eyre Basin, the lowest point on the Australian continent at 52 feet (16 m) below sea level (Fig. 3–9). The source of most of the aquifer's water is the precipitation that falls on the moist western slopes of the Great Dividing Range.

The Black Hills (44°N, 104°W)—low, isolated mountains in west-central South Dakota and eastern Wyoming—cover an area of 6000 square miles (15,500 km²) (Fig. 3–10). They were formed when an intrusion of igneous rock raised a 50-mile (80-km) wide dome. Erosion has worn the dome into rock stubs or ridges, known as **hogbacks.** The Dakota Hogback, which stands 500 feet (150 m) above the adjacent valley, is formed of sandstone. The Dakota Sandstone is a famous artesian aquifer that is an important source of water in areas of the northern Great Plains, including parts of Colorado, Wyoming, South Dakota, Nebraska, and Kansas, for example. Water enters this aquifer in the Black Hills and along the eastern foothills of the Rocky Mountains.

The Sahara, stretching from the Atlantic Ocean to the Red Sea, is the largest desert in the world; it covers about 3 million square miles (7.8 million km²). Many of the northern Sahara **oases**—especially those in Algeria—exist because artesian aquifers transport water southeastward from the Atlas Mountains for hundreds of miles into the desert. In the eastern Sahara, water is carried northward into the desert by a sandstone aquifer that has its recharge area in the Sudan. Oases in Egypt, which are 400 to 800 miles (640 to 1290 km) from the recharge area in the Sudan, receive water via the Nubian sandstone aquifer.

SPRINGS

When a groundwater reservoir is exposed or its water table intersects the surface, a **spring** develops from the discharge. Under some cir-

This artesian well at Bedourie, in western Queensland, Australia, taps water at a depth of 1314 feet (400 meters). This artesian bore supplies 1.5 million gallons (5.7 million liters) of water per day under natural pressure. It would have been impossible to open up many parts of the Queensland outback without artesian bores. (Australian Information Bureau)

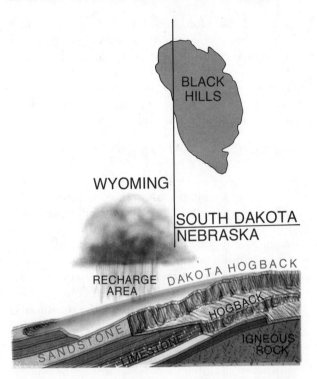

FIGURE 3–10 The Dakota Hogback, which consists of sandstone, stands 500 feet (150 meters) above the adjacent valley. The Dakota sandstone is sandwiched between impermeable rock layers and is an artesian aquifer that runs under many of the states in the Northern Great Plains. The recharge area of the aquifer is in the Black Hills.

cumstances, a surface-discharge occurs because the pressure within a confined aquifer is sufficient to force the water to the surface through the overlying rock. In the Great Artesian Basin of Australia, south and east of Lake Eyre, for example, some water rises to the surface under artesian pressure to produce mound springs, which are volcanolike hillocks. Pressure, high salinity of the artesian water, and a rapid rate of evaporation combine to produce some mound springs with 130-foot (40-m) elevations. Most, however, are only a few feet in height.

Springs vary a great deal in terms of their flow characteristics. Most springs yield small quantities of water. Some flow for a portion of the year and are called **intermittent**; others flow all year long and are called **perennial**. Seasonal change

in the height of the water table is the prime reason for an intermittent spring.

Southern Idaho has more than 100 significant springs. Some of its springs are quite large and yield a lot of water. One of the largest, Thousand Springs, is on the plain of the Snake River about 20 miles (32 km) northwest of Twin Falls. Thousand Springs yields 660 million gallons of water per day.

Florida is another state that is famous for its springs, all of which are formed in limestone. Seventeen of Florida's springs are of significant size. The largest, with a flow of 900 million gallons per day, is Silver Springs, southeast of Ocala. Wakulla Springs, near Tallahassee, has a depth of 185 feet (56 m), which makes it one of the deepest.

THE ACTION OF GROUNDWATER

Water is said to be the universal solvent. When water vapor condenses to form rain or snow, it absorbs small amounts of mineral and organic substances from the air as it falls. After falling, water continues to dissolve some of the soil, sed-

iment, and rock through which it passes. As groundwater percolates and moves through aquifers, it not only dissolves but also deposits minerals.

Carbon dioxide dissolves easily in subsurface water and forms a solution of weak carbonic acid. The acidic quality of some groundwater hastens the solution of material as the water passes through sediment and rock formations. The solution process is particularly rapid in carbonate rocks—limestones, for example—which are readily attacked by acids.

CHEMICAL QUALITY OF GROUNDWATER

The most common mineral constituents dissolved in groundwater are the chlorides, sulfates, and bicarbonates of sodium, calcium, magnesium, and potassium. In addition, small but, at times, significant amounts of manganese, iron, silicon, fluoride, nitrate, and boron are in solution in groundwater. All these chemical elements are dissolved from the soil, the sediments, and underlying rocks.

Ordinarily, groundwater that has a high mineral content—that is, more than 1000 parts per million of dissolved solids—is not fit for human consumption. Water that contains more than 2000 parts per million of dissolved solids is considered unfit for livestock to drink. High mineral content is also a deterrent to its use in irrigation and in most industrial processes.

Excessive amounts of calcium and magnesium produce **hard water** that makes washing difficult; hard water is not suitable for launder-ing clothes or for use as boiler feed water. As groundwater passes through some aquifers, it exchanges its dissolved calcium and magnesium for sodium and potassium. Through this natural process of exchange, the water becomes **soft**. Two artificial methods of water softening are used when the water available is too hard: (1) the water is treated with lime and soda ash to precipitate the calcium and magnesium, after which the water is filtered; (2) the water is passed through a porous exchanger that has the ability to substitute sodium in the exchange medium for calcium and magnesium.

Iron and manganese are a nuisance because they leave stains on clothing that is being laundered as well as on sinks or bowls in which water is held. Groundwater with a high boron or sodium content is unsuitable for irrigation. Fluorides may also be present in groundwater. If fluoride is present in concentrations not exceeding 1 part per million, it is considered beneficial because it tends to inhibit tooth decay in children who drink the water. On the other hand, in concentrations exceeding 1 part per million, fluoride can cause mottling of tooth enamel.

KARST TOPOGRAPHY

A region that is underlain by carbonate rocks is profoundly affected by acidic groundwater. A striking topography—sinkholes, streams that

It is important to know the mineral constituents dissolved in groundwater. A portable lab is being used by this scientist to test the water that accumulates at the Avdat spring in the Negev Desert. (R. Nowitz)

disappear into the ground, dry valleys, and natural bridges—is produced. The word **karst** is used to describe the topography because it resembles a region in Yugoslavia, the Kras, where this type of landscape predominates. The Kras, or karst region, of Yugoslavia comprises a series of dry limestone plateaus that range in altitude from 1000 to 5000 feet (300 to 1500 m) and stretch from the Italian border at Trieste to the Albanian border in the south (Fig. 3–11). The Dalmatian coast of Yugoslavia, which is shown in Figure 3–11, is an example of a recently drowned coastline; it is quite irregular and is broken by gulfs, bays, coves, and channels.

Underground streams and disappearing riv- ers are common in karst country. Rivers that stop without joining or flowing into any other stream are referred to as *interrupted* or **lost rivers.** A lost river that flows through Gospic is mapped in Figure 3–11. Three tributaries join to make this stream, which is almost 40 miles (64 km) long. Figure 3–11 shows that the stream ends abruptly about 18 miles (29 km) southeast of Gospic. Lost rivers usually disappear into cavelike depressions and continue as a subterranean flow.

A karst region develops slowly. At first acidic groundwater begins the dissolving of the calcium carbonate found in the subsurface rocks. Gradu- ally large empty underground areas, called **caves,** develop. As water drips from a cave's ceil-

The terrain around Postojna, Yugoslavia, is quite rugged. Postojna is located in the Kras region at 45°47′N, 14°13′E, which places it just north of Rijeka (Fig. 3–11). (Govt. of Yugoslavia)

Below the rugged terrain of Postojna, Yugoslavia, there are many huge caves which have been carved from the rock by the action of groundwater. Many of these caves are tourist attractions. A small tracked vehicle is used to move people through the caves of Postojna. (Govt. of Yugoslavia)

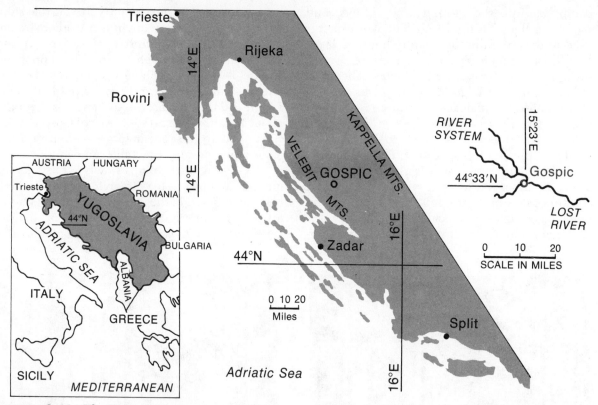

FIGURE 3–11 The Kras, or Karst region, of Yugoslavia stretches from the Italian border at Trieste to the Albanian border in the south. Disappearing streams, called *lost rivers*, are common in karst country.

ing, some of the calcium carbonate held in solution precipitates and remains behind to form iciclelike structures, called **stalactites,** that hang from the ceiling. When a drip of charged water strikes the floor, some more of the calcium carbonate it carries in solution precipitates to build a floor-based mound, which is referred to as a **stalagmite.** These structures—stalactites and stalagmites—grow toward one another; when they finally meet, they form columns of carbonate rock that extend from the floor to the roof of the cavern. Mammoth Cave (Kentucky), Carlsbad Cavern (New Mexico), and Luray Cavern (Virginia), are some of the most famous caverns in the United States produced by groundwater action.

Eventually a cave may be enlarged to such an extent that its roof can no longer support the weight of the overlying material. When the roof of a cavern collapses, it forms a **sinkhole** or open depression in the land surface. As sinkholes develop, the landscape takes on the appearance of a karst region. If a sinkhole becomes filled with water, it forms a lake. Karst topography is found in southern Indiana, Kentucky, Tennessee, Virginia, and parts of central Florida. In central Kentucky, for example, there are more than 70,000 sinkholes.

GROUNDWATER PRECIPITATES

The so-called petrified structures strewn across the landscape in the Petrified Forest of Arizona are the result of groundwater action, too. **Petrification** is, in reality, the process by which organic and inorganic materials are replaced by silica minerals deposited from groundwater solutions. For example, in the Arizona area, trees from dis-

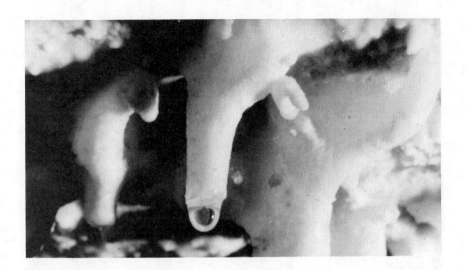

The iciclelike structure of a developing stalactite is clearly shown in the photo at the top. A bubble of charged water is about to drop from the stalactite. A floor-based mound, called a *stalagmite*, has grown toward and made contact with a thinner stalactite, which is suspended from the roof in the Carlsbad Caverns of New Mexico, shown at the left. (National Parks Service)

The Petrified Forest National Park lies in the Painted Desert in northern Arizona. The region is surrounded by a rugged terrain. Giant logs of agatized wood, which are no more than black dots at the scale of this photo, lie on the ground in the park; the trees which underwent petrification grew about 150 million years ago.

tant forests were carried by flood waters and deposited among the gravels of streams. The logs were covered by sediment, and the original wood (organic material) of the trees was gradually replaced by silica minerals deposited from groundwater. Buried materials such as wood, bone, and shell are sometimes completely replaced by precipitates from groundwater. The replacement process proceeds slowly and with such delicacy that the finest details, such as growth rings and radial structures of trees, are accurately preserved.

HOT GROUNDWATER

Groundwater is heated when it is exposed to hot rock material beneath the surface. Hot water that makes its way to the surface is designated as either a **hot spring** or **geyser**. The only essential difference between a hot and cold spring is temperature. Geysers consist of water that is blown into the atmosphere under pressure. Most geysers are hot, but some do blow cool or cold water.

THERMAL SPRINGS

Springs that average 18 Fahrenheit degrees (10 C degrees) or more above average groundwater temperatures are classified as **thermal springs** or hot springs. If a hot spring has a concentration of dissolved solids in excess of 1000 parts per million, it is called a *hot mineral spring.*

The water that supplies the flow of a hot spring is practically all meteoric in origin, coming from rain or melting snow. A small amount may, of course, be magmatic—water given off by the molten rocks below. In the United States there are only two locations in which the water supply for hot springs is almost totally magmatic with little or no meteoric source; one of these is in Katmai, Alaska, and the other is in southern Idaho.

There are numerous springs in Arkansas, especially around the foothills of the Ouachita and Ozark mountains. The famous Hot Springs National Park is a city, spa, and tourist resort in the central portion of Arkansas about 45 miles (72 km) southwest of Little Rock. There are 47 different springs at Hot Springs that yield about 1 million gallons of water per day. The temperature of the water is fairly constant at about 143°F (62°C).

Figure 3–12 shows that the water supply for the springs at Hot Springs, Arkansas, is meteoric in origin. Precipitation enters the aquifer to become groundwater along the valley just west of the mountain. The shale acts as a capping layer. The hot springs are located in a valley along the outcrop of a porous sandstone, which is identified as the Hot Springs sandstone in Figure 3–12.

These thermal springs in Arkansas were visited by Hernando de Soto in 1541. The valley was frequented by the Spanish and French throughout the 1700s as a health spa. The region was well known to the American Indians and, over the centuries, various tribes reveled in the therapeutic effect of the hot-spring water. In 1804, President Thomas Jefferson ordered the area analyzed and mapped.

GEYSERS

A geyser is really a special type of spring that throws water into the air with explosive force from time to time. The temperature of geyser water is often at its boiling point. The water Old Faithful spews forth in Yellowstone National Park (Fig. 3–13), for example, is 200°F (93°C), which is the boiling point of water at the geyser's 7500-foot (2886-m) elevation.

The conditions that produce a geyser are somewhat special: Water of meteoric origin enters the ground and moves down through the bedrock until it encounters a mass of hot rock

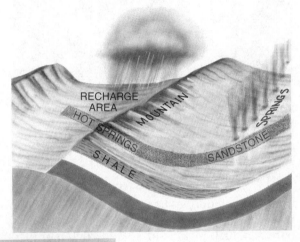

FIGURE 3–12 (Above) The water supply for the thermal springs at Hot Springs, Arkansas, is meteoric in origin.
The towering medical building at the left center of the photo stands at the north end of Bathhouse Row in Hot Springs, Arkansas. The city is built around the famous hot mineral springs at the west base of Hot Springs Mountain. (National Parks Service)

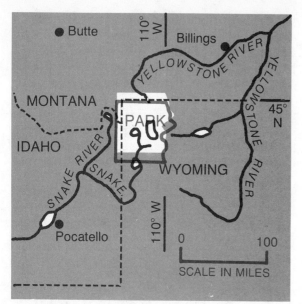

FIGURE 3–13 Yellowstone National Park is located primarily in Wyoming, but small sections protrude into Montana and Idaho. The geysers for which Yellowstone is famous are found at the end of the eastern Snake River Plain.

from which it derives heat. The hot groundwater in Yellowstone Park derives its heat at depths between 8000 feet (2440 m) and 3000 feet (910 m) from a partly molten batholith, that is, a large magma intrusion. Pressure from behind forces the heated water to the surface along joints and cracks in the bedrock. When the recharge area is higher than the point of egress, there is a constant outflow of water as a result of the hydrostatic head, and the egress-water is called a *hot spring*. There are more than 3000 hot springs in Yellowstone. On the other hand, if the point of entry, or recharge area, is lower than the point of egress, there is not sufficient hydrostatic pressure to maintain a constant flow and the egress-flow is intermittently and violently ejected in a display called a *geyser*.

Geyser action results because the meteoric water that reaches the hot rock builds its temperature and volume until the pathways of egress gradually become filled with water of a very high temperature. Some of the very hot water in the lower portion of the egress-pathways eventually reaches a position in its upward flow where the

confining pressure from above is reduced sufficiently to allow it to boil and expand. The boiling action causes some of the geyser-column water to overflow onto the surface gently at first. As the column of water becomes lighter, due to the gentle overflow onto the surface, a large quantity of the hot water flashes into steam. The hot liquid water changing to gaseous water, under the influence of reduced pressure, expands explosively. The rapidly expanding steam in the egress-pathway pushes the column upward and forces water onto the surface in a violent outpouring.

Old Faithful puts on a display that lasts about four minutes. The eruption shoots water to heights of 100 to 150 feet (30 to 46 m). About 3000 barrels of water are discharged during each display, which just about empties the egress-pathway. The column must be filled again before another discharge can occur; it takes Old Faithful about an hour to fill its egress-pathway sufficiently to bring on the display.

Geyser water contains a lot of gas and mineral matter. The mineral matter is dissolved out of the rock through which the water passes. For example, as geyser water rises through a bed of limestone, it dissolves calcium carbonate from the rock. Geyser water in Yellowstone National

Castle Geyser is part of the Upper Geyser Basin in Yellowstone National Park. The castlelike structure or cone that surrounds the geyser is a deposit of silica. The trees to the left of the cone show that the cone is rather large. (Wyoming Travel Commission)

Park contains silicate minerals, which are deposited by the hot water when it reaches the surface. The deposition of the silica around geysers in Yellowstone builds up tubes, or chimneys, as well as geyser cones.

There are at least 200 active geysers in Yellowstone National Park. There are two other well-known geyser regions: one is in Iceland 70 miles (110 km) from Reykjavik; the other is located along a zone of volcanoes in the Rotorua district of North Island, New Zealand (Fig. 3–14). The most famous of Iceland's geysers is the Great Geysir in southwest Iceland, which hurls a column of water from 180 to 220 feet (54 to 67 m) in the air. The Great Geysir has been known for many centuries and is the source of our present word, *geyser*, which we apply to all similar phenomena.

The northern part of North Island (Fig. 3–14) is a narrow peninsula with rollings hills and low mountains. The southern half of the island rises from fertile plains along the coast to Mt. Ruapehu, which is a volcanic peak near the center. The locations of two additional volcanic peaks north of Mt. Ruapehu are indicated by crosses in Figure 3–14; Mt. Ngauruhee, immediately to the north, stands at 7515 feet (2290 m), and Mt. Tongariro is 6458 feet (1968 m)

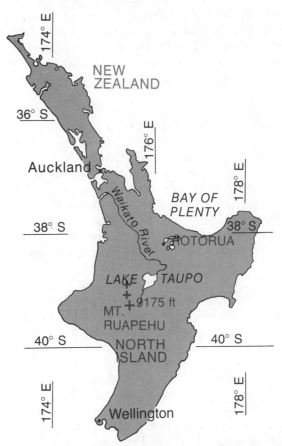

FIGURE 3–14 The Rotorua District is an active geyser area. Lake Taupo and the Waikato River are fed by water that flows from the volcanic peaks in the Mt. Ruapehu region.

The eruption of the Pohutu Geyser in the Rotorua district is shown in this photo. The terrain in the area is quite rugged. (New Zealand Consulate)

above sea level. Swift rivers, such as the Waikato which is is shown in Figure 3–14, rise in the mountains and flow to the sea. The region in and around Rotorua has a number of lakes and is like Yellowstone National Park.

GEOTHERMAL ENERGY

Useful energy can be extracted from naturally occurring steam and hot water found beneath the surface. The term **geothermal energy** is applied to heat derived from the interior of the Earth. The rate of heat conduction outward from the Earth's interior to its surface averages about 1.5 calories per centimeter per second, which over a one-year period amounts to more than 1 billion trillion calories to the total surface of the Earth.

Although the total delivery of heat to the Earth's surface is tremendous, most of it is not exploitable. To be retrieved, the heat must be concentrated in geothermal reservoirs. Underground reservoirs that are currently being exploited are found near Larderello, Italy, about 50 miles (80 km) from Pisa; in the geyser fields of North Island, New Zealand; in north-central California, at The Geysers, located about 90 miles (145 km) north of San Francisco; in the geyser fields of Iceland; in Matsukawa, Japan, which is roughly 100 miles (170 km) west of Tokyo; and in Cerro Prieto, Mexico, about 20 miles (32 km) southeast of the border town of Mexicali.

GEOTHERMAL SYSTEMS

The driving mechanism for the motion of the crustal plates is not fully understood, but it appears to be associated with convective movement within the Earth's mantle. The energy for the movement is supplied, whatever the mechanism, by the Earth's internal heat. Mass transfer of heat by magmas brings a lot of energy to shallower levels of the crust. It is from these heat sources that geothermal systems develop.

A geothermal system develops in the upper 1 to 2 miles (1–3 km) of the Earth's crust from a source of heat at some greater depth (Fig. 3–15). Groundwater, which contains dissolved minerals and salts, is heated, becomes less dense, and attempts to move upward. A rock cap of low per-

meability must overlie the system to prevent the escape of the hot groundwater to the surface. Leaks from the reservoir to the surface are manifested by steam vents, hot springs, and geysers.

USING THE SYSTEM

The steam fields at The Geysers (California), Larderello (Italy), and Matsukawa (Japan) have reservoirs that consist of highly fractured or porous rocks. The Geysers reservoir consists of fractured sandstone and volcanic rock (Fig. 3–15). Fractured volcanic rock also makes up the reservoir at Matsukawa. The Larderello reservoir consists of porous limestone and dolomite. Superheated steam at 480°F (250°C) issues from these reservoirs and is used to drive conventional steam turbines, which produce conventional electric power.

Another type of geothermal reservoir produces a mixture of water and steam when wells are drilled into it; but the steam is separated from the water and its pressure is used to operate a turbine. Examples of this kind of system are found in New Zealand, where wells are drilled into a permeable pumice-type volcanic rock, and in Cerro Prieto, Mexico. The temperature of the fluid in New Zealand is 500°F (260°C), and about 20 percent of the fluid is flashed to steam for power production. At Cerro Prieto, electric power is produced from a fluid having a temperature of 570°F (300°C).

Geothermal energy can also be used for space heating. Some buildings in Klamath Falls, Oregon, for example, are heated by hot well water. Iceland has the most elaborate system for using geothermal energy for space heating; more than 50 percent of Icelandic homes were heated by geothermal energy in 1980. Temperatures within the Icelandic reservoir range from 140 to 300°F (60 to 150°C).

The cost of energy provided by a geothermal system is low when compared to a system that utilizes fossil fuels—oil, natural gas, and coal. Another advantage to geothermal energy is that it causes minimal pollution problems; there is no smoke or discharge of oxidized material to the atmosphere. Some slight problems do result, however, from the disposal of warm saline effluents after the heat is extracted.

FIGURE 3–15A The steam fields of The Geysers are located north of San Francisco. The reservoir of The Geysers consists of fractured sandstone and volcanic rock.

EXPLORING FOR GEOTHERMAL ENERGY

Searching for a surface display of heat is the simplest way to identify a geothermal system. Some reservoirs, however, have little or no surface display. Thus, aerial and satellite photography have become increasingly more valuable in locating domed areas and other structures that have a potential for concealing geothermal resources.

The Gulf and East coasts of the United States don't steam with geysers, but they are active areas of geothermal exploration at the present time. On the East Coast—from Fort Monmouth, New Jersey, to Savannah, Georgia—50 test wells were drilled in 1978. The geothermal heat that has been found along the East Coast is not hot enough to use for electric generation. It can, however, be used for heating or cooling homes and office buildings.

The 1978 test borings were made to measure the energy available and to make heat-contour maps. The data presently indicate that the most

FIGURE 3–15B Molten magma found at great depths below The Geysers radiate heat energy through the solid rock above it. The heat comes in contact with groundwater and transforms the groundwater to geothermal steam. Some of the steam finds its way to the surface through cracks or fissures. This steam is tapped and brought to the surface by drilling wells into the reservoir area.

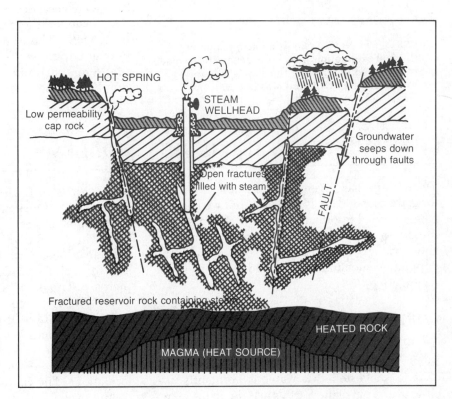

FIGURE 3–16 New Jersey is an active area of geothermal exploration. Test borings have been made at six sites (red dots) along the state's Atlantic coastline. The Sea Girt well was hot.

promising locations in the East are along the southern New Jersey coast, the Delmarva Peninsula, and around Norfolk (Virginia), Wilmington (North Carolina), Charleston (South Carolina), and Savannah (Georgia). A well in North Carolina yielded water at 140°F (60°C); another in Ocean City, Maryland, yielded water at 216°F (102°C).

Six wells were drilled in New Jersey during 1978 (Fig. 3–16). One was drilled at Fort Monmouth and another at Sea Girt. The other sites were in Lacey Township, Atlantic City, Ocean City, and Cape May. The Sea Girt well was hot; the prospects look good and hope runs high. Some predictions indicate that by 1985 geothermal hot water could be supplying energy for a variety of uses in New Jersey.

We have known for decades that hot saline water under tremendous pressure exists under the Gulf Coast of the United States. The drill bits of companies searching for oil have tapped into reservoirs of fiercely hot, pressurized salty water from time to time over the years. In the past, these strikes were considered to be nuisances; they were cast-off and written-off as useless. Today, however, there are plans to use these hot-brine wells. The water temperature of the brine in many of the wells is 300°F (150°C); this hot, high-pressure fluid can be sent through a hydraulic turbine to produce electricity.

CONTAMINATION AND OTHER PROBLEMS

Technology has given us remarkable methods for gathering and analyzing data, but we seem to have minimal information about water, our most important resource. We actually know very little about water dynamics although our survival may depend on such information and how wisely we use it.

A DRYING CYCLE?

In 1977, W. S. Fyfe of the University of Western Ontario suggested that water is constantly cycling in and out of the Earth's mantle. In his view,

the Earth is in the process of absorbing more water into its interior than is being outgassed and sent to the surface.

If the trend Fyfe identifies continues, the volume of ocean water will constantly decrease and the volume of deep subsurface water will increase. The process producing this water loss, Fyfe says, is related to plate movements and collisions. A subducting plate, for example, carries huge quantities of water into the Earth's interior where it is absorbed in various ways, including the creation of certain kinds of rock formations. The Earth, according to Fyfe, is in a drying cycle.

GROUNDWATER OVERDRAFT

When the withdrawl rate of groundwater from an aquifer exceeds the recharge rate, an **overdraft** problem develops. An overdraft can be produced by too many small wells or an industrial well with a high-volume pumping operation. The obvious consequence of the imbalance is that the water table will be lowered and eventually shallow wells will run dry. There are other consequences too: the land can subside and saltwater can move into the freshwater zone.

WATER-TABLE LEVEL

A dramatic example of the change in the level of a water table is found in North Carolina along the Pamlico River. The area affected spreads from Washington eastward to the mouth of the Pamlico River and from New Bern to Phelps Lake (Fig. 3–17). The principal source of water for this region is wells sunk into a limestone aquifer. The subsurface reservoir is a solution-cavity aquifer in which fossil shells have been dissolved out to leave a highly porous and permeable rock. In the 300 years from colonial times to the 1960s, the numerous wells sunk into the aquifer had little effect on the water table, which stood at 8 feet (2.4 m) above sea level in 1965.

An open-pit mining operation, the Lee Creek Mine (Fig. 3–17), was begun in 1965 to exploit a large phosphate deposit. The mine penetrates well below the water table. As phosphates are removed, a high-volume pumping operation is maintained to keep the pit clear of water. Within five years, shallow wells in the area had gone dry and the water table was standing at 90 feet (27 m) below sea level. In other words, the high-volume pumping operation dropped the water table almost 100 feet (30 m) in the five-year period from 1965 to 1970.

LAND SUBSIDENCE

Water moving through the pores of an aquifer helps to support the load of overlying material. When water is removed from the pore spaces of an aquifer, the soil's ability to resist compaction is reduced and the weight of the overlying rock repacks the aquifer's grains so they are closer together. Compaction reduces the water-carrying capacity of the aquifer and also allows the surface above to settle or subside.

The use of groundwater in California has been intensive. Most of the water withdrawn is used for irrigation, although some groundwater is also pumped for domestic and industrial use. The groundwater withdrawals are primarily from confined-aquifer systems. The intensive pumping has drawn water tables down by more than 100 feet (30 m). The maximum draw-down, about 500 feet (150 m), has occurred on the west side and in the southern end of the San Joaquin Valley; for each 10- to 25-foot (3- to 7.6-m) decline in the water table, there has been as much as 12 inches (30.5 cm) of surface subsidence.

Mexico City is in southeastern Mexico—in the west-central part of the Valley of Mexico, which is a closed basin at an elevation of about 7500 feet (2290 m). Sand and gravel beds constitute the highly productive aquifer underlying the city. More than 3000 privately owned wells and 230 municipal wells supplied water to a population of 2.9 million people during the 1970s. In fact, more than 140 artesian wells had been sunk and were in use by 1854. Over the years, the discharge of Mexico City's wells has exceeded the natural recharge. The compaction that has resulted from the imbalance has produced a subsidence of more than 20 feet (16 m) during the twentieth century.

Venice, built on 117 small islands and reinforced only by ancient pilings, was a sinking city for most of its history. Until the 1970s, the sink-rate was slightly more than 0.1 inch (2.5 mm) per year. A ban on pumping from the city's thousands of artesian wells was established in 1972 and water was supplied to the city by a new aqueduct. By late 1975, studies confirmed that the three-year-old ban on pumping from the artesian wells had increased the underground water pressure and had reversed the sinking process. This is a good example of a city taking affirmative action based on a scientific principle.

SEAWATER INTRUSION

In an aquifer that discharges on a shoreline, there is a distinct line of contact between fresh and salty water. The freshwater floats on top of the denser seawater, with very little mixing between the two. As long as the water table is above sea level and slopes toward the sea, the groundwater flows seaward and aquifers along coastlines and on islands yield freshwater. But, if

FIGURE 3–17 The sand dunes, reefs, sandbars, and islands beyond North Carolina's shoreline are referred to as the *outerbanks*. The lighthouse at Cape Lookout (photo above) stands lonely vigil at the southern end of the outerbanks. The Lee Creek Mine is an open-pit mine north of Moorehead City. The mine penetrates well-below the water table into the local limestone aquifer of coastal North Carolina. The water table in a wide region around the mine dropped 100 feet (30 meters) within a 5-year period and seawater began to encroach on the freshwater of the aquifer. (National Parks Service)

the water table is depressed below sea level, seawater will begin to move inland and contaminate the aquifer.

One of the earliest instances of seawater intrusion occurred in England in 1855. Overpumping in London and Liverpool caused saline water to intrude and contaminate the wells. Today, we have serious saltwater contamination of coastal aquifers in Germany, the Netherlands, Japan, and the United States.

Numerous examples of seawater intrusion into coastal aquifers of the United States can be cited: In North Carolina, the water table south of Lee Creek Mine is more than 110 feet (33.5 m) below sea level; there is a definite seawater intrusion into the aquifer. Saltwater encroachment has also occurred in Florida, especially around Miami in Dade County.

Long Island (Fig. 3–18) is really a portion of the Atlantic Coastal Plain. It consists of a wedge of unconsolidated sediment that is 2000 feet (610 m) thick on its eastern edge and thins to zero on its northwestern edge. Groundwater is held in this wedge, which overlies igneous rock. The groundwater reservoirs beneath the surface of the island contain mostly freshwater, except in Brooklyn and Queens (Fig. 3–18). There are cones of depression in the water table of Brook-

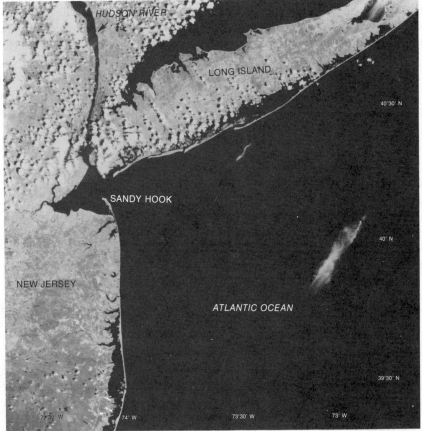

FIGURE 3–18 In the photo, rows of small, fair-weather cumulus cover Brooklyn, Queens, the lower part of Manhattan, North Jersey, and Westchester County. Long Island and New Jersey, from the Sandy Hook area south, are part of the Atlantic Coastal Plain. Both areas consist of unconsolidated sediment with barrier bars lining the ocean side of the plain. The aquifer beneath the surface of Long Island is a wedge of unconsolidated sediment that is thickest at the eastern edge, or Montauk region of the island. At the Brooklyn-Queens end, that is, western end of the island, saltwater is encroaching on the freshwater of the aquifer.

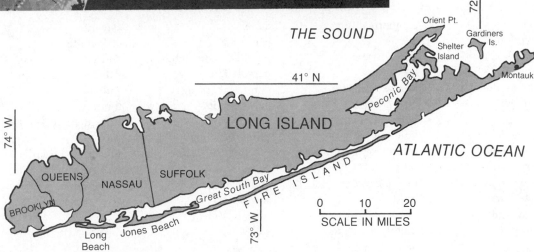

lyn and Queens caused by overpumping. Thus, in these western reaches of the island, there is a definite seawater intrusion. The problem is compounded because these areas are largely paved over and there is no effective natural recharge area. Precipitation is discharged into the sea by drainage pipes. Nassau County has a developing seawater-intrusion problem because its population is large and growing and its groundwater withdrawal rate is much higher than the recharge rate. Saltwater intrusion has not occurred as yet in the groundwater reservoirs of Suffolk County.

POLLUTION

The pollution of groundwater has always been a problem for humankind. When people used shallow dug wells, they had to be careful that material from the surface did not wash or fall into the wells. Waterborne diseases, such as typhoid, took their toll and were a source of general concern. We have not learned very much over the years because, today, we are still plagued and threatened by groundwater pollution.

There are a number of important sources of groundwater contamination: (1) wastes in the form of sewage, (2) hazardous industrial chemicals, (3) leaching from solid-waste landfills, (4) agricultural wastes, pesticide residues, and soluble chemical nutrients applied as fertilizers,

(5) radioactive wastes, (6) coal mining and processing operations, (7) oil field operations, (8) highway salting, and (9) disposal wells.

You don't need to search very far from home for examples. The newspapers in almost every state have carried reports of well and groundwater contamination. In 1979, for example, three communities in New Jersey— Fair Lawn, Mahwah, and Ramsey—had to close public wells because seven suspected cancer-causing chemicals (carbon tetrachloride, trichloroethylene, tetrachloroethylene, chloroform, trichloromethane, trichloroethane, and dichloroethane) were found in the water. Substantial contamination was spread through the area's aquifer. At the time, neither federal nor state law required testing for these chemicals so no one could say how long they were polluting the water supply. Unfortunately, these were not the first communities in New Jersey to have to close wells. Six months earlier, Allendale, New Jersey, closed half its wells; Camden, Perth Amboy, South Orange, and South Brunswick are other communities in the state that have closed contaminated wells.

One thing is eminently clear: contamination of groundwater is a pervasive phenomenon. Communities need to be alert and they should employ, on a routine basis, accurate and highly specific analytical methods to detect groundwater pollution.

——————— THE LANGUAGE OF PHYSICAL GEOGRAPHY ———————

A physical geographer uses a technical vocabulary to make explanations. Review your understanding of the vocabulary used to develop the concepts in this chapter. Among the important words and terms used are:

acre-foot	geothermal energy	oases	stalactite
adhesive force	geyser	overdraft	stalagmite
aquiclude	groundwater	perched water table	spring
aquifer	hard water	percolation	subterranean water
artesian aquifers	hogback	perennial springs	suction
artesian well	hot spring	permeability	thermal spring
bar	hydraulic head	petrification	transpiration
capillarity	hydrology	phreatic zone	tuff
capillary fringe	intermittent springs	piezometric surface	unconfined aquifer
capillary water	interstices	porosity	vadose water
caves	juvenile water	sinkhole	water table
cohesive force	karst	soft water	zone of aeration
confined aquifer	lost river	soil moisture	
connate water	meteoric water	soil water	

SELF-TESTING QUIZ

1. How much of the freshwater supply of the United States is below the surface?
2. Which countries use large quantities of groundwater to satisfy their freshwater needs?
3. Contrast the zone of saturation with that of aeration.
4. What role do adhesive and cohesive forces play in the development of soil water?
5. How are comparisons of the water held in a soil made among soils?
6. Explain the location and position of a perched water table.
7. Why is there a lower limit to the saturated zone?
8. List the primary sources for the origin of groundwater.
9. What factor determines aquifer capacity?
10. List and contrast the porosities of sand, gravel, and clay.
11. Under what conditions will an aquifer transmit water readily?
12. Why are some limestones good aquifers?
13. Explain the difference between a pump well and an artesian well.
14. Trace the development of an underground cavern in a limestone region.
15. List the common mineral constituents in groundwater.
16. What causes hard water?
17. Why are disappearing rivers common in karst country?
18. What is the source of the water flow from a hot spring?
19. Describe the conditions that produce a geyser.
20. Locate regions in which geothermal systems are being used.
21. Explain the conditions that can develop when the withdrawal rate of groundwater from an aquifer exceeds the recharge rate.
22. List some important sources of groundwater contamination.

IN REVIEW
WATER BENEATH THE SURFACE

I. SUBTERRANEAN WATER
   A. The Zone of Aeration
      1. Capillary Water
      2. Soil Water
   B. The Zone of Saturation
      1. The Water Table
      2. Origin of Groundwater

II. GROUNDWATER RESERVOIRS
   A. Aquifer Capacity
   B. Flow in an Aquifer
   C. Aquifer Types
      1. Unconfined Aquifer
      2. Artesian Aquifer
   D. Springs

III. THE ACTION OF GROUNDWATER
   A. Chemical Quality of Groundwater
   B. Karst Topography
   C. Groundwater Precipitates

IV. HOT GROUNDWATER
   A. Thermal Springs
   B. Geysers
   C. Geothermal Energy
      1. Geothermal Systems
      2. Using the System
      3. Exploring for Geothermal Energy

V. CONTAMINATION AND OTHER PROBLEMS
   A. A Drying Cycle?
   B. Groundwater Overdraft
      1. Water-Table Level
      2. Land Subsidence
      3. Seawater Intrusion
   C. Pollution

4 The earth's surface water

Surface water flows over the ground in stream and river channels. It accumulates in standing water bodies called **ponds, lakes,** and **swamps.** Significant amounts of water are also held on land surfaces as large ice sheets and glaciers.

Geographic variations in the volume of surface water are the result of differences in precipitation and evaporation rates. Areas within the continental United States with an average precipitation of more than 2 million gallons per square mile per day are primarily in the eastern portion of the country (Fig. 4–1). Except for the Pacific Northwest and some sections of California, there is a progressive decline in precipitation from east to west. The balance between surface evaporation and precipitation leaves most areas west of the 100th meridian surface-water deficient.

The greatest average volume of precipitation—4.9 million gallons per square mile per day—falls along the coast of Washington (Fig. 4–1). The Olympic Mountains, which extend along Washington's coast, force moist air flowing from the Pacific to rise, cool, and drop a lot of water. An inspection of Figure 4–1 reveals that a coastal zone of abundant precipitation—2-million-plus gallons—starts around the San Francisco Bay region and extends northward through California, Oregon, and Washington.

North Dakota is divided into four precipitation regions in Figure 4–1. The Red River of the North runs south along North Dakota's eastern border; there is a 1-million-gallon zone on the other side of the river in Minnesota. In fact, a 1-million-gallon zone extends southward through South Dakota, Nebraska, and on to

Reelfoot Lake is in northwestern Tennessee. It was formed as a result of crustal movements produced by earthquakes that struck the region in 1811 and 1812. The lake and its surrounding area are an attractive recreation area. There are significant populations of freshwater fish, waterfowl, and other game in and around the lake. (State of Tennessee)

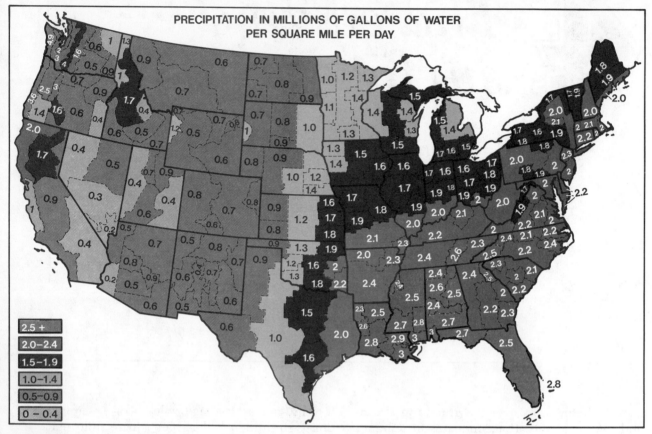

FIGURE 4-1 Precipitation zones within states are identified in this map. The figure printed on the zone is in millions of gallons of water per square mile per day. For example, New Jersey is divided into two zones: the southern zone gets an average of 2 million gallons of water per square mile per day. The northern zone receives 2.3 million gallons during the same period. According to the six-category color code, New Jersey is in the second highest category.

Texas. The western edge of this 1-million-plus zone, which extends from South Dakota to Texas, runs roughly along the 100th meridian and is generally taken as the division between the water-rich areas to the east and the water-poor areas to the west.

STREAMS AND RIVERS

Water that flows in a definite path, called a **channel,** is referred to as a **river** or a **stream.** Since the water flows under the influence of gravity, the path of a stream channel is always down a slope. From a practical point of view, the millions of rills that form with each rain and flow down each slope and hillside are part of the stream system. Thus, a thorough study of a stream must include the full surface of the surrounding region—the vegetation, the soil, and the mantle of rock debris that covers the slope.

STREAMFLOW

The water in a stream channel is called **stream-flow.** The amount of streamflow is strongly influenced by the weather and climate of the region; in other words, streams are relatively more numerous and streamflow is greater in more humid regions. For example, about two-thirds of the streamflow in the 48 contiguous states of the United States is found in the area east of the Mississippi River where we have the greatest volume of precipitation (Fig. 4–1). Most of the remaining U.S. streamflow is in the Pacific Northwest.

The surrounding topography and soil also influence the amount of streamflow, but it is the vegetation in the drainage basin, or watershed, that plays a particularly significant role in the control of streamflow. Convincing evidence of the role of vegetation was gathered from a site in the southwestern corner of North Carolina: Two experimental watersheds—each with deep, permeable soils and steep slopes—were converted from mature deciduous hardwood cover to white pine. The streamflow in this southern Appalachian region was measured over a 15-year period. The white-pine forest reduced the streamflow during every month of the experimental period; in fact, the annual streamflow was reduced 20 percent below that expected for the hardwood cover.

Throughout most of the twentieth century, white pine has been planted as a convenient and easily maintained cover on municipal watersheds in the East. If the experimental evidence cited above is conclusive, there are important implications for the management of water resources because changes in the type of forest vegetation will have a significant impact on the amount of streamflow.

SOURCES OF STREAMFLOW

Streamflow comes from a variety of sources. The most direct comes from precipitation that falls into a stream's channel, which is called **channel precipitation.** Since most streams have a relatively small surface area, channel precipitation generally makes a rather insignificant contribution to the total flow of a stream.

Another source of water for streams is **runoff,** which is the gravity-controlled flow of water draining over the surface toward a channel. Surface runoff comes into being following a

rainstorm. It is residual water that is above and beyond surface storage and infiltration capacity. Severe and torrential downpours on barren impermeable soil, road surfaces, rooftops, and hillsides produce the largest amounts of runoff.

A lateral movement of water through soil, called **throughflow,** usually exists where there is impermeable rock near the surface on a fairly steep slope. Throughflow makes a contribution to streamflow in a mountainous or hilly region where this special pattern of underlying rock exists.

In some regions, snowcover is a source of water for streams. It is an especially important source of streamflow in western North America. Precipitation in the form of snow is retained on the surface in high mountains and in cold regions. In mountainous regions there is usually a progressive melting of the snowcover from lower to higher elevations as springtime arrives. When snowcover melts, the **meltwater** behaves as though it were direct precipitation—some evaporates, some percolates into the soil, and some becomes runoff. In the United States, **snowmelt** contributes streamflow to the headwaters of the Missouri, Platte, Colorado, Rio Grande, Snake, and Columbia rivers. The Fraser River in the Canadian province of British Columbia is also fed by snowmelt from the high mountains that line its path of flow from Jasper in the east to Vancouver in the west (Fig. 4–2).

The meltwater from glaciers is another source of water for some streams. There are, for example, glaciers on Mount Rainier, Washington; in Glacier National Park, Montana; in the Jasper-Banff Park of Alberta; and in the Selkirk Mountains of British Columbia. The streams in all these areas are fed to some extent by the meltwater from the glaciers. The Fraser River in British Columbia, for example, is fed by meltwater from the glaciers in Jasper National Park (Fig. 4–2). The North Saskatchewan River is another Canadian stream that depends on glacier meltwater; it picks up its supply in the Banff area of Alberta and then moves eastward into Saskatchewan across the prairie. In fact, both the North and South Saskatchewan rivers rely heavily on glacier meltwater to sustain their flow during the summer.

The Fraser River Valley in the vicinity of Dog Creek (51°35′N, 122°15′W) spreads before us in this photo taken during the Canadian summer. (Canadian Film Board)

Although channel precipitation, runoff, throughflow, and meltwater all play a part in maintaining streamflow, the single most important source is groundwater. Most groundwater moves toward some kind of outlet such as a spring, swamp, or lake. Eventually, the groundwater flow into these holding basins becomes part of a streamflow. When the base of a stream intersects the groundwater zone, the groundwater enters the streamflow directly through the stream's banks and bed.

STREAM CLASSIFICATION

The flow of a stream that is supplied and maintained by the slow seepage of groundwater into the channel is called the **base flow.** The relationship of the water table to the bed of a stream channel can be used to classify streams. Three types of streams can be identified using this scheme of classification: (1) **perennial streams** which normally flow all the time, (2) **ephemeral streams** which carry only surface runoff, and (3) **intermittent streams** which have a seasonal flow.

PERENNIAL STREAMS
The channel bed of a perennial stream is always below the water table, and groundwater constantly seeps into the stream. Since this arrangement of water table and stream-channel bed carries groundwater away from the aquifer, the stream is termed an **effluent stream.** Perennial streams with a constant base flow are characteristic of humid regions with high year-round precipitation. When such a region experiences a drought, the entire streamflow may consist of effluent from the groundwater reservoir.

FIGURE 4–2 The Fraser River has its origin in the mountains of Jasper National Park. The streamflow of the Fraser River is fed by glacier meltwater and snowmelt. The Fraser flows northwest along the base of the Cariboo Mountains; then it turns south and flows through the coast mountains to empty into the sea at Vancouver.

The Stikine River of British Columbia is in the foreground. Meltwater from the Great Glacier adds to the streamflow of the Stikine. The river crosses the British Columbia-Alaska border at the left of the photo. The Stikine River empties into Sumner Strait at 56°40'N, 132°30'W, which is just north of Wrangell, Alaska. (Canadian Film Board)

EPHEMERAL STREAMS

These streams, which are typical of desert regions, carry water only during and after a rainstorm. An ephemeral-stream bed is dry once the runoff is carried away because the water table is well below the bed of the stream's channel. In fact, some of the water entering the channel as runoff seeps through the stream's bed and eventually works its way into the groundwater below. Since the stream actually makes a contribution to the groundwater supply, it is termed an **influent stream.**

INTERMITTENT STREAMS

Wide fluctuations in the level of the regional water table produce an intermittent or seasonal stream. Sometimes the stream's bed is below the water table. At other times, the water table is below the channel's bed. Water seeps into the channel when the water table is higher than the

bed. The channel is dry when the water table is below the level of the bed. Intermittent streams are common in regions of very porous rock or where rainfall is seasonal.

STREAM CHANNELS

The sloping trough within which the water flows is termed the **stream channel.** It is shaped by the flow of water. Some stream channels are long, wide, and deep; others are shallow and narrow. The long, deep channels are formed by great rivers like the Mississippi, which flows from its source in north-central Minnesota to its outlet in the Gulf of Mexico below New Orleans. The Mississippi River has a channel depth of 100 feet (30 m) in some places. Shallow, narrow channels are carved by brooks and creeks which flow across the countryside.

The description of a stream channel and the water flow within it is referred to as **hydraulic geometry.** The depth of the stream, d, is measured in feet (meters) at any desired point (Fig. 4–3). The distance across the stream's water surface from bank to bank is termed its width. The cross-sectional area of the stream in Figure 4–3 is outlined in red and is reported in square feet (sq m).

The angle between the water surface and a horizontal surface constructed above it is reported as the **streamslope.** In fact, slope is commonly reported as the vertical drop of the stream's water surface in feet per mile (m/km) of horizontal distance. It can also be reported as the percentage of gradient—the ratio of vertical drop to horizontal distance. A gradient of 0.05, or 5 percent, means there is a vertical drop of 5 feet

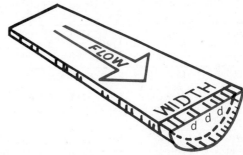

FIGURE 4–3 The sloping trough of the stream channel is shaped by the flow of water.

for each 100 feet of horizontal distance or, in metric units, a drop of 5 meters for each 100 meters of horizontal distance.

STREAM DISCHARGE

The term **discharge** is used when referring to the water-flow rate through a given channel section. In other words, it is a measure of the volume of water passing a given point per unit of time. Stream discharge is measured in cubic feet per second (cfs) or in cubic meters per sec (cms). The U.S. Geological Survey has more than 9000 stations where they monitor stream discharge.

Table 4–1 is a summary of the discharge data for the three longest rivers on five continents. The Nile River of Africa is the longest in the world. It empties into the Mediterranean Sea from the shores of Egypt at latitude 31°32′N, longitude 31°51′E (Fig. 4–4). The Nile discharges water at an average rate of 92,900 cubic feet (2630 cubic meters) per second.

The Mississippi-Missouri system documented in Table 4–1 includes the Missouri-Jefferson-Beaverhead-Red Rock rivers as part of the system. Even with this expansion of the Mississippi-Missouri system, you will note that

A boatload of tourists is approaching the Philae Temple in Aswan. Aswan is a resort city on the east bank of the Nile. The city marks the southern limit for seagoing ships and barges that move into the river. Aswan is famous for its red-granite quarries, which provided stone for its dams and for many of ancient Egypt's monuments. (Egyptian Consulate)

the average discharge of the Amazon system is ten times that of the Mississippi.

The Saint Lawrence River proper, which begins at the northeastern end of Lake Ontario,

TABLE 4–1 DISCHARGE DATA FOR LONGEST RIVERS

CONTINENT	RIVER	LENGTH (mi)	(km)	DRAINAGE BASIN AREA (mi²)	(km²)	LOCATION OF OUTFLOW Body of Water	Lat./Long.	AVERAGE DISCHARGE RATE cfs	cms
Africa	Nile	4160	6690	1,082,000	2,802,000	Mediterranean Sea	31°32′N,31°51′E	92,900	2,630
	Congo	2880	4630	1,476,000	3,822,000	Atlantic Ocean	6°04′S,12°24′E	137,000	39,000
	Niger	2550	4100	808,000	2,092,000	Gulf of Guinea	4°20′N,6°E	201,000	5,700
Asia	Yangtze	3720	5980	705,000	1,827,000	East China Sea	31°48′N,121°10′E	1,137,000	32,190
	Yenisey	3650	5870	1,011,000	2,619,000	Kara Sea	71°50′N,82°40′E	636,000	18,000
	Amur	3590	5780	792,000	2,050,000	Tatar Strait	52°56′N,141°10′E	346,000	9,800
Europe	Volga	2300	3700	533,000	1,380,000	Caspian Sea	45°45′N,47°52′E	271,000	7,680
	Danube	1770	2850	298,000	773,000	Black Sea	45°20′N,29°40′E	226,200	6,400
	Dnieper	1370	2200	194,000	503,000	Black Sea	46°30′N,32°18′E	57,000	1,620
North America	Mississippi/ Missouri	3740	6020	1,244,000	3,222,000	Gulf of Mexico	29°N,89°10′W	610,000	17,270
	MacKenzie	2630	4240	681,000	1,764,000	Beaufort Sea	69°15′N,134°08′W	410,000	11,610
	St. Lawrence	1900	3060	508,000	1,316,000	Gulf of Saint Lawrence	49°09′N,67°10′W	460,000	13,030
South America	Amazon	4080	6570	2,375,000	6,150,000	Atlantic Ocean	0°10′S,49°W	6,180,000	175,000
	Platte- Parana	3030	4880	1,197,000	3,100,000	Atlantic Ocean	35°S,57°W	809,000	22,900
	Madeira	1990	3200	463,000	1,200,000	Amazon River	3°22′S,58°45′N	770,000	21,800

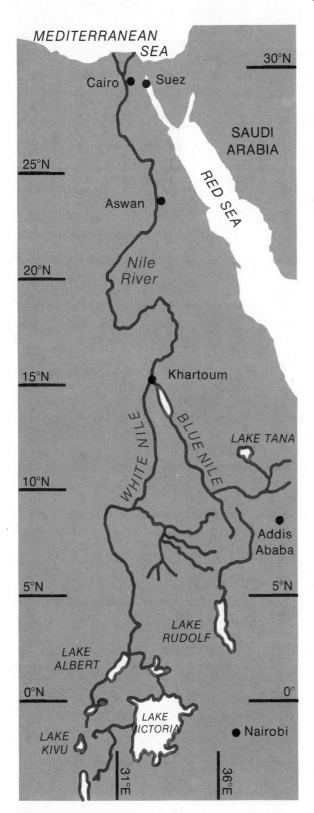

is 590 miles (960 km) long. In Table 4–1, the Saint Lawrence River system is taken to include the Great Lakes and the Saint Louis River as well. The vast Saint Lawrence River system has its source in the Saint Louis River in northeastern Minnesota.

HYDROGRAPHS

Generally, the discharge of a river varies from month to month. The plot of a stream's discharge through time is called a **hydrograph.** Stream-discharge graphs are a useful way to examine regional differences in **streamflow.**

Table 4–2 lists the average monthly discharges for three rivers: the Nile, Columbia, and Mekong. These data are plotted as hydrographs in Figure 4–5. Examine the hydrographs and note the differences in the timing and duration of **peak flow.** There are also some obvious differences in the variations between peak flow and low flow.

The peak flow along the Nile River occurs during and following the wet seasons in its watersheds. One of the watersheds for the Nile is in the Ethiopian highlands around Addis Ababa in Figure 4–4, which is an area drained by the Blue Nile. The Nile River peaks in September during and following heavy rainfall in the Ethiopian highlands (Fig. 4–5).

Lake Victoria is often referred to as the source of the Nile, but the Luvironza River at latitude 3°40'S, which is southeast of Lake Kivu in Burundi (Fig. 4–4), is the Nile's most remote headstream. Every summer for the last 7000 years of human history, the Nile has overflowed its banks, flooded the lowlands, and deposited a rich alluvial soil. The region most benefited by the yearly Nile flood lies north of Aswan (Fig. 4–4). The richest soils, however, are found in the Nile Delta, which extends from Cairo north to the Mediterranean Sea.

The Columbia River begins in southeastern British Columbia on the western slopes of the

FIGURE 4–4   The Nile's most remote headstream is southeast of Lake Kivu in Burundi. The rich soils of the Nile Delta are found from Cairo north to the Mediterranean Sea.

TABLE 4–2   AVERAGE MONTHLY DISCHARGE

MONTH	NILE RIVER DISCHARGE		COLUMBIA RIVER DISCHARGE		MEKONG RIVER DISCHARGE	
	cfs	cms	cfs	cms	cfs	cms
Jan.	42,400	1200	98,900	2800	95,300	2700
Feb.	35,300	1000	105,900	3000	70,600	2000
Mar.	30,000	850	127,100	3600	60,000	1700
Apr.	28,300	800	201,300	5700	53,000	1500
May	24,800	700	363,700	10,300	81,200	2300
June	49,400	1400	494,400	14,000	367,300	10,400
July	67,100	1900	331,900	9400	596,800	16,900
Aug.	233,000	6600	187,200	5300	925,200	26,200
Sept.	289,600	8200	123,600	3500	985,200	27,900
Oct.	183,600	5200	98,900	2800	660,400	18,700
Nov.	81,200	2300	98,900	2800	328,400	9300
Dec.	49,400	1400	98,900	2800	151,900	4300

Rocky Mountains (Fig. 4–6). It flows northward for 180 miles (290 km) along the foothills of the Rocky Mountains through a wild, forested region. Then it makes a big bend around the north end of the Selkirk Mountains and flows southward toward the state of Washington. The Canadian part of the Columbia River is 460 miles (740 km) long. The Kootenay River is the chief **tributary** of the Columbia River in Canada. The Okanagan River in British Columbia is another major tributary, but it enters the Columbia River in Washington.

The Columbia River is 1320 miles (2130 km) long. It discharges into the Pacific Ocean near Astoria, Oregon, at latitude 46°14′N, longitude 124°W with an average rate of 194,200 cfs (5500 cms). The peak flow along the Columbia occurs

in June (Fig. 4–5). The river relies heavily on snowmelt, and its peak flow corresponds to the period when snowmelt is at a maximum in the Columbia's mountainous watersheds. The period of maximum rainfall also occurs during June in these watersheds and contributes significantly to the peak flow.

The Mekong River is the longest stream in the Indochinese Peninsula and one of the greatest rivers of Asia. It rises in Tsinghai Province in the eastern Tibetan highlands and flows in a generally southerly direction for 2600 miles (4180 km) and finally discharges into the South China Sea from the coast of Vietnam at latitude 10°33′N, longitude 106°24′E. Its average discharge rate is 364,600 cfs (10,325 cms). Its hydrograph (Fig. 4–5) shows a peak flow in Sep-

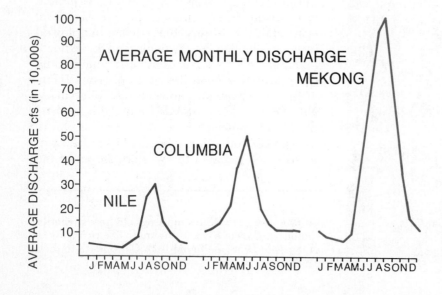

FIGURE 4–5 These hydrographs are plotted on the same scale for easy comparison. There are obvious differences in the variations between peak and low flow.

Okanogan Lake (50°N, 119°28'W) in British Columbia is surrounded by a hilly, forested terrain. The Okanogan River, which flows from this lake, is a major tributary of the Columbia River. Figure 4–6 shows that the Okanogan joins the Columbia in Washington. (Canadian Pacific Railway)

tember, which reflects the tropical rainy season of the region. Closer scrutiny of the hydrograph reveals a long period of high flow that begins in May and does not end until December. In fact, the immediate region around the Mekong Delta receives about 80 inches (200 cm) of rain between April and October. The fertile soils of the delta make this region a very productive agricultural area.

FLOODS

The condition known as a **flood** occurs when the water within a channel rises above the stream's banks and spills over. In other words, the discharge exceeds the capacity of the channel to contain it. The conditions that can cause flooding in river basins are many and varied.

The Danube River (Table 4–1), which has its start in the Black Forest near Donaueschingen (Fig. 4–7), is the second longest river system in Europe. Its basin includes southern Germany, Austria, part of Czechoslovakia, Hungary, Yugoslavia, Bulgaria, and Romania. The Danube has been the chief water route for the commerce of Central Europe since the time of the Romans. Many serious and unpredicted floods have occurred along its basin.

The hydrograph of the Danube's discharge throughout the year is remarkably uniform (Fig. 4–8). At an average flow of 226,200 cfs (6400 cms), the Danube's discharge is greater than that of the Nile or the Columbia River. In fact, the Danube's flow exceeds that of the Nile by more than 140 percent. The three hydrographs in Figure 4–5 and the Danube's hydrograph in Figure 4–8 are plotted on the same scale so simple comparisons of the vertical exaggerations can be made. The difference between the Danube's peak flow and its base flow is much less than that of the other three river systems—the Nile, Columbia, and Mekong.

The uniform flow of the Danube reflects the fact that this river rises in and moves through relatively moist areas. The Danube's source of water includes autumn-winter rainfall and spring snowmelt. The snow and glacier melt in the Austrian Alps takes place over an extended period of time and helps to maintain the Danube's flow in late summer. In addition, the Danube is controlled in many places along its course for power generation and navigation. The purpose of the control is to reduce natural variations in discharge.

Serious unpredicted flooding occurred along the Danube River in 1342, 1402, 1501, and 1830. Each of these floods resulted from an ice jam during the spring rise. An ice jam produces a natural dam that causes water from winter rains and snowmelt to accumulate and back up over hundreds of square miles. The accumulating river water very quickly spills over its banks and floods the surrounding land. When these ice jams finally break, they release torrents of water into the lower stream channels that are difficult to contain between the river's banks.

Ice jams along the Danube in 1970 produced serious early spring flooding that inundated

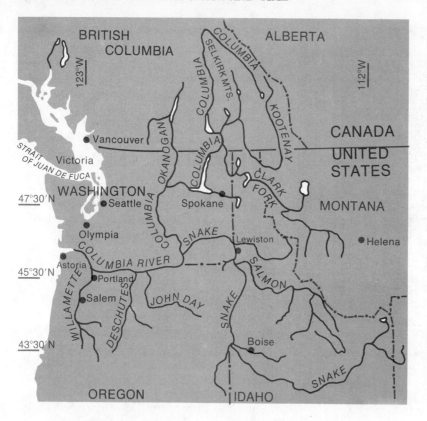

FIGURE 4–6 The Columbia River and its drainage form an important system of waterways in southwestern Canada and the Pacific Northwest of the United States. The Columbia takes shape on the western slopes of the Rocky Mountains in Canada. Its 1320-mile (2130 kilometer) journey to the sea ends at Astoria, Oregon. The mouth of the Columbia River is the main harbor between San Franciso and the Strait of Juan de Fuca.

more than 1000 villages and 1.7 million acres of choice farmland. More than 250 deaths were attributed to the rampaging waters, while

Two patrol boats are positioned across the channel in the lower loop of the Mekong River. The loops are referred to as *meanders*. There are many well-developed meanders in the portion of the Mekong that flows through Vietnam. (U.S. Navy)

thousands of persons were left homeless. By late May the flood crest had worked its way to the Danube's delta on the Black Sea (Fig. 4–7). The residents of the city of Galati fought to keep the floodwater away from factories built in low-lying areas. The Romanian government characterized the 1970 Danube floods as the worst to strike the region since Roman times.

Spring thaws take first prize for causing floods because springtime warmth melts the snows on mountain slopes and high plateaus and sends millions of gallons of water on an unrelenting journey to the sea. Certain combinations of conditions bring about rapid melting and flooding while others do not. When a warm, moist wind blows across a snow surface, for example, condensation occurs. Each time condensation occurs, there is a simultaneous release of heat. The heat produced in this way raises the temperature of the air and, as a consequence, increases its melting power. Thus, whenever a large mass of fast-moving warm, moist air moves over a snowfield, there is danger; if the melting process proceeds quickly enough, it can waterlog the

FIGURE 4-7 The Danube River is the second longest river in Europe. It rises in the Black Forest of Germany, winds its way eastward through Central Europe, and discharges its water into the Black Sea. Many hydroelectric projects have been developed along the Danube; the project shown in the photo is along the Austrian portion of the Danube. (Austrian Information Service)

ground and produce a lot of runoff that can lead to serious flooding. This, in fact, is one of the commonest causes of spring floods throughout New England, the Great Lakes, and the Great Plains region. The same combination of conditions produces the spring floods in the mountains of the Pacific Northwest.

On the other hand, when a warm, dry wind blows across a snow surface, there is little danger of flooding. Now, it is true that the snow may, in fact, disappear practically overnight under these conditions. However, the flooding risk is not as great as in the case of the warm, moist air because a warm, dry wind causes a considerable portion of the snow to vaporize directly into the dry atmosphere.

Many people believe rain causes snow to melt, but rain alone seldom succeeds in melting snow very fast. A rainfall of 5 inches (12.7 cm), for example, with a temperature of 41°F (5°C) is unlikely to release more than one-third of an inch of snowmelt. A heavy rain of 5 inches (12.7 cm), on the other hand, occurring in conjunction with warm winds causes the melting rate of the snow

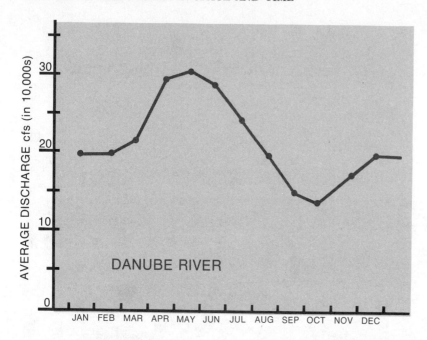

FIGURE 4–8 **This hydrograph shows that the Danube's discharge throughout the year is rather uniform.**

to rise sharply. The Red River floods of 1950 and 1979, for example, were caused by a combination of heavy rainfall and warm winds.

Another popular belief is that sunlight is an efficient destroyer of snow. In fact, most of the insolation is reflected back into the atmosphere when the snow surface is clean. A dirty snow surface, however, is a different story because each little speck of dirt acts as an absorber of the Sun's energy, which contributes to the melting process. The stored energy in a blackened layer of snow causes it to continue to melt even after it has been covered by a new snowfall.

The temperature of soil is a crucial factor in its ability to cause snow to melt. Under some conditions, soil becomes frozen before the first snows of winter. Frozen ground hidden below a blanket of snow does not contribute much heat to melt the overlying snow cover. However, if soil is not frozen at the time of the first snowfall, the ground below has a reservoir of stored-up warmth that melts the underside of the snow. The heat of the soil, under these circumstances, may be sufficient to keep the lower layer of snow in a constant state of liquefaction that saturates the soil with water throughout the winter. By the time the spring thaw comes, the risk of flooding through rapid runoff is very great because the surface below is already saturated.

During mid-April 1969, the town of Crookston, Minnesota, lived through a brutal 70-hour battering by the rampaging Red River of the North, which was swollen by the heaviest accumulation of melting snow in more than a century. Rivers throughout the region swelled, and waters gushed over their banks. In five states—North Dakota, South Dakota, Minnesota, Wisconsin, and Iowa—town after town was devastated by floodwater that reached new peaks when an ice jam broke up in Saskatchewan. In many places, tumbling, gigantic chunks of ice were carried by the torrents of water that drove more than 22,000 people from their homes. At Grand Forks, North Dakota, the Red River surged to its highest mark in 72 years. In South Dakota, rampaging waters from the Sioux River struck Sioux Falls. Throughout the heartland of the country, the Mississippi River roared past flood-control walls at more than 9 million cfs (0.3 million cms) or 15 times its normal rate.

Recognize that whenever water rises in a natural stream bed above a certain level, floods threaten the area. The danger stage is reached when the stream is bank full; in other words, the channel is completely occupied by water. Periodically during the history of any stream or river, water overflows its banks. During Feb-

This wide, meandering expanse of the Mississippi River was photographed at a point between Baton Rouge and New Orleans, Louisiana. The levee protection on each side of the river also carries a roadway. The homes and farms on each side of the river are on the Mississippi's floodplain. (Army Corps of Engineers)

ruary 1969, the experts at the National Weather Service and the Army Corps of Engineers correctly predicted the floods that struck in April. They based the prediction on the massive Canadian snowpacks that had developed over the winter. Our problem is not with our ability to predict when a flood will occur; our problem is with encroachment! The encroachment of home, town, and industry on river channels contributes to the damage caused by flooding. Part of the solution to the problem of flood control is halting this movement onto the **floodplains** of rivers. Floodplain zoning is needed along most of the rivers of the world.

LAKES AND WETLANDS

Lakes and wetlands are part of the surface-water component of the hydrologic cycle. From a hydrological point of view, lakes and wetlands serve the same purpose; that is, they are storage mechanisms that function to regulate the flow of water in streams. Each of these storage mechanisms consists of a depression that is filled or covered with water either because there is a relatively impervious structure below the surface, which captures and holds runoff, or because the floor of the depression is below the water table.

Some wetlands—bogs and swamps, for example—are the evolutionary offspring of lakes. Geologically speaking, a lake is a short-lived surface feature. As soon as it is formed three processes function to bring about its eventual destruction and conversion to a swamp: (1) All inflowing streams bring sediment into the lake, deposit it, and thus start the process of filling the depression. (2) When inflowing water fills the basin sufficiently, it overflows and the outflowing stream tends to erode a notch through the lip of the basin. Eventually, the notch is eroded sufficiently and the basin is drained. (3) The accumulation of organic deposits from vegetation causes a lake to change to a bog and then later to become a swamp.

LAKE-BASIN ORIGIN

A lake may form in any undrained depression or in any depression that has an outlet somewhat above its lowest point. Of course, an adequate supply of inflowing water is needed to keep the basin filled. These basins and depressions are produced in a number of ways.

THROUGH GLACIER ACTION

The continental glaciers of the **Pleistocene epoch** covered Canada, the adjacent northern United States, Finland, and parts of Sweden. These glaciers gouged huge depressions in the bedrock of these regions. Thousands of rock-shored lakes—whose basins were produced by glacier action—can be found in northeastern Minnesota and the adjacent parts of Canada. Ontario, for example, has a profusion of glacier-produced lakes, some are large, others are small; most are quite irregular, and many are connected by winding and twisting rivers.

Lake Winnipeg, in south-central Manitoba, is the third largest lake lying entirely within Canada (Fig. 4–9). It is 260 miles (416 km) long and from 20 to 60 miles (32 to 96 km) wide. The lake's surface is 710 feet (216 m) above sea level, but its depth does not exceed 70 feet (21 m). Lake Winnipeg and its two large neighbors—Lake Winnipegosis and Lake Manitoba—lie in the deeper portions of a glacier-produced shallow basin that was created late in the Pleistocene epoch.

Continental glaciers also create lakes by placing deposits of rock debris across preexisting drainage patterns. The debris forms a dam. Then, stream water accumulates behind the dam to form a lake. Lake Placid—about 145 miles (233 km) north of Albany—is a small beautiful lake in the Adirondack Mountains of Essex County, New York, which was formed when a valley in the Adirondacks was blocked by glacier debris. The summit of Mount Whiteface, standing 3000 feet (914 m) above Lake Placid's surface, is a good point from which to observe the rectangular outline of the lake.

METEOR CRATERS

A giant eroded crater in northeastern Quebec was formed about 210 million years ago, during

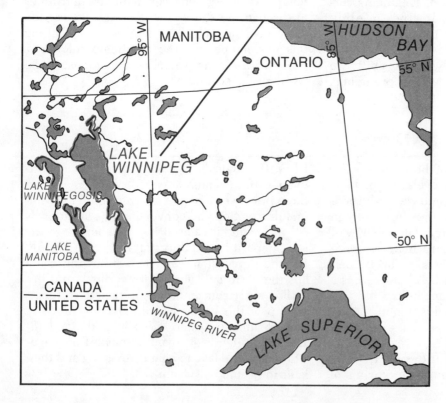

FIGURE 4–9 Lake Winnipeg, Lake Manitoba, and Lake Winnipegosis lie in the deeper portions of a glacier-produced basin.

The village of Lake Placid is in the foreground. The village spreads along the shore of Mirror Lake. The body of water known as *Lake Placid* is in the left background of the photo. These lakes owe their origin to glacier action, which placed debris across a somewhat rectangular valley. (F. MacNeill)

the Mesozoic era, by the impact of a large asteroidal body. The crater has a diameter of about 40 miles (65 km) (Fig. 4–10). The asteroid that produced the crater must have had a diameter of almost 2 miles (3 km). The water of Lake Manicouagane (latitude 51°31′N, longitude 68°19′W) fills the crater's circular depression. In the center of the lake is a peak of shocked rock that forms a small mountain, Mont de Babel. Smaller lakes dot the area of shocked rock (Fig. 4–10). The water of the lake drains southward through the Manicouagane River to the Saint Lawrence River.

The youngest-known meteor crater larger than 6 miles (10 km) in diameter is found in Ghana (Fig. 4–11). The impact that created this Ghanian crater occurred 1.3 million years ago during the Quaternary period of the Cenozoic. This Ghanian crater with a diameter of about 8 miles (13 km), provides the basin for Lake Bosumtwi. The lake is located at latitude 6°30′N, longitude 1°25′W, which is about 100 miles (162 km) northwest of Accra, which is on the Gold Coast.

VOLCANIC-REGION LAKES

The craters of extinct or dormant volcanoes commonly contain lakes. Crater Lake—located

in the crater of Mount Mazama, an inactive volcano in the Cascade Mountains of southwest Oregon—is one of the best-known examples. The lake is nearly round and 6 miles (9.6 km) across at its widest point.

Intermontane basins, which are occupied by lakes, are formed by the irregular distribution of volcanoes in the volcanic regions of central Mexico and the Armenian sections of Turkey, Iran, and the U.S.S.R. Lake Chapola, Lake Pátzcuaro, and Lake Cuitzeo—all in the Mexican highlands—lie in basins hemmed in by volcanoes. Lake Van in Turkey, Lake Urmea in Iran, and Lake Sevan in the U.S.S.R. are three shallow lakes all about equidistant from the volcanic cone of Mount Ararat (39°42′N, 44°18′E). These lakes of "Armenia" do not have any outlets and, thus, are salt lakes.

Another way in which lake basins develop in volcanic regions is by the damming of river valleys by lava flows. Snag Lake near Lassen Peak in Lassen National Park is an excellent example of such a lake.

SINKHOLE LAKES

A region underlain by highly soluble limestones develops depressions called sinkholes, which under proper circumstances may fill with water

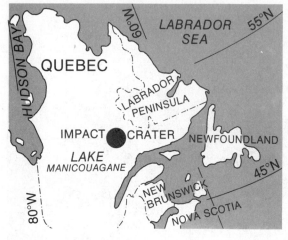

FIGURE 4–10 The impact of an asteroidal body created the circular depression of Lake Manicouagane. The lake forms a ribbon of water around the large area of shocked rock known as Mont de Babel. A number of similar lakes dot the shocked-rock area.

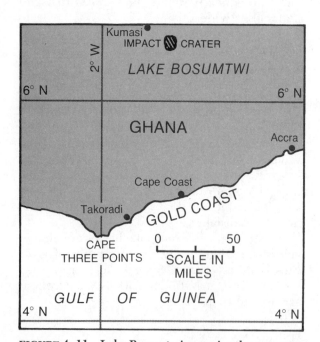

FIGURE 4–11 Lake Bosumtwi occupies the youngest known meteor crater; it is larger than 6 miles in diameter.

and become lakes. The lake region of north-central Florida is of this type (Fig. 4–12). The sinkhole-lake region of Florida extends for more than 200 miles (320 km)—from close to the northern border southward to the northern edge of the Everglades, not far from Lake Okeechobee. Lake Okeechobee is not a sinkhole lake, however. Gainesville, Orlando, and Lakeland are well within the sinkhole-lake region. Lake Apopka and Lake Kissimmee are two of the larger sinkhole lakes of Florida.

The Yunnan Plateau in southern China is a limestone region pitted with many depressions and sinks. Lake Tien Ch'ih is one of the largest of the lakes on this high plateau; it stands on the floor of K'unming Plain close to the city of K'unming. The Chinese lakes in this limestone region have a similar origin to those of Florida.

DELTA LAKES

The Salton Sea, which is located in the Imperial Valley of southeastern California, is a salt lake whose surface is 240 feet (73 m) below sea level.

This is Oregon's Crater Lake. A mighty volcanic mountain in Oregon's Cascades spewed its interior over the surrounding countryside; when the explosive activity ceased, a cauldron remained. Eventually, the cauldron filled with water; the lake maintains a level of 2000 feet (600 meters) above the crater floor. A perfect cinder cone rises 700 feet (210 meters) above the lake's surface to form Wizard Island. The cliffs in the background, which surround the lake, stand about 2000 feet (600 meters) above the lake's surface. (Oregon State Highway Dept.)

This lake was formed because the Colorado River built a delta in the Gulf of California.

At one time, the water of the Gulf of California extended much farther north and occupied a long troughlike depression between the Chocolate Mountains on the east and the mountains of southern California on the west. The Colorado River carried great quantities of silt and debris from the Rocky Mountains and from the Colorado Plateau through which it was cutting its great canyon. It dropped this debris into the quiet waters of the gulf at a point just to the west of present-day Yuma. The developing delta cut off the head of the gulf, after the gulf water evaporated, a valley remained. This valley, which is known today as the Imperial Valley, is the floor of the old gulf. Until 1905 the Imperial Valley was a salt-covered depression. Between 1905 and 1907, the Colorado River broke through irrigation head gates and burst into the basin. The floodwater formed the Salton Sea.

Lake Pontchartrain near New Orleans, Louisiana, is another example of a delta lake (Fig. 4–13). Pontchartain spreads over 40 miles (64 km) in a west-east direction and its north-south axis is 25 miles (40 km) long. It is Louisiana's largest lake. This saltwater lake was formed as a result of the building of the Mississippi Delta; that is, part of the Gulf of Mexico was cut off by ridges of sand and deposits of silt dropped by the Mississippi River as it rushed toward the sea.

STRUCTURAL DEPRESSIONS

The longest freshwater lake in the world, Lake Tanganyika (6°S, 29°30′E) in East Africa, occupies a structural depression (Fig. 4–14). It lies in the Great Rift Valley of Africa—a huge, long, down-dropped block of the Earth's crust. Lake Tanganyika is 420 miles (676 km) long and a fantastic 4825 feet (1471 km) deep. It is, however, comparatively narrow, ranging in width from about 25 to 45 miles (40 to 72 km). This lake forms the boundary between Tanzania and the Democratic Republic of the Congo. For most of its length, the land rises steeply from its shores.

LAKE WATER

Most lakes consist of freshwater but some are saltier than the ocean. Four—the Caspian Sea, the Aral Sea, Lake Maracaibo, and Lake Patos—of the 15 lakes listed in Table 4–3 are salty. Note that a number of "seas" are actually salt lakes; the Dead Sea is another example of a salt lake. All salt lakes are found in desert or semiarid climates.

Recall that typical ocean water has a salinity of 35 parts per thousand (°/oo). Salt lakes differ

FIGURE 4–12 The lakes in this figure that are in color are sinkhole lakes. The sinkhole lake region of Florida extends from north of Gainesville almost to Lake Okeechobee—a distance of about 200 miles (320 kilometers).

quite a bit in their degree of salinity. The Great Salt Lake of Utah, for example, has a dissolved salt content of 150 %. The Dead Sea is really salty; its salinity is 246 %. The Aral Sea with a salinity of 11 % and the Caspian Sea with 6 % are quite low. The composition of the salts in these lakes is related to the chemical characteristics of the rock structures in their drainage areas.

There are also marked differences in the dissolved materials found in the water of freshwater lakes. Lakes with a sluggish inflow—especially where the inflowing water has a lot of contact with marginal vegetation—tend to have water with a high organic content. A lake with a drainage that flows over igneous or metamorphic rock tends to have a low dissolved-solids content. The major drainage into Lake Superior, for example, flows over tough crystalline rock and its dissolved solids amount to a mere 0.05 %. On the other hand, lakes within limestone drainage areas have a pronounced calcium carbonate and magnesium carbonate content.

Lakes are the only major areas of relatively stationary freshwater found on the surface of the Earth. Thus, they form a specialized type of environment for both plants and animals. Most lakes are comparatively shallow with depths of less than 100 feet (30 m). Only a small number plunge to great depths like Lake Baikal (Table 4–3), which is the deepest lake in the world. In high latitudes with cold winters, a shallow lake may experience a complete double turnover of the water each year.

TABLE 4–3 THREE LARGEST LAKES ON EACH CONTINENT

CONTINENT	LAKE	CENTER-LAKE LOCATION LATITUDE, LONGITUDE	AREA		ELEVATION ABOVE SEA LEVEL		GREATEST DEPTH	
			mi²	km³	Feet	Meters	Feet	Meters
Africa	Victoria	1°S,33°E	24,300	62,940	3,721	1,134	266	81
	Tanganyika	6°S,29°30′E	12,350	32,000	2,539	774	4,825	1,471
	Nyasa	12°S,34°30′E	8,680	22,490	1,558	475	2,316	706
Asia	Caspian Sea	42°N,50°E	143,240	371,000	−91	−28	3,215	980
	Aral Sea	45°N,60°E	24,900	64,500	174	53	223	68
	Baikal	54°N,109°E	12,160	31,500	1,496	456	5,315	1,620
Europe	Ladoga	61°N,31°30′E	7,100	18,400	13	4	738	225
	Onega	61°30′N,35°45′E	3,710	9,600	108	33	377	115
	Vanern	58°55′N,13°30′E	2,160	5,580	144	44	328	100
North America	Superior	48°N,88°W	31,820	82,410	602	183	1,302	397
	Huron	44°30′N,82°15′W	23,010	59,600	581	177	750	229
	Michigan	44°N,87°W	22,400	58,020	581	177	923	281
South America	Maracaibo	9°40′N,71°31′W	5,520	14,300	0	0	197	60
	Patos	31°6′S,51°15′W	3,860	10,000	0	0	15	5
	Titicaca	15°48′S,69°24′W	3,220	8,340	12,497	3,809	997	304

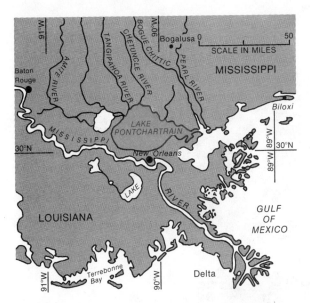

FIGURE 4–13 Lake Pontchartrain, a satellite lake, was cut off from the Gulf of Mexico by sand and silt dropped by the Mississippi River during its rush to the sea.

WETLANDS

Bogs, swamps, and marshes are considered **wetlands.** The term *swamp* is applied to a wetland where trees and shrubs are an important part of the vegetative association. The word **bog** implies a lack of solid foundation; some bogs consist of a thick zone of vegetation floating on water. In popular usage, the term **marsh** is often used indiscriminantly when referring to a swamp. It is true that a marsh has many physical features common to a swamp, but, a marsh differs from a swamp in having vegetation composed dominantly of grasses and grasslike plants such as sedges and rushes.

BOGS

A bog is a soft, spongy, waterlogged ground that is covered by vegetation, chiefly mosses in various stages of decomposition. Generally, a bog is an area which has no external drainage or almost none.

The vegetation in a bog accumulates to great thicknesses. If it is a floating mat, it thickens from both the top and the bottom. The vegetative mat can become thick enough to support the weight of a person. As the lower portion of the vegetative mat decomposes and becomes compressed by the weight of overlying material, it forms a layer called **peat.** In some cases, the peat in a bog is almost pure organic material.

The cranberry, an edible berry, grows on a small, creeping woody plant that thrives in the bogs of North America, northern Asia, and north-central Europe. The American cranberry is found wild from Newfoundland to the Carolinas and westward to Minnesota and Arkansas. New Jersey, Massachusetts, and Wisconsin are important cranberry-growing states.

SWAMPLANDS

The evolutionary process described on page 123 is not the only way in which a swamp develops. A swamp is essentially a low, flat-lying wetland that is subject to daily, seasonal, or perennial flooding. Some swamps develop on the low flat-

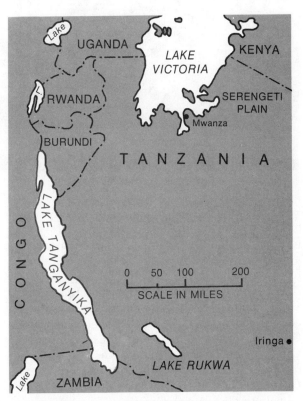

FIGURE 4–14 Lake Tanganyika is the longest freshwater lake in the world. It is in a structural depression—a huge, down-dropped block of the Earth's crust.

These cranberry bogs are in the Chatsworth area of central New Jersey, which is southwest of Toms River and west of Forked River. In the smaller photo taken in June, three men are inspecting the plants in the bog. The vegetative mat is thick and supports their weight with ease. At harvest time (below), the bog is flooded and machines are used to loosen the berries from the bushes. (N.J. Dept. of Agriculture)

lying terrain of seacoasts and on the floodplains of rivers with low drainage gradients. In Arctic regions, the thawing of frozen ground may allow a swamp to develop because the permanently frozen ground below prevents drainage.

There are some well-known swamps in the United States. We have the mangrove swamps of coastal Florida and the Okefenokee Swamp of Georgia. Then there is the densely forested Great Dismal Swamp located on the coastal plain of southeastern Virginia and northeastern North Carolina between Norfolk, Virginia, and Elizabeth City, North Carolina (Fig. 4–15).

The Dismal Swamp is one of the largest swamps in the United States. Lake Drummond lies in the center of the swamp. In the early 1700s, the north-south axis of the Great Dismal Swamp was 40 miles (64 km) long. The swamp spread over an area of 2000 square miles (5180 km²). Some of the Dismal Swamp was drained and cleared for farming. Today, its north-south axis is about 30 miles (48 km) long and the swamp only covers an area of 750 square miles (1940 km²). Over a period of 200 years, human intervention has reduced the Great Dismal Swamp's area by more than 62 percent.

The Okefenokee Swamp sits in a shallow, saucerlike depression that is roughly 20 miles (32 km) wide and 40 miles (64 km) long (Fig. 4–16). It spreads over about 800 square miles (2070 km²) of land in southeastern Georgia and northern Florida. A low, sandy ridge—Trail

FIGURE 4–15 The area of the Dismal Swamp is shown in color; it is one of the largest swamps in the United States.

Ridge—forms the swamp's eastern boundary and prevents drainage toward the Atlantic Ocean. The Okefenokee is partially drained by the Suwannee and St. Mary's rivers. Giant tupelo and cypress trees festooned with Spanish moss, brush, and vines are part of the dense vegetation of the swamp. Pine trees predominate where sandy soil is above water.

Ordinarily we do not associate swamps with deserts, but in Africa there is an extensive swampland at the very edge of a great desert. The Kalahari Desert, with an area of 200,000 square miles (518,000 km²), covers most of southern Botswana. The Okavango Swamp (latitude 18°45′S, 22°45′E) in northern Botswana sits on the northern edge of the Kalahari Desert (Fig. 4–17). The swamp, which is surrounded by thick bush and forest, is found in the 4000-square-mile (10,300-km₂) inland delta of

the Okavango River. As Figure 4–17 indicates, the great swamp tract is in a basin fed by the Okavango River. It stretches for about 170 miles (270 km) along the inland-delta region, which starts about 10 miles (16 km) below Andara at Popa Falls. Lake Ngami occupies a depression at the southern edge of the Okavango's delta region. The flow of water within the swamp of the inland delta is complex. Any outflow from the swamp normally follows the annual floods that begin in March at the head of the delta.

MARSHLANDS

We have some notable marshes in the United States. The Everglades in Florida and the tule marshes of the Sacramento and San Joaquin valleys in California are some well-known examples. Salt- and brackish-water marshes are com-

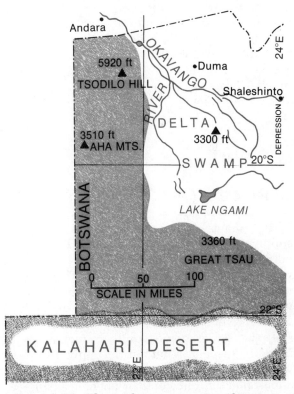

FIGURE 4–17 The northwestern section of Botswana is portrayed in this map. The great volumes of water and sediment poured into the basin of the Okavango River produced the delta and the swamp. The Okavango River is surrounded by higher land that traps and checks the outflow of water from the basin.

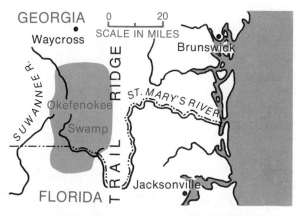

FIGURE 4–16 The area around the Okefenokee Swamp was once part of the ocean floor. The swamp sits in a shallow depression with Trail Ridge as its eastern boundary.

mon along low seacoasts, inside barrier bars and beaches, and in estuaries and deltas. There are extensive salt marshes in some deserts, too. The Humboldt Salt Marsh (39°50′N,117°55′W), which is approximately 100 miles (160 km) northeast of Reno, Nevada, is a good example of these desert marshes.

With the exception of its northeastern and southeastern corners, Nevada lies wholly within the Great Basin. The floor of the Great Basin is between 4000 and 5000 feet (1220 and 1520 m) above sea level. Buttes, mesas, and isolated mountain ranges rise 1000 to 7000 feet (300 to 2100 m) above the basin's floor. The Humboldt Salt Marsh, which is on the floor of the Great Basin, is fed by streams that flow out of the surrounding mountains. Most of Nevada's rivers are small and flow only during the wet season, from December to June. Only a few of the rivers have outlets to the sea. Most empty into the Great Basin, where they flow into lakes without outlets or into wide, shallow depressions. The low humidity causes water to evaporate rapidly from the lakes and depressions. In fact, many of the smaller lakes dry up to become salty mud flats during the summer.

There are many small and medium sized marshes—that is, areas of soft wetlands—throughout the United States. Typically, marshes are treeless but have many grasses, sedges, cattails, and lilies. The marsh photographed above is representative of those found in Wisconsin; it is a marsh of cattails and open water areas with a border vegetation of alder, cottonwood, and aspen. (USDA)

ICECAPS, GLACIERS, AND SNOWFIELDS

Today, about 1.9 percent of the world's water supply is held in **icecaps** and **glaciers.** This figure may seem insignificant when it is contrasted with the overwhelming percentage (97.6%) held in the world's oceans. But it is significant when you realize the sum total of water in all the world's rivers, freshwater lakes, saline lakes, soil, and underground aquifers amounts to a mere 0.5 percent of our entire supply. In other words, the water held as glaciers and icecaps is almost four time greater than that held as surface and subsurface water.

QUATERNARY ICE AGES

The **Quaternary period** of the Cenozoic is characterized by the advance of great glaciers or

ice sheets from a number of centers. The most important centers of glaciation in Europe were Scandinavia and the Alps. In North America the Laurentian ice sheet had its origin in the high regions of northeastern North America—in northern Quebec and Labrador, and on the islands of Newfoundland, Baffin, Bylot, Devon, and Ellesmere—and spread toward the west and south (Fig. 4–18). The Laurentian ice sheet made contact in the west with the Cordilleran ice sheet, which stretched continuously from the Columbia River to the Aleutian Islands.

The Quaternary period is divided into the **Pleistocene epoch** and the **Holocene epoch** (see Table 1–2). The Holocene embraces the last 15,000 years, and it is considered the post-glacial epoch; the Pleistocene is the epoch of the

The valley between the two peaks at the right of the upper photo is filled by the Portage Glacier, which is an hour's drive from Anchorage, Alaska. The glacier in the photo at the far left, called Mer de Glace, or Sea of Ice, is near Chamonix, France. The Mer de Glace, on Mont Blanc in the Pennine Alps, has a very irregular surface with many cracks and crevasses. Two men (photo at left) are peering into a deep crevasse on King's Glacier in Kongs Fiord on West Spitsbergen Island. (Alaska Airlines; French Government; Norwegian Government)

Quaternary in which the ice ages occurred. Minor centers of Pleistocene **glaciation** were located in Spitzbergen, Iceland, Ireland, Scotland, and northern England, the Pyrenees, the Caucasus range, the Himalayas, the mountains of central Asia, the Rockies, the Andes, the highest mountains of equatorial Africa, New Guinea, southeastern Australia, and New Zealand. Approximately one-tenth of the Earth's surface was covered by ice during the Pleistocene. The present ice sheets of Greenland and Antarctica are remnants of the glaciation that occurred during the Pleistocene.

The post-glacial period—the 15,000 years of the Holocene—has been marked by alternating rainy and dry climates. There were dry periods, for example, from 2200 to 1900 B.C., 1200 to 1000 B.C., and again from 700 to 500 B.C. With the beginning of the Christian era, the historical records give us excellent detail about climatic fluctuations: For the first 500 years of the Christian era, the climate was somewhat rainier than today, especially in northeast Africa. From A.D. 500 to 700 a period of relative dryness with accompanying droughts set in—the Caspian Sea was well below its present level, and there was a

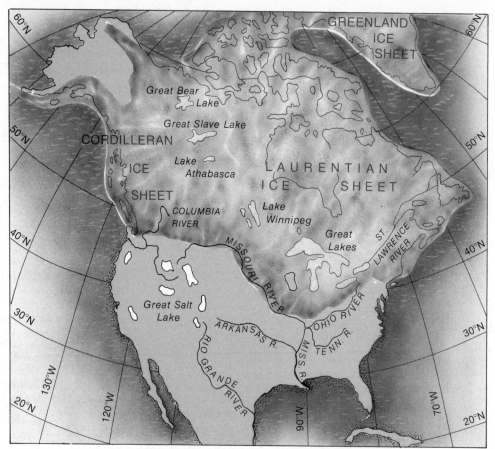

FIGURE 4–18 Canada was covered by massive ice sheets during the Pleistocene. The present channels of the Missouri and Ohio rivers are not far from what was the southern boundary of these ice sheets in the U.S. There were at least 6 minor centers of glaciation in mountain areas within the U.S.

lot of traffic over the Alpine passes. From A.D. 800 to 1250 there was an increase in rainfall, and the records indicate extensive flooding along the coasts of England and the Netherlands toward the end of the period. The period from 1540 to 1890 is known as the Little Ice Age because glaciers generally advanced during this time. Since the end of the nineteenth century, there has been a general retreat and glaciers have tended to shrink.

ICE-AGE THEORY

For more than 100 years, the scientific literature has contained a suggestion that changes in cli-

mate, which lead us into an ice age, are caused by an alteration of insolation that is brought about by variations in the Earth's orbit. Milutin Milankovitch worked out the orbital changes and developed a detailed theory of their effects in the 1930s. James Hays, John Imbrie, and Nicholas Shackelton developed a chronology of global climatic reversals and compared it with the orbital motions and cycles postulated by Milankovitch. The graph in Figure 4–19 represents a comparison between the variations in the eccentricity of the Earth's orbit and a temperature curve based on microfossils in two deep-sea cores. Figure 4–19 shows that minimum temperatures are reached when eccentricity is at a minimum.

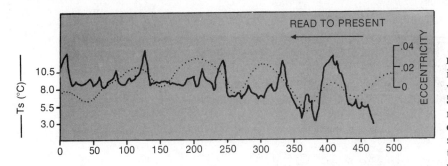

FIGURE 4–19 The solid line is the curve of temperature variation during the last 500,000 years. The dotted line represents the eccentricity of the Earth's orbit during the same span of time. Minimum temperatures are reached when the eccentricity of the Earth's orbit is at a minimum.

THE ORBITAL PATTERN

The cyclical changes in the shape of the Earth's orbit around the Sun are depicted in Figure 4–20. Starting from the left, the orbit ranges from elliptical in the first diagram to nearly circular in the fourth diagram and back again to an **elliptical orbit** in the seventh diagram. **Eccentricity** simply refers to the deviation from the circular pattern or norm. If we consider the norm to be a circle, then eccentricity refers to the the deviation from the circular pattern. The more nearly circular orbits have little or no eccentricity; the more elliptical the orbit is, the greater is its eccentricity.

Refer to Figure 4–19 again and note the peaks and troughs in the dotted line that plots the eccentricity of the Earth's orbit around the Sun. Peaks repeat in a cycle of roughly 100,000 years, indicating that the shape of the Earth's orbit around the Sun reaches its greatest eccentricity once every 100,000 years. Actually, the cycle is 93,000 years when calculated with care. The troughs, or minimum points, of eccentricity—when the orbit is most nearly circular—occur on the same cycle of 93,000 years.

The comparison made by Hays and his associates (Fig. 4–19) indicates that the coldest periods, or ice ages, occur when the orbit of the Earth around the Sun is most nearly circular. The warmest, or interglacial periods, occur when the orbit of the Earth is most eccentric or elliptical.

In the fourth diagram (Fig. 4–20) the Earth is in a nearly circular orbit around the Sun. A circular orbit means the Earth is at the same distance from the Sun at any location in the circular orbit. In the fifth diagram the Earth's orbit has become

eccentric—somewhat elliptical. With an elliptical orbit, the Earth is not always equidistant from the Sun.

Today, the Earth is in an elliptical orbit around the Sun, and its closest approach to the Sun occurs in January. As the Earth advances in its elliptical orbit—as the eccentricity of the orbit changes—its closest approach to the Sun will occur at different times of the year. Within 10,000 years from today, the eccentricity of the Earth's orbit will place it closest to the Sun in July. This cycle of change occurs every 23,000 years, and it, too, affects climate. A greater distance between the Earth and Sun in July means cooler summer temperatures in the Northern Hemisphere and less snow and ice melting from the glaciers that exist, allowing a positive growth in glaciers over the years.

A TILTED AXIS

At the present time the Earth's axis is tilted at an angle of 23.5 degrees with respect to a line erected perpendicular to the plane of the Earth's orbit around the Sun. Variations in the obliquity, or **tilt,** of the Earth's axis occur on a cycle of about 42,000 years. The tilt of the Earth's spin axis varies from 22.1 degrees to 24.5 degrees over the length of the cycle (Fig. 4–21). The peak tilt of 24.5 degrees occurred about 9000 years ago. The present trend is toward minimum tilt; the axis is moving from 23.5 degrees toward 22.1 degrees.

An increasing tilt produces hotter summers and colder winters. Hays and his associates think that cooler summers are the key to building an ice sheet because less ice melts when the summers are cooler. In other words, more of the

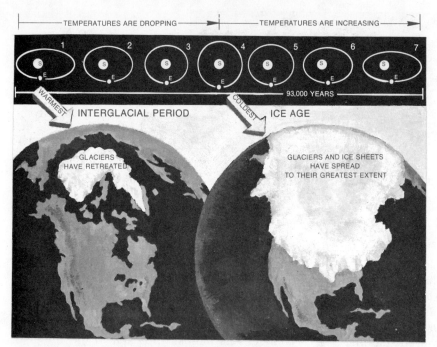

TEMPERATURES ARE DROPPING ⟶   ⟵ TEMPERATURES ARE INCREASING ⟶

93,000 YEARS

WARMEST INTERGLACIAL PERIOD   COLDEST ICE AGE

GLACIERS HAVE RETREATED

GLACIERS AND ICE SHEETS HAVE SPREAD TO THEIR GREATEST EXTENT

FIGURE 4–20 Cyclical changes in the Earth's orbit around the Sun take place at regular intervals. Orbit 4, which is circular, is taken as the norm; the more nearly circular orbits have little or no eccentricity. The more elliptical the orbit is, the greater is its eccentricity.

winter accumulation of snow and ice persists over the year when summer temperatures are lower. Thus, as the axis moves toward less tilt, the Earth should experience cooler summers. According to Hays, minimum tilt is followed in the geological record by a cold climatic period.

SUMMING UP

The last ice age occurred in the Pleistocene. For more than 11,000 years of the Holocene we have been in a warming trend. Today, the eccentricity of the Earth's orbit is high, but it is decreasing toward a circular orbit, which according to Hays means a moderate cooling trend has already set in. The cooling trend is encouraged in the Northern Hemisphere, at the present time, by the fact that the Earth is closest to the Sun in January. The tilt of the Earth's spin axis is decreasing, and this is bringing cooler summer temperatures that also encourage the development of a colder climate.

The evidence indicates that the accumulation of ice in the Northern Hemisphere lags behind a sea-surface temperature drop in the Southern Hemisphere by about 3000 years. Hays and others believe the sea-surface temperatures in the Southern Hemisphere have already cooled to the glacial stage. Therefore, he indi-

cates there is a high probability of substantially more ice in the Northern Hemisphere 3000 years from now.

GLACIAL ICE TODAY

A glacier has its origin in an extensive snowfield where more snow accumulates than melts or evaporates. The positive accumulation of snow over the cycle of a year takes place at altitudes above the permanent snowline. At locations below the snowline, snow melts completely during the warm season. As snow accumulates from year to year, it is compressed by the weight of the recently fallen and overlying snow to become **névé,** a granular ice that is sometimes called *firn.* The firn is just slightly denser than the new-fallen surface snow. Eventually, however, firn recrystallizes into solid glacial ice with a density between 0.82 and 0.87.

As glacial ice accumulates and increases in thickness, it begins to spread and to move to lower altitudes under the influence of gravity. The area of the glacier above the snowline is known as the **zone of accumulation.** The portion of the glacier that moves to an altitude below the snowline is referred to as the **zone of wastage**

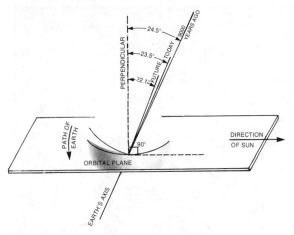

FIGURE 4–21   The tilt of the Earth's axis changes on a cycle of about 42,000 years.

because a net loss in its surface occurs through melting and evaporation.

A glacier—depending on terrain and the rate of accumulation—may move a few inches or as much as 3 feet (1 m) per day. However, it does not move in a uniform flow. The tongue of a glacier moves most rapidly at its center and slowest at its edges, a fact that is true of stream-flow, too. As a glacier moves, its upper rigid surface fractures and breaks into deep cracks, called **crevasses.** A huge glacier may have a 160-foot (50-m) crevasse; but these cracks do not usually extend beyond this depth because the flow of ice below prevents the chasm from penetrating to deeper levels.

Four kinds of glaciers can be identified on the basis of form and relationship to the surrounding topography: cirque, valley, piedmont, and ice-sheet glaciers. A cirque is a steep-walled niche shaped like an amphitheater or half-bowl; a very small glacier that occupies such a niche in a mountainside is called a **cirque glacier. A valley glacier** is simply one that is moving through a valley. When a glacier emerges from a mountain valley and spreads onto a plain at the base of a mountain, it is called a **piedmont glacier.** Broad glaciers, such as those on Greenland and Antarctica, that blanket a large land surface are called **ice sheets.**

Today glaciers cover about 5.8 million square miles (15 million km²). The Greenland ice sheet covers almost 80 percent of the island for a total area of 670,000 square miles (1.7 million km²). The area of Antarctica's ice sheet is approximately 5 million square miles (13 million km²). The Greenland and Antarctica sheets account for 97.8 percent of the world's glacier cover. The remaining 130,000 square miles (300,000 km²) of glacial ice are found in the remote valleys of high mountains such as the Alps of Europe, the Rockies of North America, the Andes of South America, and the Himalayas of Asia. Valley glaciers are also found in other mountain systems—in Africa, eastern North America, and New Zealand, for example. The Pamirs (U.S.S.R.) and the Tien Shan (U.S.S.R. and China) mountains have significant valley-glacier cover.

SYSTEMS UNDER STRESS

The Great Lakes, the Everglades, the Amazon River, and the Hackensack Meadowlands are great systems of surface water. One hundred years ago, a journey across the Great Lakes or a trip into the Everglades was an experience filled with the unexpected and the anticipation of danger. Today the Amazon Basin still requires courage and a spirit of adventure to move into and across its great expanse.

The early explorers and settlers of the Northwest Territory used the Great Lakes as their chief route of travel. Gradually the life and economy surrounding these lakes has changed.

Today the region is one of the most important industrial areas of North America. So, too, have the life and economy of the Everglades and the Hackensack meadowlands changed; for example, today, a huge sports arena stands in the Hackensack Meadowlands. The current human intrusion into the Amazon Basin is certainly affecting all in that region. The changes in the Great Lakes region, the Everglades, and the Hackensack Meadowlands have been accompanied by tremendous error and a willful neglect of the natural balances that existed; the same mistakes are being repeated in the Amazon.

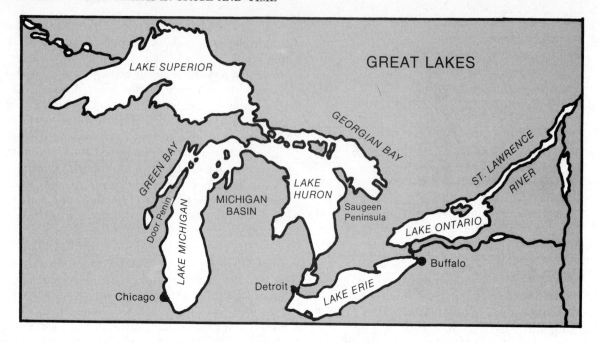

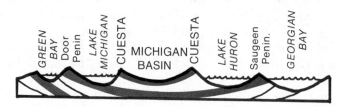

FIGURE 4-22 The major overflow from the interconnected Great Lakes moves by way of the St. Lawrence River to the Gulf of St. Lawrence. The structure of the Great Lakes region consists of sedimentary layers arranged like a series of saucers sitting one on top of another.

THE GREAT LAKES

The five large freshwater bodies, the Great Lakes, discovered by Samuel de Champlain in 1615, dominate the central section of the North American continent. These lakes—Superior, Michigan, Huron, Erie, and Ontario—are interconnected and the major overflow from them moves by way of the St. Lawrence River into the Gulf of St. Lawrence (Fig. 4–22). A portion of the outflowing water, however, passes into the Mississippi River through the Chicago Drainage Canal and the Chicago River. The course of the Chicago River was turned away from Lake Michigan and directed to the Mississippi in 1900.

The surface water in Lake Superior stands at an elevation of 602 feet (183 m) above sea level. Each of the other lakes to the east has a lower elevation. Thus, as the water moves from its maximum elevation to the sea, it descends more than 600 feet (180 m). The greatest single drop of 326 feet (99 m) occurs at Niagara Falls between Lake Erie and Lake Ontario.

Lake Superior is the largest freshwater lake in the world, covering a greater area than the state of South Carolina (Table 4–4). Superior and three of the other four lakes form part of the

TABLE 4-4 GREAT LAKES DATA

LAKE	AREA (mi^2)	(km^2)	ELEVATION OF SURFACE ABOVE SEA LEVEL (Feet)	(Meters)	GREATEST DEPTH (from surface to bottom) (Feet)	(Meters)
Superior	31,820	82,410	602	183	1302	397
Michigan	22,400	58,020	581	177	923	281
Huron	23,010	59,600	581	177	750	229
Erie	9940	25,740	572	174	210	64
Ontario	7540	19,530	246	75	778	237

natural boundary between the United States and Canada. The combined area of the Great Lakes is more than the total for the six states of New England. Lake Ontario, which is the smallest, is almost as large as New Jersey.

THE BASINS

The basins of the Great Lakes developed from major preglacial river valleys, partly by glacial excavation of the valley floors and partly by the accumulation of glacial debris, which formed dams. Note the Michigan Basin in Figure 4–22; it is one of a series of basins that form the Great Lakes region.

The structure of the region consists of sedimentary layers that are arranged like a series of saucers sitting one on top of another (Fig. 4–22). The edges of the sedimentary saucers form ridgelike features called **cuestas.** Study the vertical cross section from Green Bay to Georgian Bay depicted in Figure 4–22. Green Bay occupies a basin between two cuestas. Lake Michigan occupies the basin between the Door-Peninsula cuesta on its western shore and the Michigan Basin cuesta on its eastern shore. Lake Huron also lies between two cuestas.

These basins were formed by various stream systems. Then, the whole area was overrun by the continental ice sheet of the Pleistocene. When the glaciers melted, glacial debris, called *till*, was deposited by the outwash. Huge quantities of **glacial till** blocked many of the valleys and formed natural dams. As the lowlands behind these dams filled with water, the lake system developed.

POLLUTION

In the 1700s, the Great Lakes were beautifully blue and contained tremendous fish populations. With the buildup of the region, the lakes attained vast commercial importance. This commercialism was so irresponsible that its result was the fouling of these natural wonders. Stinking water, algae, debris, and filth became part of what was once a healthy, well-balanced environment for plants and animals.

Lake Erie is the shallowest of these lakes (Table 4–4). It is also the filthiest. The Cuyahoga River, which moves through Cleveland into Lake Erie, carried a load of sewage, detergents, and chemicals into the lake each day for more than 100 years. Throughout the 1960s there were very few fish in Lake Erie; the organic content of its water was staggering. Overwhelming amounts of sewage pouring in from Detroit and other cities sapped the oxygen from the water for years. The rapid accumulation of these wastes sent the central portion of the lake to a zero-oxygen level in the 1960s. Strict enforcement of antipollution laws in both the United States and Canada during the 1970s eliminated much of the industrial pollution by 1980. But Lake Erie's life cycle has been under stress for a long time, and it is so out of balance that we can truly say that the lake is still in danger of dying.

Sophisticated sewage treatment methods—coupled with state bans on laundry detergents containing phosphates—reduced the flow of phosphorus into the lake from 80 pounds (35 kg) per day to less than 30 pounds (15 kg) by 1980, but more comprehensive action is needed if Lake Erie is to be truly rescued from death. The four other lakes are also in need of a massive cleanup campaign if they are to be saved.

The existence of many small communities around Lake Erie has added to the difficulty of remedying its polluted condition. The lake environment is, after all, not that of the water alone, and it does not simply include a few miles of land surrounding the water. It is, in fact, composed of the land, water, plants, and animals of a rather large region. The only cleanup approach that can succeed is a continuing concerted effort by both the United States and Canada and the establishment of a comprehensive governmental structure such as a Lake Erie Authority. A clear-cut jurisdiction is needed over the entire water system and surrounding area. Individual communities cannot be depended on to maintain the necessary action to ensure a healthy lake environment.

The process that Lake Erie has been undergoing is called **eutrophication,** which means its waters were so filled with fertilizers and organic waste material that the entire character of the plant and animal population changed. The water close to the surface was filled with plankton throughout the 1960s and most of the 1970s. Light did not reach the deeper water levels, and any plants that survived there neither underwent photosynthesis nor produced oxygen. If deep-water plants existed at all, they used the dissolved oxygen in the water for their own respiration. Thus, as a result of decreasing oxygen levels in the lake, most plants and animals died and decomposed. The only organisms that survive in such an environment are those that can exist on a very low oxygen supply.

The problem with Lake Erie is, of course, not the fault of the lake; the problem is caused by people in the surrounding communities who have no large-scale social conscience or awareness of what a healthy environment means. Rather than living in harmony with nature, communities around these lakes produced stresses in the natural environment that finally caused widespread change. It is only now, with catastrophe at our door, that we realize the damage that was done. The Great Lakes, after all, contain one-half of the world's freshwater supply. Let that fact sink in and you will recognize the dimensions of the tragedy that allowed these lakes to become foul and polluted.

THE EVERGLADES

The Everglades is one of the largest marsh areas in the world. It covers more than 5000 square miles (12,900 km²) in the southern part of Florida (Fig. 4–23). During the 6-month period from

April through September, 39.3 inches (99.8 cm) of precipitation falls in the area, and the whole marsh is flooded. The marsh dries out somewhat from October through March when the total

The Everglades are south of Lake Okeechobee, which is the large lake in the lower third of the satellite photo. Sawgrass is in the foreground of the photo at the right taken in the Everglades during December. Two islands appear prominently in the background. (NASA, Celeste Navarra)

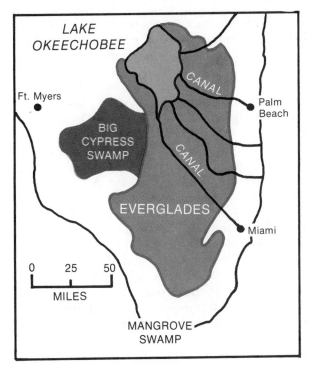

LAKE
OKEECHOBEE

Ft. Myers

CANAL

Palm
Beach

BIG
CYPRESS
SWAMP

CANAL

EVERGLADES

Miami

0   25   50

MILES

MANGROVE
SWAMP

FIGURE 4–23 The Everglades is a large marshland in the southern part of Florida. It spreads from Lake Okeechobee south almost to the tip of the Florida Peninsula.

rainfall decreases to 12.7 inches (32.3 cm). The water flow during the winter half of the year moves mostly in the sloughs. The water of the Everglades drains into the Gulf of Mexico.

The Everglades is generally south of Lake Okeechobee, which is the second largest freshwater lake wholly in the United States. Figure 4–23 shows arms of the Everglades extending around the eastern and western sides of Lake Okeechobee. The main body of marshland, however, stretches southward from the lake for about 100 miles (160 km) to the tip of the Florida Peninsula. The width of the Everglades is, on the average, about 40 miles (64 km). It merges into saltwater marshes and mangrove swamps near the Bay of Florida and the Gulf of Mexico.

The northern and eastern sections of the Everglades are covered by saw grass, which is a sedge rather than a true grass. Wax myrtles, willows, and palms grow on clumps of higher land called *tree islands,* or **hammocks.** The soils found in the Everglades are composed largely of

muck, peat, and gray marl. The muck and peat develop primarily from the remains of decayed plant life.

The state of Florida began draining large areas of the Everglades in 1906 because the rich muck made suitable land for agriculture. The drainage canals built from Lake Okeechobee to the ocean (Fig. 4–23) divert water and deprive the marsh of large volumes of water. The falling water level within the marsh has created difficulties over the years. For example, the dry muck is very susceptible to fire, and saltwater has intruded into the freshwater wells of the area. The decreasing volume of freshwater has also disturbed the natural balance of life in the marshland; without sufficient freshwater the wildlife and vegetation cannot reproduce.

A porous rock called *oolitic limestone* underlies southern Florida and holds a subterranean reservoir of freshwater. A sufficient volume of freshwater recharge is needed to keep seawater from intruding into the aquifer. The removal of great quantities of freshwater from this aquifer by a growing population has lowered the water table and is allowing seawater intrusions to occur. In addition, in some locations the withdrawal of the water from the rock causes the surface area to subside and sometimes to collapse.

THE AMAZON BASIN

The Amazon River extends across a large portion of the northern part of South America (Fig. 4–24). The mighty Amazon has its beginnings in small streams that flow from the Andes Mountains. Its basin drains an area of 2,375,000 square miles (6,150,000 km²). The established length of the Amazon is 4080 miles (6570 km) (Table 4–1). It flows into the Atlantic Ocean very close to the Equator.

At a distance of 2000 miles (1240 km) from the ocean, the river's width is about 1 mile (1.6 km). As the Amazon approaches the Atlantic, its width increases. For example, at 1000 miles from the ocean the river is 4 miles wide (6.5 km), but it is 200 miles (320 km) across where it empties into the Atlantic Ocean.

The Amazon River and its tributaries have many thousands of miles of water surface that are suitable for navigation. During the rainy season the Amazon and the lower courses of the larger tributaries overflow their banks. At these times, the surrounding country takes on the appearance of a vast inland sea.

The Amazon Basin is known to contain valuable natural resources such as minerals and lumber. Some of the most luxuriant vegetation in the world can be found in this basin; its dense forests are filled with an astounding assortment of trees. The waters of the Amazon are rich in

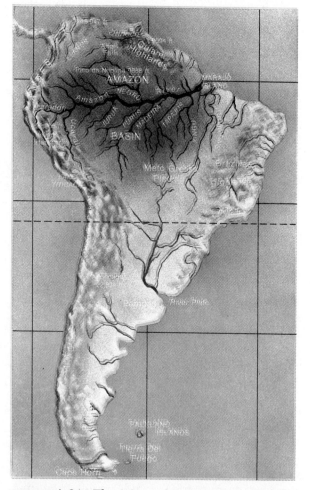

FIGURE 4–24 The Amazon begins in the lowland of northern Peru at the junction of the Marañon and Ucayali Rivers, which rise in the Andes. The Amazon flows east through the jungles of Brazil. Tributaries flow into the Amazon from north and south. The mighty Amazon empties into the Atlantic Ocean on the north side of Marajó.

fish, and large tracts of the basin's fertile land have never known the plow or the trod of human feet. It is one of the few wild frontiers that remain in the world.

During 1968 there was a proposal that nine relatively inexpensive low dams be placed along this river system at strategic points to create a system of seven navigable lakes. One dam more than 20 miles (32 km) long would have enough water behind it to form an inland sea several times the size of Lake Superior. A vast water-transportation system could be developed, with these lakes at the hub, to link the interiors of Brazil, Argentina, Paraguay, Venezuela, Ecuador, Colombia, Peru, and Bolivia. This is a daring plan that is being touted as a boon for these countries. Before we agree, however, some questions need to be raised: What will happen to the ecology of the region? What will happen to the delicate balances that exist in the plant and animal communities? What kinds of disturbances can be predicted? Will the benefits outweigh the disadvantages? How will these changes in the Amazon Basin affect the rest of the world?

Some indicate that the added weight of water at the Equator—from water retained by the series of nine dams—would add 3 seconds to the year because the additional weight at this location would cause the Earth to rotate more slowly. Common sense should tell us that this may not be the only change that will occur. The storage of large quantities of water in this region has the potential to produce changes in the daily flow of the weather as well as in the climate. The building of the nine dams also requires the relocation of people—at least one city of 100,000 people would need to be uprooted and moved. At the present time, the Brazilian government has not embraced the scheme, but there is some speculation that this proposal may be pushed at a later time.

An agency known as SUDAM (Superintendency for Amazon Development) is seeking to attract industry to the Amazon region. More than 135 industrial projects—most of them in agriculture, lumber, textiles, food products, and minerals—were operational during the 1970s. There is a great temptation for industrialists to move into the Amazon Basin; many incentives

The Brazilian government is emphasizing the need for transport of all kinds to open new lands and to move products from the interior to coastal consumption areas and ports for export. This is a highway in the Brazilian highlands of the state of Paraná. (World Bank)

have been offered, and the lure of great profits seems irresistible.

One of the obvious investment lures in the Amazon is lumber. There are billions of acres of woodland with many different kinds of hardwoods, including mahogany and rosewood. The Georgia-Pacific Corporation began operating in the Amazon region in the 1960s and has produced huge harvests from its timberland. When you realize Brazil is one of the world's largest importers of petroleum products, you can understand the compelling reason for the exploration and exploitation of this natural resource. Minerals such as bauxite, copper, industrial diamonds, and gold also invite the speculator to move into the region.

Rubber was the impetus for the first fantastic boom in the Amazon more than 50 years ago. Whether the lessons learned from the mistakes made throughout the world can be used constructively to correct our procedures in the continuing exploitation of the Amazon Basin remains to be seen. It will, of course, be tragic for the Amazon and the world if we repeat some of the blunders of the past.

THE HACKENSACK MEADOWLANDS

The region in New Jersey known as the Hackensack Meadowlands is a tidal marsh underlain by a succession of marine, freshwater, and glacial deposits. This tidal marsh, larger than the island of Manhattan, is surrounded by a ring of urban development (Fig. 4–25).

GEOLOGIC HISTORY

During the Pleistocene epoch of the Cenozoic, this region of New Jersey was scoured by advancing ice sheets. The worldwide warming trend, which ushered in the Holocene epoch,

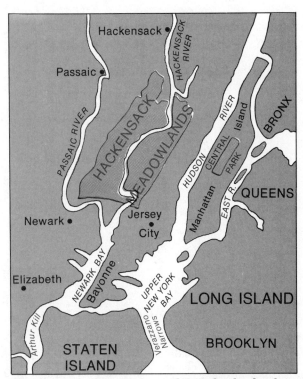

FIGURE 4–25   The Hackensack Meadowlands of New Jersey is a tidal marsh surrounded by a ring of urban development.

caused the ice sheet to retreat. Glacial till—gravel, boulders, sand, and silt carried by the ice—was deposited in the basin as the glacier melted.

A **terminal moraine** extending from Long Island across Staten Island to a location south of Elizabeth trapped the glacier's meltwater behind it. The trapped water formed a tremendous lake, referred to today as Glacial Lake Hackensack. Freshwater clays and organic silt were deposited in layers on the lake's bottom.

The continued melting of the ice sheet raised the level of water behind the terminal-moraine dam. Finally, the rising water breached the terminal moraine in two locations—at the Verazzano Narrows and the Arthur Kill (Fig. 4–25). The water of Glacial Lake Hackensack flowed

from the outlets, and vegetation took hold on the moist ground of the former lake bottom.

The meltwater from the Pleistocene glaciers flowed into the ocean and increased its volume. Eventually the rising sea entered the lower reaches of the former lake bed, which was now coursed by a winding river, the Hackensack. River and sea met, freshwater mixed with salt, grasses sprouted, and the Hackensack marsh-estuary was born.

POLLUTION

Prior to the twentieth century, the Hackensack River, which drains a 200-square-mile (518 km²) watershed, carried a rich assortment of nutrients to the Meadowlands. On the rise of the tide, nutrients were carried from the Meadowlands to

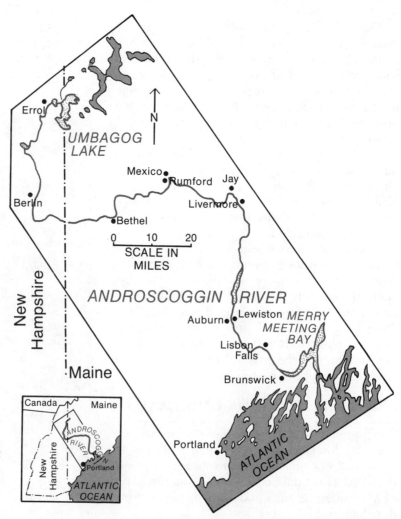

FIGURE 4–26   The Androscoggin River rises from Umbagog Lake on the border of Maine and New Hampshire. The river drops 1245 feet (380 meters) in its channel from Umbagog to Merrymeeting Bay where it empties into the Atlantic Ocean.

Newark Bay and on to the Atlantic Ocean. The Hackensack Meadowlands was the spawning, food-producing, and feeding area of a very productive estuary. Today, this function of the Meadowlands has been destroyed. The destruction of the Meadowlands was precipitated by its location at the center of a ring of urban development. One of the most pressing problems of the nineteenth and twentieth centuries has been the garbage generated by urban life. A classic example of the wrong way to approach the problem is found in the solid-waste disposal practices of the New York–New Jersey metropolitan area: Some 30,000 tons of waste from more than 100 municipalities are dumped into the Hackensack Meadowlands each week. This practice of carting garbage to the Meadowlands of New Jersey has been going on for at least 100 years. The solid waste has significantly changed the lay of the land and also the ecology of the marsh itself.

The fouling of the swamp by countless tons of garbage has destroyed the natural balances and the normal food chains. Today, we do not find a rich assortment of razor clams, oysters, periwinkles, mussels, barnacles, mud snails, and crabs. All of these animals turn algae and other organic matter into food for fish and birds. In the place of these animals, we find rats, sea gulls, and owls. The rats and sea gulls are the scavengers picking through the garbage. The owl, of course, preys on the rat. The Meadowlands—once a thriving community of life and beauty—is today the site for the castoff refuse of urban life.

THE ANDROSCOGGIN RIVER

This river rises from Lake Umbagog on the border of Maine and New Hampshire (Fig. 4–26). It has a 169-mile (272-km) run to its outflow point in Merrymeeting Bay. The Androscoggin drops 1245 feet (380 m) in the course of its journey from Umbagog to Merrymeeting, an average of 7.4 feet per mile (1.4 m/km). At several places the Androscoggin drops abruptly and spectacularly. For example, it plunges 240 feet (73 m) in 2.5 miles (4 km) at Berlin, New Hampshire, and another 180 feet (55 m) in 1.6 miles (2.6 km) at Rumford, Maine. Important falls in the river are found at Berlin, Rumford-Mexico, Jay-Livermore, and Lewiston-Auburn.

Throughout the eternity of time prior to the arrival of European settlers, the gravelly bottom of the Androscoggin was well-suited to spawning salmon. The fish made the trip upriver over the lower falls with very little difficulty. When the European settlers arrived in the region, they turned to the wealth that lay in the forests. The forests were cut, sawmills appeared along the river's banks, and more than 22 dams of various sizes were built across the width of the river's channel. The logs that were floated downriver tumbled over dams and falls. Logs were also flushed down the channels of tributary streams. The tumbling logs and flush-water scoured the channels and removed the rubble needed by fish for their spawning beds. Protective cover and aquatic plants were destroyed, too. Thus, food for young fish was drastically reduced and in some locations completely eliminated.

Textile mills, paper mills, and other factories were built at the sites of the larger falls to take advantage of the power the flow of water could produce. The waste matter from mill and factory that was dumped into the river added a poisonous load to the water that sent oxygen levels to zero at some locations.

It took a great deal of neglect to produce devastation in the waters of the Androscoggin. There is, after all, an abundant supply of water from rain and snow fed into the stream each year. For example, the Presidential Range, which is drained in part by the Androscoggin River, receives about 60 inches (150 cm) of precipitation annually, and the average snowfall in the area is slightly more than 170 inches (430 cm). Ordinarily this abundant flow coupled with the aeration provided by the rapids and falls should be sufficient to cleanse the water. The problem, of course, is the waste generated in the area. If it is handled properly, the Androscoggin will cleanse itself and its water quality will improve.

THE COLORADO RIVER

Hoover Dam extended Lake Mead into the lower reaches of the Grand Canyon in the 1930s; even so, the streamflow in the upper reaches of the canyon was not adversely affected (Fig. 4–27). In 1963, the Glen Canyon Dam at Page,

Arizona, was completed; since then, the flow of the Colorado has been altered significantly. Prior to 1963, the river water was charged with sediment. Now the streamflow is clear and almost completely dependent on the release of water from the Glen Canyon Reservoir.

The sediment load of the Colorado River is being trapped behind the Glen Canyon Dam, and the peak streamflow has been cut in half. Thus, the Colorado River is not carrying out its natural deposition and scouring regime below Glen Canyon Dam. In the absence of new depo-sition and sedimentation, former flood-stage ter-races are being eroded without being replaced. In addition, the sediment and debris deposited in the Colorado River by floods issuing from side canyons is not being scoured away because of the decreased streamflow. These changes are also producing stresses in the Colorado's ecosys-tem. The kinds of adjustments being made by the plants and animals of the ecosystem to this new river regime are complex and not completely understood. The desirability of these changes is certainly open to question.

In this aerial view, we are looking upstream to Hoover Dam and Lake Mead. The dam spans the Colorado River between Nevada and Arizona. Lake Mead extends upstream behind the dam for about 115 miles (185 kilometers). (U.S. Dept. of Interior)

FIGURE 4–27   The portion of the main channel of the Colorado River shown in this figure runs from the Glen Canyon Reservoir to Lake Mead. The Little Colorado River merges with the main channel in northern Arizona. The streamflow of the Colorado that moves through Grand Canyon National Park is now almost completely dependent on the release of water from the Glen Canyon Reservoir.

THE LANGUAGE OF PHYSICAL GEOGRAPHY

A physical geographer uses a technical vocabulary to make explanations. Review your understanding of the vocabulary used to develop the concepts in this chapter. Among the important words and terms used are:

base flow	flood	marsh	stream channel
bogs	flood plain	meltwater	streamflow
channel	glacial till	névé	stream slope
channel precipitation	glaciation	peak flow	swamps
cirque glacier	glaciers	peat	terminal moraine
crevasse	hammocks	perennial streams	throughflow
cuestas	Holocene epoch	piedmont glacier	tilt
discharge	hydraulic geometry	Pleistocene epoch	tree islands
eccentricity	hydrograph	ponds	tributary
effluent streams	icecaps	Quaternary period	valley glacier
elliptical orbit	ice sheets	river	wetlands
ephemeral streams	influent stream	runoff	zone of accumulation
eutrophication	intermittent streams	snowmelt	zone of wastage
firn	lakes	stream	

_____ SELF-TESTING QUIZ _____

1. Characterize the pattern of precipitation as you move from east to west across the United States.
2. List the various elements that are collectively referred to as a stream system.
3. In what ways do surrounding topography and vegetation affect streamflow?
4. What kind of topography and structure encourage throughflow?
5. What are the major sources of streamflow?
6. Explain how the relationship of the water table to the bed of the stream channel can be used to classify streams.
7. Describe the kinds of streams that produce shallow, narrow channels.
8. Compare the discharges of the Nile and the St. Lawrence rivers.
9. In what ways are hydrographs useful?
10. What kinds of conditions can cause flooding in river basins?
11. Why are bogs and swamps referred to as the evolutionary offspring of lakes?
12. List some of the ways in which lake basins originate.
13. How do the rock structures in a lake's drainage area affect its water?
14. In what ways do swamps and bogs differ?
15. Describe the setting of the Okefenokee Swamp.
16. What were the important centers of glaciation in North America during the Pleistocene epoch?
17. Explain how the Earth's orbital pattern affects our movement into and out of ice ages.
18. Contrast the tilt of the Earth's axis today with that of 9000 years ago.
19. Trace the evolutionary development of a glacier.
20. Describe the structure of the Great Lakes region.
21. What kind of impact have drainage canals made in the Everglades?
22. What are the advantages and the disadvantages flowing from present developments in the Amazon Basin?
23. Trace the development of the Hackensack Meadowlands from the Pleistocene epoch to the present.
24. What developments along the Androscoggin were harmful to the river's health?
25. Describe the impact Glen Canyon Dam has had on the Colorado River.

IN REVIEW
THE EARTH'S SURFACE WATER

I. STREAMS AND RIVERS

  A. Streamflow
  B. Sources of Streamflow
  C. Stream Classification
    1. Perennial Streams
    2. Ephemeral Streams
    3. Intermittent Streams
  D. Stream Channels
  E. Stream Discharge
    1. Hydrographs
    2. Floods

II. LAKES AND WETLANDS

  A. Lake-Basin Origin
    1. Through Glacier Action
    2. Meteor Craters
    3. Volcanic-Region Lakes
    4. Sinkhole Lakes
    5. Delta Lakes
    6. Structural Depressions
  B. Lake Water

  C. Wetlands
    1. Bogs
    2. Swamplands
    3. Marshlands

III. ICECAPS, GLACIERS, AND SNOWFIELDS

  A. Quaternary Ice Ages
  B. Ice-Age Theory
    1. The Orbital Pattern
    2. A Tilted Axis
    3. Summing Up
  C. Glacial Ice Today

IV. SYSTEMS UNDER STRESS

  A. The Great Lakes
    1. The Basins
    2. Pollution
  B. The Everglades
  C. The Amazon Basin
  D. The Hackensack Meadowlands
    1. Geologic History
    2. Pollution
  E. The Androscoggin River
  F. The Colorado River

PART TWO

LANDFORM PATTERNS OF THE EARTH

The variable character of the Earth's surface is nowhere more apparent than in its landforms. In the four chapters of this part we will study the processes that have produced the variations in shape and form at the Earth's surface.

The content within this part includes information about the ways in which winds, running water, moving ice, and temperature changes work together to determine the shape of a valley, round out the outline of a mountain, and over long periods of time succeed in transforming an entire landscape. The large-scale movements of the Earth's plates will be examined, too, because these movements produce the orderly distribution of initial and significant relief features.

The fourth chapter in this part looks at the North American continent as a total region. The purpose of this examination is to establish and study the spatial relationships of the North American landforms—the landforms of your continent!

A. The water that rushes over the five-story waterfall at Blackwater Falls in West Virginia cuts and erodes the land to produce deep valleys. (Department of Tourism, West Virginia)

B. Ayers Rock consisting of arkose sandstone is an inselberg—an island mountain or isolated steep-sided summit formed by erosion that rises abruptly from a surface of low relief—in the Northern Territory of Australia. (James O'Keefe)

C. The vast, awe-inspiring sand desert known as al-Rub al-Khali (Empty Quarter) is located in the southern section of Saudi Arabia. (Aramco)

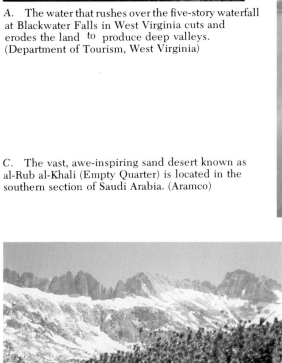

D. The sharp ridges of the Sawtooth Mountains in the Sierra Nevada of California were formed by glacial action. (James O'Keefe)

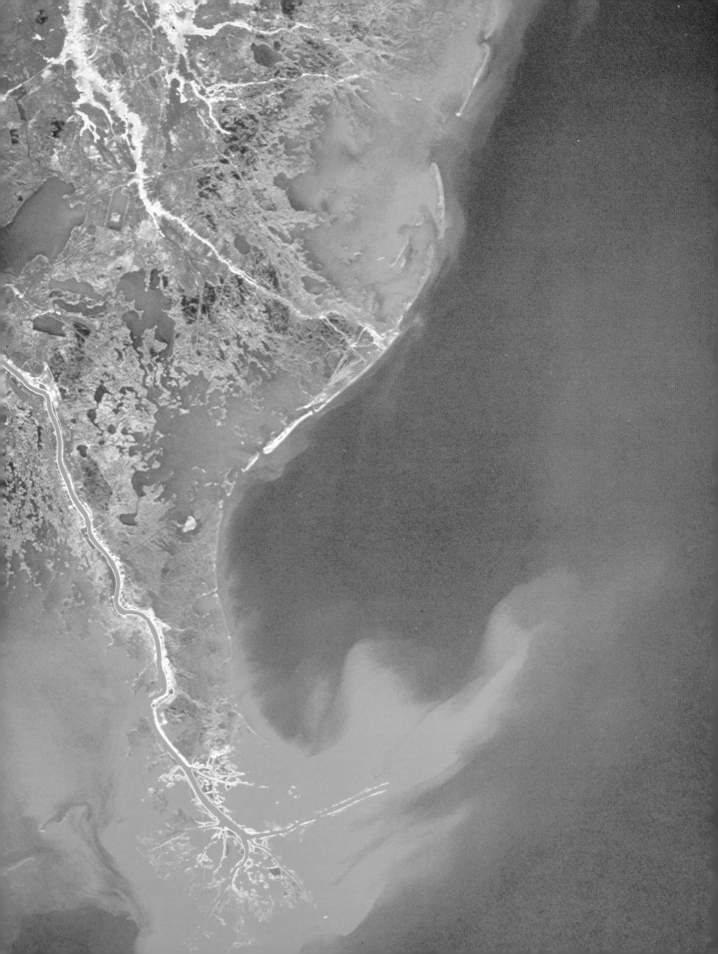

Processes and materials of landscape development

The physical geographer refers to the various shapes and patterns on the Earth's surface as **landforms.** The study of landforms includes describing and classifying the gross features of continents and ocean basins as well as the details of valleys, mountains, plains, plateaus, and coastal zones. There is, however, an underlying genesis-theme in the physical geographer's study of landforms; in other words, there is a preoccupation with how these landforms come into being.

Since the physical geographer goes beyond the mere description and classification of landforms, you will find the underlying theme of this chapter is process and mode of formation. Two basic forces—tectonic and gradational—are constantly at work shaping and reshaping the Earth's surface. The **tectonic forces—volcanism, folding,** and **faulting**—distort, buckle, and break the surface of the Earth. The **gradational forces**—running water, moving ice, wind, and waves—wear down high places on the Earth's surface and fill in low regions.

A crucial factor in the distorting buckling, and breaking of the crust is its composition—that is, the makeup of the crust's sediment and rocks. The materials of which the crust is made affect the way the tectonic forces work their will and the kind of landforms produced. The ability of gradational forces to produce different landforms also depends on the crustal materials encountered.

The accumulated alluvial deposits laid down by the Mississippi River at its mouth where it empties into the Gulf of Mexico resemble the Greek letter *delta* (Δ). Examine this satellite photo of the Mississippi's delta carefully and you will find that the delta has a distinctive bird's-foot form. It is, therefore, referred to as a *bird-foot delta.* The bird's-foot form results from the subsidence of the ocean floor in the Gulf of Mexico and the compaction of the sediments, which submerge all but the crests of the natural levees built along the distributary channels. Lake Ponchartrain can be seen in the upper left portion of the photo. (EROS Center)

VARIATIONS IN CONFIGURATION

A multitude of detail has been imprinted on the surface of the Earth. The variety of embellishments—including oceans, mountains, plains, valleys, canyons, springs, geysers, sinkholes, streams, lakes, waterfalls, rivers, deltas, and moraines—is amazing and a little confusing. The physical geographer brings some order to the variations in surface configuration by discriminating among them on the basis of scale, called **orders of relief**. Relief in this context means horizontal as well as vertical dimension.

The relief features of the Earth's surface can be organized into three orders according to scale: The largest, the continents and ocean basins, are designated as the *first order of relief*. Continental and oceanic mountain systems, large basins, and extensive plateau regions are classified as *second-order relief features*. When the four major agents of destruction—streams, glaciers, winds, and waves—go to work on the second-order relief features, they leave erosional features (valleys, canyons, and waterfalls, for example), residual features (peaks and summits), and depositional features (deltas, moraines, and sand dunes); all of these are referred to as the *third order of relief*.

CONTINENTS AND OCEAN BASINS

The Earth's outer layer consists of a mosaic of plates as described in Chapter 1 (pages 11–17). These plates are large rafts of relatively low-density rock. They are supported from below by the denser, semiplastic material that forms the upper region of the Earth's mantle. The continents and ocean basins are embedded in these plates.

EARTH'S INTERIOR

In order to understand the origin of the continents and ocean basins, we must be aware of the important features of the Earth's interior and the way they interact. By 1914, from seismic wave studies as well as the analyses of gravity data and our planet's magnetic field, scientists had subdivided the Earth into three concentric shells—the **crust**, the **mantle**, and the **core**.

The word discontinuity is a general term used to designate a surface that separates one layer from another. The lower boundary of the crustal layer is a surface of separation, which is referred to as the Mohorovičić discontinuity. This discontinuity derives its name from the Yugoslavian scientist who demonstrated its existence in 1909; his name, Mohorovičić, applied to this zone, makes it the **M-discontinuity**, or **Moho**. The Moho is found at average depths of 6 to 7.5 miles (10 to 12 km) beneath ocean basins and at average subcontinental depths of 20.5 to 21.5 miles (33 to 35 km).

Another important feature of the Earth's interior was discovered in 1936; the core, which consists of nickel and iron, was differentiated

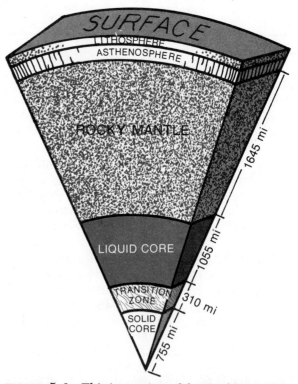

FIGURE 5–1 This inner view of the Earth's interior identifies a zone of transition between the inner and outer core. The mantle is differentiated into two zones: an upper mantle which is labeled asthenosphere, and a lower mantle which is identified as the rocky mantle. Transitional zones are found in the rocky mantle.

into two zones—a molten outer core and a solid inner core at the very center of the Earth. Figure 5–1 shows the solid inner core with a radius of 755 miles (1215 km), the molten-liquid outer core with a radius of 1055 miles (1700 km), and a transition zone of about 310 miles (500 km) between the two. The core region is 16 percent of the total volume of the Earth.

The lower region of the mantle, known as the **rocky mantle,** is rigid and constitutes the bulk of the interior; it is 80 percent of the Earth's volume. The rocky rigid mantle extends from the liquid core upward for approximately 1645 miles (2650 km) (Fig. 5–1). There are two zones of transition, one at 435 miles (700 km) and another at 240 miles (390 km) below the surface.

The **asthenosphere** is the soft, semiplastic layer of the upper mantle. This semiplastic region ranges from 35 to 155 miles (55 to 250 km) in thickness. The plates that carry the continents and ocean basins are supported by the asthenosphere.

THE OUTER ZONE

The rigid layer above the asthenosphere is called the **lithosphere** (Fig. 5–1). The lithosphere is the solid, outer region of the Earth that is broken into the mosaic of plates. The mobile lithospheric plates range from 30 to 60 miles (50 to 100 km) in thickness.

The word *crust* is a general term that means a hard layer at the surface of softer material. The term has been applied to the outer layer of the Earth in a figurative and imprecise sense. At this juncture in your study of physical geography, it is better to think of the mobile plates as the main outer layer of the Earth and to use the term *lithosphere* to designate this zone. When the term *crust* is used from this point on, it will be used to designate the portion of the lithosphere that is above the Mohorovičić discontinuity.

THE CRUST

The chemical composition of the shallow portion of the Earth's crust can be investigated directly. The eight elements listed in Table 5–1 make up 98.6 percent of the Earth's crust

TABLE 5–1 COMPOSITION OF THE CRUST

ELEMENT	PERCENT BY WEIGHT
Oxygen	46.60
Silicon	27.72
Aluminum	8.13
Iron	5.00
Calcium	3.63
Sodium	2.83
Potassium	2.59
Magnesium	2.10

by weight. All the other 92 naturally occurring elements—including such important elements as carbon, nitrogen, and phosphorus—make up only 1.4 percent of the crust.

Temperatures throughout the Earth's crust increase at a rate of 1°F per 50 feet (1°C/30 m) of depth. Some of this heat is released as the result of the tremendous compression of the rocks at great depths, but most of this heat is produced by the radioactive decay of unstable elements. In any event, the buildup of heat is so rapid that miners cannot survive without special air-conditioning installations when they excavate beyond a depth of 1 mile (1.6 km).

The crust makes up a mere 1 percent of the volume of the Earth. When we define the thickness of the crust on the basis of the Moho, the continental crust has an average thickness of 21 miles (34 km), and the oceanic crust averages 7 miles (11 km).

THE PROTO-CONTINENT

In the 1930s, Vening Meinesz, a Dutch scientist, suggested there was a single system of flow within the Earth long before the core developed. The overturning produced by the single convection cell began to separate the components of the interior and deposited a "froth" of light components on the surface. This patch of light rock accumulating on the surface formed the primordial continent called *Pangaea*.

The convective overturning caused heavy material—primarily nickel and iron—to sink to the center of the Earth, where it began to accumulate and form a core. As the volume of the core grew, it began to interfere with the single-flow pattern and broke the incipient convection current into a number of separate cells of circulation. Some of the newly formed currents under the proto-continent produced tears, or rifts, in Pangaea. The original rift separated Pangaea into two fragments—Laurasia and Gondwana (Fig. 5–2).

As magma flooded into the rift and separated the continental masses of Laurasia and Gondwana, a new ocean basin was produced. Along the centerline of this basin, the rising magma produced a ridge, known today as a **mid-ocean ridge.**

The core continued to grow and the convection patterns were further constrained. New convection systems were brought into being by the growing core, and these developing convection cells produced new rifts in Laurasia and Gondwana. The patterns of continental drift changed and new mid-ocean ridges evolved to produce additional first-order relief features.

THE TWO-STAGE PROCESS

After the formation of the initial continental structure and the breakup of Pangaea, a two-stage process for producing continental material developed: The first stage begins with the extrusion of new ocean floor along mid-ocean ridges; the rock material of the ocean floor is lighter than the rock-forming material from which it derives, but it is not light enough to avoid subduction and recapture at the ocean trenches. The second stage of the process is initiated when friction along a subducting oceanic plate causes some of the plunging plate's material to melt and rise to the surface through volcanic action; the lava ejected from a volcano onto an overriding plate is light enough to qualify as continental material. Thus, the two-stage development of continental material includes the reprocessing of an ocean floor in a trench and ends with the eruption of material that is light enough to float as continental rock.

The result of this second-stage melt and volcanic action is to make continental rocks that are rich in silicon and aluminum—sometimes referred to as **sial,** which is a word derived from the first two letters of *si*licon and *al*uminum. These continental rocks are appreciably less dense than the rocks of the ocean floor and the rock-forming material of the Earth's interior. New continental material is being formed in the Andes of South America from the plunging Nazca Plate and on the Japanese islands from the plunging Pacific Plate. Continental material is also being formed close to the borders of other overriding plates.

MOUNTAINS, BASINS, AND PLATEAUS

Second-order relief features consists of (1) the continental and oceanic mountain systems, such as the Andes, Appalachians, Rockies, Alps, Himalayas, and the Mid-Atlantic Ridge; (2) the subcontinental plateaus—the Tibetan and South African plateaus, for example—which are very large but smaller than a continent, (3) large depressions, or subcontinental basins, such as the West Siberian Lowland and the Mississippi Valley. All second-order relief features have been produced by either vertical or horizontal displacements of the Earth's crust and may be explained in terms of plate tectonics.

TECTONICS IN THE SEA

Oceanic mountain systems were described in Chapter 1 (pages 13–15) and Chapter 2 (pages 47–49). The Mid-Atlantic Ridge is an excellent example of an oceanic mountain that is a source region for new ocean floor material.

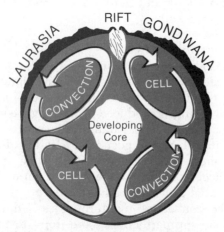

FIGURE 5–2   The developing core within the Earth constrained old patterns of flow and brought new convection patterns into being. The changing pattern of convection currents produced rifts in the continental mass of Pangaea.

This panoramic view was taken from Mount Hamilton, outside San Jose, California, at an elevation of 4260 feet (1300 meters). Mount Hamilton is in the Coast Ranges. The mountains in the foreground are the Coast Ranges, a second-order relief feature. We are looking east toward Yosemite National Park in the Sierra Nevada, another second-order relief feature. The Sierras, the mountains in the background, are approximately 120 miles (190 kilometers) from us. The valley between the Coast Ranges and the Sierras is the San Joaquin Valley, which is also considered to be a second-order relief feature. (Lick Observatory)

These undersea mountain chains rise to heights of nearly 10,000 feet (3050 m) above the ocean's floor.

Magma is being injected from below into the central rift valley of the Mid-Atlantic Ridge. The newly emplaced oceanic crust moves away on both sides of the injection zone. The divergence or spreading away from the oceanic ridges has taken place at rates from 0.4 to 4 inches (1 to 10 cm) per year, averaging about 1 inch (2.5 cm) in the Atlantic and about 2 inches (5 cm) in the Pacific.

The active lifetime of a mid-ocean ridge is only as long as the lifetime of the convection current that sustains it—possibly 200 or 300 million years. As a result of the divergence along present ridges, which has been continuing at least since the Jurassic period of the Mesozoic era, young crustal material has spread entirely over ocean basins. Oceanic crust older than the Jurassic has been pushed into oceanic trenches and recycled.

CONTINENTAL TECTONICS

Continental mountain systems, subcontinental plateaus, and subcontinental basins have developed through processes involving large-scale deformation of the Earth's crust. The term **diastrophism** is often used to refer to these processes of large-scale deformation.

Orogeny is another name for mountain building; most of the continental mountain systems, for example, have been or are presently being created by the convergence of lithospheric plates. The Andes—young, rugged mountains in South America—are currently being uplifted as the Nazca Plate plunges under the western edge of the South American Plate; the subduction of the oceanic Nazca Plate is taking place in the Peru-Chile Trench. A magmatic belt and folded mountains are features that develop on an overriding continental plate when an oceanic plate plunges under it. The Andes are a volcanic chain that lie in the magmatic belt of the overriding continental plate.

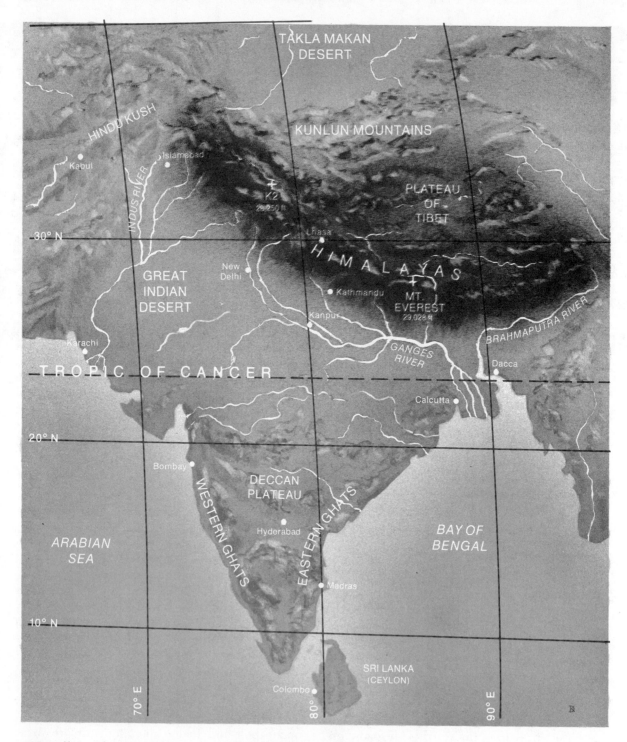

FIGURE 5–3 The Himalayas and the Tibetan Plateau are the result of the collision of the Indian Plate and the Eurasian Plate along the Indus-Brahmaputra zone.

Two continents are in collision along the Indus-Brahmaputra zone in India (Fig. 5–3). The Himalaya Mountains and the Tibetan Plateau were produced when the lithospheric plate carrying India collided with LePichon's Eurasian Plate. The Tibetan Plateau has an average elevation of 16,000 feet (4880 m) above seal level. The Himalayas rise higher than any other mountain chain in the world; Mount Everest, the highest peak in the chain, has an elevation of 29,028 feet (8848 m) (Fig. 5–3). The Himalayas are a prime example of mountains that result from a two-continent collision. One possibility during this type of collision is that subduction ceases at the continent-to-continent suture and a new subduction zone is started elsewhere; in this case, the suture is marked by a mountain range made up of either folded or thrusted rocks, and the mountain range is either in or adjacent to a magmatic belt.

In the scenario of plate tectonics, ocean basins are impermanent features of the Earth's surface and continents are permanent. The term *permanent* is used because over a period of hundreds of millions of years continents may be torn apart, deformed (to produce mountains, plateaus, and basins), welded together, and torn apart again, but they survive! On the other hand, an ocean is swept clean and replaced by new crustal material every 300 to 400 million years.

LANDFORMS

Streams, glaciers, winds, and waves sculpt second-order relief features into third-order relief features—the individual hills, valleys, canyons, waterfalls, deltas, moraines, and sand dunes that are generally regarded as landforms. The magnitude scale of a third-order relief feature is much smaller than a first-order or second-order scale. With many landforms we can personally observe them and see them in their entirety. Canals, dams, and reservoirs produced by human effort are also on the scale of third-order relief features.

Although the variety of landforms is enormous, most of them are produced by gradational processes rather than by diastrophism or tectonic forces. The best way to discuss the origin of any landform is in terms of the mechanics of gravitationally produced stresses acting on materials of varying resistance. Four factors affect the sculpturing of second-order relief features into landforms of third-order relief: (1) the type and arrangement of rock materials composing the relief feature, (2) the gradational processes at work, (3) the period of time the gradational processes have operated, and (4) the nature of the local movements in the Earth's crust that result from tectonic activity.

Glaciers are at work on Atlin Mountain (59°35′N, 133°42′W) in northern British Columbia. The rock of the mountain is being carved by the glacial ice into small niches or amphitheaters, called *cirques*, as well as valleys. (Canadian Film Board)

MINERALS OF THE CRUST

A **mineral**—in the common sense—is any chemical element or compound that occurs naturally. The word, however, takes on special meanings when used by different scientists, including different physical geographers. Therefore, rather than get involved in the complications of a "simple" definition, the discussion over the next few pages will be devoted to some basic mineral properties.

The obvious properties of minerals such as size, shape, and color are not very informative. In fact, these obvious properties tend to obscure the facts that are essential in defining a mineral species. Two items of information are the most reliable bases for identifying and describing a mineral: chemical composition and crystal structure.

Before proceeding with the discussion of crystalline structure and chemical composition, however, we should note that various physical properties of minerals are often used as guides to identification. Minerals, for example, can be classified according to specific gravity, color, streak, luster, hardness, and other special physical characteristics; the tests for these properties must be used together for reliable identification.

CHEMICAL COMPOSITION

A mineral may consist of a single element or of compounds made up of more than one element. Most rock-forming minerals are chemical compounds. Quartz, for example, is a rock-forming mineral composed of silicon and oxygen; its chemical formula is SiO_2. Calcite consists of the elements calcium, carbon, and oxygen arranged in the form of a chemical compound called *calcium carbonate*, $CaCO_3$. Graphite—

Water is the primary agent sculpting second-order relief features into third-order relief features on the island of Molokai in the Hawaiian chain. We are looking into Halawa Valley, which is at the eastern end of Molokai. The arrow at the far end of the valley points to a waterfall cascading down the steep slope. (USDA)

a soft, lustrous mineral that is used to make lead pencils—consists solely of one element, carbon.

Most of the 92 naturally occurring elements play some role in the formation and composition of minerals. In Table 5–1 the composition of the crust is given as percent by weight. If we calculate the composition of the Earth's crust in terms of the relative number of atoms of each element present, we find that ten elements predominate (Table 5–2). Oxygen, silicon, and aluminum still head the list. Hydrogen, however, has moved onto the list and into fourth place. Iron has dropped to seventh place. On the basis of relative numbers, sodium stands before calcium and magnesium is more abundant than potassium.

If we expand Table 5–2 to a list of 15 elements and include carbon (0.06%), phosphorus (0.05%), manganese (0.04%), sulfur (0.03%), and fluorine (0.03%), we can account for 99.81 percent of the total number of atoms in the Earth's crust. The remaining 77 naturally occurring elements account for a mere 0.19 percent of the relative number of atoms in the Earth's crust.

Although chemical composition is used to describe a mineral, it cannot be relied on to the exclusion of other data. For example, graphite—lustrous, black, and soft—consists of carbon; a diamond, which is one of the hardest minerals known, is also built of carbon. Chemical analyses of graphite and diamond are identical. However, the physical properties of these minerals are quite different because the spatial arrangements of the carbon atoms in graphite and diamond are different (Fig. 5–4).

Various types of cut diamonds are displayed in this photo. The round diamonds are referred to as *brilliant cut*; rectangular are called *emerald cut*; and the boatshaped diamond is known as *marquise cut*. The spatial arrangement of the carbon atoms in a diamond is a prime factor in the difference between it and common graphite, which is used in "lead" pencils. (South African Consulate)

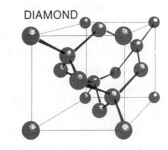

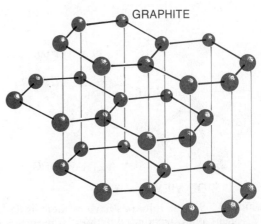

FIGURE 5–4 Carbon atoms are represented as red balls in this figure. The different spatial arrangement of the carbon atoms in diamond and graphite produces radically different physical properties.

TABLE 5–2 RELATIVE NUMBERS OF ATOMS IN EARTH'S CRUST

ELEMENT	PERCENT OF ATOMS PRESENT
Oxygen	60.5
Silicon	20.4
Aluminum	6.2
Hydrogen	2.8
Sodium	2.5
Calcium	1.9
Iron	1.9
Magnesium	1.8
Potassium	1.4
Titanium	0.2

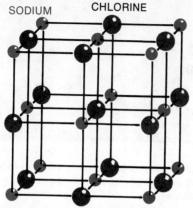

FIGURE 5–5 Two atoms form the basic building block of halite's cubic space lattice.

CRYSTAL LATTICE

The term **crystal** refers to a homogenous solid made up of an element, chemical compound, or mixture. The word **lattice** is used to refer to the regular three-dimensional spatial arrangement, or geometric configuration, of the atoms that make up a mineral. Most rock-forming minerals are crystalline solids. The crystal lattice for a mineral is a systematic, regular framework whose geometric configuration is symmetrical.

X-rays are used to analyze the crystal lattice of minerals. A crystal lattice is a fundamental property of a mineral; the mineral halite, for example, consists of a spatial arrangement of two different atoms, sodium and chlorine, in a cubic space lattice (Fig. 5–5). The silicate minerals that occur again and again in rocks have as their basic unit a tetrahedron—a geometric figure having four faces. X-ray analysis shows the tetrahedron to consist of a silicon atom surrounded by four larger oxygen atoms (Fig. 5–6).

Olivine, with a chemical formula of $(Mg,Fe)_2 SiO_4$, is an example of a silicate mineral that consists of isolated individual silicon-oxygen tetrahedra that are bound together by metallic atoms of magnesium and iron. Other silicate minerals have different arrangements of the silicon-oxygen tetrahedra. The tetrahedra can be arranged as groups, single chains, double chains, layer-type or sheets, and three-dimensional networks. Augite is a silicate mineral in which the silicon-oxygen tetrahedra form single chains with each tetrahedron sharing two of its oxygen atoms with two other tetrahedra as in Figure 5–7. The single chains of augite tetrahedra are bound together by metallic atoms of calcium, magnesium, and iron.

There are a few minerals that are not crystalline in nature; opal is an example. Minerals that are not crystalline have no regular form and are therefore called *amorphous*.

CRYSTAL FORM

Atoms at the surface of a mineral are in the same crystal lattice as those below the surface and, thus, are part of the same geometric pattern. The atoms at the surface, however, form smooth planes that intersect at sharp angles to produce crystal faces. The term **crystal form** refers to the

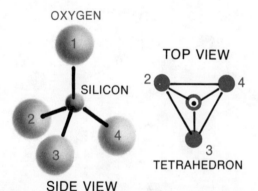

FIGURE 5–6 Silicate minerals have as their basic unit a silicon atom surrounded by four larger oxygen atoms shown here (left) in side view. When we view this arrangement from above oxygen atom 1, we can enclose the atoms within a well-formed tetrahedron, shown here (right) in top view.

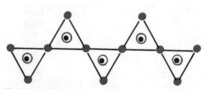

FIGURE 5–7 In a single-chain arrangement, each silicon-oxygen tetrahedron shares two of its oxygen atoms with two other tetrahedra.

FIGURE 5–8 Many minerals show distinctive geometric configurations on their surface. One-half of the geometric formation is shown for each mineral in this illustration.

HALITE PYRITE   QUARTZ   DIAMOND MAGNETITE   GARNET   FELDSPAR

external shape exhibited at a mineral's surface or face.

When developmental conditions are ideal, many minerals show distinctive geometric configurations at their surfaces (Fig. 5–8). Two very familiar minerals, halite and pyrite, for example, form cubes. A diamond exhibits the geometric form of an octahedron; so does magnetite. Omnipresent quartz, under ideal conditions, displays the form of a six-sided prism whose ends look like hexagonal pyramids. A well-developed garnet crystal, on the other hand, has 12 sides, each of which is a parallelogram. The crystal form of feldspar, which is present in many continental rocks, is an almost-rectangular block that is thin and tabular.

In actuality, however, the growth surfaces of a mineral do not always display the regular geometric configuration associated with its ideal development. The external, surrounding conditions under which a lattice grows affect the visible form at the crystal's surface. Quartz, for example, is one of the minerals that develops as magma cools, but it crystallizes late. Quartz starts its growth after other minerals have begun to form in the cooling magma; therefore, quartz, under certain conditions, may not have ample "growing room." If quartz is cramped at a critical stage in its process of crystallization, it displays an irregular form.

Interference between two developing mineral species, which causes the crystal face of one or both to develop irregularly, can also occur when the minerals are developing their crystal lattices simultaneously. On the other hand, some minerals—garnet, for example—can grow and develop regular crystal faces even in the solid state during metamorphism.* Minerals that develop while surrounded by soft mud or fluids almost always develop regular, complete crystal faces.

*Refer to pages 176–177 for a description of metamorphic processes.

CRUSTAL ROCKS

The outermost shell of the Earth is the strong, solid lithosphere, which includes the crust. The lithosphere, riding on the partially molten asthenosphere, has a relatively thin crustal topping under the oceans and a thicker continental crust. This variation in crustal thickness means about 79 percent of the total volume of the Earth's crust is continental and 21 percent is oceanic.

We generally say the crust is composed of rocks. A rock is simply an aggregate of one or more minerals. Most rocks consist of minerals of different kinds in varying proportions. Rocks can be classified in a variety of ways, but the system most physical geographers find useful is based on inferred origin. The three major rock groups, according to this system, are igneous, sedimentary, and metamorphic.

INFERRED-ORIGIN SYSTEM

Igneous rocks form as a result of the cooling of a melt of molten magma. Some igneous rocks cool and solidify below the Earth's surface, others cool on the Earth's surface. Igneous rocks, which compose about 65 percent of the total volume of the crust, are by far the most abundant rocks.

New technology allows us to hold a ship in a sufficiently stable position to allow drilling into the ocean's crustal bed. Information gathered from deep-sea drilling has contributed to a comprehensive picture of the distribution of crustal rocks. (Exxon)

THE ROCK CYCLE

Any one of the three rock types—igneous, metamorphic, sedimentary—may be derived from either of the other two. For example, igneous and sedimentary rocks, when subjected to pressure by tectonic forces or exposed to extreme heat, may become metamorphic rocks. Metamorphic and igneous rocks, in turn, when exposed to gradational forces, may be eroded to form sedimentary types. Finally, metamorphic or sedimentary rocks that are subducted or melted by other means may, when cooled, produce rocks of the igneous type.

A rock cycle is a system of rock production, with no beginning and no end. Each of the three types of rock is constantly being developed, destroyed, and changed into other rock types. The concept of the rock cycle is a way to think about the ever-changing nature of the material of the crust that results from plate movements and adjustments, erosion and deposition, and rock burial.

IGNEOUS ROCKS

Magma—the pudding from which igneous rock develops—is a molten silicate material that is

Fountains of molten lava and flame climbed to heights of more than 1500 feet (460 meters) during this eruption of Hawaii's Kilauea volcano. (Hawaii Visitor's Bureau)

Sedimentary rocks make up 8 percent of the volume of the Earth's crust. They consist of particles that were worn away from or dissolved out of other rocks and were then carried by a transporting agent such as water or wind to a new location. After deposition the sediment underwent a conversion; the process of conversion from sediment to rock is called **lithification.** During lithification, a compact mass is formed as particles are compressed and cemented together by some substance acting as a binding agent.

Metamorphic rocks, which compose 27 percent of the volume of the Earth's crust, are derived from preexisting rock types through a series of changes produced by an alteration in the heat and pressure of their environment. The heat and pressure changes leading to the development of a metamorphic rock may result from rocks being subducted and forced toward the asthenosphere in an oceanic trench, or may result from renewed deformation produced by plate movement.

Table Rock in South Carolina is an exposed pluton. It was intruded into the crust, cooled slowly to become a granite, and was exposed through a long, slow process of erosion, which removed the overlying cover. Elevation at Table Rock is 3124 feet (952 meters). Mountains in the vicinity of Table Rock are on the fringe of the Blue Ridge. (South Carolina Dept. of Parks)

located below the Earth's surface in the realm of the asthenosphere. On reaching the surface, this molten silicate material is known by the more familiar term **lava.**

When magma cools beneath the surface of the Earth, it cools slowly. Minerals that develop within slowly cooling magma usually form relatively large crystals under ideal conditions. Large-grained, coarse-textured rocks derived from magma and emplaced in older bedrock develop into a variety of **intrusive** rock bodies. Gradational forces may eventually expose and position an intrusive rock structure at the surface of the Earth, but the texture of the minerals reveals the intrusive origin of the rock.

Lava that spills onto the surface of the Earth cools relatively rapidly. The crystal forms of the minerals within rapidly cooling surface rocks are quite small. Fine-grained rocks derived from lava produce extrusive rock structures.

INTRUSIVE ROCK FORMATIONS

Any formation that has been intruded into the crust is referred to as a **pluton.** Plutons vary in size and shape, but they are all characterized by a coarse-grained texture, which is indicative of slow cooling. Erosion frequently exposes these deep-seated rocks.

The largest of these plutons is called a **batholith** (Fig. 5–9). A batholith in the central portion of Idaho, which has been exposed by gradational forces, is nearly 100 miles (160 km) wide and more than 300 miles (48 km) long (Fig. 5–10). The exposed area of the Idaho batholith is about 16,000 square miles (41,440 km²). The Sierra Nevadas—a range of granitic mountains that extends north and south for 400 miles (640 km) in eastern California—are composed of several large batholiths.

A smaller intrusive structure, similar to a batholith, is known as a **stock** (Fig. 5–9). At times, the distinction between a stock and a batholith is based on the area of exposed rock. Quite arbitrarily, a batholith is defined as a pluton that has a surface area exposure ex-

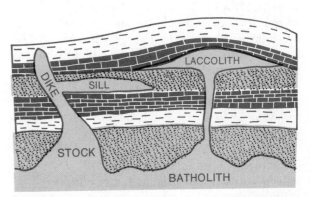

FIGURE 5–9 The term *pluton* is used to refer to any igneous formation that has been intruded into the crust, regardless of its shape and size.

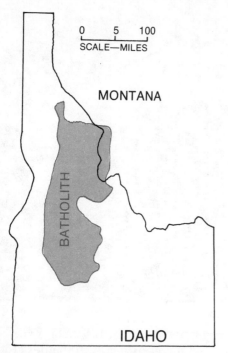

FIGURE 5-10 The batholith shown in color dominates the central area of Idaho. A small portion of the Idaho batholith protrudes into Montana.

Magma moving upward to form a dike may also find an opportunity to spread out horizontally at some point. The horizontal movement begins as magma forces its way between two preexisting rock layers. The shelflike structure produced by a horizontal flow is called a **sill** (Fig. 5-9). Sills produce a concordant contact, that is, a contact that is parallel to the layers of the preexisting intruded rock. The concordant contact is produced because the sill forces two strata apart rather than slicing through them as a dike does. The Palisades of New Jersey is a sill formation of igneous rock that was intruded into the strata below the surface during the Jurassic period of the Mesozoic. The overlying strata were eroded and the igneous sill is now exposed to our view along the west shore of the Hudson River (Fig. 5-11).

The various shapes of plutons result from different ways in which strata bend. For ex-

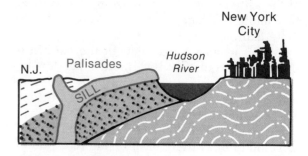

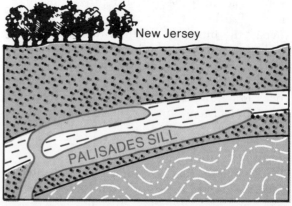

FIGURE 5-11 Intrusive igneous rock was emplaced as a sill between strata below New Jersey during the Jurassic (Bottom). Gradational forces carried away the overlying strata and, today, the Palisades sill is exposed along the west bank of the Hudson River (Top).

ceeding 40 square miles (100 km²). Stocks, on the other hand, are defined as roughly circular or elliptical plutons with an exposed area less than 40 square miles (100 km²). Stocks are found in great numbers in the mineral-rich area of the southern Rocky Mountains.

Plutons are masses of magma that intrude into the crust and cut through or push aside preexisting rock structures. Structures that cut through older bedrock are said to produce discordant contacts with the existing rock. The term **discordant,** as applied to a pluton, means the contact surface of the pluton is not parallel with the layers or other boundaries within the intruded rock.

When a fissure or crack develops in the lower region of the lithosphere, magma forces its way upward in a sharply inclined tablelike structure known as a **dike** (Fig. 5-9). Dikes are tilted tabular plutons that are often rich in mineral deposits. These tabular plutons, or dikes, cut through the older bedrock and, thus, have discordant contacts with the preexisting rock.

The Palisades of New Jersey is a sill formation that is now exposed to view and lines the west shore of the Hudson River. These cliffs along the lower Hudson River are formed of igneous rock that is arranged like massive columns. (H. Gilmore)

ample, a magma flow between two rock layers forms a dome when the preexisting top layer of rock arches upward from the pressure of the flow—the arched structure is called a *laccolith* (Fig. 5–9). A laccolith is known as a **concordant** lenticular pluton. The term **lenticular** means the laccolith has a shape somewhat reminiscent of a lens, that is, an essentially plane floor with a domed roof. The La Sal Mountains east of the Colorado River are domes that have been formed by groups of laccoliths; the highest peak of the cluster, with an elevation of 12,720 feet (3877 m), is Mount Peale just to the north of La Sal, Utah (38°19′N, 109°15′W). The pluton known as a *lopolith* has a shape that resembles a funnel (Fig. 5–12). Some indicate the lopolith's shape is produced when the intruded magma pushes down the underlying strata; others credit the funnel shape to the fact that the lopolith occupies a tectonic basin. When both the roof and floor of an intrusion are convex upward, the pluton is called a *phacolith* (Fig. 5–13). The igneous rock of a phacolith is usually associated with the highest region of a large fold.

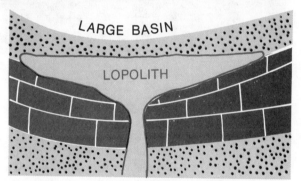

FIGURE 5–12 The pluton known as a *lopolith* has a funnel shape and occupies a large basin.

EXTRUSIVE ROCK FORMATIONS

As molten magma works its way toward the surface, the pressure at which it is confined gradually decreases. When it spills onto the surface as lava, the comparatively low atmospheric pressure allows the gases entrapped within the lava to escape. The rocks that result from the solidification of lava are called **extrusive igneous rocks.**

FIGURE 5–13 Magma injected into the highest region of a large fold develops into the pluton known as a *phacolith.*

Majestic Mount Hood (45°23'N, 121°41'W) standing with its peak above the clouds at an elevation of 11,234 feet (3424 meters) dominates this scene of Timothy Lake and the fir-forested slopes of Oregon's Cascade Mountains. The lake is behind a dam on the Oak Creek fork of the Clackamas River east of Estacada. The patches in the timber are logged-off areas where trees have been harvested under a U.S. Forest Service program. (Oregon State Highway Dept.)

Lava may emerge from a fissure in the Earth's crust or from the conduit of a volcano (Fig. 5–14). When lava issues from the vent of a volcano, it usually flows downslope as a narrow tongue. The Cascade Range—a chain of mountains that extends from northern California through western Oregon and Washington into southern British Columbia—is built of lava that issued from volcanic cores. Mount Rainier in Washington (46°52'N, 121°46'W), with an elevation of 14,410 feet (4392 m), is the highest

peak in the range. Other peaks include Mount Shasta in northern California, Mount Adams in Washington, and Mount Hood in Oregon. Most of the peaks are extinct volcanoes, but Lassen Peak in California (40°29'N, 121°31'W) is an active volcano. In March of 1980 Mount St. Helens in Washington (46°12'N, 122°11'W) after 123 years of dormancy began to erupt. See pages 170–171 for photos, maps, and descriptions of Mount St. Helens' eruptions.

FIGURE 5–14 This cross section of a composite volcano shows the typical cone shape, interbedded lava flows, and beds of pyroclastic material. Dikes and sills also develop within the large principal cone. When a dike erupts along the downslope of the main cone, a small lateral cone develops.

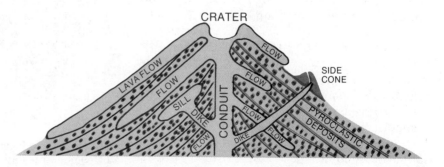

The Devils Postpile is a formation of symmetrical, gray-brown basaltic columns which fit closely together. The postpile is a remnant of a lava flow that originated in what is now Mammoth Pass in the Sierras. During a by-gone ice age, one side of the flow was quarried away by the glacier, leaving exposed a sheer-wall of columns 40 to 60 feet (12 to 18 meters) thick. Many of these columns have broken away and lie in fragments on the slope below. (National Parks Service)

A fissure flow moves as a thin sheet when the lava is fluid. More often than not, a fissue flow results in one sheet being built on another until a considerable thickness of extrusive igneous rock covers the area. Such accumulations are referred to as **plateau basalts**. The Columbia Plateau is the result of an outpouring of basaltic lavas during the middle Miocene epoch of the Cenozoic era. Figure 5–15 shows the plateau basalts cover southeastern Washington and a large section of Oregon; they also follow the sweep of the Snake River across southern Idaho. The lava of the Columbia

Plateau is more than 4000 feet (1220 m) thick in several places.

After lava has cooled and solidified, the interior of the flow may show columnar jointing, that is, joints that split rocks into long prisms or columns. Columnar joints are a very common feature in tabular igneous formations, whether extrusive or intrusive. We do, for example, find spectacular columnar jointing in exposed sills such as the Palisades of New Jersey, which is an intrusive formation. Jointing can also be observed in the extrusive igneous formations along the banks of the Columbia River in Washington. Another spectacular example of columnar jointing is found at the Devils Postpile, a national monument in the Sierra Nevadas.

Volcanic ejecta is another class of extruded igneous material. It consists of rock and mineral fragments blown out of a volcanic vent under pressure. The ejecta are classified on the basis of size: fragments larger than 1.5 inches (38.1 mm) are called *bombs* or *blocks* depending on their shape; lapilli are small particles ranging from 0.26 to 1.5 inches (6.35 or 38.1 mm) in diameter; and volcanic ash consists of particles less than 0.25 inches (6.35 mm) in diameter. Beds built of volcanic ejecta are referred to as *pyroclastic deposits* (Fig. 5–14). A rock composed of a mixture of small and large ejecta is called *volcanic breccia*.

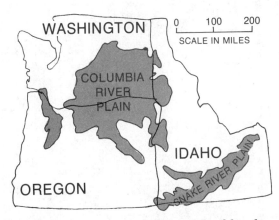

FIGURE 5–15   A vast outpouring of flood basalts during the Miocene epoch of the Cenozoic era produced the Columbia Plateau.

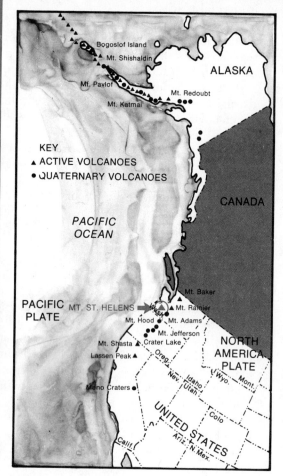

KEY
▲ ACTIVE VOLCANOES
● QUATERNARY VOLCANOES

MOUNT ST. HELENS

As indicated on the map at left, Mount St. Helens is one of a number of volcanoes at the western edge of the North America Plate. A triangle is used to designate a volcano that has been active in recent times; a circle indicates a volcano that has been active some time during the Quaternary period of the Cenozoic. All of these volcanoes are on an overriding plate, that is, the North America Plate. In the Alaska area magma rises from the subducting Pacific Plate and spills onto the surface of the overriding North America Plate as lava. In the California-Oregon-Washington area, the magma is rising from subducting plates which underlie the region. The subduction of the Juan de Fuca Plate produced the lava that issued from Mount St. Helens.

The photo at left shows Mount St. Helens on March 27, 1980, when it began to spew forth clouds of steam and ash. This snow-clad 9677-foot (2950-meter) volcano in the Cascade Range of southwest Washington had been dormant for 123 years; its last eruption was in 1857. Mount Lassen is the only other peak in the Cascades that erupted in this century—that is, from 1914 to 1917. The other peaks in the Cascades also rise snow-clad from a landscape blanketed by green pine forests.

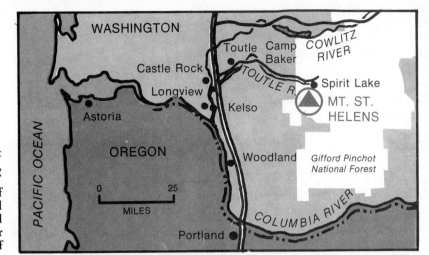

This satellite photo (at right) taken on April 1, 1980, at 16:46 GMT shows the pattern of clouds moving toward the Pacific Coast as well as the cloud pattern over the western portions of Washington and Oregon. Mount St. Helens, hidden below the cloud deck, was continuing to spew forth steam and ash that was drifting into eastern Washington. The map at right details the region immediately surrounding Mount St. Helens.

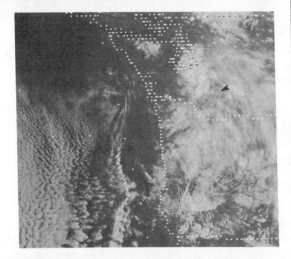

On May 18, 1980, Mount St. Helens exploded with a thud felt 100 miles (160 kilometers) away. A drifting column of steam and ash turned day into night. The eastward sweep of the volcanic ash cloud from Mount St. Helens is detailed in the map at right. The drifting ash has proved to be hazardous to crops, water supplies, and to the health of the people forced to inhale it. Forest fires and flash floods resulted from melting glaciers.

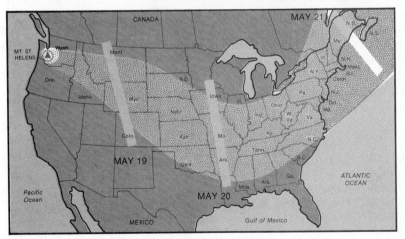

CLASSIFICATION

A simple classification of the igneous rocks can be made on the basis of the texture of the aggregate and the proportions of the various dominant minerals present. The system used below groups the igneous rocks into five categories: (1) coarse-grained, (2) fine-grained, (3) porphyritic, (4) cellular, and (5) glassy. Each of these categories is related to texture and subdivided according to the minerals present.

With respect to texture: Remember, the size of the crystal grains within an igneous rock indicates the rate at which the rock cooled. Thus, distinctions made on the basis of texture have some relationship to the rate at which the rock-forming material cooled.

With respect to the dominant minerals: The most abundant minerals found in igneous rocks are feldspar, quartz, mica, amphibole (hornblende), pyroxene, and olivine. Amphibole, pyroxene, and olivine are all dark green to black and they are often grouped together and referred to collectively as the *ferromagnesians* (ferromagnesium silicate minerals). Thus, the dominant minerals used in the five-category classification that follows are simply the feldspars, quartz, micas, and ferromagnesians.

Long drills are used to punch holes into the granite mass in this Georgia quarry at Elberton. Charges are set and slabs are blasted and split off from the huge rock mass. (Elberton Granite Association)

The Coarse-Grained Category The crystals of coarse-grained igneous rocks are visible to the naked eye. Examples of large-grained intrusive rocks are granite, diorite, and gabbro. Granite, the lightest colored member of this group, is composed mostly of feldspar (55% by volume), with another 27 percent of the rock volume consisting of quartz. The biotite mica (12% by volume) and hornblende (6% by volume) found in granite are referred to as accessory minerals. Granite is a very common and widely occurring igneous rock that is quarried in many regions of the world.

Diorite, slightly darker than granite, is another important plutonic rock. Feldspar makes up 61 percent of the volume of diorite and, thus, is its dominant mineral. There is very little quartz present—less than 2 percent, in fact; for all practical purposes you might say quartz is absent. The biotite micas (2%), horn-

blendes (17%), and pyroxenes (17%) that are present account for the dark color of diorite.

Gabbro is also a plutonic rock, and it is the darkest in this group of coarse-grained igneous rocks. The chief ferromagnesium mineral in gabbro is pyroxene, which accounts for 57 percent of the rock's volume. Feldspar, which makes up about 43 percent of the rock volume, is the second most abundant mineral in gabbro.

The Fine-Grained Category The rocks in this category are extrusive igneous rocks. As a result of fairly rapid cooling, these rocks possess crystals that cannot be distinguished with the naked eye. Rhyolite, andesite, and basalt are examples of fine-grained extrusive igneous rocks. Rhyolite, with its light gray to pink color, is the extrusive equivalent of granite and contains essentially the same minerals. Andesite, formed from volcanic lava flows, is the extrusive equivalent of diorite. Basalt is the extrusive

A lava flow on the Big Island of Hawaii built the point of land in the background of this photo. The black sand along the beach at the lower right and middle of the photo was formed when a lava flow entered the water and exploded into tiny black granules. (USDA)

equivalent of gabbro and contains essentially the same minerals in the same proportions. Basalt is a rather common rock that develops from cooling lava; it is, in fact, the predominant rock of the ocean floor. When lava flows into the rift valley of the Mid-Atlantic Ridge, it cools and forms basalt.

The Porphyritic Category The word *porphyritic* is a textural term that refers to large crystals set in a finer groundmass. The intrusive rocks in this category contain large crystals, called **phenocrysts**, which are embedded in a groundmass of fine-grained crystals. A cooling rate that varies at different points in the solidification process produces the different sizes of crystals within the same rock mass. The phenocrysts form gradually when the temperature of the rock mass is falling at a relatively slow pace; the bed of smaller fine-grained crystals

takes shape quickly at higher temperatures when the rate of cooling is proceeding at a faster pace. The rocks vary from light to dark and are described according to the relative mineral composition by the addition of a prefix: granite-porphyry, andesite-porphyry, basalt-porphyry, etc.

The Cellular Category The word *cellular* refers to small openings or cells in a rock's surface. The extrusive igneous rocks in this category are fine-grained in texture. The cellular character is produced as a result of the rapid expansion of gases as pressure is reduced when the lava is ejected from the volcano. Volcanic scoria is full of cavities formed by gas bubbles. Pumice is another volcanic rock in this category.

The Glassy Category The extrusive igneous rocks in this category result from the rapid cooling of lava, and they display a glassy texture. There are no visible crystals present. Obsidian is an example of a rock in the glassy category.

IGNEOUS ROCKS AND PLATES

The magmas that are extruded to become the lavas of mid-ocean ridges come from the upper asthenosphere. The composition of the magma in this zone of the asthenosphere is rich in iron and magnesium. Lava with this composition—that is, relatively large amounts of iron and magnesium—develops large volumes of pyroxene and feldspar. Thus, the igneous rocks developing from the magmatic mix extruded along mid-ocean ridges are basaltic. Basalts, recall, are rich in pyroxene (57% by volume) and feldspar (43% by volume).

Within a zone of subduction, the descending plate carries crustal sediments and other rocks toward the asthenosphere. Magmas are produced from the melt of the plunging plate. The plunging-plate magmas contain more silica, potassium, and aluminum than the magmas of the asthenosphere. Thus, the lava issuing from a volcano at the border of a subduction zone on an overriding plate tends to produce rhyolitic rocks. Rhyolite—the extrusive equivalent of granite—contains feldspar, quartz, biotite mica, and hornblende.

SEDIMENTARY ROCKS

Most sedimentary rocks are produced by the cementation of sediments, which are derived from the erosion products of other rocks. Sandstone, a common sedimentary rock, is a good example of the sedimentary cycle since it consists of quartz grains which, in all likelihood, weathered out of a granite. Some sedimentary rocks, however, are derived from precipitation and biologic secretion. Halite, which is composed of sodium chloride, and gypsum, which consists of calcium sulfate, are two sedimentary rocks whose origins stem from precipitation that occurred in shallow ancient seas. Chalk—simple calcium carbonate—is an example of a biologically secreted sedimentary rock since its origin is the soft minute shells secreted by what were once living organisms. Still other sedimentary rocks—bituminous coal, for example—result from the accumulation and lithification of plant matter.

DETRITUS

Any loose material removed directly from rocks and minerals by mechanical means such as disintegration or abrasion is called **detritus.** This fine particulate debris may be of inorganic or organic origin. Broken pieces of rock, minerals. and shells are detritus. In the last analysis, detritus is the basic building block for many sedimentary rocks.

The detritus from which most sedimentary rocks develop varies greatly in size and is graded according to its diameter as clay, silt, sand, pebbles, cobbles, or boulders. Particles of **clastic sediment,** for example, range from clay, which is less than 0.00016 inch (0.004 mm) in diameter, to boulders, which are larger than 10 inches (254 mm) in diameter. Between these two extremes are gradations of particle size. The word silt is used to refer to particles larger than those of clay but less than 0.00246 inch (0.0625 mm) in diameter. Sand is next on the grading scale with particle diameters up to 0.0787 inch (2 mm). Pebbles range in size from the upper diameter-limit of sand particles to 2.5 inches (64 mm) in diameter. Cobble diameters range up to 10 inches (254 mm).

This sandstone at Potsdam, New York, had its origin in sand-sized sediment that was laid down in shallow Cambrian seas. (Rosanna Rosse)

LITHIFICATION

The term *lithification* refers to the complex of processes that converts newly deposited sediment into a consolidated rock. The process includes changes in the packing of the particles, known as **compaction,** as well as cementation, and possibly some recrystallization.

When sediments are transported by water and deposited in a large, generally linear trough, called a **geosyncline,** the particles may be loosely packed and have a maximum of surrounding pore space. As additional sediments are deposited in the geosyncline, the pressure of overlying material contributes to the reduction of pore space. This process of forcing particles closer together begins to convert the sediments into a coherent mass with a dramatic reduction in the thickness of the layer or bed.

Cementation is the binding together of the clastic particles by the precipitation of a mineral from the water contained in the pore space.

Precipitates of calcium carbonate, quartz, and iron oxide are the most common mineral cements. Calcium carbonate develops into a soft, gray-colored cement; quartz (silica) forms a hard glassy bond; and iron oxide becomes a fairly soft cement of rust-red color. During the process of cementation, the clastic particles are more or less passive as the precipitated cementing agent binds them together.

Recrystallization involves the formation of new mineral grains within the solid rock material. The process includes the interlocking of sediments as they enlarge and reorganize. The new grains may have the same chemical and mineralogic composition as in the original material; the coarse-grained character of many limestones, for example, develops from a recrystallization of its calcium carbonate.

CLASSIFICATION

Clastic sedimentary rocks are built from particles that have been transported by water, compacted, and then cemented. We can classify these rocks on the basis of the size of the particles found within the rock or on the basis of the origin of the sediment. Discrimination on the basis of sediment size is the easiest way to proceed since the particles in a rock will, obviously, be all the same size or of various sizes.

Shale is a clastic sedimentary rock in which clay-sized particles predominate but some quartz and other minerals may be included. As a result of the small size of its clastic particles, shale has a fine texture in which particles are indistinguishable with the naked eye.

Sandstone consists generally of sand-sized grains of quartz. The color in a sandstone is produced partly by the color of the sand grains, but more often than not a sandstone's color is imparted to the rock by the cementing agent. The red sandstones of New Jersey, which were formed during the Triassic period of the Mesozoic, for example, get their color from the red iron-oxide cementing agent. Sandstone has been widely used as a building stone, and it is extensively quarried. Cambrian-age sandstone is quarried, for example, near Potsdam in New York and on the south shore of Lake Superior.

Conglomerate, which is another type of clastic sedimentary rock, contains a prepon-

Conglomerate is a clastic sedimentary rock. Pebbles in this sample are of various sizes; all are waterworn and somewhat rounded. (Larry D'Andrea)

derance of rounded pebbles. Breccia is similar to conglomerate, but the clastic particles of which it is constructed are sharp and angular and not rounded as in conglomerates.

The class of sedimentary rocks that originates from precipitates and has fine, dense structures falls into a chemical class. Limestone, in which calcite ($CaCO_3$) dominates, is one form of chemical sedimentary rock. Evaporites, like gypsum and salt from dry seabeds, are also found in this group.

Sedimentary rocks whose origin is plant and animal remains fall into an organic class. Calcium carbonate is a constituent of many of the rocks in this sedimentary-rock class since a lot of calcium carbonate is derived from the residue of seashells. Reef rock, which is built by lime-secreting organisms, coquina, which consists of large shell fragments bound by calcium carbonate, and chalk, which is derived from the calcium shells of one-celled sea life, are organic-class sedimentary rocks. Coal, too, is an organic-class sedimentary rock since it is rich in carbon from the remains of ancient forests. Other organic sedimentary rocks are lignite and peat.

Sedimentary strata often carry special features that have been imprinted in the rocks. For example, ripple marks consisting of alternating ridges and hollows formed by wind or water action on incoherent sedimentary materials are often found in hardened sedimentary rock surfaces. Fossil imprints of plants and animals are also common. Mud cracks, produced by the shrinking and drying of mud beds, and raindrop impressions can be found along the bedding planes of many sedimentary rocks.

A rather interesting feature found associated with some sedimentary rocks—notably limestone—is the geode. Within the cavity of a geode, a lining of crystals points inward; the crystals are usually made of calcite or quartz. A geode originates after the sedimentary rock that encloses it is formed. One explanation has the tiny cavity excavated by solution. Then, the cavity is partially refilled at a later time by mineral substances that crystallize from the water solution.

METAMORPHIC ROCKS

Metamorphism is the process by which a pre-existing rock is altered in texture and composition as a response to a greatly altered environment. The transformation of the consolidated rock takes place without melting, although the principal driving forces behind the alterations are pressure and heat. Plate movements as well as intrusive and extrusive igneous activity can produce the pressure and heat that trigger the chemical and physical changes associated with metamorphism. Some of the resulting changes are crushing, pulverization, and recrystallization.

Shellfish take calcite from water and use it to build their shells. When shellfish die, their shells settle to the bottom; if the shell fragments are bound together by a fine cement of calcium carbonate, we have a sedimentary rock. Rather large shell fragments have been bound together to produce this coquina limestone. (Larry D'Andrea)

The metamorphic rock called *gneiss* has a laminated structure which results from the segregation of crushed and pulverized minerals during metamorphism. Note the alternate black and white bands of minerals in this piece of gneiss from Manhattan Island. (Larry D'Andrea)

Mechanical strain produces crushing and pulverization of minerals within a rock. The laminated structure that results from the segregation of the crushed and pulverized minerals is referred to as **foliation** or **schistocity.** The metamorphic rocks called *schists* and *gneisses* show banding and foliation. Large zones containing great volumes of rock are usually involved when schists and gneisses are produced. Because the metamorphism is so extensive, it is referred to as **regional metamorphism.**

The sedimentary rock called *limestone* undergoes recrystallization or enlargement of its particles during metamorphism. Coarse-grained marble, a metamorphic rock, is produced as a result of the transformation. Application of high temperatures alone can trigger the transformation. For example, an igneous intrusion, such as a batholith, can apply sufficient heat to a bed of limestone to bring about its transformation into coarse-grained marble.

When metamorphism occurs in a rather narrow zone around a mass of hot magma, it is referred to as **contact metamorphism.** The sedimentary rock known as *sandstone* can be changed to quartzite, a metamorphic rock, by contact metamorphism; recrystallization of the mineral quartz is involved in the transformation. The change of a pure limestone to a glistening-white, coarse-grained marble is brought about by contact metamorphism, too; it involves only the recrystallization of $CaCO_3$. The Yule marble in the heart of the Rocky Mountains in Colorado was produced from limestone by contact metamorphism. The Tomb of the Unknown Soldier in Arlington National Cemetery is constructed of Yule marble.

Pure limestones contain only calcium carbonate. Many limestones, however, also contain mixtures of other minerals, including quartz and clay. If a limestone containing both quartz and clay minerals undergoes contact metamorphism, a new mineral, garnet, is formed. Garnet is a complex aluminum-silicate mineral. Some garnets, however, also contain iron. It is thought that the iron included in the garnet is transferred to the limestone from the magma of the intrusive igneous mass that moves into the region.

THE LANGUAGE OF PHYSICAL GEOGRAPHY

A physical geographer uses a technical vocabulary to make explanations. Review your understanding of the vocabulary used to develop the concepts in this chapter. Among the important words and terms used are:

asthenosphere	discordant	lithification	pluton
batholith	extrusive	lithosphere	regional meta-
clastic sediment	faulting	magma	morphism
compaction	folding	mantle	rocky mantle
concordant	foliation	M-discontinuity	schistocity
contact metamorphism	geosyncline	metamorphic	sedimentary
core	gradational forces	mid-ocean ridge	sial
crust	igneous	mineral	sill
crystal	intrusive	Moho	stock
crystal form	landforms	orders of relief	tectonic forces
detritus	lattice	orogeny	volcanism
diastrophism	lava	phenocrysts	
dike	lenticular	plateau basalts	

SELF-TESTING QUIZ

1. How does the physical geographer use the concept of orders of relief?
2. List some first-order relief features.
3. What agents carve second-order relief features?
4. Describe the characteristics of the three concentric shells that make up the Earth's interior.
5. Differentiate between the terms *lithosphere* and *crust*.
6. How were the tears and rifts in Pangaea produced?
7. Explain the two-stage process for producing continental material.
8. How are third-order relief features produced?
9. In what way are most of the continental mountain systems built?
10. List some of the variety of landforms that exist on the Earth.
11. Why are ocean basins considered to be impermanent features but continents are considered permanent?
12. Why don't the growth surfaces of a mineral always display the same regular geometric configuration?
13. What is meant by the inferred-origin system of classifying rocks?
14. Explain the concept of the rock cycle.
15. What accounts for the various shapes of plutons?
16. Under what conditions do fissure flows develop?
17. How is texture used to discriminate among the various kinds of igneous rocks?
18. Compare the composition of plunging-plate magmas with the magmas of the asthenosphere.

19. Describe the formation of sedimentary rock from newly deposited sediment.

20. How do plate movements figure in the development of metamorphic rocks?

IN REVIEW

—— **PROCESSES AND MATERIALS OF LANDSCAPE DEVELOPMENT** ——

I. VARIATIONS IN CONFIGURATION

    A. Continents and Ocean Basins
        1. Earth's Interior
        2. The Outer Zone
        3. The Crust
        4. The Proto-Continent
        5. The Two-Stage Process
    B. Mountains, Basins, and Plateaus
        1. Tectonics in the Sea
        2. Continental Tectonics
    C. Landforms

II. MINERALS OF THE CRUST

    A. Chemical Composition
    B. Crystal Lattice
    C. Crystal Form

III. CRUSTAL ROCKS

    A. Inferred-Origin System
    B. The Rock Cycle
    C. Igneous Rocks
        1. Intrusive Rock Formations
        2. Extrusive Rock Formations
        3. Classification
            a. The coarse-grained category
            b. The fine-grained category
            c. The porphyritic category
            d. The cellular category
            e. The glassy category
        4. Igneous Rocks and Plates
    D. Sedimentary Rocks
        1. Detritus
        2. Lithification
        3. Classification
    E. Metamorphic Rocks

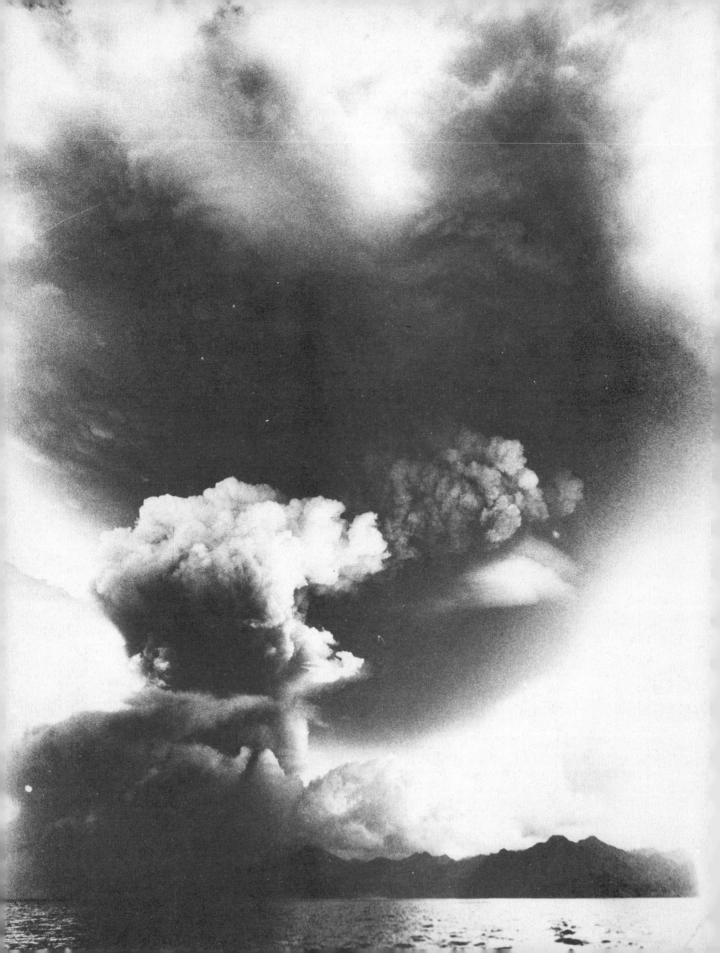

6 *Tectonic forces and landscapes*

The Earth's surface is distorted, buckled, and broken as a consequence of folding, faulting, and volcanic activity. Each of these tectonic forces leaves its imprint in the crustal surface and is an important influence in landscape development.

It is, however, the large-scale movements of lithospheric plates that compress the Earth's crust and eventually cause rock structures to fail by bending and producing folds or by rupturing and producing faults. Plate movements that stretch the crust also cause rock-structure failure—primarily through collapses, which result in faults. Each of these types of rock-structure fail-

ure is expressed in the landscape as a mountain range.

Volcanic activity, which is a major source of new rock, is also an important factor in landscape development. However, volcanism implies much more than eruption through a volcanic cone. Any invasion of the crustal zone by magma from below is properly termed **volcanism**. In Chapter 5, the discussion of intrusive and extrusive igneous rock formations is, in reality, a discourse on volcanism. Plateaus, mountains, domes, and hills are landscape features that develop as a result of the crustal stress produced by volcanism.

DEFORMATION THROUGH FOLDING

The development of a **fold system** is closely associated with the concept of global tectonics. A fold system is formed at the margin of converging

plates where an oceanic plate is plunging under a continental plate or where collision is occurring between two continental plates. In both

Mount Soufrière—shown in eruption on April 22, 1979—has a history of violent eruptions in this century and the last. This active volcano is the highest peak (4918 feet, 1225 meters) on Saint Vincent, an island of the Lesser Antilles in the Caribbean Sea. Thickly wooded volcanic mountains form the backbone of Saint Vincent. (J. Hollis, U.S. Coast Guard)

cases—oceanic-plate subduction or a two-continent suture—mountain chains are created. The process of suturing creates collision-type mountains—the Himalayas, for example—and the process of subducting creates cordilleran-type mountains—the Andes, for example.

THE FOLDING PROCESS

It seems somewhat unbelievable that layered rocks can be distorted into wavelike forms or folds, but it happens! It happens as strong lateral pressure is exerted by colliding plates on strata that are buried at some depth beneath the confining pressure of overlying layers. These compressional forces, exerted over a long period of time, gradually distort some very stiff rock layers into broad warps. Other more pliable, flat-lying strata contort into a series of parallel folds that build and eventually collapse over one another as the pressure continues.

All classes of rock—igneous, metamorphic, and sedimentary—fold under the proper set of circumstances. The folds in sedimentary rocks are the easiest to identify because folding changes the obvious particle-to-particle relationship of the original bedding plan, and the disruption in the flat-lying character of sediments is immediately apparent.

FOLD TERMINOLOGY

Any pronounced bend in rock strata is called a *fold*. The high portion of a fold or arch is referred to as the **anticline;** the lower portions or depressions on either side are called **synclines** (Fig. 6-1).

The terms *anticline* and *syncline* are applied to folds of relatively small scale—some with widths and lengths measured in inches, others with widths of a few miles and lengths in the order of tens of miles. If the same sort of folding occurs on a very large scale—involving hundreds of miles—the high portion of the fold is called a **geanticline** and the depression is known as a **geosyncline**. The prefix *geo* from the Greek word *geos*, meaning Earth, emphasizes the size of a geosynclinal structure.

A plane that intersects the crest or the trough of a fold in such a manner that the limbs or sides of the fold are more or less symmetrically arranged with reference to it is called the **axial plane** (Fig. 6-1). The simplest folds are said to be symmetrical, that is, the limbs on either side of the axial surface are of equal length. A limb extends from the axial surface of one fold to the axial surface of the adjacent fold. The region of curvature is called the *hinge*.

The term **dip** means the angle at which a stratum or any planar feature is inclined from the horizontal. Using the axial plane and the concept of dip, we can describe three different kinds of folds as illustrated in Figure 6-2: (1) An upright fold has an axial surface that dips from 81° to 90° (Fig. 6-2A). (2) Inclined folds have axial surfaces that dip from 10° to 80° (Figs. 6-2B and C). (3) Recumbent folds have axial surfaces that dip less than 10° from the horizontal (Fig. 6-2D).

When the terms **symmetrical, asymmetrical, overturned,** and **isoclinal** are used to describe folds, reference is being made to the attitude or relative lengths of the limbs. Symmetrical folds

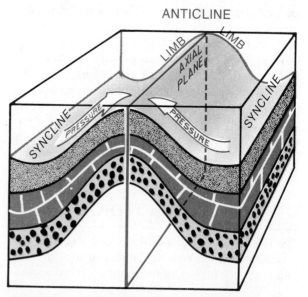

FIGURE 6-1 When rock strata are subjected to a compression in the direction of the arrows, the strata bend to produce a fold. The complete fold is composed of two parts: a high part, the crest, and a low part, the trough. A plane drawn vertically through the upper part of the anticline cuts the fold in half and is called the *axial plane.*

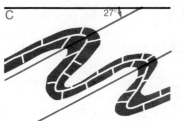
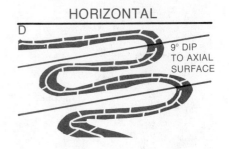

FIGURE 6–2 Four different folds are depicted in this illustration. The axial planes of these folds assume different positions with respect to a horizontal drawn at the Earth's surface. Depending on the pattern of the fold, the angle between the horizontal and the axial plane can range from 0° to 90°.

have limbs of equal length; asymmetrical means the limbs are of unequal length. Overturned folds are inclined folds in which both limbs dip in the same direction. Both limbs in an isoclinal fold are parallel.

FOLDS IN THE FIELD

It does not follow that anticlines become visible hills and synclines become valleys. Erosion is at work on these structures as soon as they are exposed at the surface; and it is not uncommon to find a syncline at the top of a hill. The present-day Appalachians, which stretch from the Gaspé Peninsula in the Canadian province of Quebec to central Alabama for a distance of 1500 miles (2400 km) (Fig. 6–3), are folded mountains; but erosion has played a significant role in forming the ridges from the harder rock and carving valleys from the softer rock of the folds.

In Figure 6–4, a very schematic section is drawn across the Appalachians from the Blue Mountain area (40°20′N, 76°55′W) outside Harrisburg to Bald Eagle Mountain (41°N, 77°45′W) just beyond Bellefonte, Pennsylvania. The cross section runs in a northwesterly direction for slightly more than 60 miles (about 100 km). Shade Mountain near Lewiston is an anticline that is intact. Stony Mountain is a remnant of

rock that was at the central part of a syncline. Lewiston sits between Jacks Mountain and Shade Mountain. Jacks Mountain is the remnant of a limb; it is one side of a fold—the missing section of the arch was eroded away. Bald Eagle Mountain is just one side·of a fold, too. Shade, Jacks, and Bald Eagle mountains are all part of the same rock formation, as the color code in Figure 6–4 indicates. The valleys—Nittany, Penn, and Kishacoquillas—occupy depressions produced when the tops of the original up-arched folds were worn away.

THE FOLDED APPALACHIANS

The Appalachians—the chief mountain system of eastern North America—are a composite product of initial cordilleran activity, which included subduction, and subsequent collision tectonics. The cordilleran-to-collision sequence resulted from the fact that the Atlantic Ocean opened, closed, and then opened again. The following summary of Appalachian development is based on the research and suggestions of John Dewey, John Bird, and Robert Dietz.

RIFT AND SEDIMENT STAGE
A continuous North American–African continent existed as part of Pangaea. The present-day east-

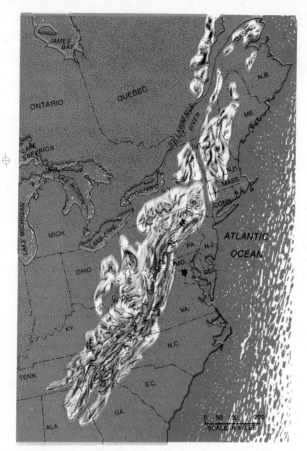

FIGURE 6–3 The folded Appalachian system runs in a northeasterly direction from Alabama to the Gaspé Peninsula of Quebec.

ern coast of North America was joined to the African continent as shown in Figure 6–5A. Then 600 million years ago a spreading rift developed and began to separate North America from Africa. Through the process of sea-floor spreading, the ancestral Atlantic Ocean opened and North America receded from Africa. The newly formed continental shelves on both sides of the proto-ocean received sediments from the continental areas (Fig. 6–5B). The geosyncline on the North America Plate had a thin, shallow-water section on the shelf, called **miogeosyncline,** and a thick, deep-water section, called **eugeosyncline** (Fig. 6–5C).

CORDILLERAN AND COMPRESSION STAGE

About 500 million years ago, the convection currents within the mantle underwent a reorganiza

tion as the Earth's core increased in size. Basalt was no longer supplied to the recently developing Atlantic Basin. New lateral pressures, which developed along rifts between other continental masses, pushed the Africa Plate toward the North America Plate and caused the **lithosphere** to rupture along the continental margin of North America. The denser oceanic portion of the Africa Plate was forced under the North America Plate by the compressive forces (Fig. 6–6A).

The trench produced by the subducting Africa Plate was adjacent to the North American continental margin. As the oceanic portion of the Africa Plate was being consumed in the subduction zone, the thick deep-sea sediments of the eugeosyncline were being uplifted and folded to create the ancient Appalachians. Magma rising from the plunging plate was intruded into these eugeosynclinal sediments and some of it spilled onto the surface as eruptions of lava. The intense lateral compression also folded the miogeosynclinal sedimentary deposits on the continental shelf into a series of ridges (Fig. 6–6B).

CONTINENTAL COLLISION STAGE

Eventually the oceanic crust of the Africa Plate was consumed in the subduction zone that developed off the North American continental margin, and the eugeosynclinal sediments on the Africa Plate plowed into the folded sediments on the North America Plate (Fig. 6–7). Thus, between 350 million and 225 million years ago the continental masses of Africa and North America were sutured together. Sediments eroded from the mountain foldbelt of the eugeosyncline and were carried into the folded miogeosyncline. Over eons of time, the lofty mountain range in the foldbelt of the North American eugeosyncline was completely eroded.

PRESENT RIFT CYCLE

Approximately 180 million years ago another reorganization of the internal convection currents occurred. A rising current that developed from this reorganization opened a new rift at the old plate-to-plate suture that joined North America to Africa. The present-day Atlantic Ocean was born when this rift, which is known

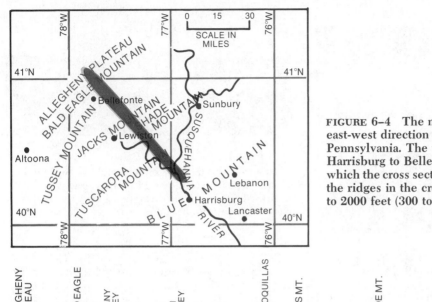

FIGURE 6-4 The map at the left spreads in an east-west direction from Lancaster to Altoona, Pennsylvania. The wide (color) line that runs from Harrisburg to Bellefonte marks the region through which the cross section was made. The elevations of the ridges in the cross section range from 1000 to 2000 feet (300 to 600 meters).

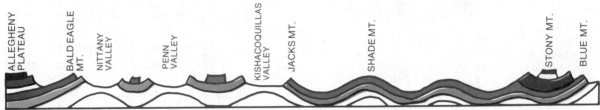

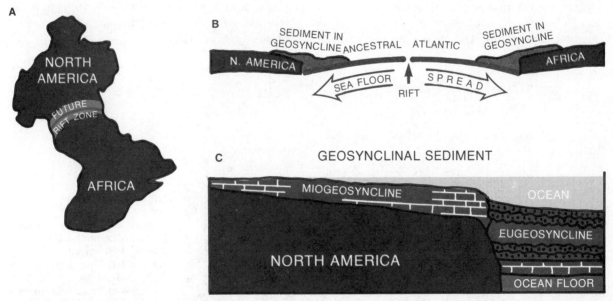

FIGURE 6-5 The ancestral Atlantic Ocean was created by a rift that opened between North America and Africa. The geosynclines produced at the continental margins by the rifting had sediments deposited into them. The shallow-water section of the deposits on the continental shelf is referred to as a *miogeosyncline*. The thicker section of deposits on the ocean floor is known as a *eugeosyncline*.

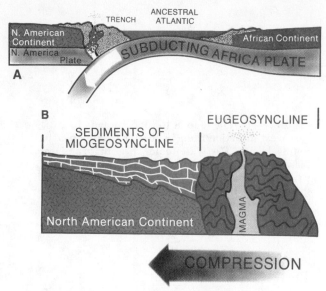

FIGURE 6–6 When the Africa Plate was forced under the North America Plate by the lateral force of compression, the eugeosynclinal and miogeosynclinal sediments were uplifted and folded.

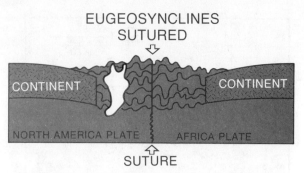

FIGURE 6–7 When the eugeosynclinal sediments on the Africa Plate plowed into the eugeosynclinal sediments on the North America Plate, the plates were sutured together.

today as the Mid-Atlantic Ridge, developed. As new ocean floor was emplaced in the developing rift, the continental masses of North America and Africa were driven apart once again. The present-day Atlantic Ocean is opening and en-

larging at an approximate rate of 1 inch (3 cm) per year.

Since the ancient Appalachians in the foldbelt of the eugeosyncline were completely eroded, geosynclines developed anew on the continental shelf and the developing ocean floor. The present coastal plain that forms the eastern shore of North America and the continental shelf to its east occupy the site of the ancient eugeosyncline (Fig. 6–8). The folded Appalachians of today—from the Blue Ridge line to the Allegheny Front—occupy the site of the ancient miogeosyncline (Fig. 6–8).

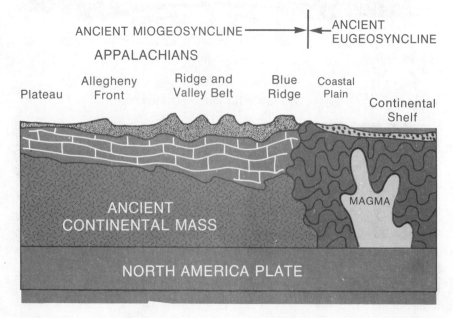

FIGURE 6–8 The present-day Appalachians occupy the site of the ancient miogeosyncline. The folding and uplift of the miogeosynclinal sediments and rocks is the result of compressional forces that develop as magma is intruded into the Mid-Atlantic Ridge.

FAULTING OF THE CRUST

The concept of moving plates is fundamental to the theory of continental drift. The response of crustal plates to compression and tension accounts for most large-scale second-order–relief features. Compression produces folding, thrusting, trenching, and thickening. Thinning, down-dropping, and rifting are produced by tension. A fracture in the Earth's crust along which the adjacent rock surfaces have been displaced is called a **fault.** Faults—the San Andreas Fault system, which forms the boundary between the North America Plate and the Pacific Plate, for example—result from the operation of tectonic forces. Both compression and tension can produce faults.

Most crustal faulting takes place beneath the surface or on the ocean floor; thus, it is hidden from our immediate view. Earthquakes and the waves generated by them signal and document movement along hidden faults. Surface faulting, on the other hand, can be observed directly, and the characteristic features produced by it can be studied.

FAULT TERMINOLOGY

The definition of the term *fault* in the first paragraph of this section includes a reference to motion. A fault is a fracture along which movement has taken place. Cracks or joints along which there has been no movement are not faults.

The trace of a fault on the Earth's surface is called a **fault line.** When a mass of rock is bounded on at least two opposite sides by faults, it is known as a **fault block.** The term *faulting* is reserved to describe the movement or sudden slippage between two rock masses.

The fracture surface between two blocks is referred to a a **fault plane** (Fig. 6–9). The inclination of the fault plane from the horizontal is known as its **dip.** When an escarpment—a cliff or steep slope—is produced by the displacement of a fault block, it is called a **fault scarp.**

Most faults occur along fractures that are inclined. Thus, a fault block is often elevated or depressed relative to the adjoining region. In Figure 6–10, the person is facing the fault line;

the block overhanging the person's position is called the **hanging-wall block,** and the block on which the person's feet rest is called the **footwall block.** Technically, the terms *footwall* and *hanging wall* are defined with respect to the fault plane itself. If the fault is not vertical, the face of rock lying above the fault plane is the hanging wall and that below it is the footwall.

GRABENS AND HORSTS

Some special relief features develop from combinations of more than two fault blocks. A fault block, for example, may be elevated relative to the region on one side and depressed relative to that on the other. On the other hand, as shown in Figure 6–11, the hanging-wall block, labeled *B*, is depressed with respect to the footwall blocks (labeled *A* and *C*) on either side. The horizontal surface of the hanging-wall block forms a trench-like structure called a **graben.** Grabens are important aspects of the lanscape in many parts of the world. For example, the valley through which the Rhine River flows in western Europe, north of Basel, is a giant graben. Russia's Lake Baikal (53°N, 107°40′E)—the deepest lake in the world—is also located in a graben.

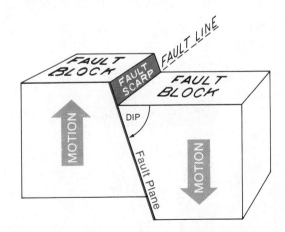

FIGURE 6–9 The fracture surface between two blocks is called the *fault plane.* The cliff or steep slope above the fault line is known as a *fault scarp.*

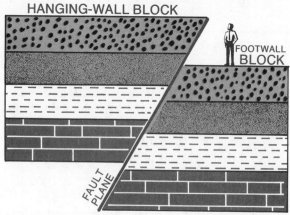

FIGURE 6-10 The fracture plane is oblique with respect to the strata and the vertical. The surface lying above the fracture plane is called the *hanging wall*; the surface located below the fault plane is called the *footwall*.

The most famous graben in the world is the Great Rift Valley of East Africa. At times, it is referred to as the *Kenya Rift Valley* because it is a few miles west of Nairobi, capital of Kenya (Fig. 6-12). Two imposing scarps that run north and south frame the valley. The scarps are separated by about 40 miles (64 km)—the average width of the valley. The ground between the scarps is a fault block that has subsided.

In the last 20 million years, the blocks on either side of the Great Rift Valley have eased apart by about 6 miles (10 km). The Great Rift Valley is a part of a system of rifts that stretches

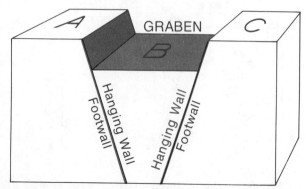

FIGURE 6-11 The trenchlike structure between two footwall blocks is called a *graben*.

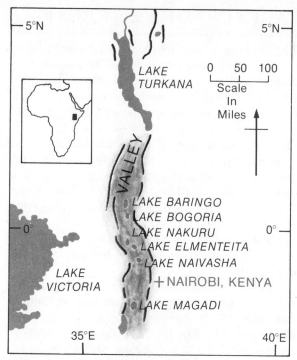

FIGURE 6-12 The Kenya Rift Valley is located in a graben in East Africa.

south through Mozambique and north through Ethiopia to the Red Sea. A western branch runs along the Congo border; Lake Tanganyika, mentioned on page 127, occupies a graben in the Western Rift Valley of Africa.

There are a number of lakes in the Kenya Rift Valley whose development is related to a wetter climatic phase that occurred approximately 10,000 years ago. The lakes, shown in Figure 6-12, lie in fault-controlled troughs. Lake Bogoria, which is located just north of the Equator, is typical of the lakes in the Kenya Rift Valley; it extends in a north-south direction for 10.5 miles (17 km) and varies in width from 150 to 1070 feet (500 to 3500 m). Lake Bogoria receives water from short ephemeral streams, the semipermanent Sandai River, and three groups of hot springs and geysers.

In Figure 6-13, a footwall block, F, is elevated with respect to the hanging-wall blocks, E and G, on either side. The upraised central fault block is called a **horst.** The top surface of the horst is generally long compared to its width.

The tectonic activity which gave rise to the Eastern Rift Valley and its flanking highlands, which lie mainly in Kenya, occurred primarily in the Pliocene and Pleistocene periods. Volcanism accompanied the instability, and lava extends over most of the Kenya highlands. The lava-produced soils are generally porous but fertile. The fertile soils and adequate rainfall in the highlands of Kenya allow many cash-crops to be grown. The Kenyan woman is harvesting carrots against a background of pyrethrum (used in insecticides) in the highlands. (World Bank)

The Great Basin region of Nevada displays many horsts.

As mentioned above, the Rhine River, which rises in Switzerland and empties into the North Sea, flows through a giant graben north of Basel. The Rhine's graben, however, is flanked by two huge horsts—the raised region of the Vosges Mountains on the west in France and Germany's Black Forest on the east.

FAULT CLASSIFICATION

Generally, faults are classified with reference to either apparent or relative movement. In the system that follows, downward, upward, and horizontal movements are used as the basis for classifying faults into three basic groups.

NORMAL FAULTS

If the apparent condition can be accounted for by a downward displacement of the hanging-wall block with respect to the footwall block, then the fault is said to be a **normal fault.** The faults in this category are generally caused by tension. Since the relative downward movement of the hanging wall suggests that gravitational forces are at work, such faults are sometimes called **gravity faults.**

The dips of normal fault planes vary over a wide range, but the fault planes in most normal block-faulted regions dip at angles greater than 45°. In fact, numerous studies indicate that the dip angle for these faults averages between 45° and 70°.

The Sierra Nevada mountain range of California is created by normal block faulting within the North America Plate. The hanging-wall

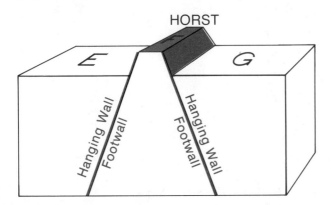

FIGURE 6–13 The elevated region of the footwall block is referred to as a *horst.*

block—known as Owens Valley—moved down (Fig. 6–14). The Sierras sit on the footwall block. In fact, faulting in this region has occurred in recent time; about 13 feet (4 m) of vertical slippage occurred in 1872 and resulted in a violent earthquake. The west side of the footwall block is depressed, and this makes for a mountain range that is steeply cliffed on the east and gently sloped on the west. The steep eastern face of the footwall block stands more than 10,000 feet (3000 m) above Owens Valley. The range extends 400 miles (644 km) north-northwestward from Tehachapi Pass to Lake Almanor. The Sierra Nevada is about 40 to 70 miles (64 to 113 km) wide. Mt. Whitney (36°35′N, 118°18′W), standing at 14,494 feet (4118 m) above sea level, is the highest peak in the range. The Sierra Nevada range, of course, is an example of a normal fault on a very large scale!

REVERSE FAULTS

The faults in this category are produced by compression. If the apparent condition can be explained by an upward movement of the hanging wall with respect to the footwall, the situation is described as a **reverse fault.**

Reverse faults can dip at very low or very steep angles. Generally, however, they dip at somewhat less than 45°. In fact, some of the greatest reverse faults dip at very low angles. Often, the lateral displacement in a low-angle reverse fault is so pronounced that it is termed a **thrust fault.**

In the southern Appalachians, the entire Blue Ridge—from Roanoke, Viginia, southwestward across North Carolina, South Carolina, Georgia, and into Alabama—has moved westward on a great thrust sheet. The dip of the thrust fault is very low—almost approaching horizon-

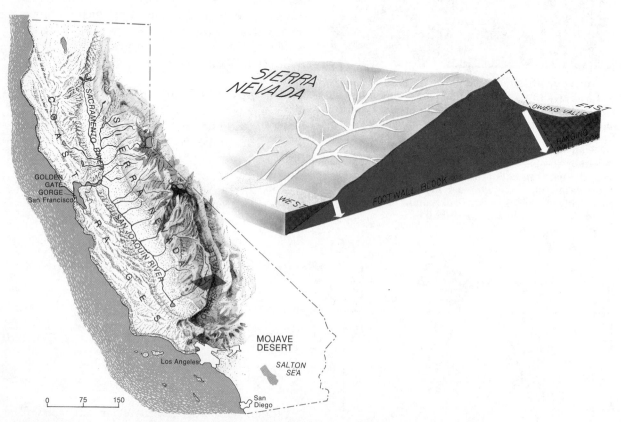

FIGURE 6–14 Normal block faulting is responsible for the creation of the Sierra Nevada.

This portion of the southern Appalachians is in North Carolina, just south of the Virginia border. This is an area of reverse faulting. Rolling bluegrass pastures terminate in precipitous bluffs. The flowers in the foreground are those of rhododendron, which bloom in June. (National Parks Service)

tality. The westward movement of the thrust sheet is as much as 78 miles (125 km).

STRIKE-SLIP FAULTS

In **strike-slip faults** the relative displacement of the blocks is horizontal or approximately so. Because strike-slip faults are readily observable where they cross the trend of folds, they have been called **transcurrent faults.** They are also known as *tear faults, wrench faults,* and *lateral faults.*

Strike-slip, or lateral, faults have a very nearly vertical dip. Lateral faults are referred to as being either left-lateral or right-lateral. Identification of lateral faults is relatively easy: simply face the fault or fracture surface and determine whether the block on the other side has moved to the left or right. If the block has moved to the left, the fault is a left-lateral fault.

Some faults—especially lateral faults—are plate-edge phenomena. California, as you know, is split by the boundary of the North America and Pacific plates and, thus, the San Andreas Fault, which runs through California, is interpreted as a plate-edge phenomenon (Fig. 6–15). The North America Plate is on the eastern side of the fault and the Pacific Plate is on its western side. The San Andreas Fault is described as a right-lateral fault. This means that if you stand on the North America Plate and face the fracture surface, the Pacific Plate has moved to the right. Strike-slip faults that are plate-edge phenomena have been given a special name—**transform faults.**

FAULT COMBINATIONS

The complex of forces operating at any location can, of course, produce combinations of vertical and horizontal motion along faults. For example, in addition to the relative downward movement of the hanging wall in a normal fault, there may also be a relative horizontal movement of the blocks. If the relative horizontal offset of the blocks is to the left, the combination is identified as a left-lateral normal fault. There are also left-lateral reverse faults, right-lateral normal faults, and right-lateral reverse faults.

EARTHQUAKES

A trembling or shaking of the ground is identified by the average person as an **earthquake.**

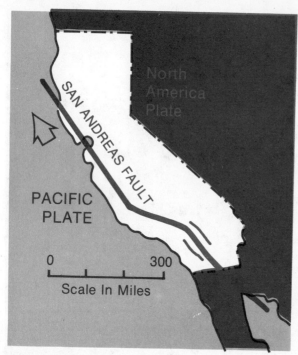

FIGURE 6–15 The San Andreas Fault system is outlined in color. It is one of the world's major lateral faults. There is a general northwestward movement of the ocean side of the fault relative to the continental side.

Since movement along faults tends to be abrupt rather than a steady sliding, faulting is one of the prime causes of earthquakes. As the forces of deformation build within the crust, friction temporarily locks the two sides of a fault in place and causes stresses to build. When the developing force of deformation is sufficient, the locked rock fractures under the stress and abrupt slippage occurs along the fault zone. It is this sudden fracture and release of pent-up energy that causes earthquakes. If the forces of deformation persist, the whole cycle of developing stress and frictional locking followed by abrupt slipping repeats itself.

The San Andreas Fault represents a zone of powerful collision where two great plates are sliding past each other. The sliver of California that is west of the fault is located on the Pacific Plate, which is moving toward the northwest. The portion of California to the east of the fault is on the North America Plate, which is moving

westward. Any sudden movement along the San Andreas Fault causes an earthquake in California.

In 1972, scientists at the Goddard Space Flight Center in Maryland developed a new system for measuring motion along the San Andreas Fault. Two laser-device stations were established on opposite sides of the fault. One station is at Quincy; the other is at Otay Mountain outside San Diego (See Fig. 1–9). From each station, a laser beam is directed at a satellite orbiting 600 miles (965 km) above the Earth. The information provided by this technique allows the distance between the stations to be measured with great accuracy.

Over a four-year period—from 1972 to 1976—Quincy and Otay moved closer by 14 inches (35.6 cm). This is an average rate of 3.5 inches (8.9 cm) per year, which is more than anyone had expected. These findings indicate far more activity along the fault than previously suspected. It may mean that stress is accumulating at a rapid rate. In addition, because more energy than anticipated is being locked into the rock fractures, its release might produce an enormous earthquake.

The principal zones of earthquake activity coincide with major plate boundaries. In fact, 90 percent of the world's earthquakes occur along plate boundaries. Although the strongest earthquakes occur along plate boundaries, small earthquakes occur practically everywhere. The earthquakes that are not associated with plate boundaries are less frequent and weaker. Some of these nonplate-boundary earthquakes are caused by volcanic activity, gravitational stress, and rebound of the crust following a period of severe loading that might have been produced by an ancient ice sheet.

EARTHQUAKE FOCUS

The point within the Earth where an earthquake begins is called the **focus**. The focus generally lies below the surface. The point on the Earth's surface directly above the focus is referred to as the **epicenter**. The location of an earthquake focus is identified by giving the geographic position of its epicenter and the depth at which the disturbance was initiated. The focus of the

The abrupt violent nature of the crustal movement produced by the Alaskan quake of 1964 can be imagined when you view the offset of the railroad tract (*far left*) and the displacement of this Anchorage street (*left*). (NOAA)

Alaskan earthquake of March 27, 1964, for example, was reported as latitude 59°48'N, longitude 148°42'W at a depth of 24.8 miles (40 km).

Earthquakes are classified as being shallow- or deep-focus. A **shallow-focus earthquake** is one in which the focus is less than 31 miles (50 km) from the surface. An earthquake whose focus lies below 31 miles (50 km) is identified as a **deep-focus** disturbance. The Alaskan shock of 1964 was a shallow-focus earthquake. The foci of some earthquakes may be at depths of several hundred kilometers.

Every year more than a million earthquakes shake the Earth strongly enough to be felt. The foci of most earthquakes lie at depths between 1.9 miles (3 km) and 18.7 miles (30 km). By definition, then, most earthquakes are shallow-focus disturbances. Shallow-focus earthquakes abound along all the ocean ridges, although they are not restricted to these belts. In fact, at shallow depths, where the edges of two plates are converging and pressing against each other, there is intense earthquake activity.

Deep-focus earthquakes, on the other hand, cannot occur except in descending plates. They occur in the lower portion of the plunging plate. Recent studies indicate that deep-focus earthquakes are centered in the coolest region of the descending plate because it is here that the stresses generated by gravitational forces acting on the dense interior of the slab are most intense. The resistance of the surrounding mantle to the slab's penetration also develops the greatest stress in the coolest region of the descending plate. The cool and rigid interior of the slab acts as a channel to transmit stress and, thus, gives rise to deep-focus earthquakes.

EARTHQUAKE WAVES

As with any release of energy, vibrations or disturbances spread outward from the focus of an earthquake. The vibrations are called **seismic waves** after the Greek word for earthquake. Seismic waves move out in all directions from the focus, just as sound waves move out in all directions when a gun is fired.

Seismic waves are the shakers and wreckers that accompany the release of energy at the focus. There are two general types of seismic waves produced by earthquakes: surface waves and body waves. Surface waves, which produce most of the destruction because they actually make the ground roll, are so named because they travel along the Earth's surface. Body waves are sometimes referred to as *preliminary waves* because they arrive before the surface or rolling waves. There are, however, two different and distinct types of body waves: the **compression wave** and the **shear wave**.

The compression waves produced by an earthquake travel at great speeds and arrive at the surface like a hammer blow. The blow of a compression wave travels in the same way that a bump from a locomotive on one freight car travels clear through a whole train. Since compression waves are the first to arrive, they are referred to as the **primary waves,** or P waves.

P waves—like sound waves—move through both liquids and solids by compressing the material directly ahead. Each compressed particle, in turn, springs back to its original position as the energy moves on. This event—compression of a particle to its spring back—is called a *cycle.* The time within which such a cycle is completed is referred to as the wave's period.

The P wave is the swiftest seismic wave. Its speed, however, varies with the material through which it passes. P-wave velocity in the crust of the Earth usually is less than 4 miles (6.4 km) per second, or 14,000 miles (22,500 km) per hour. But just below the crust in the mantle, the speed of a P wave jumps to 5 miles (8 km) per second. As a P wave passes deep into the Earth and moves below the mantle through the core, its speed increases to 7 miles (11 km) per second. Thus, it travels through the Earth's core at more than 25,000 miles (40,200 km) per hour.

When a P wave strikes an object embedded in the ground, it produces a series of sharp pushes and pulls. These pushes and pulls are in a direction parallel to the wave path. The second type of body wave, on the other hand, produces a shearing effect or a side-to-side shaking of an object embedded in the ground. The shear waves produced by an earthquake are referred to as **secondary waves,** or S waves. One reason why the shear waves are called secondary is that they ordinarily reach the surface after the P wave. The shear, or S, waves displace an object at right angles to their direction of travel and are thus sometimes called **transverse waves.**

The S wave must have a rigid medium through which to move. These transverse or shear waves do not travel below the mantle. The outer portion of the Earth's core, which is just below the mantle, is liquid and S waves cannot travel through it.

RECORDING SEISMIC WAVES

The vibrations produced by earthquakes are recorded and measured by instruments called **seismographs.** A variety of different seismographs—some sensitive to short-period waves and others sensitive to long-period waves—have been designed. Remember, the period of a

The California earthquake of April 18, 1906, is commonly known as the *"San Franciso earthquake."* Until 1906, it was presumed that fault movements were predominantly vertical; but horizontal displacements of 20 feet (6 meters) in this 1906 quake shattered that notion. The horizontal motion caused widespread lurching in areas. The buckled trolley tracks, displaced sidewalks, and gaping fissures in the streets shown in this 1906 San Francisco photo were rather common. (California Palace, Legion of Honor)

seismic wave is nothing more than the time within which it completes a cycle. The seismic waves generated at a focus vary over a wide range of periods. Some have periods that are extremely long; others have periods of less than a tenth of a second.

The zigzag lines made by the recording mechanism of a seismograph are called **seismograms.** In Figure 6–16 you see a seismogram recorded by a seismograph located at Florissant, Missouri, a few miles northwest of St. Louis. The seismogram is marked with a time scale. Each unit on the time scale is one minute. The points labeled "P" and "S" on the seismogram indicate when the P-wave trains and the S-wave trains begin. The continuous record of exceedingly small waves made by the seismograph prior to the arrival of the P-wave train—known as *microseisms*—represent the background vibrations that exist at Florissant. The microseisms are made by disturbances other than earthquakes.

P and S waves, which leave the earthquake focus at the same instant, travel outward in all directions. The fast moving P waves reach the seismograph first. Some time later, the slower moving S waves arrive. The delay in arrival time is proportional to the distance traveled by the waves. In other words, the farther away the center, or focus, of the earthquake shock, the longer is the spread of time between the arrival of the P and S waves.

The seismogram is used to describe the earthquake. The recorded heights, or amplitudes, of the P, S, and surface waves, for example, indicate the amount of energy released. By combining data from selected seismograph stations, the epicenter and the focal depth of an earthquake can be located.

SCALING EARTHQUAKES

In the case of a violent dislocation that produces a severe earthquake, the energy is released in one large wrench followed by smaller tremors. The tremors, referred to as *aftershocks*, are produced by continuing collapse and movement of crustal blocks. Sometimes the violent wrench is preceded by small structural failures that produce foreshocks, which are really small tremors. At the present time, two scales are used to describe the severity of an earthquake: the Richter scale and the modified Mercalli scale. The Richter and Mercalli scales are completely separate in intent.

The **Richter scale,** which is based on instrument records, measures an earthquake in terms of its energy or magnitude at the moment of creation. Magnitude is expressed as a number by the Richter scale. The Richter number is derived from the amplitude of the recorded seismic waves. The **Mercalli scale,** which is based on subjective observation, describes the actual effect or intensity of an earthquake at a particular location in terms of the damage to life and property.

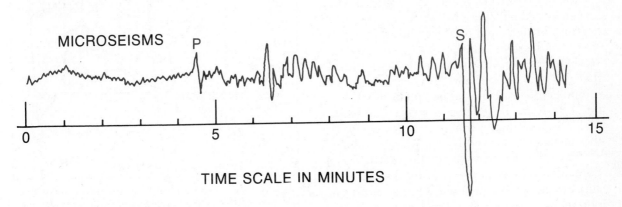

FIGURE 6–16 Microseisms, the P-wave train, and the S-wave train are shown in this seismogram. The S-wave train arrived at the station 7 minutes after the first P-wave was recorded.

In order to assign a number on the Richter scale, the measurement must be based on a seismogram made at a distance of 62 miles (100 km) from the epicenter. Most stations that record earthquakes, however, are at some distance other than 62 miles (100 km). This means that seismograms from several different stations are studied. Then, complex conversion tables are used to arrive at the final or standard number.

The Richter scale actually has no fixed lower or upper limit. It does not rate the size of an earthquake on a scale of 0 through 10. Small earthquakes are measured at figures around zero and some are recorded as minus numbers of magnitude. An earthquake of magnitude 1 normally can be detected only by a seismograph. The weakest disturbances noticed by people are magnitude 2 earthquakes. Any earthquake with a Richter value of 6 or more is commonly considered to be a major disturbance. The Alaskan earthquake of March 27, 1964, was described as having a magnitude of 8.5 on the Richter scale.

According to the Mercalli scale, an earthquake may vary in intensity from Degree I to Degree XII. Degree I earthquakes are not felt—except by a few people under especially favorable conditions. A Degree II earthquake causes delicately suspended objects to swing. An earthquake that cracks walls and produces the sensation that a heavy truck has just struck a building is rated as Degree IV. A Degree IX rating on the Mercalli scale is given when the earthquake shifts buildings off foundations and conspicuously cracks them. An earthquake that leaves few, if any, masonry structures standing, destroys bridges, and produces broad fissures in the ground is given a rating of XI on the Mercalli scale. A rating of XII is reserved for those quakes that produce visible and large-scale waves in ground surfaces. The Alaskan quake of 1964 was rated as X on the Mercalli scale. Landslides, rockfalls, and slumps in river banks were part of the destruction caused by the Alaskan quake.

The effect as measured by the Mercalli scale depends to a large extent on local surface and subsurface conditions. For example, an area underlain by unstable ground such as sand or clay is likely to experience more noticeable surface effects than an area that is equally distant from the epicenter but underlain instead by granite or firm ground. Thus, an earthquake of large magnitude on the Richter scale does not necessarily cause intense surface effects in a particular area; in such a circumstance, the earthquake is given a very low rating on the Mercalli scale.

PREDICTING EARTHQUAKES

A devastating earthquake (39°02′N, 70°22′E) struck the Garm region (Fig. 6–17) of Siberia in 1949. The earthquake triggered an avalanche that buried the village of Khait, which was about 30 miles (50 km) northeast of Garm. More than 12,000 people were killed. The Russians were shocked and stunned by the enormity of the disaster. The Russian government organized a scientific expedition and sent it into the quake-prone area to discover any early warning signals that might be occurring. By 1971, they had, indeed, learned how to recognize the signs of an impending earthquake.

The Russians found, through a study of seismograms, that the difference in the arrival times of P and S waves began to decrease markedly for days, weeks, and even months before an earthquake occurred. The lead time mysteriously returned to normal, however, shortly before the earthquake struck. The study also indicated that the longer the period of abnormal

FIGURE 6–17 The enormity of the earthquake that devastated the Garm region of Siberia in 1949 spurred the Russian government to search for early signals which might be used to alert the population to an impending disaster.

wave velocity before the shock, the larger the magnitude of the quake was likely to be.

The Russian hypothesis was confirmed through a study of seismograms gathered in upper New York State by scientists at Columbia University's Lamont-Doherty Geological Observatory. Two American scientists, Christopher Scholy and Amos Nur, suggested the following explanations for the Russian findings: When rock is subjected to great pressure and approaches its breaking point, a myriad of tiny microscopic cracks develop. As the cracks open in the rock, the P waves slow down because they do not travel as fast through the open spaces as they do through solid rock. Eventually, ground water begins to seep into the openings created in the

rock, and the P wave velocity quickly returns to normal. But the water has another effect, too: it weakens the rock until it suddenly gives way and causes the earthquake.

Crustal uplift and tilting that precede some earthquakes are probably related to the fact that the cracking of the rock increases its volume. The Japanese, for example, observed a 2-inch (5-cm) rise in the ground for about five years before the Niigata earthquake of 1964 (Fig. 6–18).

Variations in the local magnetic field prior to an impending earthquake have also been noted. The strength of the magnetic field between two monitoring stations, for example, may suddenly rise and then gradually subside over a protracted period. In some cases, it takes one week or more

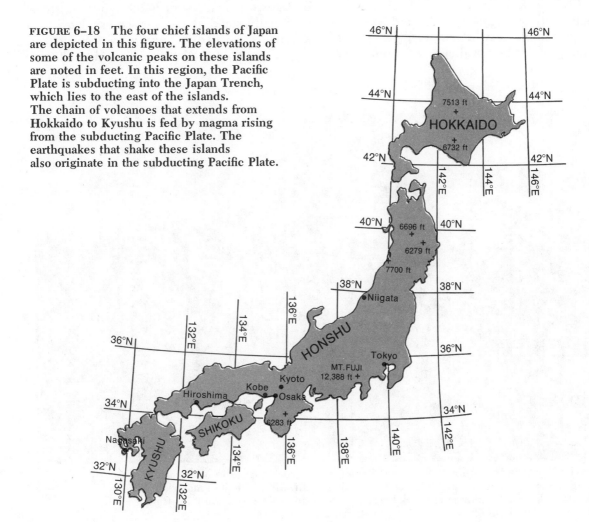

FIGURE 6–18 The four chief islands of Japan are depicted in this figure. The elevations of some of the volcanic peaks on these islands are noted in feet. In this region, the Pacific Plate is subducting into the Japan Trench, which lies to the east of the islands.
The chain of volcanoes that extends from Hokkaido to Kyushu is fed by magma rising from the subducting Pacific Plate. The earthquakes that shake these islands also originate in the subducting Pacific Plate.

for the magnetic field to settle back to normal. At this stage in the studies, it is not certain how the variations in the local magnetic field occur, but the effect may be related to changes in the rock's electrical resistance. As cracks open in the rock, for example, the rock's electrical resistance may rise because air is not a good conductor of electricity.

With this new knowledge, predictions of impending earthquakes have been made with greater frequency and greater accuracy. The Chinese have been particularly active in this field and have had some notable success. For example, they correctly predicted a major earthquake in densely populated Mukden, an industrial area of northeastern China. The quake, which occurred on February 4, 1975, registered 7.4 on the Richter scale and was severe enough to have caused extensive casualties if the population had not been warned in time to take precautionary measures. Prediction and early warning save lives when an earthquake occurs.

VOLCANISM AND LANDSCAPES

Intrusive and extrusive volcanism play an important role in the development of landscapes. Massive intrusions of magma into the Earth's crustal zone, for example, may lift and warp the surface layers to produce hills, valleys, and mountains. Magmatic intrusions, which are slowly exposed by the wearing away of softer overlying strata, eventually stand at the surface as plateaus or highlands. Extrusive eruptive volcanism that pours lava onto the surface through faults and

The Devils Tower is an 865-foot (264-meter) stump-shaped rock cluster in northeastern Wyoming. The tower was formed by the cooling and crystallization of molten materials into an igneous rock upon a sedimentary rock base. This towering rock formation rises from the hills bordering the Bell Fourche River. (National Parks Service)

vents can build plateaus as well as huge piles of volcanic rock. Truly, then, volcanism once set in motion, is a potent landscape-building force.

CHARACTERISTIC PLUTONIC LANDFORMS

Magma, like any molten material under pressure, tends to move to the area of least pressure. The direction of magmatic movement is, thus, dominantly upward. The upward movement is induced by the relief of pressure caused by the removal of overlying rock by erosion, upward buckling, and faulting. Upward movement is also accomplished through assimilation—the magma may melt the overlying rock and assimilate its way upward, aided by gas pressure within itself.

In its upward movement, magma may pry off blocks of overlying rock, it may wedge apart weak strata, or it may be squirted into fractures and openings to form laccoliths, sills, dikes,

phacoliths, or dike swarms. When the magma is solidified at great depths, it forms large intrusives—batholiths, stocks, and lopoliths.

Dikes tend to occur in swarms or groups. Some swarms consist of a series of parallel dikes. In other groupings the dikes radiate outward in all directions from a common center. A parallel dike system is found in the Teenaway River dike swarm of central Washington northwest of Cle Elum (47°12′N, 120°56′W). An example of radial dike swarms is found in the Crazy Mountains of Montana (46°8′N, 110°20′W).

Laccoliths of various shapes and sizes are found in Utah, Montana, South Dakota, and Colorado. The laccoliths around Mount Ellen (38°06′N, 110°47′W) in the Henry Mountains of Utah were fed from magma that rose from the depths through a centrally located stock (Fig. 6–19). Good examples of **phacoliths** can be found in the Adirondack Mountains of New York and the European Alps.

Funnel-shaped igneous bodies that occupy a tectonic basin can be found in many parts of the

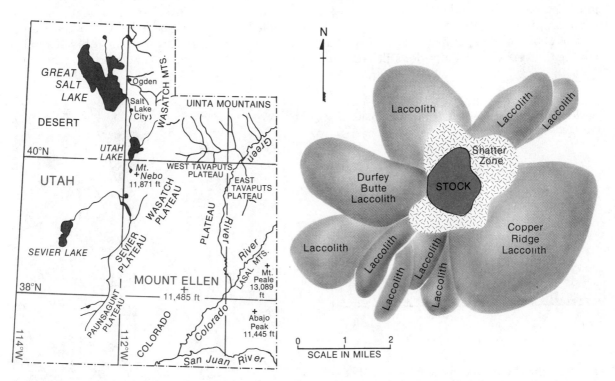

FIGURE 6–19   Mount Ellen in the Henry Mountains of Utah consists of radial laccoliths that were built from magma which worked its way upward through a centrally located stock.

world, including Minnesota. The Duluth **lopolith,** for example, is 150 miles (240 km) across and is estimated to have a maximum thickness of 10 miles (16 km).

THE PALISADES SILL

A line of near-vertical cliffs, called the *Palisades*, extends more than 40 miles (64 km) along the west bank of the Hudson River from Haverstraw, New York, to Bayonne, New Jersey (Fig. 6–20). The cliffs are at their maximum elevation of 530 feet (161.5 m) above sea level at Point Lookout, about 0.5 mile (0.8 km) south of the New Jersey–New York state boundary. They gradually decrease in elevation and finally disappear beneath the surface in Bayonne.

The Palisades **sill** is about 100 feet (305 m) thick. It is an igneous rock of intrusive origin that cooled and consolidated well below the Earth's surface. The magma of the sill solidified between sedimentary strata of sandstone and shale. The adjacent sandstones and shales were altered into metamorphic rock along the contact margins. The sandstones were metamorphosed to quartzite, and the shales were baked to a metamorphic rock called *hornfels*.

The action of erosion has exposed less than half the thickness of the Palisades sill. The cross section in Figure 6–20 shows that the dip of the Palisades sill steepens abruptly and cuts across the strata as a dike. The location of the dike is at the eastern edge of the Hackensack Meadowlands. The Granton quarry is the remnant of a sill that rose from the Palisades. The base of the Granton still was separated from the top of the Palisades sill by 350 feet (106 m) of sandstone and shale. The Watchungs are also remnants; these mountains are the remains of more extensive lava flows that covered large surface areas of New Jersey's Triassic landscape.

SIERRA NEVADA BATHOLITH

The rocks of the Sierra Nevada **batholith** form a large part of a nearly continuous belt of plutonic rock that extends from Mexico's Baja Peninsula into California and adjacent parts of Nevada (Fig. 6–21). Of this huge mass of plutonic rock, the Sierra Nevada portion is exposed over a north-south distance of about 400 miles (640 km).

The Sierra batholith was intruded into Paleozoic and Mesozoic sedimentary rocks. The collision of the North America Plate and the Pacific Plate in the area to the west of the Sierras caused faulting, folding, and the relief of surface pressure. The orogeny that occurred at the close

FIGURE 6–20 The Palisades are a line of near vertical cliffs that extend along the west bank of the Hudson River.

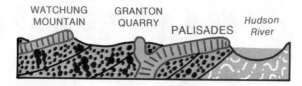

WATCHUNG MOUNTAIN

GRANTON QUARRY

PALISADES

Hudson River

This generalized cross-section represents a slice—along a line that stretches from the Hudson River westward—which is positioned north of Newark.

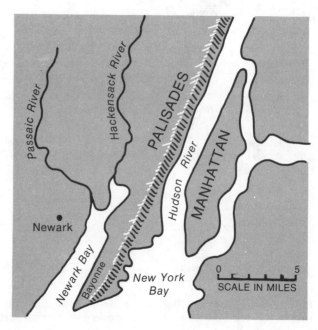

FIGURE 6–21 A nearly continuous chain of plutonic rock (color areas) extends from Mexico's Baja Peninsula into California.

of the Jurassic period of the Mesozoic era caused the principal folds in the rock to the west of the batholith. The main plutons of the Sierra Nevada batholith were intruded immediately before, during, and after the Jurassic disturbance.

The major plutons in the western part of the Sierra Nevada batholith are older than those along the crest of the range. In the Yosemite region, for example, the plutons appear successively younger toward the east. The pattern of intrusion is far from simple, however. There were probably five major episodes of intrusive activity, each of which continued for about 15 million years. The first intrusive episode occurred during the Triassic, the second and third occurred during the Jurassic, and two occurred during the Cretaceous.

ERUPTIVE VOLCANISM

Any opening in the crust of the Earth that allows molten rock-forming material to reach the surface is properly called a *volcano*. The deposits of lava surrounding the vent are also considered to be part of the volcano. Lavas are poured onto the Earth's surface along both divergent and convergent plate boundaries.

Along mid-oceanic ridges, which are divergent boundaries, rock-forming material from the mantle rises and pours into rift valleys as an underwater lava flow. The lava being injected

and accumulating between the plates causes the plates to diverge. The newly emplaced lava builds new oceanic floor.

Volcanism is also a prominent feature of subduction zones that form at convergent boundaries. Volcanoes, for example, are commonly found parallel to submarine trenches on the overriding plate. The exact source of the lava that pours from these volcanoes is not definitely known, but most of the evidence seems to indicate that the source is from the crust of the subducting plate. In other words, as the plunging plate reaches deeper and hotter portions of the mantle, some of its material melts and rises as a plume to the surface, where it breaks through as a volcano.

There is an active zone of subduction along the Java Trench (Fig. 6–22A). The Indo-Australian Plate is plunging under the China Plate in this area. Figure 6–22A shows that the Indonesian island chain is sitting on the China Plate. Krakatoa (6°7'S, 105°24'E) is a small volcanic island in the Sunda Strait on the China Plate (Fig. 6–22B); it is one of more than 100 active and recently active volcanoes that dot the Indonesian islands. The 1883 eruption of Krakatoa, which discharged nearly 5 cubic miles (20 km³) of rock fragments into the air, was fueled by a melt of crustal material from the subducting Indo-Australian Plate.

COMPOSITION OF MAGMA

The character of volcanic eruptions is governed primarily by the composition of the magma involved. The quantity of silicon dioxide (SiO_2) and water (H_2O) in the magma affects its viscosity, that is, its ability to flow. High-silica magma does not flow well and is said to be very viscous. In other words, the viscosity of magma tends to increase with increasing silica-content.

Basaltic magmas are generally erupted as very fluid lavas that spread easily over considerable distances. Andesitic magma is more siliceous and hence more viscous than that of basalt. Andesitic magma is often erupted with explosive violence. Thus, in addition to the lava, fragmental volcanic debris, loosely called *volcanic ash*, may also be ejected. Rhyolitic magma is the most siliceous of the three. It is so viscous that it is

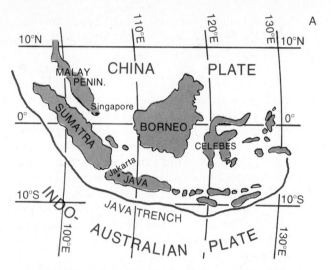

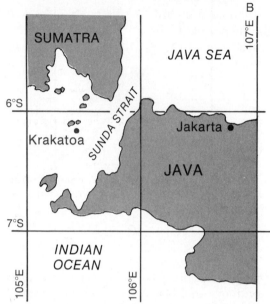

FIGURE 6–22 The Indo-Australian Plate is plunging under the China Plate along the Java Trench. The eruptions of Krakatoa and other volcanoes on the China Plate are fueled by lava that rises from the subducting plate.

incapable of forming a flow and the lava tends to consolidate in or near the vent.

FISSURE FLOW
Iceland—an island in the North Atlantic centered at latitude 65°N, longitude 18°W—is really a section of the Mid-Atlantic Ridge that stands above sea level. The processes of fissure eruption and "sea-floor" spreading are at work on the surface of this island. The eastern part of Iceland is moving with the Eurasian Plate; the western part is moving with the North American Plate.

Iceland is growing and spreading through repeated lava flows from long narrow fissures. The rate of movement is about 0.25 inch (0.64 cm) per year. The rather fluid lava with a low-silica content flows away from the fissures, floods rather extensive areas, and builds lava plains. In 1783 on Iceland, for example, the Laki fissure—a crack almost 20 miles (32 km) long—opened and 3 cubic miles (12 km³) of basaltic lava flowed onto the adjacent area.

In April 1963, in the Atlantic south of Iceland, lava poured onto the ocean floor from a fissure along the Mid-Atlantic Ridge. By November 1963, the accumulation rose above the surface of the sea to form a new island, which was named

Surtsey. Surtsey is centered at latitude 63°16′N, longitude 20°32′W. The lava flowed intermittently for more than 15 months. The accumulation built the island to a height of 558 feet (170 m) with an area of about 0.8 square mile (2 km²). Extrusive activity continued in the area on an almost daily basis for more than 4 years.

CENTRAL-VENT ERUPTIONS
The Mediterranean Sea lies over the boundary of the colliding African and Eurasian plates. Mount Etna (37°46′N, 15°E) at an elevation of 11,122 feet (3390 m) on Sicily, the island of Stromboli (38°48′N, 15°13′E) with its peak at an elevation of 3032 feet (924 m), and Vesuvius (40°49′N, 14°26′E) at an elevation of 4190 feet (1277 m) above sea level, are volcanoes on the Eurasian Plate. The lava that pours from each of these volcanoes issues from a central vent, or pipe. The source of the extrusive igneous material, in each case, is thought to be from the African Plate, which is plunging under the Eurasian Plate.

Some volcanoes erupt only once; others erupt repeatedly. Huge volcanic mountains are generally built through numerous eruptions. In fact, eruptive episodes—especially those that

A chain of 14 islands follows the southwest trend of the Mid-Atlantic Ridge off the coast of Iceland. The volcanic summit of Helgafell dominates Heimaey which, at 5 miles (8 kilometers) long, is the largest of these islands. On January 23, 1973, a fissure began to open on the slope of Helgafell above the town; the eruption that ensued is documented in these photos. Volcanic ash and fiery lava bombs fell on the town. An attempt was made to install corrugated metal sheets in windows facing the explosion to prevent lava bombs from entering homes and causing fires. The accumulation of ash exceeded 16 feet (5 meters). (Icelandic Consulate)

build giant volcanic mountains like Etna, Vesuvius, and Stromboli—may be spread over hundreds or thousands of years.

Volcanic eruptions differ widely in character. Some are violently explosive; others are much less so and might even be characterized as "gentle." In fact, eruptions of the explosive and gentle variety may even take place at the same volcano. Generally, however, each volcano has its own eruptive "style." The eruptive style depends largely on the viscosity of the lava and the relative amounts of gas and liquid rock that reach the surface. A thinly fluid basaltic lava, for example, allows moderate amounts of gas to bubble out with little more than minor spattering. In a very viscous rhyolitic lava, on the other hand, it is difficult for the gas to work upward and break through the surface of the liquid. Thus, the gas accumulates until its pressure is high enough to burst free. A major explosion or a series of explosions can occur when the lava is very viscous and there is a large amount of gas at high pressure.

VOLCANIC CONES

The hill or mountain built by the material issuing from the central vent of a volcano commonly takes the shape of a cone that has the small end cut off. Usually there is a crater at the summit. Hills and mountains with this type of configuration are referred to as *volcanic cones*.

The profile of a volcanic cone depends on the eruptive style of the volcano. Very little explosive action is involved in the building of the Hawaiian volcanoes, for example. The Hawaiian flows are so fluid that they spread out to great distances from their vents and build broadly rounded, dome-shaped mountains, known as shield volcanoes (Fig. 6–23A). Mauna Loa is a perfect example of a broad shield-shaped volcano. From its base on the Pacific floor, this huge

volcanic mountain rises well over 31,000 feet (9450 m)—13,680 feet (4170 m) of its gentle 6°-to-12° slopes are above sea level. The diameter of Mauna Loa at its base on the sea floor is about 60 miles (96 km).

Eruptions of viscous lava belch forth a lot of volcanic ejecta* with considerable explosive activity. The shape of the hill built by the fragments falling around the vent depends to an extent on the size of the ejecta. A cone formed of unconsolidated volcanic ash, referred to as an *ash cone*, usually has a very broad saucer-shaped crater (Fig. 6–23B). The explosions that produce ash cones occur mostly at ground level and many of the fragments are thrown away from the vent at a low angle, which accounts for the broad cone and the wide shallow crater. Volcanic ash tends to become cemented together to form a rock called *tuff*. Diamond Head, on the island of Oahu, is a tuff cone that was consolidated from volcanic ash.

When the fragments making up a volcanic cone are solid but are not cemented together, the cone is called a *cinder cone*. Cinder cones develop from explosions that occur within the feeding conduit. The vent acts like a gun barrel, which causes the ejecta to be shot upward. The cinders fall back close to the vent and build a narrow cone with a relatively small crater (Fig. 6–23C).

Eruptions of volcanic ejecta that are accompanied by lava flows build composite volcanoes. These are the most beautifully shaped of the volcanic mountains. They consist of interbedded cinder and lava, which produce very symmetrical steep-sided cones (Fig. 6–23D). The volcanic

*See page 69 for a definition and discussion of volcanic ejecta.

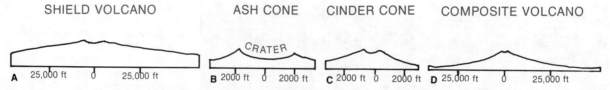

FIGURE 6–23 The hill or mountain built by the material issuing from the central vent of a volcano takes the shape of a cone. Four classic profiles are depicted in this illustration. The profile assumed by a volcanic cone depends on its eruptive style.

Mount Fuji, at 12,388 feet (3776 meters), is the highest mountain in Japan. This dormant volcano is on the island of Honshu, about 60 miles (95 kilometers) west of Tokyo. Mount Fuji has long, symmetrical slopes. (Japanese Information Service)

ejecta falling back around the central vent build a regular conical shape; then, the ribs of interbedded lava strengthen and preserve the graceful shape. Mount Shasta in California, Mount Fuji in Japan, and Vesuvius and Stromboli in Italy are all composite volcanoes.

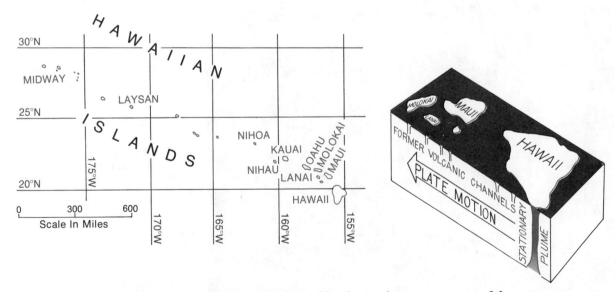

FIGURE 6-24 The Hawaiian Islands have been formed by the northwest movement of the Pacific Plate over a hot spot in the Earth's mantle. The plate rides over the plume, which punches up volcanoes in a sequence. The older, extinct volcanoes, then, are carried on to the northwest.

VOLCANIC CHAINS

Although most of the Earth's output of lava is associated with either diverging or converging plate boundaries, there are many active volcanic fields far from any plate boundary. The Hawaiian islands, for example, are an intraplate volcanic chain that has been built by a stationary plume of superhot, high-pressure material from a magma source below the lithosphere (Fig. 6–24). To understand how this chain is being built, you must realize that the main part of the Pacific Plate has been migrating northwestward for at least 100 million years. A convection plume that originates near the core-mantle boundary breaks through the Pacific Plate periodically as the sea bottom slides over the hot spot.

Since the Pacific floor is migrating northwestward, it is natural to find the oldest volcanic islands at the northwestern end of the Hawaiian chain. The rate of northwestward migration of the Pacific Plate cannot be fixed with any degree of certainty; it seems to vary between 0.8 and 2.4 inches (20 to 60 mm) per year. Midway Island, at the far end of the chain, was built by a volcano 27 million years ago when the part of the plate the island currently occupies was over the hot spot. The eruptions of the stationary plume that built the island of Oahu started about 3.5 million years ago and continued for 1.5 million years. At the southeastern end of the Hawaiian chain, we find the big island of Hawaii, which is the only island with currently active volcanoes. In other words, the present-day eruptions occurring on the big island of Hawaii are fueled by the stationary hot spot. The oldest lava on the big island of Hawaii was poured onto the surface about 800,000 years ago and is found in the Kahala volcano.

In the early 1970s, W. Jason Morgan of Princeton University suggested that there are some 20 plumes rising beneath key volcanic areas of the world. By 1979, Kevin Burke of the State University of New York and Tuzo Wilson of

Diamond Head in the background and Punchbowl in the foreground of this photo are two well-known landmarks on the island of Oahu in the Hawaiian chain. Both are extinct volcanoes. The cones are made of tuff, which makes these craters very broad and saucer-shaped. Diamond Head is 760 feet (230 meters) high. Punchbowl Crater is 500 feet (150 meters) high. (Hawaii Tourist Bureau)

the Ontario Science Center had mapped 120 world hot spots. S. Thomas Crough of Princeton, however, believes the number of hot spots is closer to 50. The disagreement as to number seems to stem from how neighboring volcanoes are classified—do they originate from the same hot spot?

Burke and Wilson suggest hot spots vary across a broad range of diameters. Morgan originally suggested that each plume is perhaps 60 miles (100 km) in diameter. In any event, all seem to agree that a **plume** constitutes a kind of pipeline that carries deep mantle material up-

ward toward the surface. At the top of the mantle, these plumes spread out; it is this spreading motion that propels the crustal plates. The eruptive island of Tristan da Cunha (37°15'S, 12°30'W) is believed to be over one of these plumes that is maintaining the Mid-Atlantic Ridge and is helping to push the South America and Africa plates apart. The two chains of submerged islands that extend from Tristan da Cunha are believed to have been formed from this same plume. The Rio Grand Rise is the chain that extends to the northwest, and the chain that reaches to the northeast is the Walvis Ridge.

Although Mount St. Helens—shown here in eruption on May 18, 1980—is within the North America Plate, its output of lava rises from the plunging Juan de Fuca Plate. (Photograph by Austin Post, courtesy of the U.S. Geological Survey.)

THE LANGUAGE OF PHYSICAL GEOGRAPHY

A physical geographer uses a technical vocabulary to make explanations. Review your understanding of the vocabulary used to develop the concepts in this chapter. Among the important words and terms used are:

anticline	fault plane	lithosphere	seismogram
asymmetrical folds	fault scarp	Mercalli scale	shallow-focus earthquake
axial plane	focus	miogeosyncline	shear wave
batholiths	fold system	normal fault	sill
compression wave	footwall block	overturned folds	strike-slip fault
deep-focus earthquake	geanticline	phacoliths	symmetrical folds
dikes	geosyncline	plume	synclines
dip	graben	plutons	transcurrent faults
earthquake	gravity fault	primary wave	transform faults
epicenter	hanging-wall block	reverse fault	transverse waves
eugeosyncline	horst	Richter scale	thrust faults
fault	isoclinal folds	secondary wave	volcanism
fault block	laccolith	seismic wave	
fault line	lopolith	seismograph	

SELF-TESTING QUIZ

1. Explain why the development of a fold system is closely associated with the concept of global tectonics.
2. Identify and describe the various portions of a fold.
3. Trace the development of the Appalachians through time.
4. Compare folding and faulting. How do they differ?
5. What special relief features develop from combinations of more than two fault blocks?
6. How are faults classified?
7. Classify the San Andreas Fault.
8. What do present-day measurements along the San Andreas Fault indicate?
9. Explain what is meant by a deep-focus earthquake.
10. Compare the different kinds of seismic waves that accompany the release of energy at an earthquake focus.
11. Describe the different methods of scaling and comparing earthquakes.
12. List the early warning signals that herald the coming of an earthquake.
13. Describe some of the landforms produced by plutons.
14. How does the composition of magma control the character of a volcanic eruption?
15. Trace the development of the Hawaiian islands as an intraplate phenomenon.
16. Why is the island of Surtsey considered to be a plate-edge phenomenon?

IN REVIEW
TECTONIC FORCES AND LANDSCAPES

I. DEFORMATION THROUGH FOLDING

  A. The Folding Process
  B. Fold Terminology
  C. Folds in the Field
  D. The Folded Appalachians
    1. Rift and Sediment Stage
    2. Cordilleran and Compression Stage
    3. Continental Collision Stage
    4. Present Rift Cycle

II. FAULTING OF THE CRUST

  A. Fault Terminology
  B. Grabens and Horsts
  C. Fault Classification
    1. Normal Faults
    2. Reverse Faults
    3. Strike-Slip Faults

  D. Fault Combinations
  E. Earthquakes
    1. Earthquake Focus
    2. Earthquake Waves
    3. Recording Seismic Waves
    4. Scaling Earthquakes
    5. Predicting Earthquakes

III. VOLCANISM

  A. Characteristic Plutonic Landforms
    1. The Palisades Sill
    2. Sierra Nevada Batholith
  B. Eruptive Volcanism
    1. Composition of Magma
    2. Fissure Flow
    3. Central-Vent Eruptions
    4. Volcanic Cones
    5. Volcanic Chains

 # Landforms shaped by erosion and deposition

The processes that work to wear down, destroy, and at times to reveal the landforms produced by tectonic forces are collectively termed **gradational forces.** The gradational forces—running water, moving ice, wind, and waves—wear down high places on the Earth's surface and fill in low regions. Through the processes of grading, new and distinctive landforms are produced.

There is one additional process, known as **mass wasting,** that I am including in this discussion of gradation. The energizing force in mass wasting is gravity. The distance through which rock material is moved by mass wasting is lim-

ited, but the landscapes produced by this process are distinctive and deserve mention at this point in the discussion.

Although **weathering** is not one of the gradational forces, it acts as a preliminary to gradation. Weathering involves the decomposition and disintegration of rocks, with little or no movement of the particles produced when the rock structure breaks down. Thus weathering, which aids and abets gradation but does not grade, is considered an essential preliminary element in an informed discussion of gradational forces and is used as the basis for the opening remarks in this chapter.

THE WEATHERING OF ROCK

The processes of weathering are constant and continuous everywhere on the Earth's surface and even, at times, below the surface. It is through these processes that massive rock formations are gradually transformed into **regolith**— the noncemented rock fragments and mineral

grains that overlie bedrock. The initial phase of the conversion of a rock formation into regolith begins in a process of cracking that starts long before the rock is exposed at the Earth's surface. As soon as the rock has a surface exposure, however, the conversion continues by physical

The dry, rocky Atlas Mountains spread across most of Morocco, which lies only 9 miles (14.5 kilometers) from Spain across the Strait of Gibraltar in North Africa. This stream is flowing through the High Atlas Mountains outside Marrakech (31°49′N, 8°W). The rush of its water is sufficient, at times, to move rather large cobbles and boulders. (Moroccan Government)

weathering, or disintegration, and by chemical weathering, or decomposition, of its minerals.

Generally, most massive rock formations develop joints or cracks long before the overlying material is fully removed and the formation assumes a position at the Earth's surface. Regional stresses brought on by plate movement or even the loading effect of a glacier can cause jointing that cracks a formation into long fragments. Faulting is another phenomenon that fragments massive rock formations into huge blocks. When folding, faulting, or volcanism places a sedimentary rock formation under stress, the rock tends to break along its bedding plane, which is a zone of weakness. Thick intrusive formations, which cool slowly below the surface, tend to fragment into blocks that develop perpendicular to the horizontal. Examples can be found in the columnar jointing of the Devils Postpile in California and the Palisades sill of New Jersey.

Sheeting is the process by which great slabs of igneous rock crack away in onionlike fashion from the surface of a parent formation. The separated sheets may be flat or curved, paper thin or meters thick. The cause of the sheeting is related to the fact that an intrusive igneous mass—batholith or laccolith, for example—has its upper surface under pressure from the overlying strata. When the overlying formations are eroded away, the pressure on the upper surface of the intrusive rock is relieved. The minerals, which were initially compressed, are free to expand in a vertical direction. As the minerals expand, slabs detach themselves and the resulting phenomenon is referred to as *sheeting*.

MECHANICAL WEATHERING

The words *mechanical* and *physical* are used interchangeably in many discussions of weathering. In the past, the mechanical disintegration of a rock formation into smaller particles was explained as alternate expansion and contraction that resulted from either the heating and cooling of the rock itself or the freezing and thawing of water that infiltrated the pore spaces of the formation. In the first instance, it was assumed that the input of energy from the Sun is sufficient to produce the mechanical expansion and contraction necessary for physical weathering. However, experiments have cast doubt on this idea and there is really no evidence to support the concept.

Water is the prime agent responsible for **mechanical weathering.** Water penetrates into the cracks and pores of a rock's surface, collects, and when the temperature drops sufficiently, it freezes. When water freezes, it expands. The tremendous pressure exerted by water in the process of freezing pries open and widens cracks and pores and leads to the physical fragmentation of the rock structure. It is, for example, responsible for the removal of quartz grains—commonly called *sand grains*—from granite.

The water flowing through the Plitvice Lakes region (44°53′N, 15°38′E), which is northeast of Gospic, Yugoslavia, is quite acidic. The chemical weathering that results produces the caves and well-known karst topography of Yugoslavia. (Yugoslav Government)

CHEMICAL WEATHERING

Water is also the chief agent of **chemical weathering.** Acidic water is an especially powerful solvent. The role of acidic water—specifically carbonic acid, H_2CO_3—in the formation of caverns and karst topography was described on pages 94–96. Carbonic acid, which is formed by the action of carbon dioxide with water, is a very common and active chemical weathering agent.

Weak solutions of nitric acid (HNO_3) and sulfuric acid (H_2SO_4) are also important chemical weathering agents. Both nitric and sulfuric acids can have an atmospheric or biologic origin. Sulfuric acid, for example, can form by the reaction of rainwater with sulfur dioxide (SO_2), which may be released into the atmosphere by the burning of coals and oils with a high sulfur content. Pollution produced through the use of fossil fuels can and does increase the chemical weathering of rock.

Feldspar, which is an important mineral in most igneous rocks, reacts readily with acidic water to form clay and silica. Acidic water also reacts with other mineral groups in igneous rocks. The ferromagnesians, for example, react with acidic water to produce clay and silica. And iron, which is abundant in the ferromagnesians, combines with oxygen to form iron oxides.

Exfoliation, a common mode of rock disruption, is a process that resembles sheeting. The resemblance, however, is superficial because the shells of rock produced by exfoliation are, in fact, separated mainly by mineral decomposition during chemical weathering. The swelling and separation of a layer of rock—in the process called *exfoliation*—is due to the formation of clay through the chemical alteration of feldspar. The creation of clay minerals in the rock is accompanied by a volume increase, which sets up tensional stresses that cause shells of rock to separate from the main formation.

MASS TRANSFER OF ROCK AND SEDIMENT

Rockfalls, landslides, and mudflows are spectacular and destructive movements that take place on sloped landscapes. These movements are referred to as **surficial,** or surface movements. Gravity is the driving force that propels the material down the slope; but, some event—an earthquake, heavy rains, or the undercutting of a stable slope—plays a role in triggering the action and upsetting the equilibrium that exists.

ROCKFALLS

The processes of mechanical weathering rupture fragments of rock from the surface of cliffs. The fall of rock fragments from a cliff face is referred to simply as a **rockfall.** If the face of the cliff is nearly vertical, the rock fragments can accelerate substantially as they fall under the influence of gravity.

Funnellike ravines develop in a cliff face undergoing mechanical weathering. The loose rock fragments that accumulate at the base of a cliff are called **talus.** The slope established by the accumulation of rock fragments is known as a **talus slope.** A talus slope establishes itself and remains fairly stable when its angle is about 35°. The somewhat conical shape that the pile of fragments establishes at the base of a cliff is referred to as a **talus cone.**

LANDSLIDES

The event referred to as a **landslide** involves the sudden mass movement of bedrock. There are two fundamental factors that must be studied in determining whether a landslide is likely to be triggered in a particular area. The first factor is related to the topography of the region; the second involves the physical characteristics of the rock formation.

THE TOPOGRAPHY

Three types of landforms are especially conducive to the development of landslides: a valley with exceptionally steep walls, a vertical fault scarp, and a headland along a seacoast. Glaciers,

In Morocco—in the area of Al Hoceima (35°15′N, 3°55′W)—the Rif Range faces the Mediterranean and runs parallel to the coast. The rock fragments at the base of this cliff, below the two men at the top, have been ruptured from the face of the cliff by the processes of mechanical weathering. (Moroccan Government)

large-scale excavations, and highway construction often leave in their wake oversteep slopes and precipitously steep valley walls. Waves that cut into headlands also tend to create steep cliffs that are landslide prone.

THE GEOLOGIC STRUCTURE

A massive rock formation that dips toward a valley and is underlain by a layer of weak shale is an unstable landslide-prone condition. If the rock formations become saturated with water, the shale becomes slick and slippery. The rock formation above it may slide down the lubricated incline into the valley below. The rock strata above any valley should be studied for layered combinations that can lead to trouble.

Another situation that is landslide prone involves shattered rock on a steep slope. When heavy rains fall on a mass of shattered rock, the rock can absorb great quantities of water, which may make it pasty and unstable. The shattered-rock mass is, in fact, lubricated by the water, and its internal friction is reduced to a minimum. The lubricated mass may break away suddenly from its position on the slope and slide into the valley if a jolt or shock jars it sufficiently to upset the equilibrium that holds it in place.

A tremendously destructive landslide occurred on October 9, 1963, at the Vaiont Reservoir in Italy (46°16′N, 12°21′E). As Figure 7–1 indicates, the reservoir is located north of Venice in the Venetian Alps. The rock formation above the dam is a classic landslide structure in which jointed and fractured limestone filled with solution cavities and clay dip steeply toward the reservoir. The mass of material above the dam had been creeping downhill at the rate of 0.4 inch (1 cm) per week for a long time before the tragedy of October 9. This fact was a signal that dangerous slope conditions existed; if the formation became saturated with water, the downhill movement would accelerate. In fact, by September 1963, the downhill creep rate had increased to 0.4 inch (1 cm) per day. There was a tenfold increase in the creep rate during the month of September, and by the beginning of October the unstable formation was moving at 4 inches (10 cm) per day. On October 9, the downhill movement culminated in a landslide of awesome proportions that sent 314 million cubic yards (240 million m³) of rock and soil into the reservoir. The water that spilled over the dam flooded the valley below and killed more than 2000 people.

SOIL AND EARTH FLOWS

A very slow downhill movement of the soil cover is called **creep**. Such movements generally go undetected, and there may not seem to be any immediate danger to people living on the downhill side of the creep. But a creep—no matter

This cliff face on the eastern slope of the Canadian Rockies, near Banff in Alberta, is undergoing mechanical weathering. Funnellike ravines have developed in the cliff face. The talus cones spread from left to right at the base of the cliff. (National Film Board of Canada)

how slow it is—is indicative of potentially dangerous slope conditions that require positive action to prevent a future disaster from occurring.

In June 1974, for example, a water pipeline in Manti, Utah, was cracked open by the pressure of moving soil. For more than two weeks the town was without its regular water supply. The fracturing of the pipeline, however, was only a symptom of a larger problem that exists in the area (Fig. 7–2). For almost two years, from 1974 to 1976, the residents of Manti watched a 2-mile- (3-km-) long, 0.5-mile- (0.8-km-) wide stretch of aspen and fir-covered mountainside slowly and surely move downhill toward a creek that runs into the town.

Viewed from a distance, the area in which the movement was occurring looked like a beautiful, forested canyon wall with a creek at its base. On closer examination visitors found 15-foot- (5-m-) wide cracks in the soft soil, 3-foot- (1-m-) high escarpments formed as portions of the soil shifted position, and trees snapped in half by the pressure of the moving soil. The same type of movement, in fact, plagues large sections of Utah where hillsides are composed of rather loose sandstone, limestone, and shale. Snowy winters in the region make these hillsides wet and unstable.

MUDFLOWS

Mud is a wet mixture of soft, unconsolidated sediments such as silt and clay. The consistency of mud varies from that of a semifluid to that of a soft plastic sediment. The basic requirement for silt and clay to become a slimy, sticky fluid-mixture, called *mud,* is that the proportion of water to mineral matter must be large. A **mudflow** develops when a huge mass of unconsolidated sediment on a slope is rapidly satu-

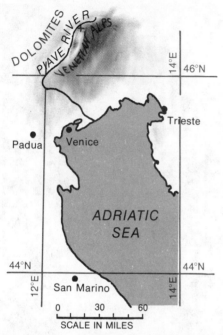
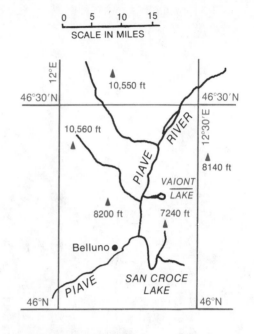

FIGURE 7–1 The Piave River in northern Italy flows between the Dolomites and the Venetian Alps. The Vaiont Reservoir (X on the map at left) is about 60 miles (100 kilometers) north of Venice. The peaks (elevations on the map at the right) of the Dolomites and the Venetian Alps tower above the Piave River. The runoff from the Venetian Alps supplies water to Vaiont Lake.

rated with water during a heavy rainfall. A sudden thaw can also place a large charge of water in unconsolidated sediments that are either beneath or on the downhill side of a snowfield.

As a result of the normal processes of erosion in humid mountainous regions, unconsolidated sediments—clay, silt, and sand, for example—accumulate in the upper regions of valleys. When these upper-valley sediments become saturated with water, they begin to flow and move down the valley. Mudflows also occur in desert regions where unconsolidated sediments, which have been eroded from the high ground, accumulate in the canyons and become saturated as the result of heavy rains falling in the mountains.

Mudflows that move at 6 miles (9.7 km) per hour or faster are a definite hazard to life and property. There are numerous examples of mudflows that have caused widespread havoc and destruction. For example, Choloma, a small town in Honduras, perished under a wall of mud in 1974. Hurricane Fifi, which moved over Honduras in late September, was the source of the heavy rains that caused thousands of tons of mud to move down from the mountains. The initial mudflow formed a dam across the normally tiny Choloma River. The river quickly swelled behind the dam. When the river's rising water finally ruptured the dam, the town was overwhelmed by a huge wall of water and mud that killed more than 2800 people.

SURFACE MODIFICATION BY STREAMS

To the average person, the hills, mountains, and other features of the landscape appear to be everlasting. But as soon as they appear, the gradational forces set to work to destroy them. Weathering prepares the way by breaking and softening the rocks, and gravity transfers the fragments

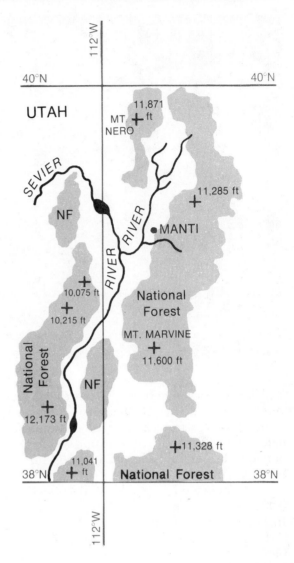

FIGURE 7–2 Manti, located in a river valley, is surrounded by forested mountain peaks. The hillsides are composed of sandstones, limestones, and shales.

downhill. Rainwater and streams running over the surface also carry loose material downslope. In addition, the running water of streams cuts into and etches new profiles in the surface of the landscape. In fact, at one time or another, all landscapes are modified by running water.

STREAM EROSION AND DEPOSITION

Drawn by the force of gravity, the water within each stream runs down the steepest available slope toward the sea. The speed of flow is governed largely by the stream gradient.* The

*Refer to pages 112–118 for the initial discussion of streams and stream gradient.

steeper the gradient, the more rapid the flow.

As the water flows in its channel, it tends to move any loose rock debris along with it. Transport is accomplished by **traction,** that is, the rolling, sliding, and dragging of large boulders and particles along the channel bottom; by **saltation,** carrying the material in small jumps; by **suspension** of materials that will eventually settle out; and by **solution,** the process through which material is actually dissolved in the water.

EROSION VELOCITY

The size of the fragment a stream can erode depends largely on the speed of its flow. Material

The material under and to the left of this roadway—north of Cloverdale in Sonoma County, California—gave way suddenly and moved downhill. (U.S. Geological Survey)

is picked up, that is, entrained, when the **erosion velocity** for particles of that size is reached. Generally, for a fine sand,* with a diameter of 0.004 inch (0.1 mm), the erosion velocity is between 8 and 20 inches (20 to 50 cm) per second.

Both finer and coarser materials than the 0.004-inch (0.1-mm) sand are less easily eroded. In other words, the erosion velocities for clay and silt, which are finer than sand, and for pebbles, which are coarser than sand, are greater than 8 to 20 inches (20 to 50 cm) per second. Clay and silt require higher erosion velocities because of their cohesiveness and the relatively smooth and frictionless surfaces they present to the moving water. Silt, with a diameter of 0.0025 inch (0.06 mm), requires an erosion velocity between 25 and 34 inches (65 and 85 cm) per second. Pebbles, on the other hand, require a higher erosion velocity because of their greater mass. It takes, for example, a velocity of about 100 inches (250 cm) per second to erode and entrain a pebble with a diameter of 0.4 inch (10 mm).

*See page 174 for the size ranges of clastic sediments.

TRANSPORT VELOCITY

Once fine particles are picked up, or entrained, they can be transported at very low velocities. Silt, which as mentioned above requires an erosion velocity of more than 25 inches (65 cm) per second, can be transported at stream velocities as low as 0.08 inch (0.2 cm) per second. Coarser particles, of course, require higher stream velocities.

Fine sand, with erosion velocities of more than 8 inches (20 cm) per second, will remain in suspension and be transported at velocities just above 0.4 inch (1 cm) per second. Coarser sand, with a diameter around 0.04 inch (1 mm) and a minimum erosion velocity above 25 inches (65 cm) per second, can be transported at stream velocities just above 4 inches (10 cm) per second.

DEPOSITION VELOCITY

The velocity at which entrained material is dropped from moving water and deposited is sometimes called the **sedimentation velocity**. In fact, the **deposition velocities** for various sediments have been identified in the prior paragraphs that detail the minimum velocity required for transport.

The deposition velocities for the sediments described above are as follows: silt, less than 0.04 inch (0.1 cm) per second; fine sand, under 0.4 inch (1 cm) per second; and coarser sand, less than 4 inches (10 cm) per second. Pebbles with a diameter of 0.4 inch (10 mm) and an erosion velocity of about 100 inches (250 cm) per second will be deposited when the stream velocity drops just below 40 inches (100 cm) per second.

STREAM LOAD

All the material moved by a stream is referred to as its **load.** Some is entrained and carried in suspension as a **suspended load.** The material transported by saltation and traction is known as the **bed load;** the portion held in solution is designated as the **dissolved load.**

The amount of the load that can be transported is directly dependent on the volume of the stream but varies as the square of any change in the stream's velocity. Thus, when the velocity of a stream doubles, the amount of the load it can transport does not double—it increases as the square of 2, which is 4. If the velocity of water flow in a stream triples, its load-carrying capacity increases 9 times.

The moving water within a stream channel can move much larger rock fragments as bed load than it can as suspended load. As the velocity of the stream increases, the size of bed-load fragments that can be moved increases far more than you might expect. Doubling the stream velocity increases the size of the fragments that can be moved by a factor approaching the sixth power of the change in velocity—the sixth power of 2 is 64. That's quite an increase!

The load of a stream consists largely of rock material that has been loosened and transformed into regolith by weathering. The removal of this weathered material is characterized as the first stage in stream erosion. The second stage is initiated as the load rolls, slides, and bumps along the stream bed. During this second-stage process, the load erodes the stream bed by abrasion and impact. As the fragments wear away the rock of the stream bed, they, in turn, are rounded, reduced in size, and generally worn.

Through the processes of this second stage, the rock formation of the stream bed is undermined and loosened to become part of the stream's load and transported away. Thus, the stream gradually cuts its way downward. At first, it cuts a gully, but eventually it scours out a valley.

THE FLUVIAL GEOMORPHIC CYCLE

A landscape that is being sculptured by stream erosion goes through a series of somewhat definite stages. Since the word **geomorphic** pertains to the form of the Earth's surface, we can refer to the stages of stream, or fluvial, erosion as a **geomorphic cycle.** Four successive stages are identified in Figure 7–3 as part of the ideal geomorphic cycle of stream erosion: youth, submature, mature, and old age. The cross section of ridges and valleys (Fig. 7–3) depicts the profiles characteristic of these stages.

YOUTHFUL STAGE

The major action of a geologically young stream is erosion—the removal, transport, and deposition of rock material. A youthful stream is always marked by a steep longitudinal profile, or gradient, because the stream is cutting downward into its own underlying bedrock. Turbulence is especially common in the swiftly flowing water of a youthful stream since impediments have not yet been removed or eroded. Boulders and outcrops of rock produce rapids and a variety of swirls and eddies in the water.

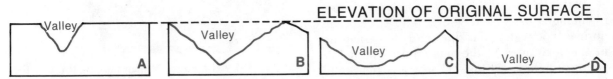

FIGURE 7–3 These four cross sections of ridges and valleys represent successive stages in the ideal geomorphic cycle of stream erosion: A. youth, B. submature, C. mature, and D. old age.

This is Avalanche Creek in Glacier National Park, which lies in northern Montana on the boundary between the United States and Canada. The rock through which this creek is cutting is part of the Rocky Mountain chain. The Continental Divide—the line separating areas drained to opposite sides of the continent—runs north and south through Glacier National Park. Avalanche Creek is on the west side of the Continental Divide. (National Parks Service)

A youthful stream is rather straight and narrow with a V-shaped valley. Broad areas of the original surface (Fig. 7–3A) remain almost unchanged. The Grand Canyon of the Colorado River is an example of a young valley. The great canyons of the Andes, plunging 7000 feet (2130 m), have been cut through masses of volcanic rock. The Andes' canyons rise, almost sheer, for thousands of feet from very narrow valley floors and are good examples of valleys cut by young streams.

Waterfalls and rapids are interesting phenomena associated with young streams. A normal-type **waterfall** is due solely to variations in the resistance of the rocks into which the young stream is cutting. When water descends from a more resistant rock to a less resistant rock on the downstream side, a waterfall is produced. There are many examples of differentially eroded waterfalls in the southern Appalachians and in the Adirondack mountains.

Because the valleys of geologically young rivers are narrow, they are not well suited for the development of railway lines. The Hudson River Gorge, however, is an example of a young river valley that is used by railways. Young rivers, with their steep gradients, do provide favorable conditions for water power development since they are easily dammed and a good head of water develops to drive water turbines to generate electricity. Boulder Dam, near the Grand Canyon, is a perfect example of how a young river is tapped for power production.

There is an abrupt decrease of slope when a young stream flows down through a steep highland area and then suddenly finds itself on a nearly level valley floor or plain (Fig. 7–4). Under these conditions, the stream deposits that part of its load which it finds impossible to transport along the new gradient. The material that the stream drops is deposited in the shape of a fan. All sediments deposited in land environments by streams are referred to as **alluvium**. Thus, a fan developed under these conditions is called an **alluvial fan**. Vast alluvial fans have

FIGURE 7–4 The young stream moving through the highland carries a large load of sediment. When the stream reaches the valley floor, its gradient changes abruptly. The sediments deposited by the stream on the valley floor are identified as an alluvial fan.

The geologically young Hudson River cuts through the Adirondack Mountains of New York. Its narrow valley is used effectively for a railroad right-of-way. (Amtrak)

been built on the east side of the Rocky Mountains. There are also extensive alluvial fans on the north and south sides of the Alps.

THE SUBMATURE STAGE

There is a transition stage in this geomorphic cycle that is intermediate between the youthful and mature stages. The valley (Fig. 7–3B) is still V-shaped but the original surface is completely destroyed and only pointy ridge tops remain. This transition stage is sometimes called *submature*. The ridge tops, however, still mark the approximate level of the original land surface. Most of the Waianae ranges on Oahu are in this stage.

THE MATURE STAGE

The transition from the submature to the mature stage occurs as the valley floor is slowly widened and the profile of the valley changes from a sharp V to a broad V with a flat floor and eventually to a U-shape. Steep mountains still stand and the tops are not far below the original land level (Fig. 7–3C).

As a stream approaches geological maturity, the channel cross section approaches its most efficient shape—a semicircular trough, which offers the least resistance to the flow of water. A mature stream is said to have a profile of equilibrium. Changing conditions, however, can cause a mature stream to change its slope; for example, an increase in its load will trigger deposition, while a decrease in load permits erosion. The

headwaters of a mature stream gradually wear away the land and eventually this process decreases the load the tributaries supply to the mainstream. The mainstream, therefore, keeps reducing its gradient.

Meanders, looplike bends of a stream channel, are a characteristic of a geologically mature river with its load so adjusted that the river is not always depositing but is alternately cutting and filling its valley (Fig. 7–5A). Deposition of sediment takes place on the inside of the curve. Cutting away of the river bank takes place on the outside of the meander curve. During the process of its development, the whole meander migrates downstream.

The **floodplain** is that part of a stream valley that is inundated during a flood. During times of high water, a meandering stream may cut across the neck of a meander spur and shorten its course. When the water recedes to a more normal level, it may not follow the old longer-meandering path but may follow the shorter course. The old, abandoned part of the meander is cut off from the main part of the river and is called an **oxbow lake** (Fig. 7–5B). These oxbow lakes eventually become swampy areas. The lower Mississippi River is a good example of a mature river with well-developed meanders, broad floodplains, and oxbow lakes.

When a river reaches the sea its speed is suddenly checked and it ceases to exist. The abrupt change in its velocity of flow means the river must deposit the sediment it can no longer carry. The body of sediment deposited by a river flowing into the essentially standing water of the sea is called a **delta.** The name is derived from the fact that the plan view of the deposited sediments resembles the Greek letter *delta* (Δ).

A stream that empties into a lake also has its velocity checked suddenly and deposits the sediment it can no longer carry in the form of a delta. Deltas that are built into lakes are much more perfect than deltas built into an open sea where there are tidal currents and strong wave action.

A mature river builds the level of its floodplain by deposition each time it floods and overruns its banks. Most of the deposits, of course, are dropped close to the river and form natural levees. The natural **levees** are broad, low ridges

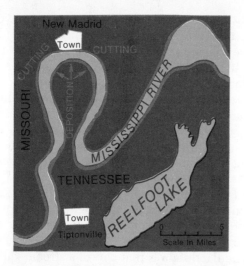

FIGURE 7-5 A: (left) There are many looplike bends, known as *meanders*, in the stream channel on the Mississippi River. The meander of New Madrid, Missouri, is well-developed. B: (right) Lake Mary and other oxbow lakes along a 20-mile (30-kilometer) stretch of the Mississippi River are identified in color.

This is a high aerial view of oxbow lakes along the Mississippi River. The top of the photo is east; south is toward the right. Lake Chico is an oxbow lake on the Arkansas side of the Mississippi River. The cutoff, Lake Ferguston, that runs by Greenville, Mississippi, was made in the 1930s; it serves Greenville as a harbor area. Lake Lee—in the upper right of the photo—is another oxbow lake. (Army Corps of Engineers)

of fine alluvium. The alluvium of natural levees can reach heights of 20 feet (6 m) above the level of the surrounding floodplain.

Because deposition also occurs along the riverbed, a mature stream gradually raises its bed above the level of its own floodplain. This action—the raising of the riverbed—along with the deposition of natural levees can block tributaries and prevent them from entering the main channel (Fig. 7–6). The blocked tributary is forced to flow alongside the mainstream until it finds a point of entry. The Yazoo River—a tributary of the Mississippi—behaves in this way and lends its name to this phenomenon of parallel-flow. Tributaries that run parallel to the main river for some distance are called *Yazoo rivers.*

OLD-AGE STAGE

This part of the geomorphic cycle is marked by the reduction of the mountains to rolling hills with broad flat-floored valleys between them. The entire surface is lowered well below the

FIGURE 7–6 The natural levees of the Mississippi River are barriers that have blocked the Yazoo's entrance to the main channel. The Yazoo River flows alongside the Mississippi until it finds a point of entry into the main channel at Vicksburg.

original surface (Fig. 7–3D). When the cycle of stream erosion is carried to its conclusion, the landscape is described as a peneplain, which is a broad, almost-plain surface with very low relief.

FLUVIAL LANDFORMS

The landforms shaped by the running water of streams are either erosional or depositional types. Valleys, ridges, hills, and mountain summits are among the fluvial landforms produced by erosion. Fans, floodplains, levees, and deltas are among the depositional landforms produced by fluvial processes.

STREAM VALLEYS

In the youthful stage of the fluvial geomorphic cycle, downcutting is greatly predominant over lateral cutting and the stream tends to produce a vertical-sided notch. Generally, mass wasting widens the upper part of the notch concurrently with its deepening by the stream, and the valley becomes V-shaped. Occasionally, however, downward cutting is much more rapid than widening, and the stream occupies the bottom of a very steep-walled trench, or gorge, which at times is several hundred feet deep.

A vertical-sided gorge may be maintained for some time if the stream is cutting through hardrock strata. However, as the stream continues its downward cutting, it may expose softer layers. When softer layers are exposed, weathering and mass wasting go to work immediately. As the softer layers are eroded away, the overlying harder rock is undercut and eventually it collapses. Gravity eventually moves the debris produced by weathering, mass wasting, and collapse into the stream channel. The spectacular gorges in the Colorado Plateau, which are commonly 0.5 mile (0.8 km) in width across the top but can run as much as 15 miles (24 km) wide, have been developed in this way (Fig. 7–7).

STREAM TERRACES

A **terrace,** which extends along a valley, is a relatively flat surface with a steep bank that separates it from a lower terrace or from the present floodplain of the river. An alluvial terrace develops in a river valley when alluvium that was

It has taken the Colorado River from 3 to 5 million years to carve the Grand Canyon. A series of microclimates were produced in the process—the inner gorge is hot and dry with desert plant and animal life; forest areas predominate along the canyon rim. (National Parks Service)

formerly deposited in the stream valley is partly removed by erosion (Fig. 7–8). In other words, the flat surface of the terrace represents a remnant of a former floodplain, which was deposited when the stream was more nearly mature.

Terraces, then, are the product of a mature stream's rejuvenation—a resumption of some youthful characteristics. The stream, in a sense, gets a new burst of energy and begins to erode downward. The rejuvenation may be brought about by increased gradient because of tilting, by increased water volume, or by a decrease of load.

If the rejuvenation is the result of a decrease in load, the meandering stream maintains most of the characteristics of a mature stream (Fig. 7–8A). It persists, for example, in its habit of swinging from one side of the floodplain to the other. The meanders continue to migrate downstream; now, however, the stream picks up alluvium and reduces the level of its floodplain and, thus, creates terraces.

WATERFALLS

Temporary checks to a youthful stream's downcutting result from unusually hard rock along its course. When a rock ledge is cut down less rapidly than the bed upstream from it, the stream's gradient is flattened and the velocity of its flow is decreased; such changes decrease the stream's transporting and eroding power. Since downcutting may be interrupted at numerous locations along a stream's course, the longitudinal profile of most streams consists not of a smooth slope to the sea but, rather, of a series of steps—each of which is maintained by a more-than-ordinarily resistant ledge of rock.

When the water flow passes beyond the resistant rock ledge onto less resistant material, it erodes more rapidly. The erosion of the softer rock increases the gradient in that location which, in turn, further increases the stream's eroding power. The cutting power of the water flow, as it passes onto the softer rock, is also augmented because some of the stream's load is dropped in the flattened part of the riverbed that is on the upstream side of the resistant ledge. As a result, just beyond the resistant ledge—on the downstream side—cutting is very rapid and a steep cliff forms over which the stream plunges in a waterfall.

Erosion of softer rock at the base of the water-

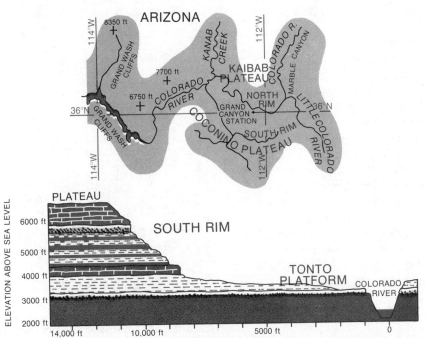

FIGURE 7-7   The cross section of the Grand Canyon is of the south wall at Grand Canyon Station, Arizona. The section runs from the Colorado River southward toward the Coconino Plateau. This portion of the canyon lies opposite the Kaibab Plateau. This is one of the widest parts of the canyon—widths of 10 to 15 miles (16 to 24 kilometers) are common. The northern brink of the canyon is much farther—more than three times—from the river at this point than the southern brink. The slope of the plateau on the south side is away from the river. Tonto Platform is a terrace about 3000 feet (900 meters) below the brink; it has a gently sloping surface. Thus, the canyon appears in cross section to have a flat bottom of considerable width that is trenched by a steep-sided gorge in which we find the river.

FIGURE 7-8   Terraces are the product of a mature stream's rejuvenation: in A, the rejuvenation is the result of a decrease in load; in B, the lowering of the base level, that is, an increased gradient, results in a trenching of the flat alluvial floor.

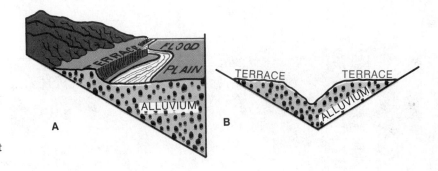

fall gradually undercuts the resistant rock ledge. Chunks and blocks fall from the resistant ledge as the softer rock is eroded away. The process of undercutting causes a waterfall to migrate slowly upstream.

The Niagara River, which connects Lake Erie and Lake Ontario, forms part of the boundary line between New York and the Canadian province of Ontario. The famous Niagara Falls are not quite midway between the two lakes (Fig. 7–9). The Falls are maintained by a thick layer of resistant dolomite, which overlies softer shale. The water drops in two streams. The larger falls over the rocky dolomite ledge on the Canadian side and forms the Horseshoe Falls; the smaller stream drops over the eastern shore and forms the American Falls. The ledge at Horseshoe Falls is being undercut at an average rate of 3 feet (1 m) per year. The upstream migration is slower at the American Falls; its shelf moves back at an average rate of 6 inches (15 cm) per year.

DELTAS

In this discussion three types of deltas are described: arcuate, estuarine, and bird-foot deltas. The initial discrimination among these types is based primarily on shape. The form of a delta, however, depends on two sets of factors: One set has to do with the quantity and character of the load carried by the river. The other set of factors has to do with the body of water in which the delta is built.

Most of the deltas of the world are relatively simple in outline—they are more or less fan-shaped or arcuate (Fig. 7–10). An **arcuate delta**

FIGURE 7–9 The Niagara River flows northward from Lake Erie. The river divides and passes on either side of Grand Island. Goat Island (black) separates the river into two streams just before the Falls. Portions of the Horseshoe Falls and the American Falls are marked in color. At the northern end of the river, the water flows into Lake Ontario.

has its outer edge curved like a bow so it is convex toward the sea. Generally, arcuate deltas are built of silt, coarse sand, and gravel, which tends to make them porous. The deltas of the Nile (30°10′N, 31°6′E), the Niger (5°N,6°30′E), the Indus (24°20′N, 67°47′E), the Ganges (23°22′N, 90°32′E), the Mekong (10°33′N,

We are looking upstream. American Falls is at the left; Horseshoe Falls on the Canadian side is at the right of the photo. Goat Island is the tree-covered area between the falls. (Niagara Convention Bureau)

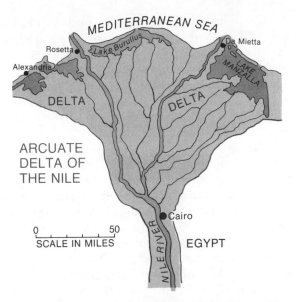

FIGURE 7–10   The Nile Delta, which begins at Cairo, is north of the color lines in the figure. The shape of this arcuate delta roughly resembles the Greek letter delta (Δ). The Nile separates into two main branches north of Cairo. There are many smaller distributaries that flow from these main branches in the delta region.

105°24′E), and the Po (44°57′N, 12°4′E) rivers are all arcuate.

An arcuate delta advances seaward almost uniformly around its entire margin. Cairo—situated at the apex of the Nile Delta, approximately 100 miles (160 km) from the coast—"moves" inland each year as the Nile builds its delta at the rate of 12 feet (3.7 m) per year. Some 2000 years ago, on the Italian Peninsula, the town of Adria (45°3′N, 12°3′E) was a seaport. Today it is about 14 miles (22 km) from the sea because the Po River has pushed its delta into the Adriatic by that amount.

An estuary is a drainage channel that is adjacent to the sea and in which the tide ebbs and flows. Some estuaries are the broad mouths of rivers that have been submerged and inundated by seawater. A river whose lower course is submerged may deposit its load of sediment as a narrow estuarine filling of submerged bars or an extensive floodplain of marshy land areas. This kind of estuarine filling is called an **estuarine delta** (Fig. 7–11). The following rivers build estuarine deltas: the Mackenzie (69°15′N, 134°8′W), the Elbe (53°50′N, 9°E), the Susquehanna (39°33′N, 76°5′W), the Seine

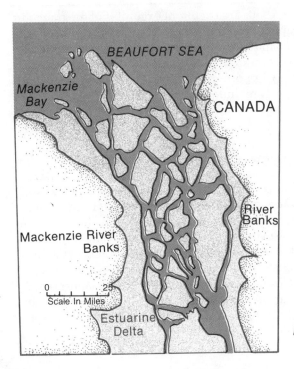

FIGURE 7–11   The mouth of the Mackenzie River is more than 40 miles (65 kilometers) wide. The filling of submerged bars and marshy land (color) that makes up its delta region is more than 110 miles (180 kilometers) long.

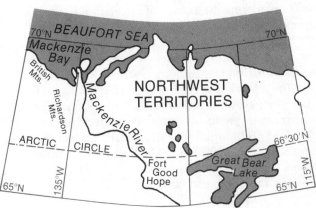

(49°26′N, 0°26′E), and the Hudson (40°42′N, 74°2′W).

The third type of delta—the **bird-foot delta**—is built by streams carrying large amounts of extremely fine material in solution, including a lot of lime. The delta formed by the Mississippi (29°N,89°15′W) is the best known example of the bird-foot delta (Fig. 7–12). The Mississippi River advances its delta by means of four primary distributaries, called *passes,* which radiate from a common point and give the bird-foot form to the delta.

The Mississippi River is unique in that no other river of its size drains so vast an area of limestone and other fine-grained sedimentary rocks. More than 75 percent of the material deposited at the mouth of the Mississippi is silt or clay. The very fine impervious material does not permit any subterranean flow and, as a result, the water is concentrated in a few large channels. The bird-foot delta of the Mississippi River is advancing at an average rate of 250 feet (76 m) per year.

The delta (42°37′N, 82°31′W) built into Lake Saint Clair by the Saint Clair River is another bird-foot delta (Fig. 7–13). This delta is unique, however, because it has been built in two stages. The eastern side was built first. Then there was a period during which delta-building ceased. When delta-building was reestablished, the western side began to take shape.

The water that flows from Lake Huron via the Saint Clair River to Lake Saint Clair is on its way to Lake Erie. This was the pattern of water flow when the eastern side of the delta was built, and it is the pattern of flow today. However, after the eastern side of the delta was built, the flow changed for a while and the upper Great Lakes discharged by way of Georgian Bay into the Saint Lawrence and very little water flowed through the Saint Clair River. Then the region to the north was uplifted and the lakes renewed their former course of discharge; as a result, the present western section of the delta is being actively formed.

THE EOLIAN LANDFORMS

The word **eolian** is applied to deposits—sands and other loose materials along shores and in deserts—that are arranged by the wind. The eolian processes are very similar to the fluvial processes described in the preceding section, that is, the wind changes landscapes by erosion, transportation, and deposition. Geographically, the eolian processes are most significant in hot, dry regions and in cold regions because the absence of a protective cover of vegetation in these areas exposes the surface to assault by the wind.

The geomorphic effect of the wind depends on wind velocity in much the same way that fluvial erosion depends on water velocity. The erosional work of the wind is carried on through the processes of deflation and abrasion. The transport capacity of wind is astounding; as much as 100 million tons of material are carried more than 2000 miles (3220 km) during some wind storms. The huge dust storms that were common in the American Midwest Dust Bowl in the 1930s are excellent examples of the transport capacity of the wind. The interruption of the transportation process and the consequent deposition of material is related to wind velocity. Two important and striking depositional landforms are dunes and wind-deposited soil, known as *loess.*

EROSION AND TRANSPORTATION

The wind carries material in suspension as well as along the surface of the ground. Fine particles are picked up primarily through the turbulent flow of the air and are then carried in suspension. **Deflation** is a general term that refers to the picking up, removal, and transportation of loose particles by the wind. Deflation occurs most readily in areas where particles are loosely packed and where there are few barriers to the flow of the wind.

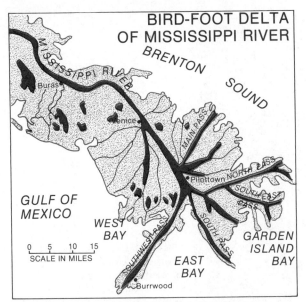

FIGURE 7-12 The delta of the Mississippi River is advanced by means of deep primary channels, referred to as *passes*. Each pass is bordered by narrow banks of unyielding clay. The bays between the fingers of the delta are rather deep and are filled in very gradually by wave and tidal actions.

There is a definite relationship between wind velocity and the size of the material picked up and transported. The material most easily moved is fine sand with a diameter of 0.004 inch (0.1 mm), which happens to be the particle most easily moved in the fluvial system. Fine sand is entrained at wind velocities around 6 inches (15 cm) per second. Higher velocities are required to entrain both finer and coarser material. Mass is a factor in the higher velocities required to pick up coarser material; with respect to finer material, higher velocities are required because of the smooth surface presented to the moving air and the moisture that allows the finer material to cohere and resist entrainment.

When strong winds blow, two layers of particles are usually present in the air. The lower layer, which extends a few inches to a maximum of 6 to 10 feet (2 to 3 m) above the ground, consists of sand grains; the upper layer often extends to great heights and generally consists of silt and clay particles. Coarser, heavier particles are also bounced along the surface by the process known as *saltation*, which was described on page 217 in the section on fluvial processes.

In addition to the erosive action of deflation, wind-carried particles collide with landforms

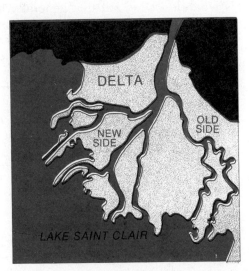

FIGURE 7-13 The St. Clair Delta has two distinct sections. The eastern side is the older of the two and represents an earlier stage of delta development. The section of this bird-foot delta that is being actively formed today is on the western side.

This is the Canyonlands area in southeast Utah. Wind is a powerful eroding and transporting agent in this semiarid region. Wind abrasion has helped to sculpt Angel Arch in the background and the Tooth in the left foreground. (National Parks Service)

and abrade their surfaces. The process, called **abrasion,** involves the mechanical wearing away of exposed rock surfaces by grains of sand and other particles carried by the wind. In fact, sandblasting is an artificial technique similar to natural abrasion. In sandblasting, a fast moving stream of air drives sand particles against a stone building to wear away a thin layer of the stone. This is the most common method by which stone is cleaned of stains or defacing marks. Natural winds, with their load of material, produce the same effect on stone—although at a much slower rate than artificial sandblasting techniques.

Any surface exposed to the constant sandblast effect of wind-carried debris will inevitably be cut, polished, and reshaped. In deserts, rocky surfaces and cliffs that face the prevailing winds are honeycombed, etched, and carved. Top-heavy pedestals and natural bridges are formed where wind wears away a rock formation at its base but not at its top.

EOLIAN DEPOSITION

Two major types of deposits are put in place by the wind: dunes and loess. **Dunes,** which consist of mounds or ridges of sand, not only occur in the interior of continents under desert conditions but also along river floodplains—the Columbia River floodplain, for example—and along seacoasts—Cape Cod's as well as others. **Loess** is wind-deposited silt that generally has some clay and fine sand mixed in with it. Some of the world's great loess belts are worked as very productive farmlands. The fertile farm valleys of the Mississippi, for example, are underlain by loess that was derived from river floodplains. Some very productive farmlands of central Europe consist of loess derived from glacial outwash. The rich soil of northern China is loess that was blown in from the Gobi Desert (Fig. 7–14).

SAND DUNES

Most major hot deserts of the world are found between the latitudes of 35°N and 35°S. These areas generally have a high rate of evaporation and less than 10 inches (25 cm) of precipitation per year. A desert surface consists of rock and sand. Although water is not a constant agent of change, alluvium can be found where mountain streams enter the desert.

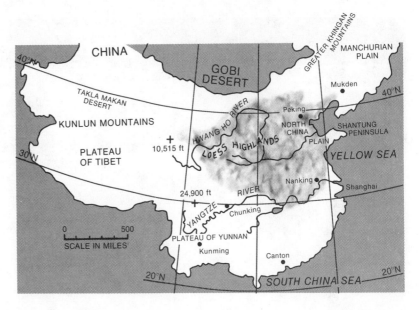

FIGURE 7–14 The main loess regions of China are shown in color. An upland region known as the *Loess Highlands* lies west of the North China Plain. The Hwang Ho River has cut deep gorges through the soft, yellow loess. The Hwang Ho, which is also called the *Yellow River*, gets its muddy yellow color from the soil it carries downstream from the Loess Highlands. The best farmland in China is found in the North China Plain in the valley of the Hwang Ho River. The Manchurian Plain of northeastern China also has good farmland.

There are large deserts on nearly every continent: Africa has three well-known deserts—the Kalahari (24°S,21°30′E), which stands at 3000 feet (900 m) above sea level and spreads over 200,000 square miles (518,000 km²) in Botswana and South West Africa; the Sahara (26°N,13°E), the largest desert in the world, covers about 3 million square miles (7.8 million km²) and stretches across North Africa from the Atlantic Ocean to the Red Sea; and the Arabian Desert (25°N,45°E), which spreads across 500,000 square miles (1,295,000 km²). The Atacama Desert (22°30′S, 69°15′W), which stretches north-south for more than 700 miles (1100 km) in northern Chile, is one of the world's driest areas. Australia's Great Victoria Desert (28°30′S, 127°45′E) is an area of shifting sand dunes that covers an area of 250,000 square miles (647,500 km²). The Mojave (35°N, 117°W), which is between the Sierra Nevada and the Colorado River in southeastern California, is considered a large desert by North American standards even though it only covers 25,000 square miles (64,800 km²).

Wind does not always move over a region from a single direction or at a constant rate. Often, however, in desert regions and along coastal areas the winds are rather constant from one direction and the sandy surfaces are formed into dunes. Dunes may be either transverse or longitudinal (Fig. 7–15). The ridges of longitudinal dunes run parallel to the direction of the prevailing wind; those of transverse dunes are at right angles to the prevailing wind.

Sand dunes are independent of any fixed structure. Therefore, they move in the direction toward which the wind is blowing. A dune has a gently sloping side, the windward side, which faces the prevailing winds. The leeward side of a dune is a steep slip-off face. As sand particles are

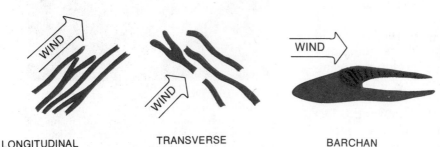

LONGITUDINAL

TRANSVERSE

BARCHAN

FIGURE 7–15 Wind blowing steadily from one direction over a sandy surface produces ridges that are either longitudinal or transverse. A barchan dune shows both transverse and longitudinal characteristics.

The White Sands area in southern New Mexico is near Alamogordo. This arid wilderness contains great deposits of white, wind-blown gypsum sand. (National Parks Service)

blown over the crest and fall down the steep leeward face, the dune creeps forward, ever so slowly, in the direction of the wind.

Transverse dunes are formed under moderate wind velocities. When several stand together, transverse dunes appear to be ripples or waves in the sand. Longitudinal dunes are formed by winds moving at high velocities. Strong winds exert enough force to move all the sand that stands before them. Smaller, crescent-shaped dunes, called **barchans,** are produced by winds of intermediate velocities. Barchans—with their crescents to leeward, that is, facing in the direction toward which the wind is blowing—show both transverse and longitudinal characteristics (Fig. 7–15).

Generally speaking, in areas where the sand supply is abundant and wind forces are moderate, transverse dunes tend to develop and dominate. In those regions where the wind has more mastery over the sand, longitudinal ridges develop and dominate. In areas where wind velocity and direction are highly variable, a complex series of different dune types usually develop.

LOESS DEPOSITS AND LANDFORMS

The loess deposits of the United States are generally derived from three different sources: floodplains of rivers, semiarid regions, and outwash plains of glaciers. The loess that caps the

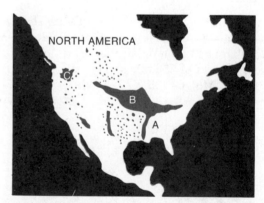

FIGURE 7–16 The loess deposits of North America are identified in color. The largest continuous area of loess stretches across the heartland of the U.S. and covers major sections of Nebraska, Kansas, Missouri, Iowa, Wisconsin and Illinois.

FIGURE 7–17   The loess deposits of South America (color) are derived from the arid land to the west.

uplands adjacent to the large rivers of the Mississippi Valley is derived from the floodplains of the rivers. Since westerly winds generally prevail in this region, it is natural to find the more extensive deposits of loess on the eastern side of a valley. A long belt of loess caps the eastern bluffs of the Mississippi River almost to the Gulf of Mexico (Fig. 7–16A).

Extensive deposits of loess are found from western Kansas and Nebraska on into Iowa and Missouri (Fig. 7–16B). These deposits were derived largely from the semiarid regions of Colorado. Since the material was transported by prevailing westerly winds, the deposits in Kansas and Nebraska are thicker than those of Iowa and Missouri.

The loess deposits in the Palouse area of Washington and Idaho (Fig. 7–16C) are great stationary dunes of silt. These deposits form deep, fertile soil. In fact, there is a 75-foot (23 m) covering of loess on the plateau between the Columbia River and the Bitter Root Mountains. These deposits were originally derived from glacial outwash plains. But, today, they are being increased by material blown from the dry eastern slopes of the Cascades and the adjacent Columbia River valley.

Loess makes a very fertile soil in which crops can be grown in great abundance, especially where water is available for irrigation. If a loess region is one of active ongoing deposition, as in the Palouse area of Washington and Idaho, there is constant renewal and enrichment of the soil. The black soils of southern Russia and northern China are of loess origin. All loess regions are productive agricultural areas. The South American loess deposits cover much of the Pampa of northern Argentina (Fig. 7–17); the source of these deposits is largely the arid regions to the west.

The flat upland surfaces of youthful loess plateaus frequently exhibit natural wells or pits, which usually occur near the rims of gullies, ravines, and canyons. These loess wells result from seepage movement along bedded layers that are somewhat more porous than the surrounding material. Between developing wells, a subterranean channel may develop to such an extent that a natural bridge is formed.

As the active destruction of a loess upland proceeds, a multitude of complex ravines with sharply cut walls develops. The divides between the heads of opposing streams are like walls that form natural causeways between one plateau and another. In the later stages of erosion, a loess upland is finally reduced to numerous pinnacles, some of which may be conical, separated by wide, flat-floored valleys.

THE WORK OF GLACIAL ICE

At least four glaciations are generally recognized as having occurred during the Pleistocene epoch in North America. The oldest of these glaciations is called the *Nebraskan*. The *Kansan* came next and the *Illinoian* followed it. The last and most recent is the *Wisconsinan*. The southernmost penetration of the Wisconsinan ice into the United States is depicted in Figure 4–18 (page 134). A ridgelike accumulation of drift, called a **moraine**, was deposited along the front margin of the Wisconsinan glaciers.

In the eastern United States, the moraine of the Wisconsinan glaciers extends in a line from Cape Cod and Nantucket Island southwest to the northside of Long Island. Figure 7–18 provides some detail on the extent of the terminal moraine in the New York–New Jersey region where it cuts across Long Island from Forest Hills to the Narrows of New York Harbor. The terminal moraine extends across Staten Island through Dongan Hills to New Jersey. From the Perth Amboy region of New Jersey, the moraine moves northwest toward Plainfield where it then turns in a more northerly course to Short Hills.

The Hudson River cut a channel—the Narrows of Upper New York Bay—through the moraine that stretched in an unbroken belt from Long Island to Staten Island (Fig. 7–18). The moraine has been breached in several other places, too. The Elizabeth Islands off the coast of Massachusetts, for example, were formed by a chain of glacial moraine, which is presently breached by passageways such as the one at Woods Hole, Massachusetts. There is another interruption in the belt of moraine between Long Island and Block Island. In the New York–New Jersey area, a strait between Staten Island and New Jersey, known as the *Arthur Kill,* is actually another breach in the moraine (Fig. 7–18).

The present locations of the Ohio and the Missouri rivers, which cut across the heartland of the United States toward the Mississippi River, are shown in Figure 4–18 (page 134). The channels of the Ohio and the Missouri rivers coincide roughly with the southernmost margin of the glacial ice during the Pleistocene. The Missouri, which flows eastward from the Rocky Mountains for 2315 miles (3726 km) to the Mississippi, was

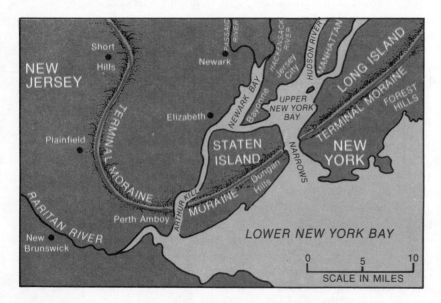

FIGURE 7–18 The terminal moraine of the Wisconsinan glaciers form a belt that encircles the metropolitan area extending from Short Hills in New Jersey to Forest Hills in New York. The Narrows, a strait in New York Harbor, is a breach in the moraine. The Arthur Kill is another strait which breaches the terminal moraine.

This is the ship-holding area of New York Bay. We are looking toward the south. The Verrazano Bridge (in the background) spans the Narrows from Staten Island (at the right) to Long Island (at the left). (Port Authority of New York–New Jersey)

blocked time and again by the advance of glacial ice. Each time it was blocked, it reworked its course to flow along the ice front. The Ohio, flowing westward for 981 miles (1579 km) from the Appalachian Plateau, encountered the same difficulty and also adapted its flow along the ice front.

GLACIAL EROSION

A **cirque** is a niche cut into a mountainside by a combination of frost action and ice plucking. During the course of a warm day, some of the snow that has fallen on a rock formation may melt and the water may seep into the pore spaces of the rock below. At night, the temperature may drop sufficiently for the water within the pore spaces to freeze, expand, and pry fragments from the rock formation. Through a very slow process of millions of fragmentations, a depression eventually forms in the rock. If the region's snow-cover builds into a glacier, the glacial ice that develops will pluck off larger pieces from the bedrock in a scouring action that serves to enlarge the depression in a cirque.

When the developing glacier builds in size and advances into a V-shaped stream valley along its path, it picks up whatever regolith is in its way. This debris, called **glacial till,** is dragged

along beneath or within the ice (Fig. 7–19). The moving glacial ice abrades the bedrock and plucks blocks from it. Scratches, called **striations,** on bedrock are indications of glacial movement over the area.

A valley glacier forms a typical U-shaped trough from a preglacial V-shaped stream valley by deepening the valley and by plucking and abrading its sides (Fig. 7–19). Crawford Notch in

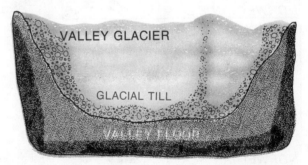

VALLEY GLACIER

GLACIAL TILL

VALLEY FLOOR

FIGURE 7–19 This cross section depicts the U-shaped trough of a valley glacier. A valley glacier forms a U-shaped trough from a preglacial V-shaped stream valley. Glacial till is a heterogenous mixture of clay, sand, gravel, and boulders carried within a glacier.

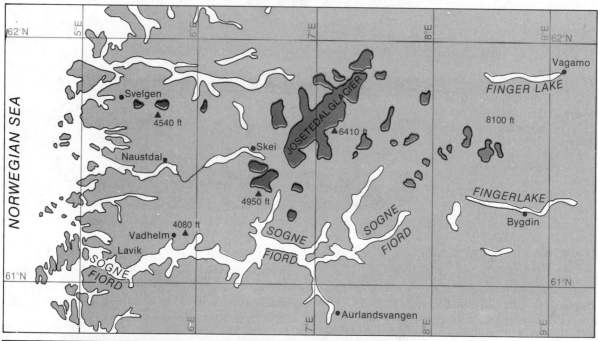

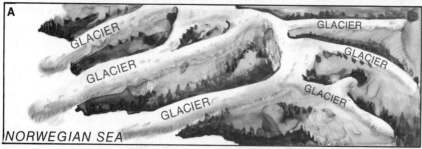

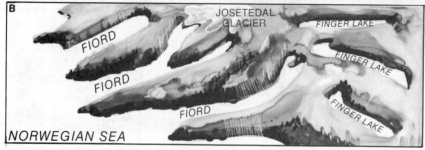

FIGURE 7–20 Fiords and finger lakes have a similar origin—both are the result of glacial action. The Josetedal Glacier is a remnant of the Pleistocene Glaciers that scoured the highlands of Norway. The fiords of Norway are legendary. The Sogne Fiord is quite long.

The rounded, U-shaped profile of Crawford Notch in New Hampshire's White Mountains is apparent in this view of the valley's left wall. That's a highway threading its way through the central area of the valley's floor. The second thread—a bit higher along the left wall of the valley—is the roadbed of the Maine Central Railroad. (State of New Hampshire)

the White Mountains of New Hampshire is an example of a U-shaped glacial trough.

A glacial trough that is inundated by the sea after the ice recedes is called a **fiord**. There is an extensive fiord region along the Alaskan coast near Juneau. The glacier-produced fiords of Norway make its coastline one of the most jagged and rugged in the world. The Sogne Fiord (61°9′N,6°E) extends inland for more than 100 miles (160 km) and is the longest on the Norwegian coast (Fig. 7–20). The Jostedal Glacier centered at latitude 61°40′N, longitude 7°E, which is in the same general area as the Sogne Fiord, is the largest present-day ice field on the European continent. The 300-square-mile (700-km²) Jostedal Glacier is a remnant of the Pleistocene glaciers that covered this area and produced the fiords. The fiords along the southwestern coasts

This is an aerial view of Milford Sound at latitude 44°40′S, longitude 167°54′E. This is a glacier-produced fiord on the southwest coast of South Island, New Zealand. Mount Tutoko, the snow-covered peak poking through the cloud at the upper right of the photo, is 9042 feet (2756 meters) above sea level. (New Zealand Government)

of New Zealand and Chile are less well known, but in some ways just as lovely.

The A and B portions of Figure 7–20 show that fiords and **finger lakes** have a similar origin. Both are the result of glacial action. The only difference is that the glaciers that worked their way to the west reached the sea; those that moved to the east terminated in a land environment as they reached lower levels. The troughs scoured out by the tongue-like glaciers that moved toward the east became finger lakes when the glaciers retreated. The fiord region near Skagway and Juneau in Alaska has a finger-lake region to the east. Finger-lake regions are also found to the east of Chile's fiords as well as on South Island, New Zealand.

When a glacier recedes and the land is uncovered, the erosional features produced by the ice are there for us to see. Glacial action in cirques that erodes a peak from at least three sides leaves in its wake sharp-pointed, pyramid-shaped peaks called **matterhorns** (Fig. 7–21). Aretes, sharp-crested divides, are the product of two cirque glaciers cutting into a ridge from opposite sides. Another spectacular feature found in glaciated regions is the **hanging valley**. During the period of glaciation, the hanging valley was a tributary glacial trough that was not eroded as deeply as the main trough. When a stream flows through a hanging valley, it descends to the floor of the main valley as a waterfall. Bridalveil Falls (37°43′N, 119°39′W) in Yosemite National Park, California, is a beautiful example of a hanging valley with a waterfall; this area of the Sierra Nevada underwent extensive glaciation.

GLACIAL LANDFORMS

The debris that remains after a glacier recedes assumes many patterns and shapes. Moraines, drumlins, eskers, and plains are some of the depositional landforms produced. In addition, when the ice sheet finally withdraws, the coun-

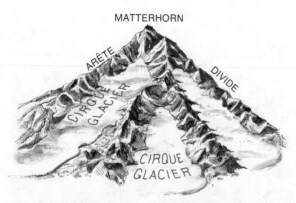

FIGURE 7–21 Several glaciers are moving outward from the summit of the mountain peak. At their heads the glaciers gouge out cuplike hollows, called *glacial cirques*. The summit resembles a sharp-pointed pyramid, known as a *matterhorn*.

Bridalveil Falls in California's Yosemite National Park is a perfect example of a glacier-rounded hanging tributary valley with a waterfall. (National Parks Service)

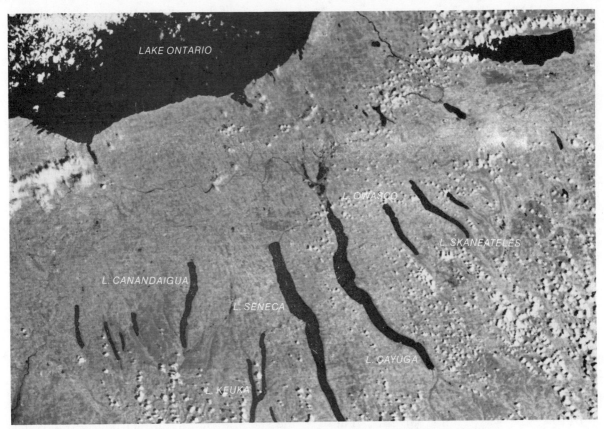

LAKE ONTARIO

L. OWASCO

L. SKANEATELES

L. CANANDAIGUA

L. SENECA

L. CAYUGA

L. KEUKA

The Finger Lakes are a group of long, narrow lakes in the western part of New York State. There are 11 Finger Lakes; the 6 largest have been labeled. Seneca is 37 miles (60 kilometers) long and 4 miles (6.5 kilometers) wide at its broadest point. The Finger Lakes were dammed by glacial debris at their southern ends; they drain north into Lake Ontario. (NASA)

tryside is covered with lakes, marshes, valleys, and channelways.

MORAINES

As a glacier advances, it makes deposits of till, called **lateral moraines,** along its boundaries on either side. The end point, or terminus, of a glacier is marked by another mound of till known as a **terminal moraine.** Orient Point and Montauk Point on Long Island, New York, were formed by terminal moraines; both points are at the eastern end of Long Island.

When a terminal moraine blocks the end of a valley, it forms a closed trough which, if it is filled with water, becomes a finger lake. The many finger lakes between Whitehorse in the Yukon territory of Canada and Skagway in Alaska are in troughs that were scoured out by advancing glaciers and then blocked by the terminal moraine of receding glaciers. The Finger Lakes, southwest of Syracuse, New York, have gone through the same developmental cycle. Recall that Norway, New Zealand, and Chile have extensive finger-lake regions just behind the fiords along their coasts.

DRUMLINS

Smooth, oval hills composed mainly of glacial till are referred to as **drumlins.** The long axis of a drumlin is parallel to the direction in which the glacier was moving. Generally, drumlins are less than 100 feet (30 m) high with a long axis of less than 2600 feet (790 m). Most drumlins seem to consist of deposits made beneath the glacier by debris that worked its way to the bottom along fissures in the ice. The most famous drumlin in the world is, without doubt, Bunker Hill in Charlestown, Massachusetts, which has been written into the history of the American Revolution. There are numerous drumlins in Massachusetts; for example, most of the elliptical islands in Boston Bay are the remains of drumlins that have been eroded by wave action, and even the State House in Boston stands on a drumlin, known as *Beacon Hill.*

OUTWASH PLAINS

The stream of meltwater that emerges from a receding glacier carries gravel, sand, and silt as part of its load. This material is carried beyond the glacier for varying distances, depending on the slope of the land and the size of the debris. When the slope is steep, the load can be carried downstream for many miles. However, as the velocity of the stream decreases, portions of the load that no longer can be transported at that velocity are dropped. The deposits made by a stream of glacial meltwater build a plain, which is referred to as an **outwash plain.** On Long Island there is an outwash plain south of the terminal moraine that stretches from Montauk westward to the Narrows. Hempstead and Freeport are located on the outwash plain of Long Island.

ESKERS

An esker is a ridge deposited by meltwater in an ice tunnel at the bottom of the glacier. These deposits are most common on low, swampy plains and can be found throughout the Mississippi Valley in Wisconsin, Minnesota, and Michigan.

SHORELINES SCULPTURED BY MARINE ACTION

The primary geomorphic agents acting on a shoreline are waves and currents. These geomorphic agents function in ways that are similar to the action of wind, streams, and glaciers when they pick up, transport, and deposit sediments. Through various processes—including the onshore action of waves and various currents that carry materials to, from and along the shore—new landforms evolve.

When wind moves over the ocean, a certain amount of energy is imparted by friction and pressure fluctuations to the water surface. The size of a water wave is determined primarily by the force of the wind—the greater the force, the larger the wave. There are two other factors, however, that contribute to the size of a water wave: (1) the duration of time the wind blows, and (2) the distance over which it blows. The term *fetch* is used to refer to the distance over which wind blows in the same direction and at the same speed. The height of a wave is increased as fetch and the duration of time the wind blows increase.

Water from a breaking wave is carried toward the shore; then, a return current develops that takes the water from the shore back toward the sea. If a wave comes in at an angle to the beach, the water spilling forward moves parallel to the beach in what is called a **longshore** or **littoral current.** The longshore current travels along the shoreline until it meets another longshore current coming from the opposite direction. The two longshore currents join and move back to sea in what is called a **rip current.** There are, of course, other currents in the sea; some are caused by tides, others result from the rotation of the Earth.

New shorelines are carved from existing landforms primarily by the relentless action of waves. Waves cast themselves against an unprotected shoreline in such massive assaults that the sheer pressure of the water is a powerful agent of gradation. For example, an 18-foot (6-m) wave exerts a pressure of more than 2000 pounds per square foot (9800 kg/m²). Such pressures can shift blocks of exposed rock weighing 10 tons

(9100 kg). The grading action of waves upon the land occurs in two zones: the zone of breakers and the surf zone. The breakers and the surf carve and sculpt the shoreline to produce new landforms.

Basically, there are two kinds of coastlines that waves carve, sculpt, and modify: the embayed coast and the gently-sloping coast. The drowned or **embayed coast** really consists of valleys, which are the water-filled bays, along with divides and hilltops, which are the peninsulas and islands. The **gently-sloping coast** is usually the inner edge of a continental shelf that is raised above sea level. The embayed and the gently-sloping coasts are at the immediate mercy of ocean waves as soon as they have been formed.

THE GENTLY-SLOPING COAST

A coastal plain, which is often a shoreline of emergence, is worked upon by small waves. Large waves, because of the shallow water covering the gentle slope, cannot reach the shoreline and break far offshore. The waves breaking offshore scour the bottom and throw up sand to form a barrier bar, which constantly grows in height as a result of sand cut from the sea bottom or drifted there by longshore currents (Fig. 7–22). A lagoon develops behind the barrier bar and is gradually filled with silt carried from the mainland by streams. After the bar has formed, the waves eventually attack it and gradually push it back onto the lagoon. Finally, the bar comes to rest on the original shore, and a newer and deeper shore profile is established with the waves now able to attack the mainland.

THE EMBAYED COAST

The embayed coast results from the partial submergence of a landmass that is dissected by normal river valleys. There are many different kinds of shorelines of submergence. There may be, for example, an embayed plain or plateau shoreline, such as the Chesapeake Bay region. Embayed mountain shorelines can be of many structural types. There is an embayed fold-mountain shoreline in the Dalmatian region of western Yugoslavia (43°N,17°E); in northwestern Spain (42°30′N,9°W), there is an embayed complex-mountain shoreline.

The embayed coast is characterized by an exceedingly irregular shoreline (Fig. 7–23). There are numerous islands, branching bays or drowned valleys, peninsulas, and a very irregular sea bottom that is fairly deep. The shoreline of submergence is immediately attacked by all sizes of waves because deep water is adjacent to it. Huge waves, exerting great pressure, can beat upon the headlands and the exposed portions of islands. Picturesque cliffs are produced and fre-

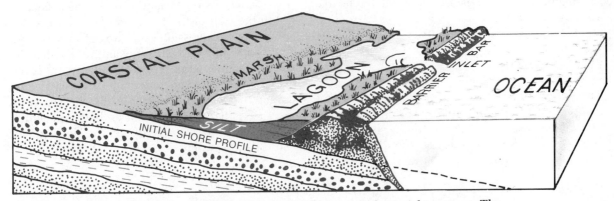

FIGURE 7–22 The initial shore profile of a coastal plain is a nearly straight contour. The barrier bar is formed by waves that cut into the initial shore profile and cast the material from the sea bottom forward into a long bar that parallels the shoreline. The offshore bar shown has two crests. The inner bar was formed first by a weaker set of waves. The outer bar is shown as partially eroded and in the process of being pushed back.

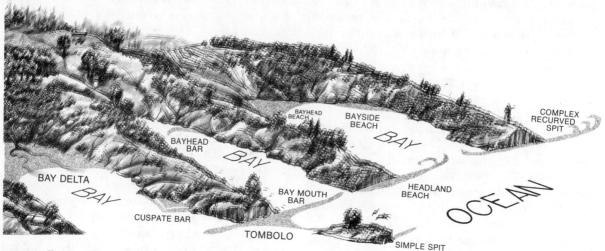

FIGURE 7–23 The embayed coast has an exceedingly irregular shoreline. Since the water depths close to the shoreline are fairly deep, large waves can beat upon the headlands and produce extensive erosion. Beaches and spits result from the accumulation of eroded material along the shore.

quent landslides result from the action of the waves along the bases of these cliffs. A great array of headlands, spits, and beaches is produced. The names given to the beaches indicate their location: headland beach, bayhead beach, and bayside beach. The term **bar** is applied to spits—accumulations of debris—built from one headland toward or to another. A **tombolo** is another feature that develops when debris joins an offshore island to the mainland. Gradually, the headlands are cut back beyond the bayheads and eventually the wave action is directed against a straightened shoreline.

OTHER SHORELINES

The variety of landforms, of course, is much too complex to fit any simplified classification system. There are, for example, delta shorelines, volcano shorelines, coral-reel shorelines, and shorelines produced by faulting that do not conform to the two categories established above. In addition, certain shorelines—the coast of Maine, for example—have gone through a complex history; the coast of Maine is essentially a shoreline of submergence, but recent emergence has resulted in the development of small coastal-plain features around certain islands.

The volcanic islands of Hawaii serve as excellent examples of the way in which ocean waves gnaw away at island shores. These islands are in the belt of the trade winds and, thus, are subject to the incessant pounding of large waves. The island of Molokai, for example, has had its volcanic dome almost completely worn away. Oahu represents the remnants of two volcanic domes that have been eroded by the sea (Fig. 7–24). Pearl Harbor is in an embayment which lies in the depression between the remnants of Oahu's domes.

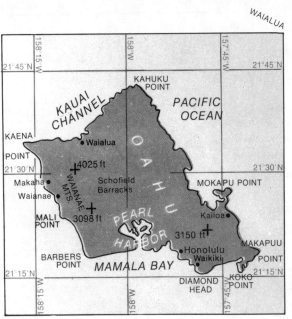

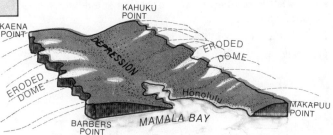

FIGURE 7–24 The island of Oahu is the remnant of two eroded volcanic domes.

THE LANGUAGE OF PHYSICAL GEOGRAPHY

A physical geographer uses a technical vocabulary to make explanations. Review your understanding of the vocabulary used to develop the concepts in this chapter. Among the important words and terms used are:

abrasion	drumlin	lateral moraine	saltation
alluvial fan	dunes	levee	sedimentation velocity
alluvium	embayed coast	littoral current	sheeting
arcuate delta	eolian	load	solution
bar	erosion velocity	loess	striations
barchans	eskers	longshore current	surficial
barrier bar	estuarine delta	mass wasting	suspended load
bed load	exfoliation	matterhorn	talus
bird-foot delta	finger lakes	meanders	talus cone
chemical weathering	fiord	mechanical weathering	talus slope
cirque	floodplain	moraine	terminal moraine
creep	gently-sloping coast	mudflow	terrace
deflation	geomorphic	outwash plain	tombolo
delta	glacial till	oxbow lake	traction
deposition velocity	gradational forces	regolith	transport velocity
dissolved load	hanging valley	rip current	waterfall
drift	landslide	rockfall	weathering

SELF-TESTING QUIZ

1. List the various gradational forces.
2. Explain what is meant by mass wasting.
3. Why is weathering considered to be a preliminary to gradation?
4. List and contrast the two processes of weathering.
5. Compare and contrast the processes of exfoliation and sheeting.
6. How is the mass transfer of rock and sediments accomplished?
7. Describe some geologic structures that are landslide-prone.
8. Discuss the difference between erosion and transport velocity.
9. Trace the ideal sequence in the fluvial geomorphic cycle.
10. How does a mature river build the level of its floodplain?
11. What does the flat surface of a stream terrace represent?
12. What conditions need to exist for a waterfall to develop?
13. Describe and compare the development of the various kinds of deltas.
14. What depositional landforms are produced by wind?
15. Trace the southernmost penetration of glacial ice into the United States during the Pleistocene epoch.
16. What kinds of erosional features are produced by glaciers?
17. Describe the characteristic shorelines carved by the action of waves.

The city of Fez (34°5′N, 4°57′W) is nestled in the foothills of the Middle Atlas Mountains of Morocco. The climate in these mountains is cool with cold, snowy winters. About 32 inches (81 centimeters) of precipitation falls on these slopes annually. The runoff that develops from the precipitation moves over the surface and carries loose material downslope. The running water also cuts and etches new profiles in these mountains of Morocco. (Moroccan Government)

IN REVIEW
LANDFORMS SHAPED BY EROSION AND DEPOSITION

I. THE WEATHERING OF ROCK

    A. Mechanical Weathering

    B. Chemical Weathering

II. MASS TRANSFER OF ROCK AND SEDIMENT

    A. Rockfalls

    B. Landslides

        1. The Topography

        2. The Geologic Structure

    C. Soils and Earth Flows

    D. Mudflows

III. SURFACE MODIFICATION BY STREAMS

    A. Stream Erosion and Deposition

        1. Erosion Velocity

        2. Transport Velocity

        3. Deposition Velocity

        4. Stream Load

    B. The Fluvial Geomorphic Cycle

        1. Youthful Stage

        2. The Submature Stage

        3. The Mature Stage

        4. Old-Age Stage

    C. Fluvial Landforms

        1. Stream Valleys

        2. Stream Terraces

        3. Waterfalls

        4. Deltas

IV. EOLIAN LANDFORMS

    A. Erosion and Transportation

    B. Eolian Deposition

        1. Sand Dunes

        2. Loess Deposits and Landforms

V. THE WORK OF GLACIAL ICE

    A. Glacial Erosion

    B. Glacial Landforms

        1. Moraines

        2. Drumlins

        3. Outwash Plains

        4. Eskers

VI. SHORELINES SCULPTURED BY MARINE ACTION

    A. The Gently Sloping Coast

    B. The Embayed Coast

    C. Other Shorelines

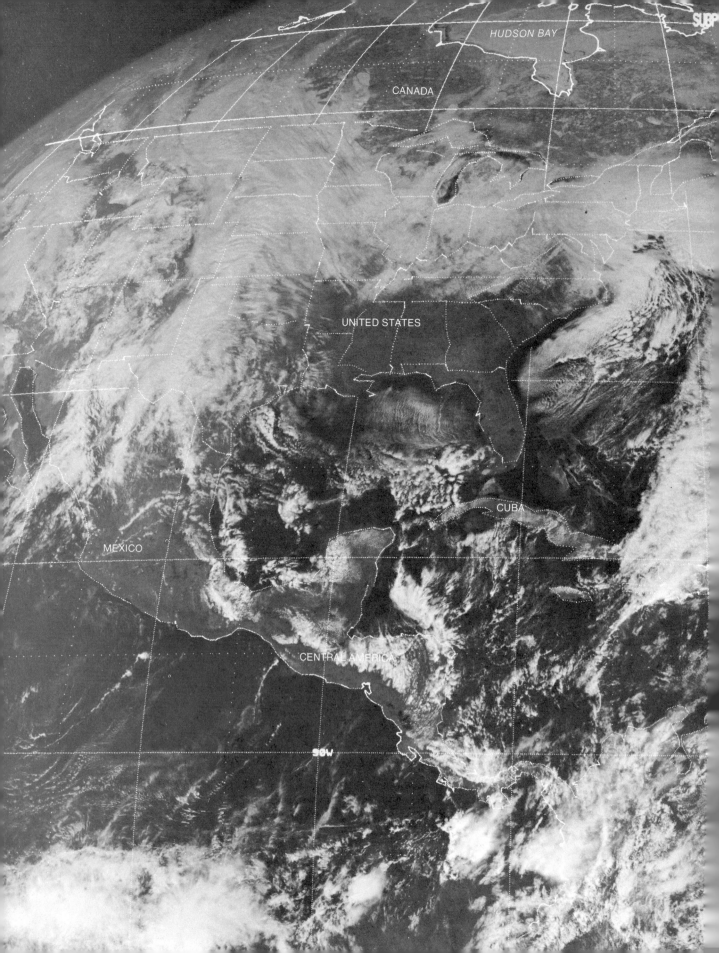

8 *A North American survey*

In the first three chapters of Part 2, we have been discussing and studying the materials from which and the processes by which landforms are developed. It seems appropriate at this point to examine the North American continent as a total region for the purpose of establishing the spatial relationships of the North American landforms.

In common geographic terms, North America is the third largest of the seven continents—only Asia and Africa have larger areas. Canada, the United States, Mexico, and Central America are thought of as occupying the main continental area. If we invoke the concept of plate tectonics, however, we find that most of Central America is on the Caribbean Plate. A part of the boundary between the North America Plate and the Caribbean Plate is marked by a great canyon in the sea floor, the Cayman Trench (Fig. 8–1). The Cayman Trench extends from between Cuba and Haiti to Guatemala's eastern coast. The landward extension of this boundary is the Motagua Fault, which moves westward across Central America to meet the Cocos Plate.

The tectonic forces that shape the North American landforms operate on the landmass north of the Cayman Trench and the Motagua Fault. Thus, the survey of North American landforms in the pages that follow will be confined to Canada, the United States, Mexico, and the small area of Central America that is north of the Motagua Fault. Canada, with an area of 3,851,809 square miles (9,976,140 km²), and the United States, with an area of 3,675,633 square miles (9,519,846 km²), make up approximately 90 percent of the landmass north of the Motagua Fault.

MAJOR TECTONIC DIVISIONS

The broad sweep or grand scheme of the spatial relationships of North American landscapes reveals itself in second-order relief features. In Fig. 8–2 the outlines and boundaries of nine major **tectonic divisions** are depicted: the western orogenic belt, the Gulf-Atlantic coastal

Most of the North American continent is under constant satellite surveillance and is spread before us in this photo. A low-pressure zone off New England spreads a distinct cloud pattern over the ocean off the East Coast from Florida to Maine. The southern tier of states from Mississippi to South Carolina is relatively free of cloud cover. If you examine the photo carefully, you will be able to identify Lake Okeechobee in Florida. (NOAA)

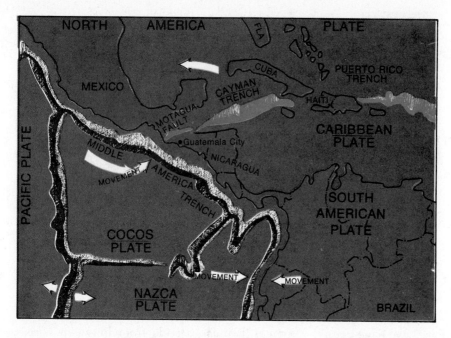

FIGURE 8–1 The Motagua Fault, the Cayman Trench, and the Puerto Rican Trench are at the boundary between the North America Plate and the Caribbean Plate.

plains, the Appalachian orogenic belt, a central stable region, a shield region, the Arctic stable region, the Arctic orogenic belt, the Arctic Coastal Plain, and the Greenland orogenic belt. The physical map (Fig. 8–3) can be consulted to get a better concept of the elevation and topography of the second-order relief features in each of these nine divisions. In the classification that follows, these nine divisions are grouped into four tectonic types—shields, stable regions, orogenic belts, and plains.

THE SHIELD REGIONS

The oldest rocks of the continental crust are found in the continental **shield regions,** which are extensive uplifted areas that are essentially bare of recent sedimentary deposits. Shields constitute the large stable blocks, or **cratons,** that are the nuclei of present-day continental masses. Although most of the rocks exposed on the shields were metamorphosed during ancient episodes of orogeny, that is, mountain building, they have remained undisturbed for very long periods—typically a billion years or more.

The Canadian Shield, an area marked by low relief (Fig. 8–3), has been the great stable craton of the North American continent since the Pro-

terozoic. The shield stretches across half of Canada and extends into northern Minnesota, Wisconsin, Michigan, and New York (Fig. 8–2). The Canadian shield consists of pre-Cambrian rock, except along the southern margin of Hudson Bay where 1000-foot- (300-m-) thick strata from the Ordovician, Silurian, and Devonian occur.

The Paleozoic strata along the southern margin of Hudson Bay continue northward under much of the bay. In the past, Paleozoic formations were much more widespread over the shield than now. Long intervals of erosion during the Mesozoic and Cenozoic eras stripped sediments from the underlying basement of Archeozoic and Proterozoic rocks.

The Greenland shield is also a stable pre-Cambrian craton. Undoubtedly, it is continuous with the Canadian shield. Greenland is covered with an ice sheet, but pre-Cambrian outcrops along the west and south shores are used to indicate a crystalline complex in the interior region.

THE STABLE REGIONS

The central region (Fig. 8–2) consists of a foundation of pre-Cambrian crystalline rock covered by a veneer of sedimentary rock. The crystalline rock is a continuation of the Canadian shield southward and westward. During certain periods of the Paleozoic and Mesozoic, shallow seas invaded this region and the streams of the Canadian shield and of the flanking mountains poured out their sand and silt into this central region. The veneer of sediment varies greatly in thickness from place to place.

Sometimes this central **stable region** is referred to as the *Great Interior* or *Interior Lowlands.* The terms **prairie** and **plains** are also used when referring to certain sections of the Great Interior (Fig. 8–3). By whatever name it is called, it is a vast region of low relief that stretches from above the Arctic Circle at the edge of the Canadian Rockies to the margins of the Appalachian Mountains.

The **Arctic stable region** (Fig. 8–2) is the northern counterpart of the central stable region; it is, however, much smaller. The stable region in the Arctic spreads from Victoria Island and the southern portion of Banks Island to the southern half of Ellesmere Island and northern Greenland. Thus, the Arctic stable zone includes some of the Arctic islands and the shallow sea-covered areas between them and the mainland. The region consists of a basement of crystalline rock covered by a veneer of nearly horizontal Paleozoic sedimentary rocks.

THE OROGENIC BELTS

Four **orogenic belts** are identified: the western, the Appalachian, the **Greenland,** and the **Arctic belts.** The orogenic belt of eastern Greenland consists primarily of folds and thrust faults of late Silurian age. The Arctic orogenic belt, which stretches from northern Banks Island to northern Greenland, consists essentially of folded Paleozoic strata.

THE APPALACHIAN BELT

The **Appalachian orogenic belt** stretches from northern Alabama to Newfoundland (Fig. 8–2). A rather extensive discussion of the role of plate tectonics in the development of the Appalachians can be found on pages 183–186. The belt consists essentially of two divisions: The inner western-division, which is folded and thrust-faulted, runs from Alabama to New York. The

FIGURE 8–2   Four tectonic types—shields, stable regions, orogenic belts, and plains—are represented by the nine divisions outlined in this figure.

outer eastern-division is compressed, metamorphosed, and intruded by great batholiths; it extends from Alabama to Newfoundland. The outer eastern-division can be recognized in the piedmont province of Georgia and South Carolina.

THE WESTERN BELT

The **western orogenic belt** extends from Alaska through Mexico to the Motagua Fault (Fig. 8–2). The belt consists essentially of three divisions: the Pacific Coastland, the Intermountain Region, and the Rocky Mountains (Fig. 8–3).

The Pacific Coastland, spreading from Alaska into Mexico, is built by two roughly parallel mountain chains that are separated by valleys (Fig. 8–3). The western outer-chain, which rises steeply from the Pacific Ocean, includes Washington's Olympic Mountains and the coastal ranges of Oregon, California, and Baja California. The eastern inner-chain is made up of the Alaska Range, Canada's Coast Mountains, the Cascade Mountains in Washington and Oregon, California's Sierra Nevada, and Mexico's

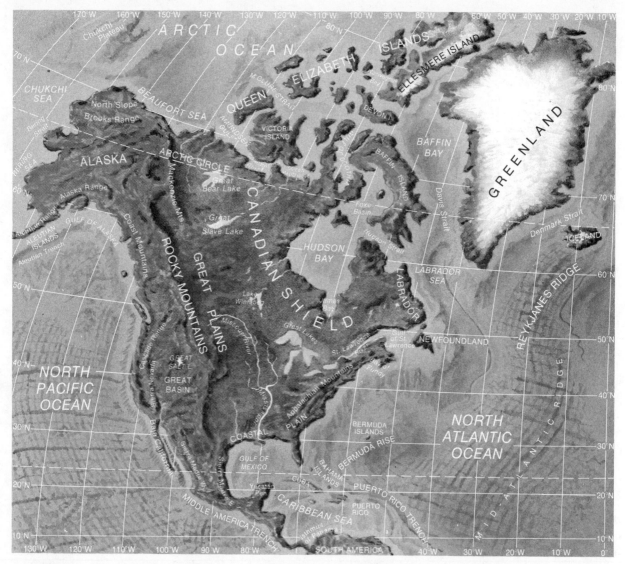

FIGURE 8–3 Second-order relief features are identified and depicted in this physical map of North America.

Sierra Madre Occidental. The highest mountain peak in North America, Mount McKinley at 20,320 feet (6194 m), is in the Alaska Range (Fig. 8–3).

The mountains at the eastern border of the western orogenic belt are known as the Rockies. The Rocky Mountains, with numerous peaks rising to heights of 14,000 feet (4270 m) or more, give the continent a high inland spine. The Rockies extend from northern Alaska into Mexico where they are known as the Sierra Madre Oriental (Fig. 8–3).

The Intermountain Region, an area of basins and high plateaus, sits between the Pacific Coastland and the Rockies (Fig. 8–3). Starting in the north, this region includes the Yukon River Basin, British Columbia's Interior Plateau, the Columbia Plateau, the Colorado Plateau, the Basin and Range Province, and the Mexican Plateau.

PLATE MOTION IN THE WEST

The story of the western orogenic belt is complicated and not fully unraveled. A number of diverse ideas—all rooted in plate theory—have been proposed. The proposals are in their infancy and, no doubt, many revisions will be made as evidence is gathered and the ideas mature. There is, however, agreement that the rocks in the Coast Ranges and the Sierras are related to the descent of an oceanic plate, referred to as the **Farallon Plate,** under the western edge of the North America Plate (Fig. 8–4). The descent of the Farallon Plate via a trench along the coast created an effect similar to that of the present-day plunging Nazca Plate, which is producing the Andes of South America. The volcanic activity that shaped much of the West probably began around 70 million years ago when the North America Plate started overriding the Farallon Plate especially rapidly as the opening of the North Atlantic Ocean went into full swing.

About 30 million years ago, the westward movement of the North America Plate brought it into contact with the ridge separating the Pacific and Farallon plates (Fig. 8–4). This encounter separated the Farallon Plate into three fragments. The middle fragment was completely consumed in the trench. The northern remnant

of the Farallon Plate is the Juan de Fuca Plate, which lies northwest of Cape Mendocino. The southern remnant is the Cocos Plate (Fig. 8–4C).

A new situation developed when the middle fragment of the Farallon Plate was consumed in the trench and the Pacific Plate came up against the North America Plate. The Pacific Plate did not plunge under the western edge of the North America Plate; it slid northwestward relative to the North America Plate. Thus, a new zone of collision, called the *San Andreas Fault,* was created. There was a gradual transformation of the tectonic forces shaping the West from those associated with descent of a plate to those associated with drag and stretching.

The northwestward dragging of the Pacific Plate against the North America Plate distorted the entire region from the Rockies westward. The continental crust stretched and released basalt eruptions which produced features like the Columbia Plateau. The tension also produced the Great Basin centered in Nevada (Fig. 8–3).

The shift of gears, which produced tension and stretching in the West, caused the Los Angeles Basin to subside and initiated a prolonged series of basalt eruptions that built the nearby Channel Islands. When this activity

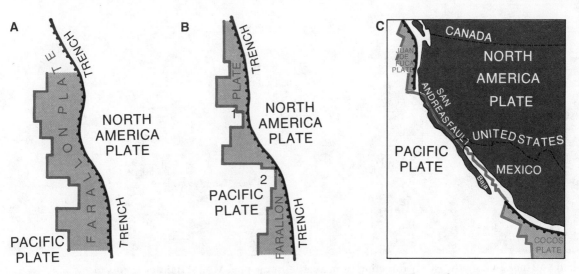

FIGURE 8–4 A. The Farallon Plate is plunging beneath the North America Plate. The zone of subduction is marked by a trench. B. Portions of the Farallon Plate have been completely consumed and the Pacific Plate is making contact with the North America Plate at locations 1 and 2. C. The Juan de Fuca and Cocos plates are remnants of the Farallon Plate.

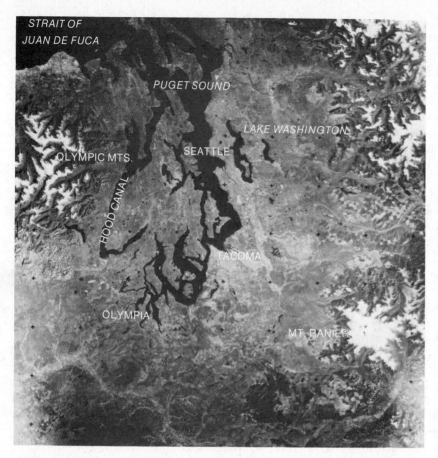

The present volcanic activity in the state of Washington is related to stresses and strains produced by movement of the Juan de Fuca Plate. This satellite photo gives us a view of the Washington region from the Strait of Juan de Fuca to Mount Rainier. Other snow-clad mountains in the Cascades can be seen along the right border of the photo. As you move from the left of the photo along the Strait of Juan de Fuca, you encounter two small indentations—Sequim Bay and Port Discovery Bay. When you move around Point Wilson, two small islands—Indian Island and Marrowstone Island—are at the mouth of Puget Sound. (NASA)

When you look in a southeasterly direction from Seattle, snow-clad Mount Rainier seems to hover over the city—although it is at a distance of about 50 miles (80 kilometers) from Seattle. The peak of Mount Rainier is 14,410 feet (4392 meters) above sea level. (Seattle Visitors Bureau)

moved up the coast, it produced the Berkeley Hills overlooking San Francisco Bay and the Sonoma volcanic field.

The batholiths that constitute much of the exposed granite in western North America were derived from the plunging Farallon Plate. The Rockies, except in their northern part, were built primarily by vertical uplift rather than crumpling from the direction of the sea. The volcanic activity in the Rockies and the uplift of these mountains is attributed to the presence of an oceanic slab—the middle segment of the Farallon Plate—beneath the region. The depth of the slab is at least 200 miles (300 km) below the surface.

THE COASTAL PLAINS

Basically, there are three major coastal plains on the North American continent: the Atlantic, Gulf, and Arctic coastal plains (Fig. 8–2). The **Gulf-Atlantic coastal plain** is bounded on the landward side by the foothills of the Appalachians, the Ouachita Mountains, the Sierra Madre Oriental, and the Central American mountain ranges (Fig. 8–3). The **Arctic Coastal Plain** spreads across northern Alaska into Canada and the lower Mackenzie River Valley.

THE ATLANTIC COASTAL PLAIN

Following the Appalachian orogeny of the Paleozoic, the eastern metamorphosed division of the orogenic belt was broken, during the Triassic period of the Mesozoic era, by high-angle faulting, which extended from South Carolina to the Bay of Fundy. The long and narrow downfaulted basins that were produced trapped sediments flowing into them from the highlands. The Triassic lowland of Maryland, New Jersey, and Pennsylvania, and the central lowland of Connecticut are the best known of these basins.

Much of the eastern shore of North America is a subsiding coast. The continental margin had begun to subside at least by early Cretaceous, if not before. The crystalline rocks of the outer division of the Appalachian orogenic system have been traced eastward under a Cretaceous and Tertiary sedimentary cover to a depth of 10,000 feet (3050 m), which is near the margin of the present continental shelf. This zone of sedimentary overlap on the older crystalline rocks of the eastern continental margin is known as the *Atlantic Coastal Plain.*

The sedimentary cover dips gently and thickens like a wedge oceanward. Since the same sediments continue out beyond the present

Martha's Vineyard is an island 4 miles (6.5 kilometers) off the southeastern coast of Massachusetts. This aerial photo shows a portion of the south shore of Martha's Vineyard in the vicinity of Tisbury Great Pond. The ocean has been very active in shaping this coastline—sealing off the many inlets along the south shore by building long beaches and spits. (NOAA)

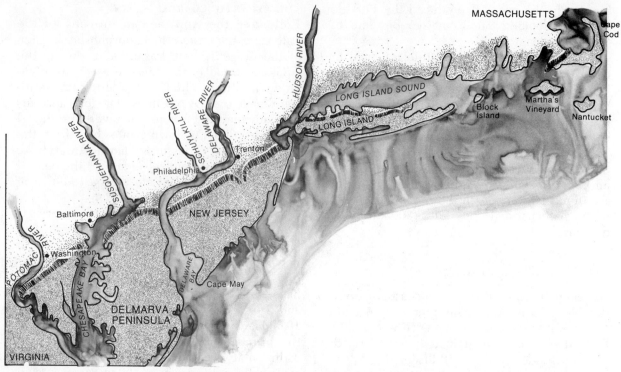

FIGURE 8–5 The Atlantic Coastal Plain between Virginia and Massachusetts is identified by color. A portion of this part of the coastal plain is submerged. The water to the north of Long Island, Block Island, and Martha's Vineyard covers a submerged portion of the Atlantic Coastal Plain. Delaware Bay, Chesapeake Bay, and the lower Potomac River also occupy drowned portions of the Atlantic Coastal Plain.

shoreline, the submerged portion is part of the same geologic province. Coastal plain sediments, for example, exist in Georges Bank (41°N, 67°W). The same type of sediment probably makes up part or all of the shallow continental shelf to the Grand Banks (47°N,52°W) off Newfoundland.

Portions of the Atlantic Coastal Plain from Massachusetts to Virginia have been submerged (Fig. 8–5). Martha's Vineyard (41°25′N,70°40′W), Block Island (41°11′N,71°35′W), and Long Island (40°50′N,73°W) are isolated portions of the Atlantic Coastal Plain that are still above sea level. The rising sea level has produced some excellent bay harbors along the eastern coast of the United States; Baltimore, Maryland, for example, is a port city due to the submergence of the Atlantic Coastal Plain.

The Atlantic Coastal Plain is a region of low relief with gently rolling to hilly land—all of which is below an elevation of 1000 feet (300 m). There is a series of falls and rapids on the rivers that move from the highlands to the coastal plain. The falls of the East Coast mark the junction between the hard rocks of the Appalachian orogenic system—represented in some areas by the Piedmont Region—and the softer deposits of the Atlantic Coastal Plain. This boundary is known as the **fall line.**

Since the relief of the land in the Atlantic Coastal Plain below the falls is very slight, the rivers are navigable up to the falls. At one time during the eighteenth and nineteenth centuries, the falls were a valuable source of power and thus became the sites at which cities were founded. Trenton, New Jersey is located at the fall line on the Delaware River (Fig. 8–5). All the following cities are located on the fall line: Philadelphia, Pennsylvania is on the Schuylkill River; Washington, D.C. is on the Potomac;

Baltimore was founded in 1729 around a protected harbor on the northwest branch of the Patapsco River at the edge of the Piedmont Region about 14 miles (23 kilometers) west of Chesapeake Bay. Today tall buildings of the central business district overlook the inner harbor and Baltimore is one of the largest port cities in the world. (Inner Harbor Management)

Richmond, Virginia is on the James River; Raleigh, North Carolina is on the Neuse River; Columbia, South Carolina is on the Congaree River; and Augusta, Georgia is on the Savannah River.

THE GULF COASTAL PLAIN

The Gulf of Mexico came into existence through the process of subsidence sometime after the Appalachian orogeny of the Paleozoic. The oldest known sediments at its marginal areas are from the Permian period, which is very late in the Paleozoic. The Gulf Coastal Plain, which is continuous with the Atlantic Coastal Plain, nearly encloses the Gulf of Mexico (Fig. 8–2).

The two most important rivers of the Gulf Coastal Plain are the Mississippi and the Rio Grande. These rivers and others draining the interior of the continent have deposited a great mass of sediments at their mouths. The depth at some locations is in excess of 25,000 feet (7600 m). The weight of these sediments has caused the crust along the Texas, Louisiana, and Mississippi coastal areas to subside faster than the crust in the middle of the Gulf.

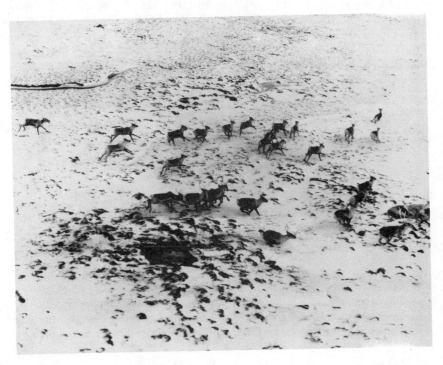

A caribou herd on Alaska's flat, bleak, snow-covered coastal plain. (U.S. Navy)

THE ARCTIC COASTAL PLAIN
The region north of the Brooks Range in Alaska is part of the **Arctic Coastal Plain** (Figs. 8–2 and 8–3). The upper sediments are from the Tertiary period of the Cenozoic. Mesozoic sediments—Cretaceous, Jurassic, and Triassic—are found below those of the Cenozoic. The cretaceous sediments found in this area range from 15,000 to 25,000 feet (4600 to 7600 m) in thickness.

From the air, the Arctic Coastal Plain appears to be a bleak, flat, snow-covered wasteland of frozen lakes and rivers. The surface dips gently northward and extends out under the Arctic Sea. If you start from the sea, the land rises gradually as you move southward toward the Brooks Range. The maximum elevation attained at the southern end of the Arctic Coastal Plain is 600 feet (180 m).

COASTAL SURVEY

Almost everyone recognizes intuitively that the shoreline is where land and water meet. The word **shore,** however, has a somewhat restricted meaning; it is the zone from mean low tide to the inner edge of wave-transported sand. The term **coast** refers to a broad zone directly landward from the shore. Thus, a coastal zone includes sea cliffs and elevated terraces as well as the lowlands landward of the shore.

The shoreline of mainland North America is about 39,000 miles (62,800 km) long. The Atlantic coastal zone is steep and rocky in the north, but south of Cape Cod it slopes gently to the sea as the Atlantic-Gulf coastal plains. Along most of the Pacific coastal zone, high mountains rise abruptly from the sea. Numerous gulfs and bays are found along the coastal zone; the largest are the Gulf of Mexico and the Hudson Bay. Many islands, which lie seaward of the coastal zone, are physically related to the mainland; Greenland is the largest of these offshore islands.

BEACHES OF NORTH AMERICA

A **beach** is the zone of unconsolidated material extending landward from mean low tide to the zone of permanent vegetation, or a zone of dunes, or a sea cliff. The upper limit of a beach marks the effective limit of ordinary storm waves. The important nomenclature of beach features is illustrated in Figure 8–6.

The backshore consists of a **berm,** which is a nearly horizontal portion of the beach formed by wave-deposited material. Some beaches have no berms, others have one or more. The foreshore area is the sloping part of the beach lying between the berm and the low-water mark.

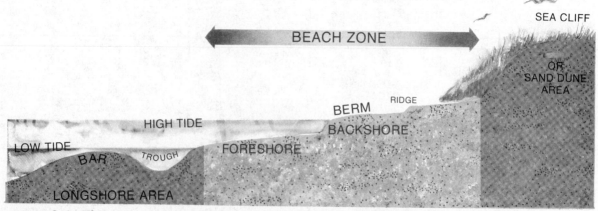

FIGURE 8–6 There are two major divisions in the beach zone: the foreshore and the backshore. Disputed ownership of beaches and problems of erosion often need to be resolved in the courts. Lawsuits involving the beach zone need to be based on a very precise nomenclature of beach features.

Sea oats grow at the upper edge of this gently sloping sandy beach along South Carolina's Atlantic seashore. The photo was taken at Edisto Beach, just south of Charleston. (South Carolina Dept. of Parks)

In the shallow-water area adjacent to the beach, there is an elongate depression referred to as a **longshore trough.** At times, the longshore trough consists of a series of depressions that form between the lower beach and the zone of breakers. Immediately seaward of the longshore trough is a sand ridge or ridges, known as the **longshore bar,** which is often exposed at low tide.

The principal unconsolidated sediments that go into the building of the beaches of North America are gravel, coarse sand, and fine sand. A typical gravel or **shingle beach** has a beach ridge—a low, lengthy ridge of beach material piled up by storm waves landward of the berm. In some areas—along the Gulf Coast in particular—a beach ridge usually consists of shells. Coarse-sand beaches may have berms that slope landward, often at considerable angles. Fine-sand beaches usually have very gentle foreshore slopes built with hard-packed sand.

The most common type of beach is located along the seaward side of barrier bars since approximately 47 percent of the United States coastline is bordered by some type of barrier. **Barrier beaches** are almost continuous along the Gulf Coast. Along the Atlantic coastline, barrier beaches are found from Florida to the New York area. Arctic Alaska is another area with long barrier bars.

CUSPATE SHORELINES

The word *cusp* simply means a point or pointed end. Some shorelines display a pointed profile that is formed by the meeting of two arcs. Since the shorelines are shaped like a cusp, they are referred to as **cuspate shorelines** (Fig. 8–7).

Specifically, however, in common terms, the word **cape** is used to refer to a point of land that projects prominently into a sea or lake. Sometimes other terms—such as **point, head,** and

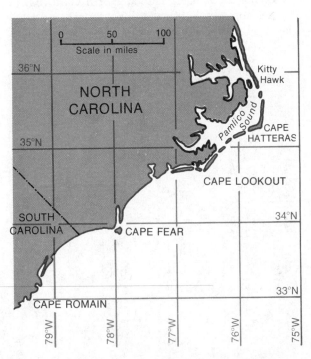

FIGURE 8–7   The pointed profile of the shoreline along the Carolina coast is referred to as a *cuspate shoreline.* The coastal area from Kitty Hawk to Cape Romain forms a huge cuspate foreland that juts into the sea.

promontory—are used to express the same idea. Capes may be formed in several ways: (1) Offshore bars or barrier beaches may form points as a result of converging currents depositing sandy material in shallow water; for example, Cape Hatteras, North Carolina has been formed in this way. (2) River deposits may accumulate until a point of land is created; Cape Gracios a Dios (15°N, 83°8′W) in Honduras is an example of this kind of development. (3) The initial outline of a cape may be established by glacial deposits that are then reworked by waves—as happened at Cape Cod. (4) Massive, resistant rock cliffs, which withstand the attack of ocean waves, may project away from the shoreline as headlands; Point Dune (34°N, 118°48′W), just west of Malibu, California, is an example of this type of developmental history.

CUSPATE FORELANDS

An area of high land that juts into the sea is referred to as a **cuspate foreland.** The Carolina coastline from Kitty Hawk, North Carolina, to Cape Romain, South Carolina, is a huge foreland. The area, shown in Figure 8–7, juts promi-nently into the sea.

A series of cuspate forelands stretches from Cape Hatteras to Cape Romain. The intervals between these capes are from 70 to 100 miles (110 to 160 km). If a straight line is extended from Cape Lookout to Cape Fear, for example, the distance from the center of the line to the nearest point on shore is about 20 miles (32 km), which can be taken as the distance these capes protrude seaward.

Cape Canaveral (28°24′N, 80°37′W), on the east coast of Florida, is a cuspate foreland that covers an extensive area. On the west coast of Florida, there are a series of smaller cuspate fore-lands that culminate to the northwest at Cape San Blas (29°36′N, 85°21′W), which is a rather extensive cuspate foreland.

There are comparable cuspate forelands in Alaska, north of Bering Strait. Working north from Cape Espenberg (66°30′N, 164°W), there is Cape Krusenstern (67°10′N, 163°42′W) and Point Hope (68°10′N, 166°50′W). Icy Cape (70°17′N, 162°W), Point Franklin (70°55′N, 158°54′W), and Point Barrow (71°20′N, 156°32′W) constitute a group of cuspate forelands similar to those of the

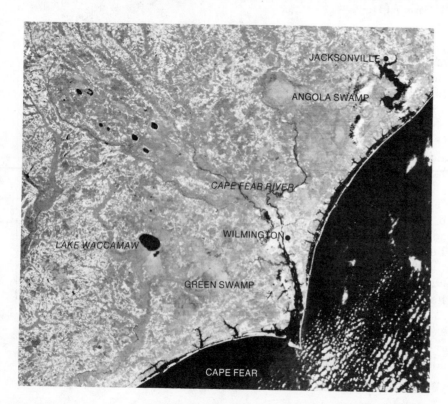

This satellite view of eastern North Carolina from Jacksonville to Cape Fear and on to the South Carolina border at North Myrtle Beach gives a perfect view of the cuspate nature of the coastline. Angola Swamp drains by way of the northeast Cape Fear River to the inlet at Cape Fear. (NOAA)

Carolinas. These last three Alaskan capes are separated by about 80 miles (130 km).

Most cuspate forelands seem to be built by deposits in the slack-water zone between two coastal eddies. The eddies commonly develop inside a major current that flows along the coast. The Gulf Stream, north of the Straits of Florida, is an example of a major current that gives rise to cuspate-producing coastal eddies.

CUSPATE SPITS

A small, narrow point of land, consisting of beach deposits, that extends out into open water is called a **cuspate spit.** Spits are produced by longshore drift where an embayment occurs. Generally, the incoming tidal currents are stronger than the outgoing currents and the longshore current producing the spit is too weak to maintain a continuous straight line and swings into the embayment. Thus, the ends of most spits projecting from headlands have a landward curvature.

Rockaway Beach on Long Island, Sandy Hook in New Jersey, and Cape Cod in Massachusetts are examples of spits. All of these spits curve back on their ends to produce hooks. On the end of the hook, on its inner side, there may be an additional little hook. This additional hook, in the case of Cape Cod, is big enough to have its own name, Long Point (Fig. 8–8). Long Point forms a natural breakwater for the port of Provincetown.

Sandy Hook is a compound spit that has evolved slowly over a long period of time. In Figure 8–9, five stages in the growth of Sandy Hook are depicted. During the first four stages, Rumson Neck and the North Long Branch land areas extended farther into the Atlantic Ocean. As the land was worn back by wave erosion, the spit gradually assumed its present position. The hook at Spermaceti Cove is the end of the third-stage spit.

GIANT CUSPS

Along many open beaches there are times when the beach develops a series of shallow bays and points, called **beach cusps.** These projecting points may be spaced at intervals of 300 feet (100 m) or more. Sometimes these cusps appear as solitary points. At times, cusps develop in a period of a few days and then disappear.

We do not have very adequate explanations as to why beach cusps appear and disappear. They do, however, seem to be related to the direction in which waves converge on a beach and then diverge from it. The approach of waves diagonally to the shore sets up currents inside the breakers; outward-flowing rip currents develop at places, and back eddies from these seaward-moving masses of water may cause deposition that produces a point, or cusp.

DELTAIC COASTS

Three types of deltas with some specific examples, including the Mississippi and Saint Clair deltas, were described on pages 226–229. A **deltaic coast** is a zone of very active deposition, particularly in bays protected from the attack of waves; the Colorado River in Texas, for example, empties into a well-protected lagoon and displays a spectacular growth. Along the unprotected Gulf Coast, which is buffeted by low-energy waves, deltas build seaward wherever a

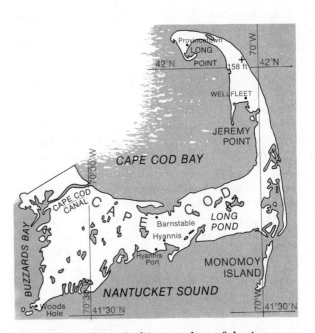

FIGURE 8–8 Cape Cod is a product of the ice sheets of the Pleistocene. It consists largely of deposits of sand, gravel, clay, and boulders brought by the ice. Over the centuries, glacial debris—sand, gravel, and cobbles—cut from the scarp of Cape Cod by storm waves has been moved along the shore to produce the present spit.

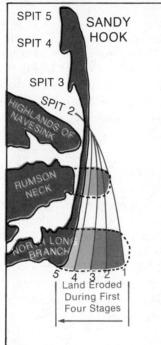

FIGURE 8–9 Five steps in the growth of Sandy Hook are shown. In the earlier stages the land area of New Jersey extended farther into the Atlantic Ocean. Today, Sandy Hook is a geologic adolescent. It is still growing. Sandy Hook's area has quadrupled since it was first surveyed in 1685. The old lighthouse on the Hook was erected in 1764 and stood near the lapping waves on the spit's end. Presently the end of the Hook is about 1.5 miles (2.5 kilometers) from the light.

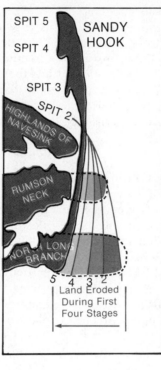

large amount of sediment is being supplied. What is generally not recognized, however, is that especially large deltas are building along the open Arctic coasts of North America where ice-pack protection allows forward building.

YUKON-KUSKOKWIM DELTAS

The Bering Sea coast of Alaska—roughly between latitudes 60° to 63°N and longitudes 162° to 166°W—is one big deltaic coast, with a shoreline of about 500 miles (800 km) (Fig. 8–10). The sediments are carried to this region by the Yukon and Kuskokwim rivers.

This deltaic coast appears to have developed a series of different lobes that frequently changed position. Initially, the main mouth for the drainage was at the present Kuskokwim River's outlet. Then the mouth shifted to Baird Inlet and the Kuskokwim's mouth sank to form an estuary. The mouth of principal drainage continued to shift northward until it reached its present position on the south shore of Norton Sound.

Several mountainous islands—Nelson Island at Baird Inlet, for example—have been incorporated into the deltaic plain as a result of its seaward movement. The numerous lakes in the delta are related largely to permafrost and thawing.

NORTH-SLOPE DELTAS

One of the largest deltaic coasts in the world is found on Alaska's North Slope. A series of com-

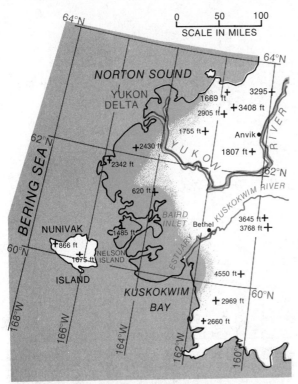

FIGURE 8–10 A big deltaic coast (color) stretches from the mouth of Kuskokwim River to the mouth of the Yukon River. This deltaic coast of Alaska is a shoreline that stretches for about 500 miles (800 km) along the border of the Bering Sea.

pound deltas stretch from longitude 147°W to 162°W. The area consists of channels interlaced with sand bars and numerous delta lakes.

Pack ice is piled against this portion of the Alaskan coast for most of the year, the only exception being the few months of summer. The source for most of the rivers that flow into the delta is in the Brooks Range. The streams flow only a few months of the year, however; during part of that time, the river water flows under the ice pack into the Arctic Ocean.

COASTAL LAGOONS

A shallow body of water that has a restricted connection with the sea is called a **lagoon**. Almost half of the U.S. coastline is bordered by barriers with lagoons on the inside. As a result of

streams flowing into the lagoons, they are in various stages of filling and generally contain large marsh areas.

GULF-COAST LAGOONS

Lagoons are common throughout the coastal area that surrounds the Gulf of Mexico (Fig. 8–11). Starting along the south coast of Florida, there is Florida Bay. Florida Bay, inside the Florida Keys, is essentially a lagoon in its upper end even though it opens out to the west (Fig. 8–11). Charlotte Harbor, which is protected by barriers, is typical of Florida's west coast lagoons (Fig. 8–11). There is a group of lagoons that extends along the northwest coast of Florida from Pensacola east to Carrabelle (Fig. 8–11).

The Mississippi and Alabama coastlines are also protected by barriers. Mobile Bay and Mississippi Sound nestle behind these barriers (Fig. 8–11). Barataria Bay, Louisiana, on the west side of the bird-foot delta of the Mississippi, is an example of a lagoon inside the barrier islands around a submerged delta lobe.

Along the Texas coastline—from Boca Chica, which is east of Brownsville, north to Galveston—a series of lagoons is found inside the almost continuous barrier islands (Fig. 8–11). One of the best known of these lagoons is Laguna Madre, which extends from Corpus Christi Bay south for 125 miles (200 km) to the Boca Chica area.

ALONG THE EAST COAST

Florida's Atlantic coastline is protected by extensive barriers, which here and there touch the mainland. Daytona Beach, Vero Beach, and Miami Beach are all built on barrier bars. Behind each of these beaches is a long narrow lagoon. Biscayne Bay is a lagoon and so is the Indian River behind Vero Beach (Fig. 8–11).

Georgia's coast is composed of short stubby barriers, which are separated by many large tidal inlets (Fig. 8–12). This kind of situation is typical of tidal current-controlled coasts. On the other hand, North Carolina's coast consists predominantly of long, thin, barrier islands with a few tidal inlets, which is typical of a wave- and current-controlled coast. Pamlico Sound, North Carolina, is bounded seaward by barriers and is the largest lagoon on the East Coast.

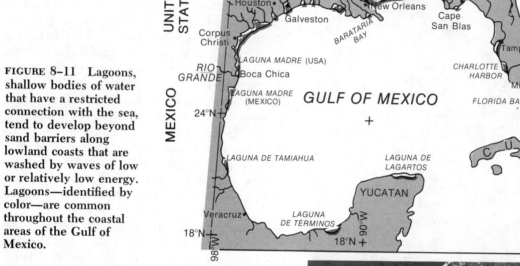

FIGURE 8–11 Lagoons, shallow bodies of water that have a restricted connection with the sea, tend to develop beyond sand barriers along lowland coasts that are washed by waves of low or relatively low energy. Lagoons—identified by color—are common throughout the coastal areas of the Gulf of Mexico.

FIGURE 8–12 The coast of Georgia, from the mouth of the St. Mary's River to the mouth of the Savannah River, consists of short, stubby barrier islands. The islands are separated by large tidal inlets.

Galveston, below us, lies on the eastern end of Galveston Island, which is a long barrier-bar island in the Gulf of Mexico, about 2 miles (3.2 kilometers) off the Texas mainland. Galveston has a deepwater harbor channel on the protected side of its barrier bar. The channel runs between Galveston and Pelican Island at the left of the photo. (NOAA)

FIGURE 8–13 South Carolina's coast represents a transition between the short stubby barriers of Georgia's coast and the long thin barriers of North Carolina's coast.

South Carolina's coast is a transition between that of North Carolina and Georgia. The geomorphology of the South Carolina coast results from a mixture of wind-generated and tidal-generated forces. The cuspate foreland of the northern section of South Carolina's coast gives way to a cuspate delta south of Winyah Bay and then to a barrier-island zone, which dominates the central and southern portion of the coast (Fig. 8–13). The cuspate delta, which extends 18.5 miles (30 km) along the South Carolina coast, is built by Santee River sediments and is the largest wave-dominated deltaic complex on the East Coast. Few tidal inlets breach South Carolina's coastline in its northern section, but the number and size of tidal inlets increases south of Winyah Bay.

The eastern shore of the Delmarva Peninsula from Cape Charles to Dewey Beach, the eastern shore of New Jersey from Cape May to Point Pleasant Beach, and the southern shore of Long Island from Brooklyn to Shinnecock Bay are barrier-bar coastlines with associated lagoons (Fig. 8–14). Each of these coastlines is the same length—about 100 miles (160 km)—and each is undergoing wave erosion just beyond the location where the barrier bar touches the mainland.

On eastern Long Island in the area of the Hamptons, in New Jersey in the area from Asbury Park to Long Branch, and on the Delmarva Peninsula from Rehoboth Beach to Cape Henlopen, waves are cutting against high bluffs and actively eroding them.

Southward-moving longshore currents in the New Jersey and Delmarva areas and westward-moving longshore currents in the Long Island area drag great quantities of sand along the coast (Fig. 8–14). The quantity of sand carried diminishes as the longshore current moves farther from its source because the current is continuously dropping sand, which contributes to the building of the bars. The **longshore bars** that are close to the source of sand are almost continuous and unbroken; in fact, they form offshore islands that are separated from the mainland by a wide lagoon. On Long Island this section of the coastline is represented by Fire Island and the lagoon back of it is Great South Bay (Fig. 3–18 provides detail of the Long Island coastline). In New Jersey there are two long island segments: Island Beach and Long Beach Island (Fig. 8–14). Barnegat Bay is the lagoon behind Island Beach and the northern section of Long Beach Island. The lagoon be-

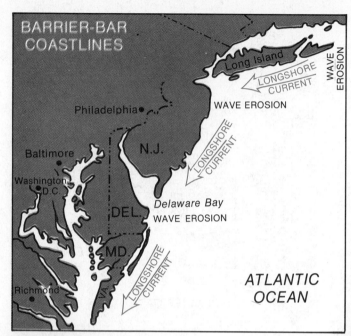

FIGURE 8–14 Longshore currents sweep southward along the New Jersey and Delmarva coasts. The longshore currents of Long Island sweep westward. The currents drag great quantities of sand along these coasts. The sand is derived from the zones of wave erosion. The pattern of barrier islands along these coasts is created by the combination of wave erosion and longshore currents.

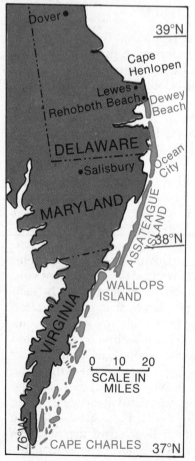

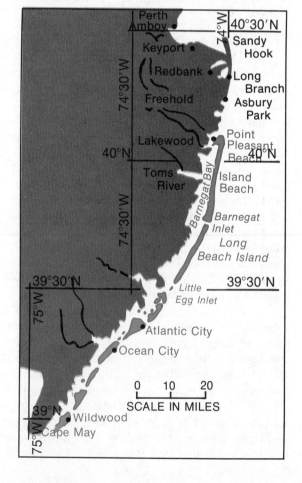

hind the central portion of Long Beach Island is known as *Manahawkin Bay;* at its southern end the lagoon is called *Little Egg Harbor.* Along the Delmarva Peninsula the long unbroken barrier is Assateague Island; its associated lagoon is Chincoteague Bay.

Since the amount of sand carried by the longshore currents is constantly diminishing as the current increases its distance from the erosion zone, the bars farthest from the source are small and fragmentary. In each of these cases— the Delmarva Peninsula, New Jersey, and Long Island—a number of much shorter bars and islands are found beyond the main unbroken segment. The lagoons in these fragmentary sections tend to contain a lot of marshland intersected by bayous and channels. The zone of small segments on Long Island includes Jones Beach, the Rockaways, and Coney Island. In New Jersey, these fragmentary barriers are found at Atlantic City, Ocean City, and Wildwood. On the Delmarva Peninsula, the fragmentary section is in the southern portion occupied by Virginia.

WEST-COAST LAGOONS

The largest lagoon along the California coast, Humboldt Bay (40°47′N, 124°11′W), is shown in Figure 8–15. Humboldt Bay is 14.3 miles (23 km) long and 4.4 miles (7 km) wide. There are three sections in the bay: South Bay, Arcata Bay, and Entrance Bay. The South Bay and Arcata Bay sections are at either end and they consist of marshlands with many winding tidal channels, which have gravel and sand bottoms with finer and more muddy material in their upper reaches. Entrance Bay is the navigable part of the lagoon along which the port facilities of Eureka are located.

The Mad River empties directly into the Pacific Ocean at latitude 40°57′N, longitude 124°7′W. Figure 8–15 shows that the mouth of the Mad River is just north of Arcata and south of McKinleyville. The Eel River also empties directly into the open Pacific; its mouth is located south of Humboldt Bay at latitude 40°40′N, longitude 124°20′W. The suspended sediment placed by these rivers into the open Pacific Ocean is carried back along the coast and swept into Humboldt Bay through its entrance channel. Thus, most of the mud in the bay is contrib-

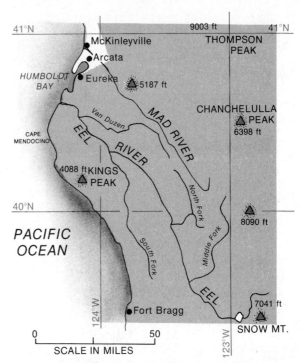

FIGURE 8–15 Humboldt Bay is the largest lagoon along the California coast. Although the Mad and Eel rivers empty directly into the Pacific, their loads of suspended material—carried from the peaks of the coast ranges—make an important contribution to the sediments in the lagoon.

uted to it by two rivers that do not empty into the bay.

Many of the lagoons along California's southern coast have been reworked and reshaped by human intervention. Harbor-building projects, especially in the Long Beach-San Pedro area, have had a decided impact on the lagoons as well as the erosional processes. The Newport Bay area has been reworked as a yachting center, but the harbor is still essentially a lagoon with broad tidal flats on the inside. The Coronado barrier at San Diego has produced an elongated lagoon, San Diego Bay, with a natural opening at its north end that is maintained largely by the strong tides. Some of the marsh areas on the north and east ends of the bay have been filled to provide expansion space for the city of San Diego.

There are a number of lagoons along the western coast of Baja California (Fig. 8–16). Cabo San Lázaro (24°48′N, 112°16′W) stands at

FIGURE 8–16 Hot, dry deserts cover the western half of Baja California. A barren mountain range runs along the eastern side of the peninsula. Some of the lagoons of western Baja have great economic importance because of the mining of salt that accumulates in the salt flats at their heads.

the point of a huge cuspate foreland with an extensive lagoon behind its barriers. Further north, Laguna Ojo de Liebre (27°50′N,114°15′W) and its immediate neighbor Laguna Guerrero Negro are protected from the open sea by sand barriers; but tidal inlets allow a large supply of seawater to enter these lagoons. Rapid evaporation in this arid region of western Baja produces wide salt flats at the heads of Laguna Ojo de Libre and Laguna Guerrero Negro.

RIVER ESTUARIES

Technically, any drainage channel adjacent to the sea in which the tide ebbs and flows is called an **estuary.** Thus, some estuaries are the lower courses of rivers or smaller streams, and others are no more than drainage ways that lead seawater into and out of coastal swamps and lagoons. The following discussion is confined to river estuaries—the typical, long, relatively narrow estuary that tapers inland toward its head and results in an embayment.

Chesapeake Bay, the longest estuary in the United States, is primarily the drowned river valley of the Susquehanna and Potomac rivers (Fig. 8–17). There are, of course, other rivers and streams that enter the bay, notably the Patuxent, Rappahannock, York, and James rivers. From the mouth of the Susquehanna to the entrance of the Chesapeake is a distance of approximately 190 miles (300 km). The flat floor of most of the bay is from 16 to 33 feet (5 to 10 m) deep. The sediments are mud in the deep areas and sand in shoal water, especially near land.

At the mouth of the Chesapeake, there is a regular predominant pattern of current flow; the current flows in on the north side and out on the south side. Studies show that the rate of erosion on the eastern shore of Chesapeake Bay is greater than that occurring on the western shore. More than 17,000 acres (6800 hectares) of land were eroded from Maryland's Chesapeake shoreline during the course of a 90-year-measurement period. There have also been substantial erosional changes in the islands of the Bay. Some of the sediment eroded from the shoreline and islands has filled depressions and contributed to the development of the Chesapeake's flat floor; other portions of the sediment have formed shoals in the lee of eroded islands.

Delaware Bay is another large estuary that is somewhat similar to that of the Chesapeake (Fig. 8–17). Other significant East Coast rivers with estuaries that are clearly drowned river valleys are the Hudson of New Jersey–New York, and Connecticut's Connecticut and Thames rivers. The Merrimack River in northern Massachusetts has a well-developed estuary that penetrates the coast north of Plum Island.

On the West Coast, the mouths of the Columbia and Yaquina rivers are well-developed estuaries. The length of the Columbia's estuary is difficult to determine, however. The tide goes up the Columbia as far as Bonneville Dam, which is

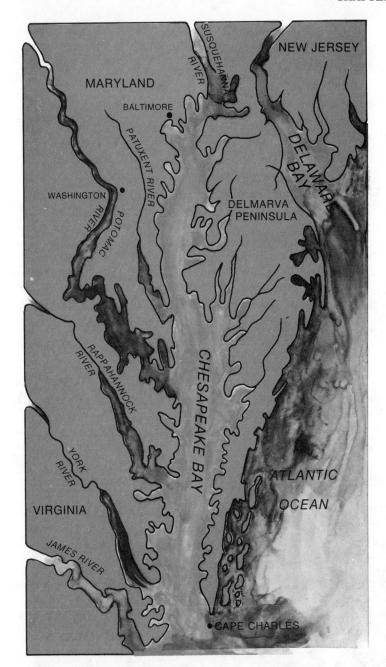

MARYLAND

BALTIMORE

WASHINGTON

VIRGINIA

NEW JERSEY

DELMARVA
PENINSULA

CHESAPEAKE BAY

ATLANTIC

OCEAN

CAPE CHARLES

SUSQUEHANNA
RIVER

POTOMAC

PATUXENT RIVER

RAPPAHANNOCK
RIVER

YORK
RIVER

JAMES RIVER

RIVER

DELAWARE
BAY

FIGURE 8–17 Chesapeake Bay is a long narrow arm that runs north from the Cape Henry—Norfolk area; the bay is almost 200 miles (320 kilometers) long and from 4 to 40 miles (6 to 64 kilometers) wide. Delaware Bay is about 50 miles (80 kilometers) long and about 35 miles (55 kilometers) at its widest point. Philadelphia lies 100 miles (160 kilometers) above the entrance of the bay on the Delaware River.

145 miles (230 km) from the estuary entrance. The lower course of the river is a maze of point bars; the sandbar channels that appear to represent the estuary extend up river at least 80 miles (130 km). The mouth of the Columbia is the main harbor between San Francisco and the Strait of Juan de Fuca.

SURVEY OF PHYSIOGRAPHIC REGIONS

The classification used in the first section of this chapter was based on four tectonic types: shields, stable regions, orogenic belts, and plains. The function of this section is to review

Beyond the twin towers of the World Trade Center lies New York Bay, an estuary that is clearly the drowned river valley of the Hudson. Liberty Island (formerly Bedloe's Island), on which the Statue of Liberty stands, lies close to the Jersey City shoreline; the island and statue can be seen at the right of the photo. (Port Authority of New York and New Jersey)

the land regions of Canada and the United States in more traditional terms and to add to your understanding by identifying additional physiographic provinces within the broader classification previously developed.

CANADA

The landmass of Canada ranges from Pelee Island (41°46′N,82°39′W) in Lake Erie to Cape Columbia (83°7′N, 70°W) on Ellesmere Island. The official physiographic map of Canada identifies eight major regions (Fig. 8–18).

THE CORDILLERAN REGION

This western mountain region of Canada covers most of British Columbia, a very small section of western Alberta, the Yukon Territory, and a small part of the Northwest Territories. The system is made up of eastern, central, and western sections, each with distinctive characteristics.

The rugged front range of the Richardson-Mackenzie-Rocky Mountains, which are high fold mountains, form the eastern boundary of the region. The Richardsons are in the north, then come the Mackenzies, and the Rockies are in the south of this eastern section. Some of the highest peaks and most beautiful scenery on the conti-

Lake Louise, called the *Pearl of the Canadian Rockies,* is located in Banff National Park in southern Alberta at latitude 51°26′N, longitude 116°11′W. The lake is about 1.5 miles (2.4 kilometers) long. The peaks of the surrounding mountains are over 9800 feet (3000 meters) above sea level. (National Film Board of Canada)

PHYSIOGRAPHIC REGIONS OF CANADA

1. CANADIAN SHIELD

2. GREAT LAKES—ST. LAWRENCE LOWLAND

3. INTERIOR PLAINS
   a. First Prairie Level
   b. Second Prairie Level
   c. Third Prairie Level
   d. Mackenzie Lowlands

4. HUDSON BAY LOWLANDS

5. CANADIAN APPALACHIANS

6. CORDILLERAN REGION
   a. Rockies
   b. Interior Plateaus
   c. Coast Ranges
   d. Inner Passage
   e. Insular Mountains

7. INNUITIANS

8. ARCTIC LOWLAND REGION
   a. Foxe Basin Lowland
   b. South Archipelago Lowland
   c. Arctic Ocean Lowland

FIGURE 8–18 There are eight principal physiographic regions in Canada.

nent are in the Canadian Rockies. Mount Robson (53°5′N,119°7′W), the highest peak in the Canadian Rockies, towers 12,972 feet (3954 m) above sea level.

The central section of this region is marked off from the eastern by a sharp break, the down-faulted depression known as the *Rocky Mountain Trench,* which is one of the most remarkable valleys of its kind. The valley is up to 15 miles (24 km) wide and several thousand feet deep. Through the trench's valley flow the waters of the Kootenay, Columbia, Fraser, Peace, and Liard rivers. West of the trench are plateaus, deep valleys, basins, and low mountains, which appear as an intricate web of deeply trenched uplands and long, narrow basins. In the north,

The Peticodiac River of New Brunswick is flowing from left to right across the top of the photo. The river empties into Chignecto Bay at 45°50′N, 64°33′W. Chignecto Bay leads into the Bay of Fundy just north of Saint John. This photo was taken at low tide in the Peticodiac—its tributary in the foreground is drained dry. (National Film Board of Canada)

the Yukon Plateau nestles between the Dawson Range on the west and the Selwyn Mountains on the east. The Cassiar Mountains form the southern boundary of the Yukon Plateau. South of this area, we find the Nechako Plateau (54°N,124°30′W) and the Fraser Plateau (51°30′N,122°W), which make up the Central Upland of British Columbia. Southward the upland is squeezed out between the Columbia and Cascade mountains. The Columbia Mountains are a series of massive ranges—the Cariboo, Monashee, Selkirk, and Purcell Mountains— with deep-faulted or glaciated troughs between them. Kootenay Lake lies in a trough between the Purcell Mountains on the east and the Selkirk Mountains on the west. The trough between the Selkirk Mountains and the Monashee Mountains is occupied by Arrow Lake.

The western section of the Cordilleras consists of a triple structure made up of the Coast Mountains, the Inner Passage, which is a drowned, coastal trench or downfold, and an outer range of mountains identified as the Insular Mountains, which are largely submerged and form Vancouver Island and the Queen Charlotte Islands. The Coast Mountains begin in the Saint Elias Mountains (60°30′N,139°30′W), which divide the Alaskan panhandle from Canadian territory. Mount Logan (60°34′N,140°30′W), the highest peak in Canada, towers 19,850 feet (6050 m) and thrusts out of glistening icefields. Glacial activity in the area is intense.

THE CANADIAN APPALACHIANS

This region extends northeastward from the eastern townships of the province of Quebec to encompass the Gaspé Peninsula, the Maritime Provinces, and the island of Newfoundland (Fig. 8–18). The folded Appalachians in this area have been eroded into low, rounded mountains that are dissected by valleys and broken by broader lowland areas developed on belts of weak rock.

Three broad groups of highlands and uplands can be recognized in the region. In Quebec and the Maritimes, the region consists of a series of southwest-to-northeast-trending ridges—for example, the Notre Dame Mountains, which terminate in the Shickshock Mountains that form the core of the Gaspé Peninsula. Mount Jacques-Cartier in the Shickshocks has an elevation of 4160 feet (1268 m). Farther to the east, the land dips down to a broad bench that consists of the Chaleur Uplands and the Miramichi Highlands. The third and most easterly of the highland groups is made up of two branches—the New Brunswick and Nova Scotia highlands, which merge to form the Long Range Mountains on the island of Newfoundland.

The few lowlands of the region—Prince Edward Island, the Cumberland Lowland, and the Magdalen Islands—generally border the Gulf of St. Lawrence. On the island of Newfoundland, however, there is a fairly broad lowland on the northcentral coast facing Notre Dame Bay and the Labrador Sea.

THE INNUTIAN REGION

This is another system of marginal mountains; they are found on the islands of the Northwest Territories (Fig. 8–18). Paleozoic rocks are intensively folded in the Parry Islands, Axel Heiberg Island, and central and northeastern Ellesmere Island. Much of the **Innutian Region** is covered with fields of permanent ice and snow through which mountain peaks protrude. The high folded mountains in this region have high peaks; on Ellesmere, for example, Barbeau Peak stands at 8051 feet (2454 m) above sea level.

THE CANADIAN SHIELD

This is by far the largest of Canada's physiographic regions. The **Canadian Shield** is centered around Hudson Bay and occupies about 49 percent of the total area of the country (Fig. 8–18).

Most of the shield is less than 2000 feet (610 m) above sea level. In the north, however, the rim of the shield is about 7000 feet (2100 m) above sea level, and fiords with walls 3000 feet (900 m) high extend many miles into it. South of Hudson Strait, the Torngat Mountains (59°N,64°W) have at least one peak with an elevation of 5320 feet (1620 m). Along the north shore of the St. Lawrence River in Quebec, the rim of the shield is a 2000-foot (9610 m) escarpment; in Ontario, the rim rises about 1500 feet (460 m) above the north shore of Lake Superior.

Starting at a point in Manitoba and running northwestward, the edge of the shield is marked by a series of lakes (Fig. 8–3). Lake Winnipeg (52°N,97°W), Lake Athabasca (59°7′N,110°W), Great Slave Lake (61°30′N,114°W), and Great Bear Lake (66°N,120°W) are all found at the shield's edge. Some of these lakes are very large. Great Bear Lake, for example, developed in a combination of faulted and ice-gouged depressions on the shield's margin. On the flatter western sides of these lakes there are extensive beds of clay and gravel beaches, which probably means they were all wider at one time. The rivers running into the lakes have developed large sandy deltas.

Glaciations during the Pleistocene had a striking effect on the shield's surface. Much of the effect was erosive. By stripping off the top layer of weathered material, the glaciers roughened the rock surface into a grained landscape of ridges and troughs. The troughs between the ridges are occupied by enormous numbers of lakes.

HUDSON BAY LOWLANDS

The largest section of this region is on the southwest shore of Hudson Bay; it is between 100 and 200 miles (160 and 320 km) wide (Fig. 8–18), and extends about 800 miles (1290 km) from the Churchill River in Manitoba to the Nottaway River in Quebec.

The region is a low, wet, rocky plain wedged between the shield and the water of Hudson Bay. All of the area is undergoing uplift as a result of the removal of the heavy load of the Pleistocene ice sheet. In fact, some sections have risen 600 feet (180 m), which has left a number of raised beaches far from water.

GREAT LAKES-ST. LAWRENCE LOWLAND

This region forms a narrow belt that stretches from southern Ontario into Quebec and on to Anticosti Island (Fig. 8–18). There is more than one part to this region: The first section, with its eastern border at the Niagara escarpment, is a rich domelike peninsula between Lake Huron and Lake Erie. The second section stretches along the lowland of the St. Lawrence River from Lake Ontario to the Gulf of St. Lawrence. Anticosti Island and the Mingan Islands at the mouth of the St. Lawrence form the third natural division in this region.

This region was covered by glacial ice during the Pleistocene. As the ice receded from around the Huron-Erie dome, it halted periodically and at each "rest stop" the glacial meltwater deposited sand and gravels around its edges in deposits known as *recessional moraines.* Today, these deposits ring the central portion of the dome. The lowlands around Ontario, east of the Niagara escarpment, consist mostly of till plains.

ARCTIC LOWLAND REGION

This region is somewhat segmented, too (Fig. 8–18). It includes the Foxe Basin Lowland, which is very similar to the Hudson Bay Lowland. Another portion of this region, called the *Southern Archipelago Lowland,* is found on the islands of the Northwest Territories. The whole of Banks Island, for example, is in this southern archipelago segment. The third segment of this region is known as the *Arctic Ocean Lowland.* Patrick Island and Mackenzie King Island are included in this third segment.

INTERIOR PLAINS REGION

This region, commonly called the **Great Plains,** comprises 18 percent of the land area of Canada.

Broad grasslands in the south of the region are called **prairies.** The region extends southward into the Great Plains of the United States from Alberta, Saskatchewan, and Manitoba. Canada's Great Plains extend northwestward from the U.S. border to the Mackenzie Delta.

There are four segments to the Great Plains. The Manitoba portion, with elevations from 600 to 900 feet (180 to 275 m) above sea level is referred to as the first level. The second-prairie level, at about 2000 feet (600 m), covers western Manitoba and eastern Saskatchewan. The western Saskatchewan and Alberta portions, with elevations up to 4300 feet (1300 m) are known as the *high plains.* The fourth segment consists of the Mackenzie Lowland, which drops from elevations of 4000 feet (1200 m) in the south to sea level in the north.

The Red River Valley, in the first-prairie level, is covered with clays and silts that accumulated on the floor of glacial Lake Agassiz. Today this silt-covered land is a rich farmland area. The Saskatchewan Plain is also covered by clay and silt laid down by glacial lakes. In addition, thick accumulations of glacial moraine form occasional isolated ranges of low hills in the second-prairie level.

This is the prairie-plains of Canada. The photo was taken between Regina, Saskatchewan, and Winnipeg, Manitoba. The dark soils of this region developed beneath prairie grasses and are the basis of Canada's most extensive agricultural development. (National Film Board of Canada)

THE CONTERMINOUS UNITED STATES

An accurate analysis of the 48 states of the conterminous United States divides the country into eight major physiographic regions. In turn, these eight major regions are subdivided into 25 divisions and 86 subdivisions (Fig. 8–19).

LAURENTIAN UPLAND

This region surrounds the western section of Lake Superior and covers portions of northeastern Minnesota, northern Wisconsin, and northwestern Michigan (Fig. 8–19). Sometimes the **Laurentian Upland** is referred to as the *Superior Upland.* It is best known for its iron mines. The rocks of the area have been upraised and eroded by running water and continental glaciers. Today, it is a maturely dissected peneplain.

THE COASTAL REGIONS

This region includes the Gulf-Atlantic Coastal Plain as well as the continental shelf along our three coasts. Note in Figure 8–19 that the coastal plain is sectioned into six subdivisions. A portion of the embayed section surrounds the Chesapeake and Delaware bays. The Chesapeake Bay was reviewed in our discussion of estuaries on pages 266–267. The sea-island section stretches from North Carolina to Georgia and discussion of these regions is found on pages 257–259 and pages 261–263. Various sections in the four chapters of this part of the book have been devoted to the three remaining subdivisions of the coastal plain—the East Gulf Coastal Plain, the Mississippi Alluvial Plain, and the West Gulf Coastal Plain.

THE APPALACHIAN HIGHLANDS

The development of the **Appalachian Highlands** region is associated with the opening of the Atlantic Ocean as discussed on pages 184–187. Previous portions of our discussion have used a rather broad brush to sketch the area. The classification in Figure 8–19 lists 7 divisions and 21 subdivisions. Obviously the region could be studied in great detail if we had the inclination and the space available.

The Piedmont, which extends from central Alabama northeastward to the Hudson River area, is an interesting region. The mountainous topography of the Blue Ridge division is to its west. The maximum elevations along the inner boundary of the Piedmont are about 3000 feet (900 m); however, this division is generally a low plateau sloping seaward.

The Appalachian plateaus are seven in number. Considered as one plateau, it stretches from northern Alabama into central New York and makes up the westernmost section of the Appalachian region. Parts of this plateau might be described as low mountains because the rivers have cut valleys 1500 feet (460 m) deep, and summit elevations reach 2500 feet (760 m) and more—especially in West Virginia. The width of the plateau division varies from about 35 miles (55 km) in the south to more than 200 miles (320 km) in the north.

THE INTERIOR PLAINS

This region sweeps westward from the low eastern highlands of the Appalachian region to the towering snow-capped mountains of the Rockies (Fig. 8–19). The western part of the region, known as the *Great Plains,* has an average width of about 400 miles (640 km). Its average elevation at the base of the mountains is about 5500 feet (1680 m) and at the eastern edge approximately 1500 feet (460 m), which makes for an average eastward slope of about 10 feet per mile (2 m/km). A part of this great division is quite literally plain; but streams, especially near the mountains, have eroded deep valleys and their tributaries have also dissected the upland areas.

There are isolated mountains in the Great Plains, the largest of which is the Black Hills. This mountain range in western South Dakota and eastern Wyoming stretches in a north-south direction for about 88 miles (140 km) and in an east-west direction for 50 miles (80 km) (See Fig. 3–10 for location and cross section). The Black Hills are the result of an igneous intrusion of granite that upwarped overlying strata of limestone and sandstone. Additional uplift and a lot of erosion finally exposed the underlying igneous rock. Today, in the central area of the range there is a core of exposed granite with its highest peak, Harney, at 7242 feet (2207 m) above sea level. Harney Peak stands about 3000 feet (914 m) above the surrounding Great Plains. The two sedimentary formations that were upwarped by

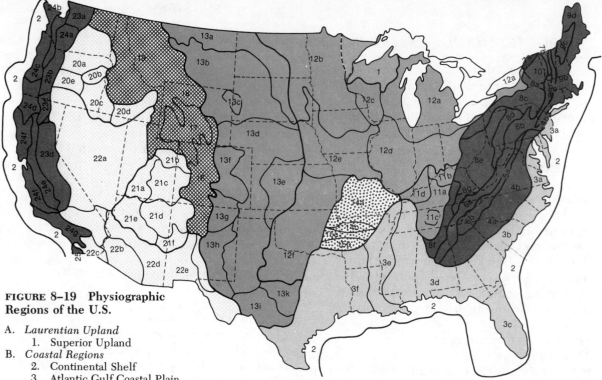

FIGURE 8–19 Physiographic Regions of the U.S.

A. *Laurentian Upland*
    1. Superior Upland
B. *Coastal Regions*
    2. Continental Shelf
    3. Atlantic Gulf Coastal Plain
        a. Embayed section
        b. Sea-Island section
        c. Floridian section
        d. East Gulf Coastal Plain
        e. Mississippi Alluvial Plain
        f. West Gulf Coastal Plain
C. *Appalachian Highlands*
    4. Piedmont Province
        a. Piedmont Upland
        b. Piedmont Lowland
    5. Blue Ridge Province
        a. Northern section
        b. Southern section
    6. Ridge and Valley Province
        a. Tennessee section
        b. Middle section
        c. Hudson Valley
    7. St. Lawrence Valley
        a. Champlain section
        b. Northern section
    8. Appalachian Plateaus
        a. Mohawk section
        b. Catskill section
        c. Glaciated Allegheny Plateau
        d. Allegheny Mountain section
        e. Unglaciated Allegheny Plateau
        f. Cumberland Plateau section
        g. Cumberland Mountain section
    9. New England Province
        a. Seaboard Lowland
        b. New England Upland
        c. White Mountain section
        d. Green Mountain section
        e. Taconic section
    10. Adirondack Province

D. *Interior Plains*
    11. Interior Low Plateau
        a. Highland rim section
        b. Lexington Plain
        c. Nashville Basin
        d. Shawnee section
    12. Central Lowland
        a. Great Lakes section
        b. Western Young Drift section
        c. Wisconsin Driftless section
        d. Till Plains
        e. Dissected Till Plains
        f. Osage Plains
    13. Great Plains Province
        a. Glaciated Missouri Plateau
        b. Unglaciated Missouri Plateau
        c. Black Hills
        d. High Plains
        e. Plains border
        f. Colorado Piedmont
        g. Raton section
        h. Pecos Valley
        i. Edwards Plateau
        j. Central Texas section
E. *Interior Highlands*
    14. Ozark Plateaus
        a. Springfield-Salem Plateaus
        b. Boston Mountain
    15. Ouachita Province
        a. Arkansas Valley
        b. Ouachita Mountains
F. *Rocky Mountain System*
    16. Southern Rocky Mountains
    17. Wyoming Basin
    18. Middle Rocky Mountains

    19. Northern Rocky Mountains
G. *Intermontane Plateaus*
    20. Columbia Plateaus
        a. Walla Walla section
        b. Blue Mountain section
        c. Payette section
        d. Snake River Plain
        e. Harney section
    21. Colorado Plateau
        a. High Plateaus of Utah
        b. Uinta Basin
        c. Canyon lands
        d. Navajo section
        e. Grand Canyon section
    22. Basin and Range Province
        a. Great Basin
        b. Sonora Desert
        c. Salton Trough
        d. Mexican Highland
        e. Sacramento section
H. *Pacific Mountain System*
    23. Sierra-Cascade Mountains
        a. Northern Cascade Mountains
        b. Middle Cascade Mountains
        c. Southern Cascade Mountains
        d. Sierra Nevada
    24. Pacific Border Province
        a. Puget Trough
        b. Olympic Mountains
        c. Oregon Coastal Range
        d. Klamath Mountains
        e. California Trough
        f. California Coast Ranges
        g. Los Angeles Ranges
    25. Lower California Province

Grand Forks County, North Dakota, is in the Great Plains. Windbreaks of trees and shrubs have been used extensively to protect the soil from wind erosion. (USDA)

the intrusion and later eroded now form a ringlike pattern around Harney Peak. Each formation produces a ridge, with a sharp summit and steeply sloping sides, known as a **hogback**. Moving from Harney Peak, the first hogback encountered is made from the outcropping edge of a tilted limestone stratum. Just beyond the limestone hogback is a circular lowland of red rocks called the *Red Valley*. Moving across the Red Valley, you encounter the second ridge formed by sandstone, which is called the *Dakota Hogback*. The Dakota Hogback stands 500 feet (150 m) above the Red Valley.

The central-lowland division of the Interior Plains is often referred to as a *prairie*. Many features of this region were created by glaciers of the Pleistocene. Most of Illinois, Indiana, and western Ohio, for example, have a glacial till cover that is remarkable for its low relief and distinguished by an absence of lakes. The beds of the Great Lakes, except Lake Superior, follow belts of weak strata that were lowlands in preglacial time and were deepened by glacial erosion. The mighty Mississippi system cuts through this region and drains it.

THE INTERIOR HIGHLANDS

This region is sometimes referred to as the *Ozark-Ouachita Highlands*. It covers eastern

Oklahoma, northwestern Arkansas, southern Missouri, and a very small section of southern Illinois (Fig. 8–19). The area is honeycombed with many rapidly moving streams, big springs, and underground caverns.

The northern part of the Ozark Plateau consists of nearly horizontal though slightly domed rocks. In its southern part, the rocks are folded along east-west axes. The elevations in the plateaus of the Ozark are generally less than 1600 feet (490 m).

The Ouachita Mountains in Arkansas and Oklahoma are essentially fold structures. They seem to be the product of two cycles of erosion with ridges in the stronger outcrops and intervening lowlands on the weaker rocks. The rocks of the Arkansas Valley are only mildly folded.

THE ROCKY MOUNTAIN SYSTEM

The plate motions that produced the Rockies are described on pages 251–253. As you will note in Figure 8–19, this system consists of four divisions in the conterminous United States. The continental divide, which is a line separating the streams that flow to the Atlantic from those that flow to the Pacific, passes through the Rockies.

The Wyoming Basin, which is a plateau with an average elevation of 7000 feet (2100 m), breaks the continuity of the Rocky Mountains and separates the southern from the middle Rockies. The floor of this basin, which is largely without mountains, covers an area of 42,000 square miles (108,800 km²). The Green, Bighorn, and North Platte rivers all leave the basin by way of canyons that are 1000 to 3000 feet (300 to 900 m) deep.

INTERMONTANE PLATEAUS

This vast region lying between the Rockies and the Pacific mountain system consists of three divisions: the Columbia Plateau, the Colorado Plateau, and the Basin and Range Province. This region extends from northern Washington south to the Mexican border (Fig. 8–19).

A lava flow, recall, produced the surface of the Columbia Plateau. Deep river canyons—for example, Hells Canyon (42°20′N,116°45′W) on the Snake River between Idaho and Oregon—have been cut through the plateau in places and reveal that the lava flow covered an uneven surface of erosion. The lava literally filled in the valleys between mountains. This is shown clearly in narrow Hells Canyon where the rushing waters of the Snake River have cut to an average depth of more than 5000 feet (1500 m). The greatest depth cut by the river in this canyon is 7900 feet (2400 m).

The Colorado Plateau covers an approximate area of 130,000 square miles (336,700 km²). The most spectacular feature of this division is the Grand Canyon subdivision, which is discussed on pages 223–225. The conditions that lead to canyon development are all present on the Colorado Plateau—great elevation above base level, swiftly flowing young streams, recent uplift of the region, and an arid climate that preserves steepness.

The Basin and Range Province is a region marked by numerous small, roughly parallel mountain ranges separated by nearly flat plains. The northern half of this province is known as the **Great Basin**, which is a large area of internal drainage discussed on pages 250–251. The floor of the basin is generally about 5000 feet (1500 m) above sea level and many of the mountain peaks stand 4000 feet (1200 m) or more above the basin's floor. Mount Jefferson (38°47′N,116°55′W) in the Toquima Range, for example, stands at an elevation of 11,949 feet (3642 m) above sea level.

THE PACIFIC MOUNTAIN SYSTEM

This system and its tectonic development are discussed on pages 249–253. The Sierra Nevada and the Cascade Range form an almost unbroken wall of mountains at the eastern border of this

These are the rugged southern Rocky Mountains in Colorado. Dream Lake is in the foreground. The peaks in the area are about 13,000 feet (4000 meters) above sea level. (National Parks Service)

GULF OF FARALLONES

PACIFIC OCEAN

GOLDEN GATE BRIDGE

SAN FRANCISCO BAY

MARIN PENINSULA

Looking south from the Marin Peninsula, the Golden Gate Bridge crosses the Golden Gate Gorge, a mile- (1.6-kilometer-) wide channel, to connect San Francisco on the south with the Marin Peninsula on the north. The topography of the California Coast Ranges is clearly visible on the Marin Peninsula. (NOAA)

region (Fig. 8–19). The Columbia River Valley is one of the few easy routes through this mountainous chain.

A series of three valleys is found to the west of the Sierra-Cascade chain. The 350-mile- (560-km-) long Puget Trough lies between the Cascades and the Coast Range. Puget Sound is the submerged northern end of this trough. The lowlands around Puget Sound constitute the first valley in the west-coast series. The second—the valley of the Willamette River—is found at the southern end of Puget Trough. The San Joaquin-Sacramento River Valley in California, which is sometimes called the *Great Valley*, is the third in the series.

A chain of coast ranges rises west of the valleys. The most northerly of the coast ranges is the Olympic Mountains. The Oregon Coast Range, a

gentle anticline, is succeeded on the south by the Klamath Mountains, which consist of deeply eroded metamorphosed Paleozoic rocks. California's Coast Ranges run along most of its coast. Since the coastal ranges of the West Coast tend to rise sharply from the Pacific, there is very little coastal plain, except around Los Angeles where the coastal plain is fairly broad.

A Lower California Province, which is quite small, is treated as a third division in this region. The Lower California division extends from the San Diego area east to the Salton trough; it stretches north from San Diego almost to the east-west trending mountains known as the *Transverse Ranges*, which are sometimes referred to as the *Los Angeles Ranges*. This third division is really the famous fruit-growing lowland of California.

THE LANGUAGE OF PHYSICAL GEOGRAPHY

A physical geographer uses a technical vocabulary to make explanations. Review your understanding of the vocabulary used to develop the concepts in this chapter. Among the important words and terms used are:

Appalachian Highlands	coast	Greenland orogenic belt	plains
Appalachian orogenic belt	cratons	Gulf-Atlantic coastal plain	point
Arctic Coastal Plain	cuspate forelands	head	prairie
Arctic orogenic belt	cuspate shoreline	hogback	promontory
Arctic stable region	cuspate spits	Innutian Region	stable region
barrier beaches	deltaic coasts	lagoon	shield region
beach	estuary	Laurentian Upland	shingle beach
beach cusps	fall line	longshore bar	shore
berm	Farallon Plate	longshore trough	tectonic divisions
Canadian Shield	Great Basin	orogenic belts	western orogenic belt
cape	Great Plains		

SELF-TESTING QUIZ

1. Why should most of Central America not be considered a part of the North American continent?
2. List the major tectonic divisions of North America.
3. Where are the oldest rocks of the continental crust found?
4. How many orogenic belts are found on North America?
5. Explain the development of the Sierras and Rockies using the theory of plate tectonics.
6. How has the northwestward dragging of the Pacific Plate distorted the region of North America from the Rockies westward?
7. List and compare the major coastal plains of North America.
8. What conditions produced the fall line of the Atlantic Coastal Plain?
9. How is the upper limit of a beach marked?
10. What is the most common type of beach along the United States coastline?
11. How are capes formed?
12. What conditions are necessary to produce a cuspate foreland?
13. Trace the development of Sandy Hook from its beginning to its present condition as a compound spit.
14. Where is one of the largest deltaic coasts in the world found?
15. Compare and contrast the coastlines of North Carolina, South Carolina, and Georgia.
16. How has human intervention reworked the lagoons along California's southern coast?
17. What is the longest estuary in the United States?
18. List the major physiographic regions of Canada.
19. How many groups of highlands can be recognized in the Canadian Appalachians?
20. What effect did glaciation have on the Canadian Shield?
21. What is the largest isolated mountain range in the Great Plains of the United States?

IN REVIEW
A NORTH AMERICAN SURVEY

I. MAJOR TECTONIC DIVISIONS

  A. The Shield Regions
  B. The Stable Regions
  C. The Orogenic Belts
    1. The Appalachian Belt
    2. The Western Belt
    3. Plate Motion in the West
  D. The Coastal Plains
    1. The Atlantic Coastal Plain
    2. The Gulf Coastal Plain
    3. The Arctic Coastal Plain

II. THE COASTAL SURVEY

  A. Beaches of North America
  B. Cuspate Shorelines
    1. Cuspate Forelands
    2. Cuspate Spits
    3. Giant Cusps
  C. Deltaic Coasts
    1. Yukon-Kuskokwim Deltas
    2. North-Slope Deltas
  D. Coastal Lagoons
    1. Gulf-Coast Lagoons
    2. Along the East Coast
    3. West-Coast Lagoons
  E. River Estuaries

III. SURVEY OF PHYSIOGRAPHIC REGIONS

  A. Canada
    1. The Cordilleran Region
    2. The Canadian Appalachians
    3. The Innutian Region
    4. The Canadian Shield
    5. Hudson Bay Lowlands
    6. Great Lakes–St. Lawrence Lowlands
    7. Arctic Lowland Region
    8. Interior Plains Region
  B. The Conterminous United States
    1. Laurentian Upland
    2. The Atlantic Plain
    3. The Coastal Regions
    4. The Appalachian Highlands
    5. The Interior Plains
    6. The Interior Highlands
    7. The Rocky Mountain System
    8. The Intermontane Plateaus
    9. The Pacific Mountain System

PART THREE

GLOBAL CLIMATE

In general, the word *climate* refers to a somewhat enduring regime of the atmosphere that takes into account extremes as well as averages. But, in simplistic terms, the climate of a place is the mean or average course of the weather for that place over a specific interval of time such as a month, a season, or a year. In a sense, then, climate is the weather we expect for summer, fall, winter, or spring at a specific location. These expectations are predicated on the results of the statistical characterization of weather data collected over a relatively long period.

Climate is not an abstraction; it applies to places and regions. The geography of climates is an integral part of physical geography. Climate is described in terms of certain elements in combination. These climatic elements are temperature, pressure, humidity, sunshine, and the conditions that derive from them such as winds, clouds, and precipitation. The climatic elements are the variables out of which the many climatic types are compounded and classified. Each of the primary climatic elements—solar energy, humidity-precipitation, and winds—also functions as a climatic control and influences each of the other elements.

Climates, their world distribution, and their effect on landforms are of great interest to the physical geographer. Since some knowledge of the individual climatic elements is essential to an understanding of global climates, we will begin the discussion of Part 3 with a study of the primary climatic elements before detailing the patterned arrangement of global climates and their effects on landforms.

A. Winds, clouds, and precipitation are conditions that develop from the primary climatic elements. This pattern of clouds developed within maritime air flowing over Beaufort, South Carolina. (John Navarra)

B. The Big Island of Hawaii lies well within the belt of northeasterly trade winds. Over Hawaii's windward slopes rainfall occurs within ascending moist trade winds. The mean annual rainfall at Hilo is 133 inches (338 centimeters). Leeward (southern and western) areas are sheltered from the trades and are drier. These prickly-pear cacti grow leeward of the southern portion of the Kohala Mountains, which receives less than 10 inches (25 centimeters) of rainfall. (John Navarra)

C. The West Indies are a long chain of islands that separates the Caribbean Sea from the rest of the Atlantic Ocean. This chain consists of three major groups of islands: the Bahamas in the north, the Greater Antilles near the center, and the Lesser Antilles to the southeast. In this photo, heavy clouds obscure the peak of St. Vincent's active volcano. The tropical climate of this island in the Lesser Antilles produces a lush vegetation. (John Navarra)

D. Mount Kosciusko is Australia's highest peak. It rises 7316 feet (2230 meters) in the Australian Alps of New South Wales. The snowfield as shown in this view is able to survive the summer period. (John Navarra)

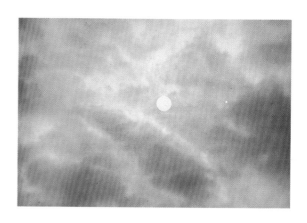

A. Cloud cover dramatically affects the amount of incoming solar radiation that reaches the Earth's surface. The combination of middle family clouds in this photo cuts the direct energy reaching the surface almost to zero. (John Navarra)

B. Antigua is in the leeward-island section of the Lesser Antilles. An absence of mountains and thorough deforestation distinguish Antigua from the other Leeward Islands. There are no rivers on the island and prolonged droughts occur despite a mean annual rainfall of 44 inches (112 centimeters). (John Navarra)

9 *Climatic elements and controls*

Certain factors are crucial in producing climatic regimes. Heat from the Sun and the rotation of the Earth are preeminent among the factors that cause climate. The Sun's radiant energy and the Earth's rotation on its axis set up circulation patterns that carry heat and moisture across land and sea. These patterns of wind, cloudiness, and temperature that spread over flat land and mountains from the Equator to the poles determine the climate of a place.

The variables—insolation and humidity-precipitation—out of which climate is compounded are referred to in this book as the *climatic elements*. The factors that cause climatic regimes are known as *climatic controls*. Among the climatic controls are latitude, continentality and sources of moisture, prevailing winds, ocean currents, elevation (altitude) and mountain barriers. The operation of each of the controls will be discussed as we consider each of the climatic elements. Wind, however, is so very significant as a climate control that a separate chapter will be devoted to it; the spatial distribution of climatic elements by wind is discussed in Chapter 10.

INSOLATION AS A CLIMATIC ELEMENT

The solar radiation that reaches the Earth and is involved in heating its surface and atmosphere is referred to as **insolation,** which is an abbreviation for *in*coming *sol*ar rad*iation*. All the weather on the Earth is, of course, ultimately powered by the Sun, and the distribution of this energy over the Earth determines to a great extent the location of the major climatic zones—tropical, temperate, and polar.

When radiant energy from the Sun is analyzed, it is found to be distributed among a continuous range of frequencies, called the **electromagnetic spectrum.** The spectrum extends from the low-frequency radio waves to the extremely high-frequency gamma rays. Members of the family include gamma rays, x-rays, ultraviolet, visible light, infrared, microwaves (radar), and radio waves.

SOLAR RADIATION

The Sun and every other star in the Universe pours tremendous amounts of radiant energy into space every second. The source of the Sun's energy is a thermonuclear reaction in which hydrogen atoms are fused together to produce helium atoms. Of the vast amount of energy radiated by the Sun, the Earth intercepts only a minute fraction—and of this, not all is absorbed.

THE SOLAR CONSTANT

The rate at which the Earth is receiving energy from the Sun is referred to as the **solar constant.** Technically defined, the solar constant is the quantity of energy that is received in unit-time on a unit-area placed perpendicular to the direction of the Sun at the top of the atmosphere. The best measurements of the solar constant are made in the upper reaches of the atmosphere, where little absorption and reflection have taken place.

The solar constant is measured in a standard unit of radiation called the **langley,** which is equal to 1 gram-calorie of heat received by 1 square centimeter of surface. The gram-calorie is a standard unit for measuring heat; it is that quantity of heat needed to raise 1 gram of water through 1°C. The solar radiation received at the outer limits of the atmosphere is approximately 2 langleys per minute. The exact figure that is used in climatological tables is 1.94 langleys per minute. It should be pointed out that there is evidence that the solar constant fluctuates as much as ± 1.5 percent. In addition, the varying distance between the Earth and Sun through the year produces a change in the solar radiation at the top of the atmosphere of ±3.5 percent from the average value.

Today, the total radiative output of the Sun is estimated to be about 100,000 langleys per minute. Based on an average solar constant of 2 langleys per minute, the Earth intercepts less than 0.002 percent of the Sun's present radiative output. John A. Eddy of the National Center for Atmospheric Research believes that there are long-term fluctuations in the total output of the Sun. If this is the case, then we must consider the possible effects of such variations on the Earth's climate. In fact, the period of minimum sunspot

TABLE 9–1 TOTAL DAILY SOLAR RADIATION

LATITUDE	Mar 21	Apr 13	May 6	May 29	Jun 22	Jul 15	Aug 8	Aug 31	Sep 23	Oct 16	Nov 8	Nov 30	Dec 22	Jan 13	Feb 4	Feb 26
					LANGLEYS (AT TOP OF ATMOSPHERE ON DATES SHOWN)											
90°N		423	772	999	1077	994	765	418								
80	155	423	760	984	1060	980	754	418								7
70	307	525	749	939	1012	934	742	519	303	129	24				24	131
60	447	635	809	934	979	929	801	629	442	273	146	72	49	73	146	276
50	575	732	867	958	989	954	859	725	568	414	286	204	176	205	289	419
40	686	807	910	972	991	967	901	798	677	545	429	348	317	350	434	553
30	775	865	929	967	975	960	921	856	765	663	564	492	466	494	568	670
20	841	894	923	935	935	930	916	884	831	760	685	627	605	630	691	769
10°N	882	897	893	881	873	877	886	887	871	835	789	748	733	752	795	845
0	895	873	837	804	790	800	830	863	885	886	870	851	843	855	878	896
10°S	882	824	760	707	687	704	753	814	871	910	927	931	933	936	936	921
20	841	750	660	593	567	590	654	741	831	907	959	988	999	993	968	918
30	775	654	543	465	436	463	538	646	765	877	964	1020	1041	1025	973	888
40	686	538	413	329	297	328	409	533	677	819	944	1027	1059	1032	953	828
50	575	408	276	193	165	192	274	404	568	743	901	1014	1056	1018	909	752
60	447	269	140	68	47	68	139	266	442	644	840	987	1046	992	847	652
70	307	127	23				23	126	303	532	778	993	1081	998	785	539
80	155	7						7	153	429	790	1041	1132	1046	796	434
90°S										429	801	1056	1149	1062	809	434

activity from 1645 to 1715 coincides almost precisely with what we now refer to as the **Little Ice Age.** Eddy believes that the climate curve looks a lot like the curve of variability in solar activity.

THE LATITUDE EFFECT

One of the most important climatic controls is latitude. Latitude has a very decided effect on the way insolation is received and spread across the Earth's surface on any one day. The total solar radiation falling during one day on a horizontal surface of one square centimeter at the top of the atmosphere is reported in langleys in Table 9–1. Trace the variation in the energy received at latitude 40°N. The energy input reaches a peak of 991 langleys per day on June 22 and a minimum of 317 langleys per day on December 22.

Each year the Earth completes a journey around the Sun. During the course of this sweep, the Earth's axis maintains a nearly constant direction in space—the North Pole points toward the relatively bright star Polaris, called the *North Star.* The Earth is a sphere, and thus the inclination of its axis causes the relative position of the Sun with respect to a given latitude to vary over the year. At 40°N latitude, for example, the Sun appears lower in the sky at noon during December than during June because of the orientation of the Earth's axis (Fig. 9–1). The noon altitude of the Sun at this latitude is approximately 25.5° on December 22 and 72.5° on June 22. As a result, at the top of the atmosphere and on the Earth's surface at 40°N, a solar beam spreads over a larger area during December than during June.

Recall the discussion on pages 29–32 and the fact that the orientation of the Earth's axis produces different lengths of daylight and darkness throughout the year at a particular latitude. The duration of daylight on June 22 at 40°N is 15 hours. This means that an average of 1.1 langleys per minute is received at this latitude during the daylight hours. On December 22, there are 9.33 hours of daylight at this latitude, which means that an average of 0.57 langley per minute is received throughout the daylight hours at the top of the atmosphere. There is a significant difference between the summer and winter input at

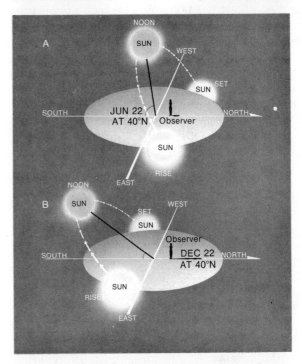

FIGURE 9–1 The Sun appears lower in the sky at Noon during December than during June. In A, on June 22, the Noon elevation of the Sun is 72.5° above the observer. In B, on December 22, the Noon elevation of the Sun is 25.5° above the horizon of the observer.

this latitude on a minute-by-minute basis because of the differences in the area covered by an incoming beam of solar radiation.

The data in Table 9–1 indicate that the peak daily input at the top of the atmosphere on June 22 occurs at latitude 90°N—with an input of 1077 langleys per day. There are 24 hours of sunshine at the North Pole on June 22. Divide the langleys received (1077) by the length of daylight in minutes and you will find that 0.75 langley per minute is received at the North Pole on June 22. This is somewhat less than the 1.1 langleys per minute received at 40°N on the same date. The total is larger at 90°N because the input goes on for an additional 9 hours.

On December 22, the peak daily input (1149 langleys) is at latitude 90°S. The South Pole receives input for 24 hours on this date. The input per minute at the South Pole is 0.8 langley. The length of daylight at latitude 20°S is 13.35 hours, and at 30°S it is 14.08 hours—which means the rate per minute is 1.25 langleys at 20°S and 1.23 langleys at 30°S. The highest per minute input is at latitude 23.5°S, which is the location at which the direct rays of the Sun are impinging on the upper atmosphere on December 22. Similarly, the highest per minute input is at 23.5°N on June 22.

The direct rays of the Sun are over the Equator on March 21 and September 23, and the length of daylight is about 12 hours at all latitudes. Sunbeams of the same diameter cover the smallest area at the Equator and cover increasingly larger areas as you move to higher latitudes. Therefore, a maximum amount of solar radiation is received at the Equator on these dates—895 langleys on March 21, and 885 langleys on September 23. Note in Table 9–1 that the langleys received fall off at higher latitudes. The radiation received at the top of the atmosphere over the North Pole and over the South Pole is zero on these dates. This symmetrical decrease of radiation with increasing latitude toward both poles occurs only during the transition seasons of spring and fall, however.

DEPLETIONS AND DIVERSIONS

As the incoming solar radiation penetrates the atmosphere, a series of depletions and diversions of the energy occurs. Within the region of the ionosphere, for example, gamma rays and x-rays are almost completely absorbed along with small amounts of ultraviolet radiation. Then reactions within the ozonosphere remove most of the remaining ultraviolet radiation. As the insolation continues its journey and penetrates into denser atmospheric layers, gas molecules turn aside visible light in all possible directions; this process is called **Rayleigh scattering.** The scattering of visible light by dust particles in the troposphere is called **diffuse reflection.** Scattering and diffuse reflection return about 6 percent of the total insolation to space.

Carbon dioxide, water vapor, and dust within the troposphere are capable of directly absorbing infrared radiation. Such absorption results in direct heating of the atmosphere. Since the atmosphere's load of water vapor and dust varies considerably from day to day, infrared absorption can also be expected to vary. A good global average for all forms of direct energy absorption is about 15 percent of the incoming total.

The combined depletion due to scattering and diffuse reflection with the depletion due to direct absorption is an average loss of 21 percent. This means that when skies are clear, about 79 percent of the insolation reaches the ground. On June 22 at latitude 40°N, for example, approximately 783 langleys would reach the Earth's surface with a 79 percent transmission, and about 624 langleys would reach ground level at the Equator.

Cloud cover, however, can dramatically affect this picture. Reflection from the upper surfaces of clouds can turn back to space 30 to 60 percent of the incoming solar radiation. Another 5 to 20 percent of the energy can be absorbed within the clouds. Thus, on a very cloudy day, the direct energy reaching the Earth's surface can be cut almost to zero.

The global average for the reflection of insolation from clouds is 21 percent; average absorption within clouds accounts for another 3 percent. Of the insolation reaching the Earth's surface, about 6 percent is reflected back into the atmosphere. Adding all the averages mentioned above and subtracting from 100, we find that the absorption of energy by the Earth's land-and-water surface averages about 49 percent of the incoming solar radiation. This means that at latitude 40°N, on June 22, 49 percent of the 991 langleys at the top of the atmosphere—roughly 485 langleys—would reach and be absorbed at the Earth's surface. An actual value will be more or less than the figure cited and will depend on how conditions depart from the averages cited.

The insolation losses due to scattering are in the shorter wavelengths. This means that blue light, for example, is scattered about 15 times as much as red light. The blue color of the sky is due to the scattering of blue light. Ozone, you have learned, absorbs very strongly in the ultraviolet region of the electromagnetic spectrum. And

The albedo of snow-covered Mount Rundle in the Canadian Rockies of Alberta, Canada, is quite different from the green, forested slopes in the foreground. (National Film Board of Canada)

water and carbon dioxide absorb very strongly in the infrared range of the spectrum.

About 40 percent of the solar radiation that finally reaches the Earth's surface is in the infrared range. Since liquid water absorbs strongly in the infrared, the world ocean "sops" up more than half of the insolation that reaches the Earth's surface. And about half the solar energy entering the hydrosphere is immediately used for evaporation, which leaves about half the energy to maintain the temperature of the world ocean.

ALBEDO

When radiant energy strikes an object, some of the radiation is absorbed and some is reflected. The term **albedo** is used to indicate the reflecting power of an object. Technically defined, albedo is the ratio of the radiation reflected from an object to the total amount that falls on it. For example, the albedo of ocean water ranges from 0.03 to 0.07, which means that an ocean surface reflects from 3 to 7 percent of the solar radiation that strikes it (Table 9–2).

Since approximately three-fourths of the Earth's surface is covered by water, which has a fairly low albedo, it is not surprising to find the Earth's surface albedo is in the 6 percent range. Considering the Earth as a whole, however—its solid surface, water surface, as well as its gaseous atmosphere—the albedo profile from pole to pole as measured from outer space by satellites is between 29 and 34 percent.

The concept of albedo means that a sandy beach, a grass-covered park, and a body of water—all within the same locale—will not be heated to the same temperature even though the

TABLE 9–2   ALBEDOS OF VARIOUS SURFACES

SURFACE	ALBEDO (PERCENT)
Water	
Bay	3–4
River	6–10
Ocean	3–7
Forest	
Green	4–6
Snow-covered	10–25
Ground	
Bare	10–20
Grass-covered	15–25
Wheat field	7
Fresh snow cover	87
Old snow cover	46
Sand	
Dry	18
Wet	9
Clouds	5–84

MEAN DAILY SOLAR RADIATION (LANGLEYS) DECEMBER

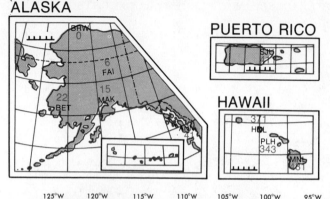

FIGURE 9–2 The mean daily solar radiation for December at different locations in the United States can be read from the Langley lines. (See Appendix VII to interpret the code used to identify each city.)

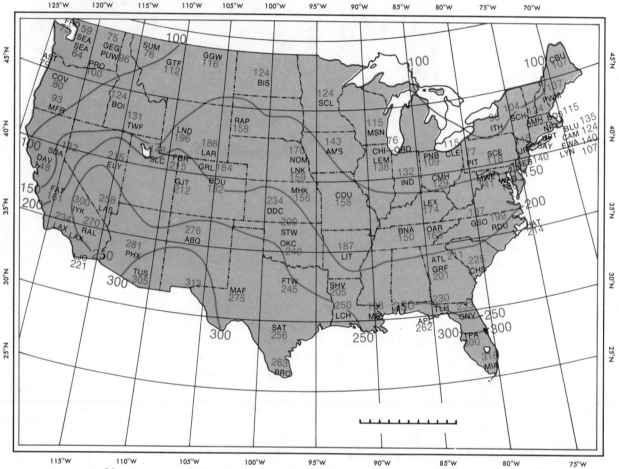

same amount of heat energy is heading toward each surface. Each of these surfaces has different albedos and different thermal properties. First of all, each surface will reflect different percentages of the total energy falling on it; thus, the amount of energy absorbed in each case will be different. Then, in addition, the differing heat capacities of each surface will prevent a given unit of energy from producing the same temperature change in each surface.

The climate control known as **continentality**, which refers to the fact that land and water sur-

faces react differently to incoming rays of the Sun, is based on this concept of albedo as well as on differing heat capacities. In any event, the uneven heating of the Earth's surface eventually produces an uneven heating of the air in contact with the surface. The solar radiation absorbed by the Earth's surface is reradiated and conducted to the air above, provided that the surface is at a higher temperature than the air in contact with it. Air over a hot sandy beach will be brought to a higher temperature than the air over a cool body of water, for example.

MEAN DAILY SOLAR RADIATION (LANGLEYS) JUNE

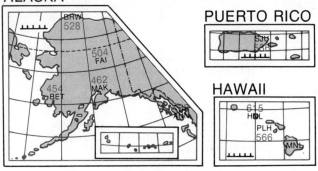

FIGURE 9–3 The mean daily solar radiation for June at different locations in the United States can be read from the Langley lines. (See Appendix VII to interpret the code used to identify each city.)

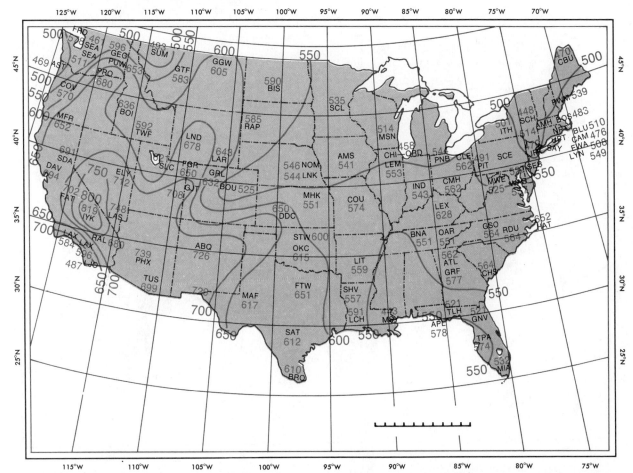

RADIATION CYCLES AND NORMS

The daily cycle of incoming solar radiation begins at sunrise. The maximum intensity of incoming radiation is reached at noon—noon according to local solar time, which occurs when the Sun is at its apparent highest point in the sky for the day. The maximum point is followed by a symmetrical decline and a cessation of incoming radiation at sunset.

The mean daily solar radiation reaching the ground is plotted for December (Fig. 9–2) and June (Fig. 9–3) for the 50 states and Puerto Rico. In December, the 300 langley line runs through Tampa and central Florida. The surface at El Paso, Texas, in December receives 313 langleys per day; North Head, Washington, gets 77 langleys per day; and Fairbanks, Alaska, gets a mere 6 langleys per day. Study the mean daily solar radiation reaching the surface during June at each of these same locations. A dramatic increase occurs at Fairbanks, which receives 504 langleys per day, and North Head, which goes up to 487 langleys per day, during June.

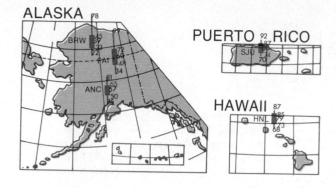

FIGURE 9-4 The temperature records of the normal daily maximum, average, minimum, and extreme temperatures for July depicted for this station on this map were compiled by averaging the results of many hundreds of days of observation. (See Appendix VIII to interpret the code used to identify each city.)

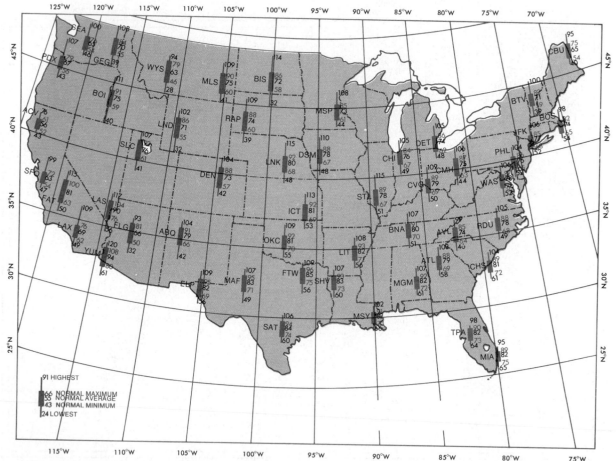

Compare the data in Table 9–1 with the data contained in Figures 9–2 and 9–3. El Paso, for example, is at latitude 31°45′N. About 978 langleys per day (Table 9–1) are received at the top of the atmosphere over El Paso in June; 729 langleys per day (Fig. 9–3) reach El Paso's surface. This amounts to a transmission of 74.5 percent of the energy to the surface, which means that the skies over El Paso are quite clear and free of cloud cover. On the other hand, the transmission during June at Gainesville, Florida, which is at 29°40′N, is 53.4 percent.

Cloud cover in the Gainesville area greatly reduces the energy reaching the ground.

Long-term records do not show an appreciable net heating or cooling of the Earth and its atmosphere. Therefore, it is believed that the Earth emits an amount of radiation equal to that absorbed. Since the Earth radiates at a low temperature—compared to the Sun's—the short-wave energy absorbed is reradiated as longer wavelengths. The penetrating solar radiation is effectively limited to wavelengths of 0.3 to 3 micrometers. The Earth's emissions, however,

290

occur through a broad range of wavelengths from 4 to 40 micrometers. The maximum emission of the Earth is at the 12-micrometer range. In this range of wavelengths, the atmosphere is not transparent and, thus, water vapor, ozone, and carbon dioxide absorb significant amounts of these long-wave radiations. During hours of darkness, energy loss from the ground continues by long-wave radiation.

The transport of energy upward from the Earth's surface into the air above constitutes the major energy supply for the atmosphere. This transfer of energy close to the Earth's surface occurs through evaporation of water, through heat conduction, and through long-wave radiation. A large portion of this energy is referred to as *latent* because it is in the form of water vapor, and it raises atmospheric temperature only when condensation takes place. This concept is described in greater detail on pages 299–303.

TEMPERATURE CYCLES

A single day's observations of air temperature in late fall and early spring will show that the lowest temperature of the day occurs just before sunrise at about 6 A.M. The highest temperature of the day, in most cases, occurs between 2 P.M. and 4 P.M. There is a lag between the time of greatest input of energy (solar noon) and the time of highest temperature. It takes time for the surface to absorb the short-wave radiation, send it out as longer waves, and then for the air to internalize the energy and increase its temperature.

Temperature records are built by averaging the results of many days of observations. Average figures usually apply to an entire calendar month. Individual station charts for a number of cities in the 50 states and Puerto Rico are indicated on the map in Figure 9–4. The July record for El Paso, Texas, is built on 78 years of observation and shows that the average or mean daily maximum is 95°F (35°C), the average daily minimum is 69°F (21°C), and the daily average is 82°F (28°C). During July, the highest temperature on record for the 78 years is 109°F (43°C); the lowest on record is 56°F (13°C).

The average hourly temperature readings for the month of July at El Paso, Texas, have been plotted in Figure 9–5. The El Paso area is a dry

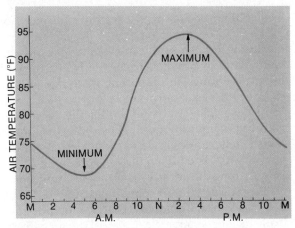

FIGURE 9–5   This plot of the hourly air temperature during July at El Paso, Texas, shows maximum temperatures are reached at 3 P.M. and minimums are reached at 5 A.M.

interior desert region. The dry soil of El Paso's desert can store very little energy; consequently the short-wave radiation absorbed is rapidly emitted as long-wave radiation. The air above this desert area heats and cools rapidly, as shown by the smooth curve of its temperature variations throughout the day. The average daily minimum temperature occurs at 5 A.M. in July. El Paso experiences its average daily maximum temperature of 95°F (35°C) at 3 P.M.

The average hourly temperature readings at Eureka, California, for the month of July have been plotted in Figure 9–6. The maximum of

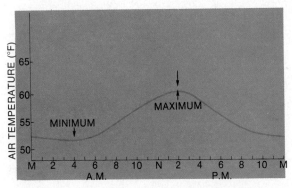

FIGURE 9–6   This plot of the average hourly air temperatures during July at Eureka, California, shows maximum temperatures are reached at 2 P.M. and minimums are reached at 4 A.M.

292 PART III GLOBAL CLIMATE

61°F (16°C) occurs at 2 P.M.; the minimum of 52°F (11°C) occurs at 4 A.M. At Eureka, the moderating influence of the huge body of cold sea water close by prevents excess daily heating and nightly cooling; the temperature range is 9°F (5°C) at Eureka compared to a 26°F (14°C) range at El Paso. The sky along the Pacific coast near Eureka is generally cloudy. The soil in the area is moist; it stores energy and radiates this energy very slowly. The high moisture content of Eureka's air tends to reduce heat loss at night, too.

A perusal of the air-temperature curves (Figs. 9–5 and 9–6) will indicate that the temperature cycle is not symmetrical, since the maximum occurs about 10 hours after the minimum. The curve, generally, rises more steeply in the morning hours than it falls in the late afternoon and early evening. This aspect is best seen in the El Paso curve (Fig. 9–5). The fact that the curve tends to flatten during the nighttime hours can best be seen in the Eureka curve (Fig. 9–6).

THE WORLDWIDE PATTERN

The distribution of temperature over the Earth and its variations throughout the year depend primarily on the insolation received at the surface in different regions. This, in turn, depends mainly on latitude but is greatly modified by the distribution of continents and oceans, prevailing winds, and oceanic currents, as well as topography and elevation. The isotherms drawn on the maps in Figures 9–7 and 9–8 show the average distribution of temperature over the Earth for January and July.

Note the pattern of isotherm distribution over the landmass of North America and Asia in January (Fig. 9–7). The lines are drawn very close together. This isotherm pattern shows that the rate of fall in temperature, called the **temperature gradient,** is very steep over the interiors of North America and Asia (north of latitude 20°). The temperature gradient in western Europe is much more gradual because it is east of an

JANUARY ISOTHERMS (F°)

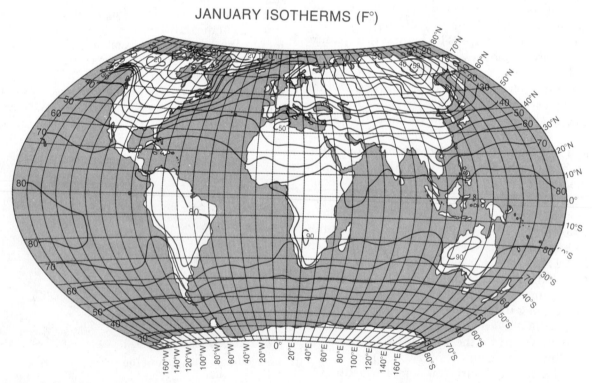

FIGURE 9–7 This pattern of isotherms is based on the average January temperatures (F°) for stations throughout the world.

oceanic current, the North Atlantic Drift, and in a region of prevailing westerly winds that flow off the Atlantic Ocean. Note the isotherm pattern of the Northern Hemisphere in July (Fig. 9–8); temperature gradients are diminished because of the warming of the extensive continental interiors.

Remember, during January the Southern Hemisphere is experiencing its summer with the direct rays of the Sun striking the surface close to the Tropic of Capricorn, that is, at 23.5°S. The temperature gradients poleward in the Southern Hemisphere during January are not very steep. And even during July, which is a winter month south of the Equator, the temperature gradient toward the pole of the Southern Hemisphere is very gradual. In fact, dipping of the isothermal lines—deflections from the west-east trend—are of minor importance because of the large area covered by the oceans, which reduces the continental effect.

CONTERMINOUS UNITED STATES

The isotherms for January, drawn on the map of the conterminous United States, indicate that the winters are coldest in the extreme northern interior and in the higher elevations of the mountains (Fig. 9–9). The highest January temperatures are found in southern Florida along the Keys. Relatively high temperatures are also found along the Gulf coast and along the Atlantic coast from Charleston, South Carolina, to Miami, Florida.

The lowest and coldest temperatures and, thus, the coldest region extends from northeastern North Dakota to northwestern Minnesota. In this Dakota-Minnesota section, the midwinter afternoon highs average 10° to 15°F (−12° to −9°C). The morning lows are below zero; they average −5° to −10°F (−20° to 23°C).

Elevation is another of the climatic controls mentioned at the beginning of this chapter. It plays an important role in modifying the latitude

JULY ISOTHERMS (F°)

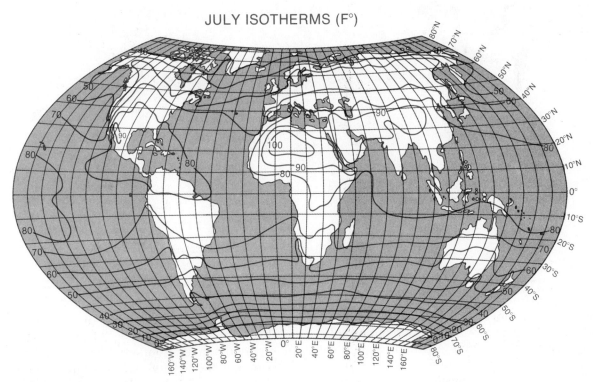

FIGURE 9–8 This pattern of isotherms is based on the average July temperatures (F°) for stations throughout the world.

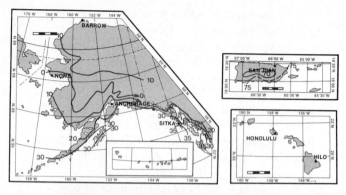

FIGURE 9–9 This pattern of isotherms is based on the average January temperatures (F°) for stations throughout the U.S.

effect of insolation. For example, in the northern and central Rocky Mountains, midwinter daily highs average 20° to 25°F (−7° to 4°C) and low temperatures run from −5°F to 5°F (−20° to −15°C). At very high elevations in the Rockies, temperatures drop to 0°F (−18°C) or below on 50 or more days each year and to 32°F (0°C) or below on 215 or more days. When 0°F (−18°C) readings are reported from the southern Sierra Nevada in California, it is elevation that is controlling the situation.

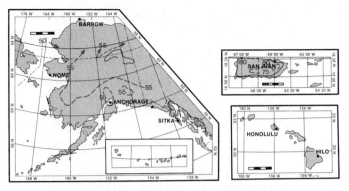

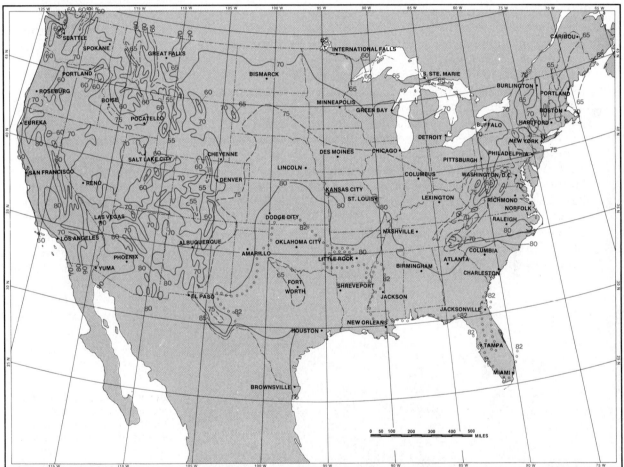

FIGURE 9–10   This pattern of isotherms is based on the average July temperatures (F°) for stations throughout the U.S.

The isotherms for the July map (Fig. 9–10) indicates the highest temperatures of the conterminous United States are found in interior valleys and desert areas of the Southwest, the southern Great Plains, and the Deep South. In southwest Arizona and southeast California, temperatures run from 100°F (38°C) to more than 110°F (43°C) in the afternoons and 75°F to 85°F (24°C to 30°C) in the mornings. Maximum daily temperatures in the southern Great Plains average about 96°F (36°C); minimums are in the range of 70°F (21°C). In the Deep South, the average daily maximum is 92°F (33°C) and minimums average 70°F (21°C).

Elevation, as a climatic control, again plays an important role in the temperature pattern over the country. Some of the lowest temperatures are found in the high mountains during July. In the East, the effect of elevation is quite apparent in the high Appalachians. The high elevations of the western mountains also show lower temperatures than the lower areas in the same latitude. At the 14,000-foot (4270-m) level in the Rockies, for example, the highest recorded temperature is 65°F (18°C). In fact, temperatures during July have reached or exceeded 100°F (38°C) in all states except in the high mountains.

Well-established ocean currents are another of the climatic controls that have an effect on the temperature pattern within the conterminous United States. For example, most northern and middle Pacific coastal areas do not run high temperatures and have never had maximums exceeding 90°F (32°C) because of the control exerted by ocean currents. The Japanese Current—a major ocean current affecting the West Coast—reaches this continent at about the latitude of the United States-Canadian border. It branches into two sections: the northern branch, the Alaska Current, brings a milder climate to coastal British Columbia and Alaska; the southern branch, the California Current, carries cooler water along the coastline of Washington, Oregon, and California. The Davidson Current moves in from the southeast during the winter and brings warmer temperatures to southern California.

The East Coast is affected by the warm Gulf Stream and the cold Labrador Current. The Gulf Stream carries warm water from the Florida Straits northward along the Atlantic coast. Winds blowing across this warm stream of water pick up heat and moisture which is then carried overland. The Labrador Current, which flows southward inside the Gulf Stream, forces the Gulf Stream to the east as it moves northward. The temperature pattern of the northeastern coast of the United States is affected by the Labrador Current.

CANADA

The temperatures during January for 66 percent of Canada, as shown in Figure 9–11, are below zero. In fact, on some of the Arctic Islands the average January temperatures are below −31°F (−35°C). Only a few small areas of Canada have average January temperatures above 20°F (−7°C); examples are the coastal area of British Columbia, southwestern Ontario, the southern coast of Newfoundland, and Nova Scotia.

The average Canadian temperatures during July are depicted on the map in Figure 9–12. The summers in northern Canada tend to be short and cool. In the northern Arctic Islands, the average July temperatures range around 40°F (4°C), and permanent icecaps cover parts of

Elevation plays an important role as a climatic control. Wharton Lake is below and to the left. We are at latitude 49°N, close to the location where the borders of Alberta, British Columbia, and Idaho come together—about 114°W. The elevations of the peaks in the region run around 8400 feet above sea level. Refer to Figure 9–12 and you will see that the mean July temperature in the region is 59°F (15°C). (National Film Board of Canada)

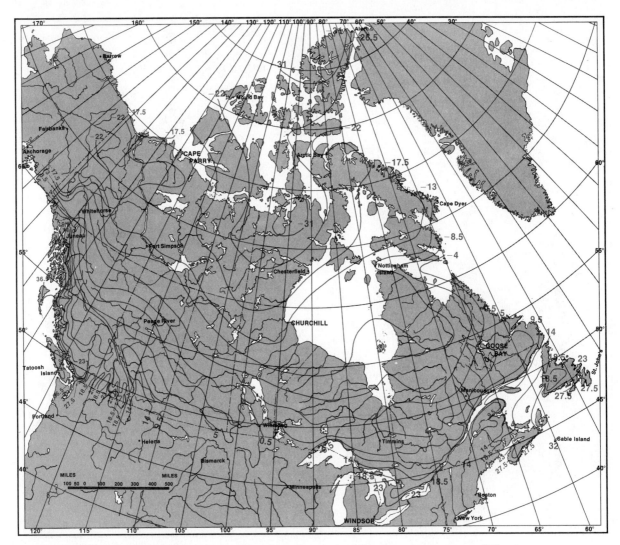

FIGURE 9–11   This pattern of isotherms based on the average January temperatures (F°) for stations throughout Canada.

Ellesmere, Axel Heiberg, and Devon islands. Bylot Island and the eastern side of Baffin Island also have permanent icecaps. The highest July temperatures, averaging around 70°F (21°C), are found along the Canada–United States border in the area of Windsor (42°18′N, 83°1′W).

MOISTURE AS A CLIMATIC ELEMENT

Water, which constantly circulates throughout the atmosphere, is the single most important component of the air from the viewpoint of a physical geographer concerned with climatic classification. The input of solar radiation evaporates large quantities of water from the Earth's surface. The water vapor that enters the atmosphere mixed with the other gases, is transported along with them as movements of air, condenses in clouds, and falls from the clouds as rain or snow. The precipitation that falls on the land eventually works its way to the sea. This whole

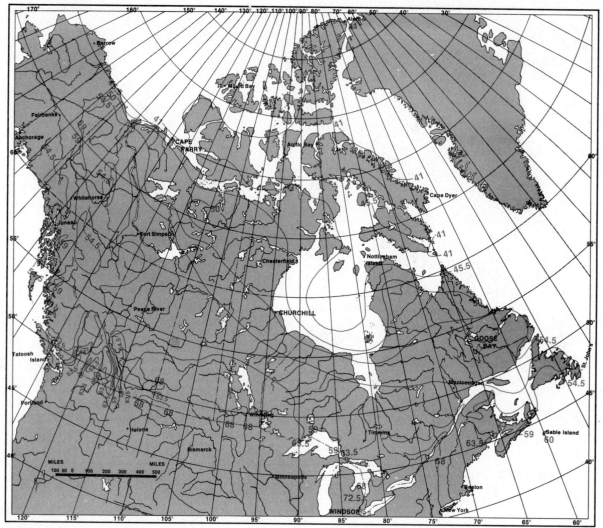

FIGURE 9–12 This pattern of isotherms is based on the average daily temperatures (F°) for July for stations throughout Canada.

process has been referred to as the *hydrologic cycle* in previous sections of this book.

At any given time, the atmosphere holds about 10 trillion tons of water. Each day approximately 1 trillion tons of water is evaporated from the surface of the sea and other bodies of water. And on the average, 1 trillion tons falls to the Earth each day as some form of precipitation. The atmosphere can be characterized as a great reservoir of water vapor—a reservoir in which the residence time of a molecule of water is about 10 days.

EVAPORATION

An interface is a surface forming a common boundary between two bodies or spaces. The common boundary between any water surface and the lower atmospheric surface is called an **air-water interface.** The boundary between the sea and the air is an interface, as is the boundary between a lake surface and the air.

At an air-water interface, water molecules are constantly passing upward from water to air and downward from air to water. The process by which any substance in the liquid state is converted into the vapor state is called **evaporation.** The difference per unit surface area between the number of molecules passing upward and those passing downward through the interface in a given unit of time is called the **evaporation rate.**

Molecules of water in a liquid—in a condensed state—are held to one another by strong forces of attraction. The molecules that escape from the liquid surface are those whose energy of thermal agitation is greater than the potential energy of attraction. In other words, the kinetic energy of an escaping molecule must be sufficient to overcome the surface-tension forces of the liquid.

Only a small proportion of the water molecules with sufficient kinetic energy to escape reach the surface in any given unit of time, and thus the rate of evaporation is limited. As water molecules of higher energy tend to escape, those left behind will have less energy, and so the temperature of the liquid is lowered. The energy of the escaping water molecules is added to the atmosphere and subtracted from the sea, lake, or river. The additional heat carried away with the evaporating molecules is called the **latent heat of vaporization,** which is the heat used to bring about the change from the liquid to the gaseous state. The rate of evaporation of water across the interface into the atmosphere is affected by temperature, vapor pressure, wind speed, and salinity.

TEMPERATURE
Since kinetic energy is a function of temperature, the rate of evaporation from the liquid is proportional to the temperature of the liquid. In other words, an increase in temperature increases the rate of evaporation, and a decrease in temperature causes a decrease in evaporation.

VAPOR PRESSURE
The water vapor that enters the atmosphere and mixes with the other gases exerts a pressure in all directions as do the other gases present. The part of atmospheric pressure that results from the presence of water vapor is called the **vapor pressure** of the air. For every open water surface, there is an equilibrium point at which the number of molecules returning to it is just equal to those escaping from it, which means the net evaporation is zero. Under these conditions, the air above the water is said to be saturated—it can hold no more water vapor.

At a constant atmospheric pressure and a given air temperature, the saturation vapor pressure has a fixed value. In Table 9–3, which is based on normal atmospheric pressure, you can find the saturation vapor pressure over freshwater and also over seawater with a salinity, or salt content, of 35 parts per thousand. Compare the saturation vapor pressure over freshwater at 32°F (0°C) with its value over seawater at the same temperature; note that salinity decreases the saturation vapor pressure slightly. The saturation vapor pressure changes rapidly with changes in temperature; for example, at 32°F (0°C) the saturation pressure is around 6 millibars, and at 86°F (30°C) it is about 42 millibars.

Now, let's consider the effect of vapor pressure on the rate of evaporation. Assume we have two parcels of air at 68°F (20°C); one parcel has a vapor pressure of 10 millibars and the other has a vapor pressure of 22 millibars. The parcel with a

TABLE 9–3 **SATURATION VAPOR PRESSURE**

AIR TEMPERATURE (°F)	(°C)	SATURATION VAPOR PRESSURE (IN MILLIBARS) Over Freshwater	Over Seawater
32	0	6.11	5.99
41	5	8.72	8.56
50	10	12.27	12.05
59	15	17.04	16.74
68	20	23.37	22.96
77	25	31.67	31.12
86	30	42.43	41.68
95	35	56.24	55.25
104	40	73.78	72.47

vapor pressure of 10 millibars is the drier of the two. The second parcel—as indicated by Table 9-3—is almost saturated. If both parcels move over a warmer body of water, the rate of evaporation or flow of water into the drier air will be greater than the flow into the moister air.

WIND SPEED

Vigorous air motions or **turbulence** in the air above a water surface carry water vapor away from the interface and increase the rate of evaporation. If there is drier air at some distance above the interface, there is usually a transport of water vapor upward to it. Turbulence intensifies with an increase in wind velocity and also gains strength as temperature increases downward toward the surface.

SALINITY

Evaporation varies with the salinity of the water body. The presence of dissolved salts in water retards evaporation. Thus, the rate of evaporation is greater from freshwater than from saltwater. Under similar conditions, freshwater will evaporate about 5 percent faster than seawater.

THE WORLDWIDE PATTERN

Since the rate of evaporation is primarily a function of temperature—although not exclusively so—it is not surprising to find that the mean annual evaporation declines from tropical to polar latitudes as shown in Table 9-4. (The figures in Table 9-4 are based on measurements and estimates made by G. Wüst.)

At most latitudes, evaporation is greater over the ocean than over a landmass. This reflects the fact that water is unlimited at the sea's surface but relatively scarce over most land areas. In the 0° to 10° band, however, evaporation is greater from the land's surface than from the ocean because of an abundance of surface water on the

land in these rainy regions and the unusual amount of water sent into the air—that is, transpired—by vegetation. In addition, the constantly high humidity in the air over the sea in the doldrums cuts down the rate of evaporation from the sea's surface. In the Northern Hemisphere, these distortions in the 0° to 10° band actually produce a lower average evaporation for its 0° to 10° band than for its 10° to 20° band (Table 9-4).

About 36.6 inches (93 cm) of water evaporates annually from the total surface of the world ocean. This corresponds to a volume of 80,100 cubic miles (333,900 km³) per year. Evaporation from inland water surfaces and from the solid ground amounts to 16.5 inches (42 cm), which is the equivalent of 14,900 cubic miles (62,100 km³). The total evaporation of 95,000 cubic miles (396,000 km³) is equal to the estimated annual precipitation of the world.

EVAPORATION MAPS

Evaporation inevitably extracts a portion of the gross water supply of reservoirs, streams, and lakes. Estimation of this loss is especially important to those concerned with water storage and reservoir design. In an arid region, the loss of water through evaporation actually imposes a ceiling on the supply obtainable through regulation.

Generalized estimates of free-water evaporation in a region are quite useful and are indicative of an aspect of climate. A rather simple system of measuring evaporation employs the standard Weather Service-type pan of 4-foot diameter. The container is filled with water, and the depth remaining at regular intervals is measured. The isopleths drawn on the map in Figure 9-13 show the pan evaporation in inches for the conterminous United States.

TABLE 9-4 MEAN ANNUAL EVAPORATION

LATITUDE BAND (in degrees)	NORTHERN HEMISPHERE		SOUTHERN HEMISPHERE	
	(in)	(cm)	(in)	(cm)
0-10	40.6	103.1	45.7	116.1
10-20	42.9	109.0	44.5	113.0
20-30	35.8	90.9	39.0	99.1
30-40	28.0	71.1	35.5	90.2
40-50	20.1	51.1	22.8	57.9
50-60	15.0	38.1	8.8	22.4

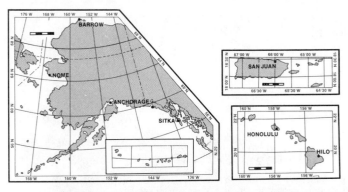

FIGURE 9–13   The pattern of isopleths drawn on this map is based on pan evaporation measurement (in inches) for many stations in the conterminous U.S.

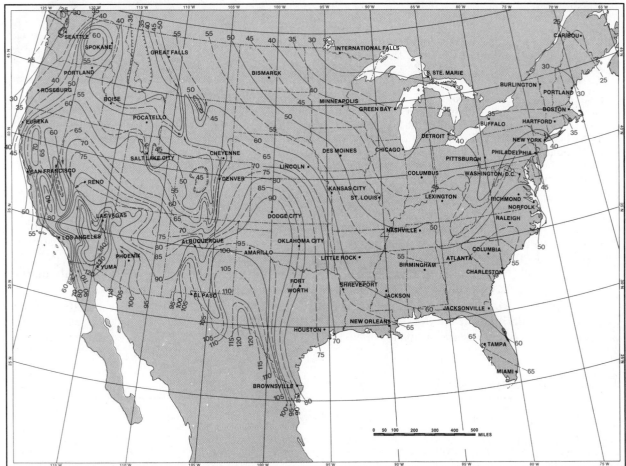

The National Oceanic and Atmospheric Administration develops pan-evaporation maps indicative of representative exposures that are reasonably free of obstructions to wind and sunshine. Topography produces a definite effect in the pan-evaporation maps because dew point and air temperatures tend to decrease with an increase in elevation; wind movement usually increases with elevation. Thus, in some portions of the map, the isopleths tend to follow the topographic features. In the deserts and semiarid regions of California, Arizona, New Mexico, and Texas, the pan evaporation ranges from 100 to more than 120 inches (254 to 300 cm) of water annually. In the Pacific Northwest, the northern Rockies, the Great Lakes region, and the Northeast, the annual pan evaporation is less than 40 inches (100 cm) per year.

Lake Jocassee in South Carolina is located at latitude 35°N, longitude 82°55′W, which is very close to the Clemson University station reported in Table 9–5. The surface area of the lake is 7565 acres. The capacity of the lake is 1,185,600 acre feet or 386 billion gallons. Depth of the lake is about 300 feet (90 meters). Jocassee adds a considerable amount of water to the air through evaporation. The evaporation from this lake would be about 70 percent of that from a pan exposed to the same temperature and wind conditions. (South Carolina Dept. of Parks)

In Figure 9–13, a 55-inch (140-cm) isopleth runs through South Carolina just to the west of Columbia. NOAA's summaries of evaporation data for four stations in South Carolina are detailed in Table 9–5. In some instances no record is reported because the water in the pan was frozen or precipitation in the pan exceeded evaporation. The two-year average for these four stations gives an evaporation of 53 inches (135 cm), which is rather close to the isopleth running through South Carolina in Figure 9–13. We must remember that any large-scale map is of necessity somewhat generalized.

In Table 9–4 the mean annual evaporation in the 30° to 40° latitude band is given as 28 inches (71 cm). This figure is not even close to the pan-

TABLE 9–5 PAN EVAPORATION IN SOUTH CAROLINA

STATION	LOCA-TION	ELEVA-TION (FT) ABOVE SEA LEVEL	YEAR	JAN	FEB	MAR	APR	MAY	JUN	JUL	AUG	SEP	OCT	NOV	DEC	TOTAL ANNUAL RE-PORTED
Clemson University	34°41′N 82°49′W	819	1975	1.79	2.19	2.82	5.21	5.57	7.17	6.48	5.69	3.88	3.15	2.71	—	46.66
			1976	—	—	3.25	5.99	5.27	6.36	7.24	6.17	3.68	2.95	2.71	—	43.62
Clark Hill Dam	33°40′N 82°11′W	380	1975	1.41	1.87	3.45	4.61	5.38	6.06	5.31	6.17	4.01	3.16	1.89	—	43.32
			1976	—	3.03	3.31	5.83	4.55	5.94	6.22	6.50	4.42	2.87	2.06	—	44.73
Sand Hill Exp. Station	34°8′N 80°52′W	440	1975	2.69	3.11	5.50	7.55	7.19	8.39	6.83	7.45	5.37	4.15	3.12	1.95	63.30
			1976	—	4.93	5.34	8.92	6.96	7.46	8.66	8.30	6.43	5.34	4.79	—	67.13
Charleston	32°54′N 80°2′W	41	1975	3.08	2.54	5.52	6.11	7.27	7.29	7.82	7.36	5.36	4.70	3.19	—	60.24
			1976	—	4.45	5.67	7.70	7.27	6.15	8.04	—	4.97	4.15	3.25	—	52.01

evaporation data reported by the four stations in South Carolina. The 28 inches (71 cm) indicated by Wüst for this latitude band, however, is an average or composite evaporation from both land and water surfaces. If we examine Wüst's calculations, we find that the evaporation from the ocean in this latitude band is 37.8 inches (96 cm) annually. Since a pan is a very shallow restricted enclosure, water evaporates faster from it than from a large water surface. In fact, evaporation from a large water surface is estimated to be about 70 percent of that from a pan exposed to the same temperature and wind conditions. If we take 70 percent of the 53-inch (135-cm) average for the South Carolina pans, we get 37.1 inches (94.2 cm), which is very close to Wüst's calculations.

HUMIDITY

The water-vapor content of the atmosphere is referred to as **humidity.** The quantity of water vapor in the air can range from almost zero to a maximum of 4 percent by volume.

RELATIVE HUMIDITY

The amount of water vapor held by air, at a particular temperature, compared to the maximum amount that it can hold, at the same temperature, is called **relative humidity.** This term is undoubtedly familiar to you; relative humidity is reported as a percentage.

Let's assume we have a parcel of air at 68°F (20°C) with a vapor pressure of 14 millibars. Refer to Table 9–3 and you will find that the saturation vapor pressure of air at 68°F (20°C) is 23.37 millibars. Since relative humidity is a comparison of the water vapor actually present in the air to the vapor pressure at saturated conditions, we divide 14 millibars by 23.37 millibars, which gives a ratio of 0.599. The relative humidity of the parcel of air at 68°F (20°C) is therefore 59.9 percent, which means the parcel is presently holding only 59.9 percent of its saturated capacity.

The relative humidity of the air can be measured by using a pair of thermometers. An arrangement of two thermometers that is used to measure relative humidity is called a **sling psychrometer.** The bulb of one thermometer (Fig.

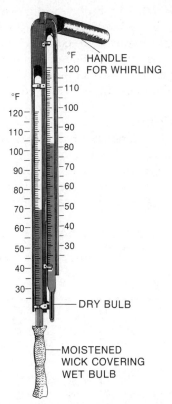

FIGURE 9–14 This arrangement of two thermometers is known as a *sling psychrometer.* The wet-bulb reading stands at 70°F. The dry bulb reading stands at 82°F.

9–14) is covered by a moistened wick, and the other bulb remains dry. Both thermometers are whirled. After a minute or two of whirling, the thermometers are read. The thermometer with the wet bulb produces a temperature equal to or lower than the dry bulb, depending on the amount of evaporation that occurs. The process of whirling the thermometers on the rotating mount is repeated until no further reduction in the wet-bulb thermometer reading can be obtained. The dry bulb indicates the current air temperature. The difference in temperature between the wet bulb and the dry bulb is proportional to the dryness of the air. The readings obtained from a sling psychrometer are used in conjunction with previously calculated tables to arrive at the relative humidity of a locality.

Assume that dry- and wet-bulb readings have been taken and were found to be 82° and 70°F, respectively (Fig. 9–14). Subtract the wet-bulb reading from the dry-bulb reading to obtain the difference between readings: 82−70 = 12. Using Table 9–6, drop down the column below 12° in the "difference" row to the number opposite the dry-bulb reading. This numeral is the relative humidity reading—in this case, 55 percent.

ABSOLUTE HUMIDITY

When the temperature of the air changes, its relative humidity changes, too. For any given air temperature, there is a fixed quantity of water vapor that can occupy any given volume of atmospheric space. An increase in temperature al-

lows more water vapor to be accommodated in the space. If no additional water vapor enters the parcel, the relative humidity will fall; whereas the **absolute humidity**—the actual weight or the amount of water vapor in a given volume of air—remains the same. If the temperature is lowered in that same parcel, the relative humidity will rise. Air temperature changes constantly; thus, relative humidity is always changing, even if water vapor does not leave or enter the parcel of air.

Absolute humidity is reported as grams of water vapor per cubic meter of air. Table 9–7 lists

TABLE 9–6   **RELATIVE HUMIDITY IN PERCENTAGES**

DRY-BULB READING (°F)	DIFFERENCES BETWEEN DRY-BULB AND WET-BULB READINGS (FAHRENHEIT)																	
	1°	2°	3°	4°	5°	6°	7°	8°	9°	10°	11°	12°	13°	14°	15°	16°	17°	18°
22°	86	71	58	44	31	17	4											
24°	87	73	60	47	35	22	10											
26°	87	75	63	51	39	27	16	4										
28°	88	76	65	54	43	32	21	10										
30°	89	78	67	56	46	36	26	16	6									
32°	89	79	69	59	49	39	30	20	11	2								
34°	90	81	71	62	52	43	34	25	16	8								
36°	91	82	73	64	55	46	38	29	21	13	5							
38°	91	83	75	66	58	50	42	33	25	17	10	2						
40°	92	83	75	68	60	52	45	37	29	22	15	7						
42°	92	85	77	69	62	55	47	40	33	26	19	12	5					
44°	93	85	78	71	63	56	49	43	36	30	23	16	10	4				
46°	93	86	79	72	65	58	52	45	39	32	26	20	14	8	2			
48°	93	86	79	73	66	60	54	47	41	35	29	23	18	12	7	1		
50°	93	87	80	75	67	61	55	49	43	38	32	27	21	16	10	5		
52°	94	87	81	75	69	63	57	51	46	40	35	29	24	19	14	9		
54°	94	88	82	76	70	64	59	53	48	42	37	32	27	22	17	12	8	3
56°	94	88	82	76	71	65	60	55	50	44	39	34	30	25	20	16	12	7
58°	94	88	83	77	72	66	61	56	51	46	41	37	32	27	23	18	14	10
60°	94	89	83	78	73	68	63	58	53	48	43	39	34	30	26	21	17	13
62°	94	89	84	79	74	69	64	59	54	50	45	41	36	32	28	24	20	16
64°	95	90	84	79	74	70	65	60	56	51	47	43	38	34	30	26	22	18
66°	95	90	85	80	75	71	66	61	57	53	48	44	40	36	32	29	25	21
68°	95	90	85	80	76	71	67	62	58	54	50	46	42	38	34	31	27	23
70°	95	90	86	81	77	72	68	64	59	55	51	48	44	40	36	33	29	25
72°	95	91	86	82	77	73	69	65	61	57	53	49	45	42	38	34	31	28
74°	95	91	86	82	78	74	69	65	61	58	54	50	47	43	39	36	33	29
76°	96	91	87	82	78	74	70	66	62	59	55	51	48	44	41	38	34	31
78°	96	91	87	83	79	75	71	67	63	60	56	53	49	46	43	39	36	33
80°	96	91	87	83	79	75	72	68	64	61	57	54	50	47	44	41	38	35
82°	96	92	88	84	80	76	72	69	65	61	58	55	51	48	45	42	39	36
84°	96	92	88	84	80	76	73	69	66	62	59	56	52	49	46	43	40	37
86°	96	92	88	84	81	77	73	70	66	63	60	57	53	50	47	44	42	39
88°	96	92	88	85	81	77	74	70	67	64	61	57	54	51	48	46	43	40
90°	96	92	89	85	81	78	74	71	68	65	61	58	55	52	49	47	44	41
92°	96	92	89	85	82	78	75	72	68	65	62	59	56	53	50	48	45	42
94°	96	93	89	85	82	79	75	72	69	66	63	60	57	54	51	49	46	43
96°	96	93	89	86	82	79	76	73	69	66	63	61	58	55	52	50	47	44

TABLE 9–7   **ABSOLUTE HUMIDITY FOR SATURATED AIR**

AIR TEMPERATURE (°F)	(°C)	WATER VAPOR (grams per cubic meter)
−4	−20	1.07
5	−15	1.61
14	−10	2.36
23	−5	3.41
32	0	4.85
41	5	6.79
50	10	9.40
59	15	12.83
68	20	17.30
77	25	23.05
86	30	30.38
95	35	39.63
104	40	51.19
113	45	65.50

the absolute humidity for saturated air at various temperatures. The data in the table indicate the impact air temperature has on humidity. Compare the absolute humidity of air at 32°F (0°C) with air at 104°F (40°C) and you will find the capacity of the air increases tenfold.

SPECIFIC HUMIDITY

Another way to describe the water-vapor content of a parcel of air is by the term **specific humidity,** which is the weight of water vapor per unit weight of moist air. Specific humidity is reported as grams of water vapor per kilogram of air. The curve in Figure 9–15 is a graphic display of specific humidity for different temperatures of saturated air when atmospheric pressure is 1013.25 millibars at sea level. For example, the specific humidity of air at 0°C is about 3.9 grams of water vapor per kilogram of air; at 12.5°C saturated air at sea level holds 9 grams of water vapor per kilogram of air.

THE WORLDWIDE PATTERN

Relative humidity is an important climatic factor that controls the amount and rate of evaporation from all surfaces, including plants, animals, land, and water. There seems to be a definite worldwide pattern to its distribution.

In a broad, general way, the zone of maximum relative humidity is centered over the Equator, with percentages averaging 85 and above. Moving poleward from the Equator, relative humidity decreases with the lowest values

occurring at 30°N (70%) and 30°S (75%). From these latitudes relative humidity increases as a result of decreasing temperature. At 60°N, the average is about 80 percent; but in the Southern Hemisphere, the averages climb back to 85 percent by the time 60°S latitude is reached.

The belts of highest and lowest relative humidity shift with the Sun—they move somewhat northward from March through June and southward from July through December. The seasonal distribution varies with latitude. For example, in the low latitudes, when seasonal variations are small, average relative humidity is higher in the wet summer than in the drier winter; in the higher latitudes, however, the highest relative humidity coincides with the cold winter. There is a worldwide diurnal varia-

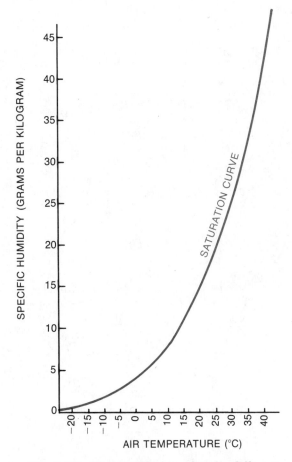

FIGURE 9–15 **The specific humidity for different temperatures of saturated air can be read from this curve. Specific humidity is reported as grams of water vapor per kilogram of air.**

tion in relative humidity, too; the maximum occurs in the early morning, and there is a decline throughout the day until midafternoon, when the lowest humidities are achieved.

THE U.S. PATTERN

The average annual relative humidity for the United States is depicted in Figure 9–16. Rela-

tive humidities are highest in coastal areas where winds blowing inland from the sea carry large quantities of moisture. Relative humidities tend to be lower at inland stations and, of course, are lowest in deserts and semiarid interiors such as the Southwest of the conterminous United States.

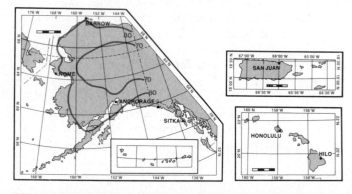

FIGURE 9–16 The pattern of mean annual relative humidity depicted on this map has been constructed from a broad study of relative humidity values at stations throughout the U.S.

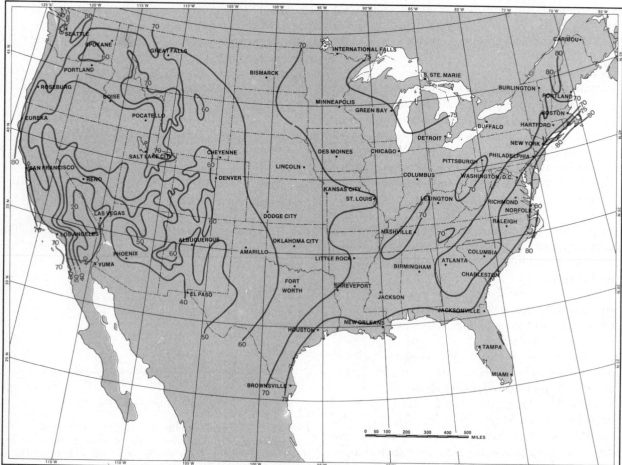

There is a definite diurnal variation in relative humidity throughout the United States. During January, for example, Tampa, on the west coast of Florida, generally has a relative humidity of 88 percent at dawn, which declines throughout the day and reaches a low of 55 percent by midafternoon. On a typical January day in Yuma, Arizona, the humidity at dawn is about 57 percent and the midafternoon humidity hovers around 30 percent.

THE DEW POINT

When a parcel of air is cooled sufficiently, its relative humidity will reach 100 percent. The temperature at which this occurs is called the **dew point.** If surface air is cooled any further, the excess water vapor may condense as dew, which is moisture that condenses on surface objects rather than on nuclei in the air above the surface. You have certainly seen blades of grass covered by sparkling droplets of water on a cool morning in autumn and spring.

When the dew point of a surface air parcel is below the freezing point of water, frost may form if the air is chilled sufficiently. In fact, when a dew point is below 32°F (0°C), it is sometimes called the **frost point.** When frost forms, the water vapor changes from the gaseous state to the solid state directly with no liquid forming; this process is called **sublimation.**

Specifically, then, the dew point is the temperature to which a given parcel of air must be cooled—at constant pressure and constant water-vapor content—to produce saturation. In other words, the dew point is the temperature at which the actual vapor pressure of the contained water in a parcel of air is equal to its saturation vapor pressure. The mean dew-point temperatures for the United States are depicted on the map in Figure 9–17. Miami, for example, has a dew-point temperature that averages 65°F (18°C) annually. Most of the dew points noted in the Alaskan portion of Figure 9–17 are, in reality, frost points.

FOGS

Fog is defined as very small drops of water condensed from and suspended in the air near the surface of the Earth; but from a practical viewpoint, the term **fog** is applied only when there are enough droplets in the air to reduce horizontal visibility. Based on visibility, fogs are classified as light, moderate, thick, and dense. A fog that is classified as dense reduces visibility to less than 0.2 mile (0.3 km). In a thick fog, visibility ranges from 0.2 to 0.3 mile (0.3 to 0.5 km). When a moderate fog exists in an area, you can expect visibility to range from 0.3 to 0.6 mile (0.5 to 1 km). In light fog, the visibility is more than 0.6 mile (1 km).

Figure 9–18 shows that heavy fog—visibility of 0.25 mile (0.4 km) or less—occurs at one time or another in most sections of the United States. It occurs, for example, on from 80 to more than 100 days annually along our western coast, in the Olympic and Cascade Mountains, in the Appalachians, and along our New England coast. Point Arguello with 140 days and Santa Catalina Island with 158 days of heavy fog have almost as many foggy days as clear days. Point Reyes, California, and some of the islands off Maine with 200 days of heavy fog per year have more foggy days than clear days.

RADIATION FOG

This is a nighttime, overland phenomenon. **Radiation fog** develops when the sky is clear, the wind is light, and the relative humidity is high near sunset. Cessation of insolation at night allows the surface of the Earth to cool by longwave radiation, mainly in wavelengths to which the water vapor in the air is transparent. The layer of air adjacent to the cooling Earth also cools because it is no longer receiving heat by conduction or radiation; the air is, in fact, losing heat by radiation in wavelengths absorbed and emitted by water vapor. The cooling, moist air loses just enough heat to cause water droplets to form in the layer close to the Earth. Autumn and winter are the most favorable seasons for radiation fog.

ADVECTION FOG

This kind of fog can form during the day or night, over land or water. It develops when warm, moist air moves in over a colder land or water surface. The warmer air gives off heat to the

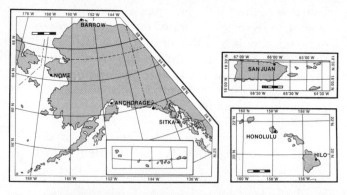

FIGURE 9–17 The pattern of mean annual dew point temperatures depicted on this map has been constructed from data collected over a 20-year period.

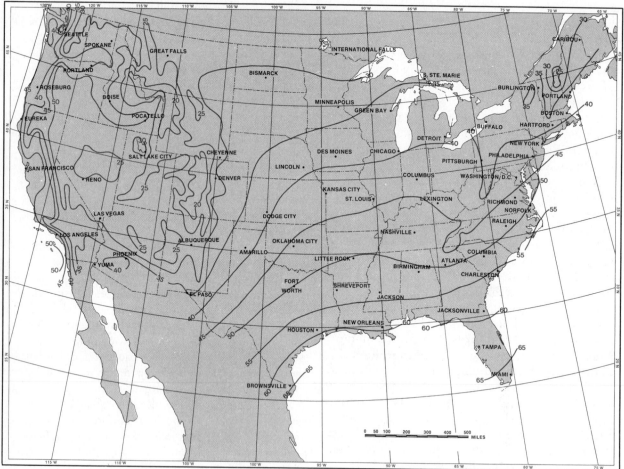

colder underlying surface. This loss of heat cools the lower layer of air to its dew point, and a fog forms. **Advection fogs** form regularly off the coast of California in the San Francisco area. The cold ocean current flowing off the coast chills the lower layer of air moving across the ocean toward the coast. The advection fogs of San Francisco Bay are legendary. The fogs of Newfoundland's Grand Banks are also advection-type fogs.

PRECIPITATION

All **precipitation**—whether in liquid or solid form—is measured and recorded either directly

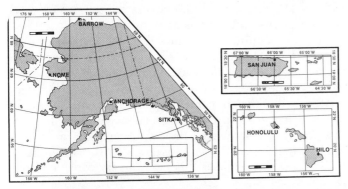

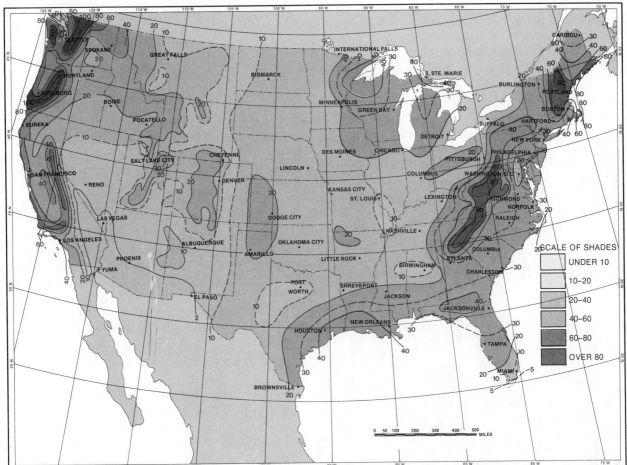

FIGURE 9–18   The isolines and shadings indicate the mean annual number of days with heavy fog—that is, visibility less than 0.25 mile (0.4 kilometer).

in inches of water or indirectly by converting other measurements, such as snow depth, to their water equivalent. In addition to rain and snow, drizzle, hail, ice pellets, and glaze ice are recorded.

Air that is free of dust may be cooled below its dew point without any condensation occurring.

Such air is said to be *supersaturated*. If smoke, salt, or other particles are added to supersaturated air, rapid condensation occurs. The presence of hygroscopic particles is essential to the condensation of moisture in the air. Hygroscopic particles are referred to as **nuclei of condensa-**

tion. Without these nuclei, the air will not yield a large volume of precipitation.

RAIN

The term **rain** refers to moisture that falls in a liquid state. The size of raindrops can vary considerably from diameters of 0.004 (0.01 cm) to 0.2 inch (0.5 cm). There is a natural limit to the size of raindrops; large drops falling through the air break up into smaller drops when they attain a velocity of 18 miles (29 km) per hour.

SNOW

This form of precipitation is produced when water vapor sublimes—changes from a gas to a solid without first going through a liquid state—at temperatures well below the freezing point. Snowflakes are most always hexagonal, that is, six-sided; in terms of their size, however, they are extremely varied. Large snowflakes, for example, are formed by the combination of many small crystals.

At low temperatures there may be very little moisture in the air, so precipitation of snow may be light; but it is never "too cold to snow." Snow is a poor conductor of heat, and when it covers the ground during the winter it serves as an insulating blanket for the plants and the soil below. A snow cover keeps the land warmer than it

otherwise would be during winter and protects plants and organisms within the soil.

A cubic foot of snow will not yield an equivalent volume of water. In any given container, a depth of 10 inches (25 cm) of snow will yield a water depth of approximately 1 inch (2.5 cm) when the snow melts. From this fact, we can say that an average ratio exists between snow and rain of 10:1. However, the condition of the snow—its texture and the closeness with which it is packed—causes this ratio to vary considerably.

HAIL

Hailstones are composed of hard pellets of ice and compacted snow granules built up in concentric layers. No one is really sure how **hail** is produced. One theory is that hailstones are formed in cumulonimbus clouds in which there are strong, circulating convection cells of cold air. The up-and-down motions of the hailstones produced by updrafts result in alternate layers of frozen ice and snow as the stones pass from one region of the cloud to another. Another theory, preferred by most meteorologists, is that hailstones form when frozen droplets pass through several layers of cold, moist air within a cumulonimbus cloud and acquire repeated layers of ice by sublimation. Hail falls only from cumulonimbus clouds. Hailstones come in various sizes; some are quite large. In fact, hailstones with diameters of 6 inches (15.2 cm) and weighing 1.6 pounds (0.73 kg) pounded Coffeyville, Kansas, in September 1970. A fast-moving hailstone can cause a lot of damage. Fortunately, the area covered by a hailstorm is usually small.

OTHER FORMS OF PRECIPITATION

Sleet, glaze, and snow grains are other common forms of precipitation. **Sleet** is partially frozen rain. **Glaze** is supercooled rain that freezes on striking cold, solid surfaces. The occurrence of glaze is usually referred to as an *ice storm*. Extensive damage is done to trees and utility company wires when they are coated by a heavy glaze. **Snow grains** are compacted ice needles.

THE WORLDWIDE PATTERN

The general pattern of annual precipitation is depicted on the world map in Figure 9–19. The

The hailstones shown here are much larger than the golf ball. (NOAA)

PRECIPITATION (INCHES)

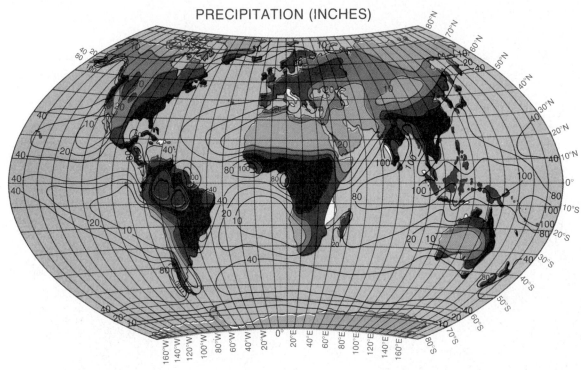

FIGURE 9–19 The general pattern of world precipitation (inches) is revealed by the isoline and the shadings.

amount of precipitation that falls on any locale is controlled by the total quantity of water vapor in the air and the nature of the process that leads to its condensation. The cooling process may occur as a result of (1) air ascending to great elevations through local convection, (2) air being forced over topographic barriers, such as mountains, to higher elevations, and (3) air being forced aloft to higher elevations over a denser parcel of air.

The areas of heaviest precipitation are generally located in tropical regions. The high tropical temperatures produce a lot of evaporation and put large quantities of water vapor into the atmosphere. Outstanding exceptions to this rule that "heaviest precipitation occurs in the tropics" are found in the high latitudes of southern Alaska, western Norway, and southern Chile where relatively warm, moist air from the sea undergoes considerable uplift as it is forced to higher elevations in the mountains.

A pine tree arches gracefully to acknowledge its burden of ice after a January ice storm outside Richmond, Virginia. (Celeste Navarra)

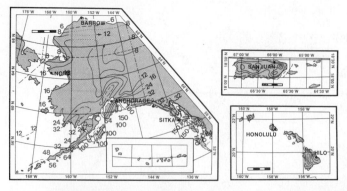

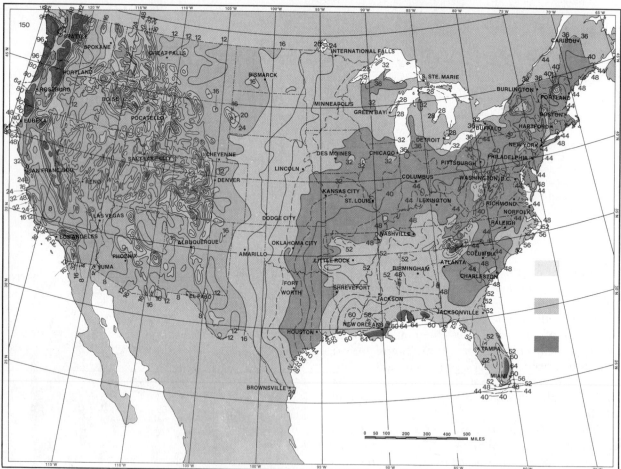

FIGURE 9–20 The conterminous U.S., as a whole, has a mean annual precipitation total of 29 inches. This total is based on a 77-year record. However, the isolines and shadings indicate the variation from this mean is quite large. The greatest mean annual precipitation on record for the conterminous U.S. is 144 inches at Wynoochee Oxbow, Washington. The least mean annual rainfall is 1.78 inches for Death Valley, California.

In marked contrast to the rainy regions are the dry regions in the vicinity of latitude 30° on practically all continents. Note, for example, the dry area that stretches from the western Sahara of Africa eastward in a broad, somewhat broken belt to the Gobi Desert of Asia (Fig. 9–19). Dry or arid regions are the result of conditions that are unfavorable to condensation of whatever water vapor may be present in the atmosphere. The polar regions, for example, are dry because low temperatures and very limited evaporation give the air a very low water-vapor content.

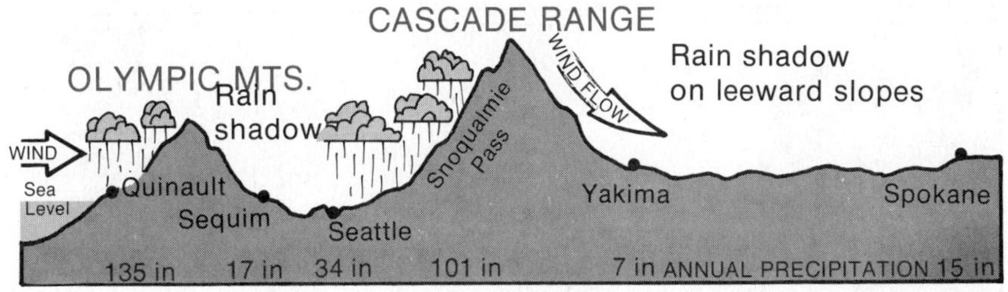

FIGURE 9–21 The windward slopes of the Olympics and Cascades are quite wet. The leeward slopes get less precipitation and are typically dry.

CASCADE RANGE

OLYMPIC MTS.

Rain shadow

Rain shadow on leeward slopes

WIND FLOW

WIND

Sea Level

Quinault

Sequim

Seattle

Snoqualmie Pass

Yakima

Spokane

135 in   17 in   34 in   101 in   7 in ANNUAL PRECIPITATION 15 in

THE UNITED STATES

The greatest mean annual precipitation on record for the entire world is 460 inches (1170 cm) at Mt. Waialeale, on the island of Kauai, Hawaii. Figure 9–20 shows that the heaviest annual precipitation for the conterminous United States falls in the mountains of western Washington. In fact, as much as 144 inches (366 cm) per year falls in the Olympic Mountains and 115 inches (292 cm) falls in the Cascades. The annual totals in the coastal regions of Washington and Oregon range from 80 to more than 100 inches (203 to 254 cm).

Much of the precipitation along the western coast of the conterminous United States is triggered by mountains. As air is forced up and over a range, the increase in elevation causes it to cool adiabatically and release precipitation on the windward slopes. Beyond the summit of the mountain, the air contains little moisture and a "rain shadow" results. In Figure 9–21, a cross section is drawn from Quinault to Spokane. The windward sides of the Olympics and the Cascades are wet; the leeward slopes are typically dry. There is a rather large difference between the amount of windward and leeward precipitation in the mountains of Washington.

The same kind of mechanism operates in Oregon and California. However, the quantity of precipitation triggered by the mountains of California is not as large as that triggered by Washington's mountains. The situation in

California from San Luis Obispo to Owens Valley in the Great Basin is depicted in Figure 9–22.

San Luis Obispo and Santa Barbara are both on the windward side of the Coast Ranges (Table 9–8). The elevations in the Coast Ranges run from about 1500 feet (460 m) beyond San Luis Obispo to more than 5000 feet (1500 m) in the higher sections. Bakersfield and Fresno are in a rain shadow on the lee side of the Coast Ranges in the San Joaquin Valley. As air from the San Joaquin ascends the lofty Sierra Nevada elevations of 14,494 feet (4418 m) at Mount Whitney, the moisture wrung from it drops an annual precipitation of 48 inches (122 cm) in the Sequoia National Forest at Giant Forest. Independence, which is in the arid Owens Valley of the Great Basin Desert, is in the rain shadow of the Sierras and receives a scant annual precipitation of 5 inches (13 cm).

The Great Basin is the driest part of the conterminous United States. Las Vegas, for example, which is in the Great Basin, receives a paltry annual precipitation of 4 inches (10 cm) (Table 9–8). The northern and central sections of the Rocky Mountain region receive from 20 to more than 40 inches (50 to 100 cm) of precipitation per year (Fig. 9–20). Precipitation in the eastern foothills of the Rockies is only 12 to 16 inches (30 to 41 cm); precipitation increases east of the foothills and reaches 30 inches (76 cm) per year over central Kansas, Iowa, and the Great Lakes.

FIGURE 9–22 Much of the precipitation in the west is triggered by mountains. The stations on the windward slopes of the mountains receive more precipitation than those locations on the leeward slopes.

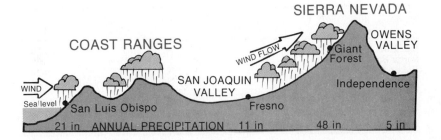

SIERRA NEVADA

COAST RANGES

OWENS VALLEY

WIND FLOW

Giant Forest

SAN JOAQUIN VALLEY

WIND

Sea level

San Luis Obispo

Fresno

Independence

21 in   ANNUAL PRECIPITATION   11 in   48 in   5 in

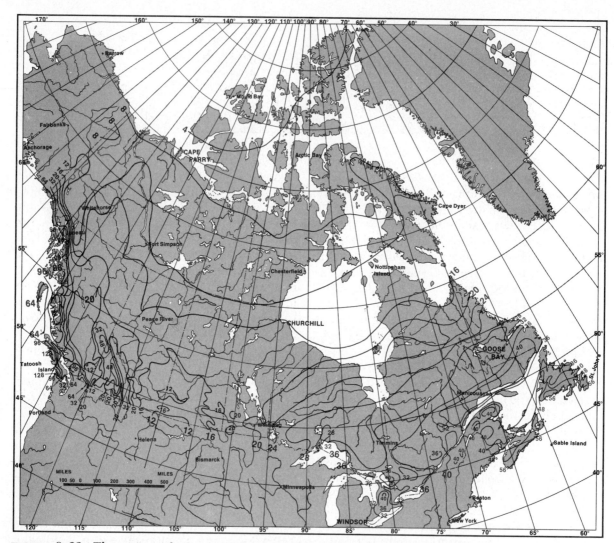

FIGURE 9–23   The pattern of mean annual precipitation (inches) throughout Canada is detailed by the isolines. This construction is based on 30 years of data.

In Tennessee, the central and eastern Gulf states, the Appalachians, and the Atlantic coastal areas, precipitation exceeds 50 inches (127 cm).

CANADA

Figure 9–23 details the average yearly precipitation throughout Canada. Some areas along

TABLE 9–8   RAIN-SHADOW EFFECT

TOWN	LOCATION (Latitude, Longitude)	ELEVATION (ft)	ELEVATION (m)	ANNUAL PRECIPITATION (in)	ANNUAL PRECIPITATION (cm)
San Luis Obispo	35°30′N,120°30′W	215	65.5	21	53
Santa Barbara	34°25′N,119°42′W	100	30.5	17	43
Bakersfield	35°22′N,119°01′W	420	128.0	6	15
Fresno	36°44′N,119°47′W	285	86.9	11	28
Giant Forest area	36°34′N,118°46′W	6400	1950	48	122
Independence	36°48′N,118°12′W	3917	1193	5	13
Las Vegas	36°10′N,115°09′W	2030	618.8	4	10

This farm is in the interior plains of Canada at 50°26′N, 105°38′W, which is near Moose Jaw, Saskatchewan. Slightly more than 14 inches (35 centimeters) of precipitation fall in the area; but more than 65 percent of this precipitation falls from May through August. (National Film Board of Canada)

Canada's western coast receive more than 100 inches (254 cm) for much the same reason that areas similarly situated in Washington receive more than 100 inches (254 cm). In the Canadian prairie, the annual precipitation ranges around 17 inches (43 cm), and most of it falls during the warm summer months. In eastern Canada, the annual precipitation ranges from 30 inches (76 cm) in central Ontario to 55 inches (140 cm) in Nova Scotia. A good portion of the precipitation in the East falls as snow. In fact, more than 100 inches (254 cm) of snow falls annually in large areas of eastern Canada.

CLOUDS AND CLOUDINESS

Clouds are accumulations of condensed liquid or frozen water, and thus their visible characteristics reflect the distribution of the temperature and the moisture content of the atmosphere in their immediate vicinity.

Clouds assume an almost infinite variety of forms, and most of them occur within the lower layer of the atmosphere, the troposphere. Only two types of clouds—nacreous and noctilucent clouds—are found above the troposphere.

The clouds of the troposphere are classified according to how and at what height they are formed (Table 9–9). Bascially there are two ways in which these clouds form: by condensation within rising air currents or by condensation within layered air that is relatively free of verti-cal currents. Rising air currents produce puffy clouds called **cumulus types**; the cooling of layered air produces sheet clouds called **stratus types**. These two cloud types—cumulus and stratus—can be further classified into four families. The family classification is based on the altitudes at which the clouds occur.

MODE OF FORMATION

A cloud is a visible accumulation of water or ice particles that are suspended in the air. The diameters of these cloud particles range from 2 through 30 micrometers; there may be less than 10 or, at times, more than 300 particles per milliliter of air within a cloud. The total amount of

TABLE 9–9   CLOUD TYPES AND ASSOCIATED CONDITIONS

ALTITUDE OVER MIDDLE LATITUDE (In Feet)	NAME, ABBREVIATION, AND SYMBOL	DESCRIPTION	COMPOSITION	POSSIBLE WEATHER CHANGES
High Clouds 18,000 to 45,000	Cirrus (Ci)	Mares tails Wispy and feathery	Ice crystals	May indicate a storm, showery weather close by
	Cirrostratus (Cs)	High veil Halo cloud	Ice crystals	Storm may be approaching
	Cirrocumulus (Cc)	Mackerel sky	Ice crystals	Mixed significance, indication of turbulence aloft, possible storm
Middle Clouds 6500 to 18,000	Altocumulus (Ac)	Widespread, cotton ball	Ice and water	Steady rain or snow
	Altostratus (As)	Thick to thin overcast; high, no halos	Water and ice	Impending rain or snow
Low-Family Clouds Sea Level to 6500	Stratocumulus (Sc)	Heavy rolls, low, widespread Wavy base of even height	Water	Rain may be possible
	Stratus (St)	Hazy cloud layer, like high fog Somewhat uniform base	Water	May produce drizzle
	Nimbostratus (Ns)	Low, dark gray	Water, or ice crystals	Continuous rain or snow
Vertical Clouds Few hundred to 65,000	Cumulus (Cu)	Fluffy, billowy clouds Flat base, cotton ball top	Water	Fair weather
	Cumulonimbus (Cb)	Thunderhead Flat bottom and lofty top Anvil at top	Ice (upper levels) Water (lower levels)	Violent winds, rain, hail are all possible Thunderstorms

water that makes up a cloud is quite variable: In some clouds there is as little as 0.1 gram per cubic meter, and in others the amount of water exceeds 5 grams per cubic meter.

All clouds form as a result of air cooling below its dew point. As the temperature of a parcel of air approaches its dew point on the saturation curve, some of its water vapor converts—condenses—to the liquid form when the hygroscopic particles called *nuclei of condensation* are present. If the temperature of the dew point is well below the freezing point of water, some of the water vapor may sublimate and convert directly to ice or snow.

The immediate cause of a decrease in the temperature of an air parcel, as mentioned earlier, is usually the ascent and consequent expansion of the air, which we have discussed as adiabatic expansion. In the free atmosphere, adiabatic expansion results from the lifting of the air (1) by its movement over terrain of increasing height, (2) over air of greater density, (3) as a result of mechanical turbulence, and (4) as a consequence of thermal instability that produces convection.

VERTICAL CLOUDS

Clouds of the troposphere can be classified by altitude into four families: low, middle, high, and vertical. The base of a vertical cloud may be as low as a typical low cloud, but its top may be at 65,000 feet (19,800 m), which is in the realm of the high clouds. Thus, vertical clouds cut across the complete sweep of cloud altitudes. The vertical-cloud family in this system consists of cumulus and cumulonimbus clouds.

CUMULUS

Cumulus are clouds produced by thermal convection. Usually, the instability, which produces the convection, develops during the morning as the ground is heated by the incoming solar radiation; the clouds form in a clear sky and grow rapidly.

Cumulus are usually clouds that are evaporating as rapidly as they are being formed. Watch a sky in which cumulus are developing and you will observe that each cloud has a life of a few minutes—"new" clouds grow as "old" clouds vanish. To casual observers, however, the scene looks generally the same because they do not focus on the "life" of an individual cloud.

family clouds are stratus, stratocumulus, and nimbostratus. In the discussion that follows some attention is given to a description of as well as to the origin of each of these clouds.

Cumulus. (NOAA)

Stratus. (NOAA)

Cumulonimbus. (NOAA)

CUMULONIMBUS

The heavy, dense, towering **cumulonimbus** cloud develops when a cumulus cloud grows to heights well above the freezing level and the upper part of the cloud becomes glaciated—the droplets of which it is composed freeze. The glaciation produces a striking change in the cloud's appearance. A cumulonimbus with a dark, flat base, and an anvil-shaped top is called a **thunderhead.**

LOW CLOUDS

At all latitudes, low clouds range from near the ground to 6500 feet (1980 m). The typical low-

STRATUS

These low, gray, sheetlike clouds cover a large part of the sky when they form. The term **stratus**—used in a strict sense—is only correctly applied to layers of fog or to clouds not far above the ground or sea. Stratus clouds are almost formless.

Stratocumulus. (NOAA)

STRATOCUMULUS

The **stratocumulus** clouds usually have a grayish tint and may appear as patchy masses or rolls in the sky. They are often arranged in long, gray, parallel bands that cover all or most of the sky.

NIMBOSTRATUS

These clouds are thick, dark gray, shapeless sheets. The undersurface of a nimbostratus has a "wet" look. There is usually irregular scud beneath and surrounding these clouds. The word **nimbus** means a cloud from which rain is falling; **nimbostratus** are associated with continuous rain or snow. When nimbostratus are present, they sometimes merge with the middle-family clouds above, and the total cloud layer is very deep. More frequently, however, there is a clear break between the two layers.

Nimbostratus. (NOAA)

Altostratus. (NOAA)

MIDDLE CLOUDS

Near the Equator, middle clouds range from 6500 to 25,000 feet (1980 to 7620 m). In middle latitudes, the members of this family range up to 18,000 feet (5490 m); at the poles, they are found at maximum altitudes of 13,000 feet (3960 m). The prefix *alto* means high, and it is used to describe both members of this family: altocumulus and altostratus.

ALTOSTRATUS

These uniform bluish or grayish-white cloud sheets consist primarily of water droplets. **Altostratus** clouds are translucent to opaque. When the Sun is viewed through a thin sheet of these clouds, it appears as though seen through a ground-glass screen. Usually, altostratus cast very little shadow.

Altostratus clouds are not produced by direct ground influence, and they may, in fact, be completely glaciated at times. These clouds are formed by the ascent of warm air over colder denser air. The lower they form, the heavier and denser they become.

Altocumulus. (NOAA)

ALTOCUMULUS

These clouds are generally layered and composed of cells or lumps, which are sometimes called *billows*. An **altocumulus** casts a shadow and varies in color from pure white to a gray shading on its undersurface. Wavy and parallel bands are characteristic of this cloud type. The woolpack clouds, commonly referred to as *sheep* *clouds,* are examples of high globular altocumulus.

Altocumulus are generally composed of water droplets, but these clouds may lie above the freezing level and be completely glaciated. Altocumulus clouds are usually formed by the widespread ascent of warm air over cooler denser air. These clouds also develop when a topographic obstacle, such as a mountain, causes the air to lift in a wave.

HIGH CLOUDS

Near the Equator, high clouds range from 20,000 feet to 60,000 feet (6100 to 18,290 m). In middle latitudes, the members of this family range from 18,000 to 45,000 feet (5490 to 13,720 m); at the poles, they are found from 10,000 feet to 25,000 feet (3050 to 7620 m). An examination of these ranges indicates that the high-family's lower ranges overlap somewhat with the middle-family's upper ranges. The word **cirrus** means curl. Cirrus is used to describe one member of the high family; it is also used in combination with cumulus and stratus to describe the other two members of the family: cirrostratus and cirrocumulus.

Cirrus. (NOAA)

CIRRUS
In their most usual form, these clouds consist of white filaments, which are many miles in length. In appearance they are thin, wispy, always whitish, and without shadows. These clouds are commonly referred to as *mares tails.* Cirrus lie entirely above the freezing level and, therefore, are composed of ice.

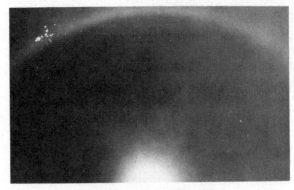

Cirrostratus. (NOAA)

CIRROSTRATUS
These clouds form a veil of more or less uniform texture. They are very thin and give the sky a milky-white appearance. **Cirrostratus** are often responsible for the appearance of a halo around the Moon and the Sun. No shadows are produced by these clouds.

Like cirrus, cirrostratus lie above the freezing level and are, therefore, composed of ice crystals. These clouds usually form where lower-density air is being uplifted over cooler, higher-density air. Under all conditions, a cirrostratus cloud is produced slowly, and it is difficult to know whether it is formed by the growth of ice particles from the start or whether it is first formed at water saturation and quickly frozen because of the low temperature—below −36°F (−38°C).

CIRROCUMULUS
The **cirrocumulus** clouds are seen as thin, white-grainy, rippled patches. They are small billowed cirrus-type clouds composed of ice crystals. The rippled appearance of these clouds has earned them the name "mackerel sky."

Cirrocumulus are quite rare and indicate high-level turbulence. The instability is in the layer at and above the cloud level. Rising currents of air form the cloud parcels and billows; descending currents create spaces in the clouds,

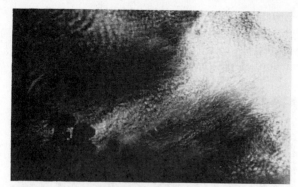

Cirrocumulus. (NOAA)

TABLE 9–10   PERCENTAGE OF SKY COVERED BY CLOUDS

LATITUDE	OVER LAND	OVER SEA	AVERAGE FOR BAND
80–90°N		63	63
70–80°N	63	70	66
60–70°N	62	72	63
50–60°N	60	67	62
40–50°N	50	66	56
30–40°N	40	52	45
20–30°N	34	49	41
10–20°N	40	53	47
0–10°N	52	53	53
0–10°S	56	50	52
10–20°S	46	49	48
20–30°S	38	53	48
30–40°S	48	57	54
40–50°S	58	67	66
50–60°S	70	72	72
60–70°S		76	76
70–80°S		64	64
80–90°S	<10		<10

which produce the overall rippled or mackerel look.

SKY COVER

Cloudiness is expressed in terms of the total area of the sky covered by clouds and may be reported as either tenths or percent. In Table 9–10, cloudiness is reported as a percent for 10-degree-latitude bands; and in Figure 9–24, the annual sky cover for the United States is reported in tenths. A cloudiness of 0, whether in tenths or as a percent, indicates a cloudless sky. A completely overcast sky is reported as either ten-tenths or 100 percent.

The glistening white surface of a cloud often has an albedo as high as 84 percent. This means that the percentage of the sky covered by clouds has an impact on the total amount of solar radiation reaching the surface. Over the landmass of Antarctica, for example, the dry atmosphere and the negligible amount of cloud cover do not appreciably reduce the total amount of solar radiation reaching the surface. In fact, the total of the incoming direct and diffuse solar radiation reaches values approaching 85 percent of the solar constant. However, as much as 90 percent of this insolation is reflected by Antarctica's snow surface; only the upper 3 feet (1 m) of the snow cover absorbs appreciable amounts of solar energy, which is quickly lost again during the dark season because the dry atmosphere has practically no blanketing effect.

Generally, the zonal distribution of cloudiness, documented in Table 9–10, parallels that of rainfall; in other words, heavy rainfall is accompanied by extensive cloud cover. There is one exception to this rule, however. The equatorial latitudes are zones of significant evaporation and heavy rainfall. However, the rainfall at these latitudes occurs primarily from convective clouds of the cumulus type, which are vertically deep but do not cover a large area of the sky. Thus, the sky cover in the equatorial latitudes is reported in the low 50s; it is not the largest percentage recorded in Table 9–10.

The zones of maximum cloudiness occur in the middle and higher latitudes. The lowest percentage of sky cover occurs in the latitude band of 20–30°N. Extensive deserts are found in the 20–30°N band and the over-land sky cover of 34 percent reflects this fact as well as that of low rainfall.

Examine the mean annual sky cover depicted for the conterminous United States in Figure 9–24. The areas of greatest cloudiness are also the areas in which the greatest amount of precipitation falls. The actual pattern of sky cover shown by the isolines and the shading on the conterminous United States map is not far from the percentages reported for these latitude bands in Table 9–10. The greater amount of sky cover over the conterminous United States probably reflects the fact that we are in a zone of air-mass collision, which produces a large quantity of spreading stratiform-type clouds.

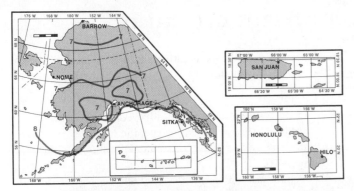

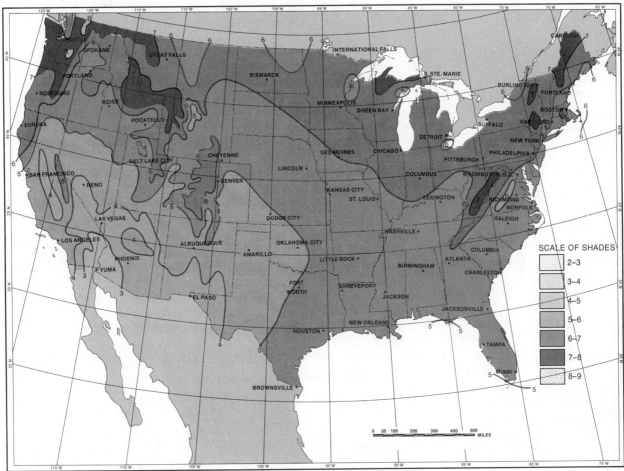

FIGURE 9–24   The mean annual sky cover from sunrise to sunset is detailed by the isolines and the shadings. The sky cover is reported in tenths.

—————— THE LANGUAGE OF PHYSICAL GEOGRAPHY ——————

A physical geographer uses a technical vocabulary to make explanations. Review your understanding of the vocabulary used to develop the concepts in this chapter. Among the important words and terms used are:

absolute humidity
advection fog
air-water interface
albedo
altocumulus
altostratus
cirrocumulus
cirrostratus
cirrus
continentality
cumulonimbus
cumulus
cumulus-type clouds
dew point

diffuse reflection
electromagnetic
    spectrum
evaporation
evaporation rate
fog
frost point
glaze
hail
humidity
insolation
langley
latent heat of
    vaporization

Little Ice Age
nimbostratus
nimbus
nuclei of
    condensation
precipitation
radiation fog
rain
Rayleigh scattering
relative humidity
sleet
sling psychrometer
snow
snow grains

solar constant
specific humidity
stratocumulus
stratus
stratus-type
    clouds
sublimation
temperature
gradient
thunderhead
turbulence
vapor pressure

—————— SELF-TESTING QUIZ ——————

1. List the factors that are important in producing climatic regimes.
2. Explain the way in which climatic controls impact and affect climatic elements.
3. Where are the best measurements of the solar constant made?
4. What is the best estimate of the total radiative output of the Sun?
5. Why is latitude an important climatic control?
6. Compare the depletions and diversions caused by carbon dioxide and clouds as incoming solar radiation penetrates the atmosphere.
7. What wavelengths suffer the greatest losses due to scattering?
8. Why are a body of water and a grass-covered park in the same locale not heated to the same temperature?
9. Describe the daily cycle of insolation.
10. Contrast and account for the temperature cycles at El Paso and Eureka.
11. How would you describe the temperature gradients in the Southern Hemisphere during January?
12. Explain the way the Gulf Stream affects the East Coast.
13. List the factors that affect the rate of evaporation.
14. Of what significance is the amount of water in the air to the climate of a region?
15. Where are zones of maximum relative humidity centered?
16. Why is radiation fog a nighttime, overland phenomenon?
17. How and under what conditions is hail formed?
18. Explain the operation of the rain-shadow mechanism.
19. List the significant clouds of the troposphere.
20. Where are the zones of maximum cloudiness in the Northern Hemisphere?

IN REVIEW
CLIMATIC ELEMENTS AND CONTROLS

I. INSOLATION AS A CLIMATIC ELEMENT

A. Solar Radiation
1. The Solar Constant
2. The Latitude Effect
3. Depletions and Diversions
B. Albedo
C. Radiation Cycles and Norms
1. Temperature Cycles
2. The Worldwide Pattern
3. Conterminous United States
4. Canada

II. MOISTURE AS A CLIMATIC ELEMENT

A. Evaporation
1. Temperature
2. Vapor Pressure
3. Wind Speed
4. Salinity
5. The Worldwide Pattern
6. Evaporation Maps
B. Humidity
1. Relative Humidity
2. Absolute Humidity
3. Specific Humidity
4. The Worldwide Pattern
5. The U.S. Pattern
C. The Dew Point
D. Fogs

1. Radiation Fog
2. Advection Fog
E. Precipitation
1. Rain
2. Snow
3. Hail
4. Other Forms of Precipitation
5. The Worldwide Pattern
6. The United States
7. Canada

III. CLOUDS AND CLOUDINESS

A. Mode of Formation
B. Vertical Clouds
1. Cumulus
2. Cumulonimbus
C. Low Clouds
1. Stratus
2. Stratocumulus
3. Nimbostratus
D. Middle Clouds
1. Altostratus
2. Altocumulus
E. High Clouds
1. Cirrus
2. Cirrostratus
3. Cirrocumulus
F. Sky Cover

10 *The spatial distribution of climatic elements by wind*

Wind functions chiefly as a control of temperature and precipitation rather than as an element of climate. Wind, for example, influences the rate of evaporation and serves two basic climatic functions that are related to the spatial distribution of the climatic elements of temperature and moisture: (1) Wind transports heat, in either sensible or latent form, from one location to another and is the principal agent that maintains the latitudinal heat balance by moving heat from lower to higher latitudes. (2) Wind transports the water vapor that is evaporated from the sea surface to continental areas where a large part of the vapor is condensed and falls as precipitation. Since the spatial patterns of the two prime climatic elements—temperature and moisture—are affected by atmospheric circulation, this chapter is devoted to unravelling and sorting out the variety of local and global winds that serve these two basic climatic functions.

WIND AS A CLIMATIC CONTROL

In this discussion, horizontal air movements are referred to as **local breezes,** or **winds.** The horizontal movement of large-scale, relatively uniform parcels of air, known as *air masses,* is discussed in Chapter 11 and should not be confused with the concept of wind. Vertical-moving air columns will be called **currents.**

Air moves when the pressure exerted on one side is greater than on the other; that is, air moves in response to unbalanced forces acting on it. A force is produced by a difference in pressure along the horizontal; it is called the *pressure gradient force.* This difference in pressure, the pressure gradient, forces parcels of air toward areas of lower pressure.

The temperature is dropping and wind-driven rain is beginning to fall as a January storm enters the Naples, Florida, area. (Celeste Navarra)

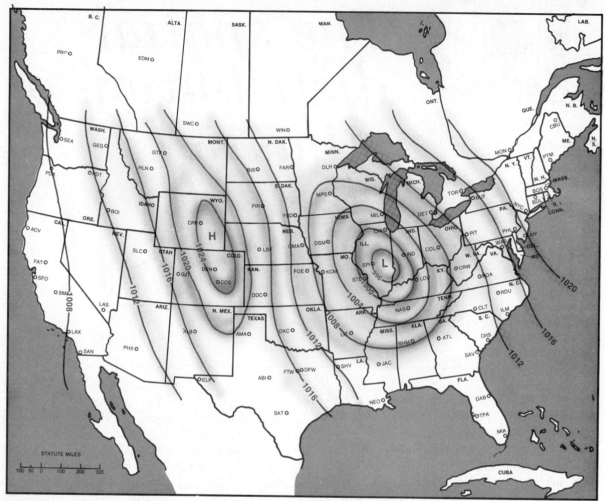

FIGURE 10–1   The horizontal distribution of pressure for sea-level conditions is depicted by the isobars. (Cities are identified by airport code; see Appendix VIII.)

THE PRESSURE FIELD

The horizontal distribution of pressure, described as the pressure field, is analyzed for sea-level conditions and for upper-air conditions. For a sea-level analysis, isobars—lines of equal surface pressure—are drawn. A different technique is used for the upper-air analysis; elevation contours are drawn on surfaces of equal or constant pressure.

THE SEA-LEVEL PATTERN

A simple technique is used to display the sea-level-pressure pattern. Points on a map showing the same pressure reading are connected with lines. Since each line is a line of constant value—a line of equal pressure—it is referred to as an **isobar**. (*Iso*, meaning "equal," is joined to *bar*, the pressure measurement unit.)

Since a large and confusing number of lines would result if all stations with the same pressure were joined, isobars are drawn at pressure intervals of 4 millibars on National Weather Service maps. Starting at 1000 millibars and going up the scale, successive isobars are drawn through points having readings of 1004, 1008, 1012, 1016, and so on. Going down the scale from 1000 millibars, successive isobars are drawn through points on the map having readings of 996, 992, 988, 984, and so on.

The horizontal distribution of pressure for sea-level conditions at one specific moment in time across the United States is depicted in pictorial form by the isobars in Figure 10–1. You will note that a continuous pressure change exists between Los Angeles, California, which has the station symbol *LAX*, and Albuquerque, New Mexico, which has the station symbol *ALB*. The pressure over this distance falls from 1016 millibars in New Mexico to 1008 millibars in California.

The rate of change in pressure for a given unit of horizontal distance is called the **pressure gradient,** a value usually expressed in millibars per 100 miles. The distance from the southeastern border of Arizona, which has a reading of 1012 millibars, to the southwestern coastline of California below Los Angeles, which has a pressure reading of 1008 millibars, is about 500 miles (800 km). The pressure gradient between these two isobars is 0.8 millibar per 100 miles (160 km).

The most rapid change in pressure is always in a direction perpendicular to the isobars. The pressure gradient is measured along the line of greatest pressure change, which is a direction perpendicular to the isobars. These lines—commonly referred to as *pressure gradient lines*—are depicted in Figure 10–2. Note the pressure gradient line that moves from the Albuquerque area to the Los Angeles area.

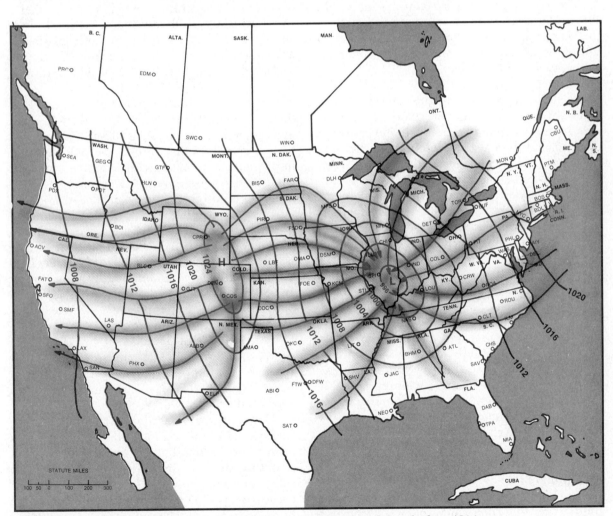

FIGURE 10–2 The pressure gradient lines move out of the high and into the low. (Cities are identified by airport code. See Appendix VIII).

HIGHS AND LOWS

In the middle latitudes, isobars display characteristic shapes; that is, they are roughly circular or elliptical. A **high** is defined simply as an area within which the pressure is high, relative to the surrounding region. Similarly, a **low** is defined as an area within which the pressure is low, compared to its surroundings.

In Figure 10–1, note the shape of the isobars surrounding the low-pressure area. The lowest pressure depicted on the map is within the 996 isobar; it never drops to 992, however, or another isobar would be drawn. The letter *L* is written within the 996 isobar, and the region is referred to as a **low-pressure area, or depression.**

The highest pressure in Figure 10–1 is within the 1024 isobar. The letter *H* is written within this isobar, and the whole region is called a **high-pressure area,** or high. The 1024 isobar has a roughly elliptical shape.

There really is no particular point where the high ends and the low begins. As the figure shows, there is a continuous decrease in pressure from the center of the high to the center of the low, and then a steady increase to the next high, which is somewhere off the eastern shore of the United States in Figure 10–1. A low-pressure configuration may cover a few states or, as depicted in the map, almost one-half of the entire nation. Similarly, a high-pressure configuration may be small or large enough to cover many states.

If you examine a series of maps and study the isobars over a large area, you will notice that the highs and lows are the principal types of pressure patterns. And further, a continuous series of highs and lows exists and moves through the atmosphere of the middle latitudes more or less from west to east, that is, in the direction of the Earth's rotation. The changes we experience in the climatic elements of temperature and moisture at these latitudes are related to the movement of these highs and lows.

THE UPPER-AIR PATTERN

The problem of pictorially representing the pressure field for upper-air conditions is handled by drawing elevation contours on a surface of constant pressure. Such a chart is referred to as a **constant-pressure chart;** it is also called a **pressure-contour chart.** In weather analysis and forecasting, pressure-contour charts are prepared for several selected pressure surfaces: 300-, 500-, 700-, and 850-millibar surfaces. A pressure of 300 millibars corresponds roughly to an altitude of 30,000 feet (9100 m). The rough altitude approximation of 500 millibars is 18,000 feet (5500 m); the 700-millibar surface is at about 10,000 feet (3000 m); and at around 5000 feet (1500 m) we find the 850-millibar surface.

Refer to Figure 10–3, which is the pressure-contour chart for the 500-millibar surface at 7:00 A.M. eastern standard time (EST) on April 25, 1979. Contour lines are drawn for each 200 feet (60 m) of difference in elevation of the 500-millibar pressure surface. This chart shows that a pressure of 500 millibars is found at an altitude of 18,900 feet (5760 m) over southern California and central Florida; but over northern California and southern Quebec the 500-millibar surface is found at 18,500 feet (5640 m).

The contour lines showing elevation are all on the 500-millibar surface in Figure 10–3. Thus, each contour line is really a 500-millibar isobar. This means, then, that the contour lines show the horizontal variation of pressure for the upper-air condition. When a contour line is at a low level, it indicates a low pressure—considered horizontally. A high level of a contour line, considered horizontally, indicates high pressure. For example, in northern Canada the 500-millibar surface is found at 16,500 feet (5030 m). At 2400 feet (730 m) above this contour line—at an elevation of 18,900 feet (5760 m)—the pressure is less than 500 millibars because pressure decreases upward. And, as was noted previously, the pressure is 500 millibars at 18,900 feet (5760 m) over southern California. Thus, these contour lines indicate high pressures over the southern United States and low pressures in the upper-air over northern Canada.

Generally, at the altitude of the 500-millibar surface, the highs and lows and the irregularity of the isobars found at sea level give way to a smooth, wavelike succession of high-pressure ridges and low-pressure troughs. The wavelike succession is shown clearly in Figure 10–3. One high-pressure ridge and four low-pressure troughs are labeled in the figure.

The terms *high* and *low* are, of course, used in

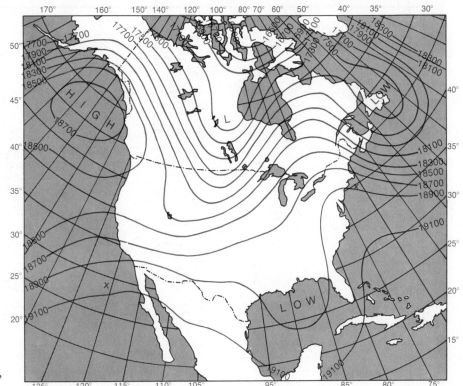

FIGURE 10–3 This is a pressure-contour chart for the 500-millibar surface. The 500-milli height contours are depicted for 7:00 A.M., EST on April 25, 1979.

a relative sense. In the upper air off the British Columbia coast, for example, an area marked as a high is enclosed by an 18,700-foot (5700-m) contour, and a low is in the trough of a 17,100-foot (5210-m) contour over northern Manitoba. With respect to the surrounding area and relative to each other, the designations do apply.

There is a relationship between the highs, lows, troughs, and ridges on the 500-millibar surface and those found at the Earth's surface. The movement of the upper-air troughs and ridges can be anticipated with fair accuracy for some days in advance. And the movement aloft is related to movement on the surface. The surface lows, for example, tend to move in directions dictated by the flowline-pattern aloft.

WORLDWIDE PRESSURE PATTERNS

Average sea-level pressures for January and July are depicted in Figures 10–4 and 10–5. The isobars are drawn at 5-millibar intervals on these maps. Even a cursory examination of these maps indicates a somewhat orderly sequence of high- and low-pressure. From this kind of data base

and other information, a worldwide pressure pattern consisting of two high-pressure zones and two low-pressure zones in each hemisphere has been constructed. The alternating belts start with a low-pressure zone in the equatorial region and culminate with high-pressure zones over the poles.

Equatorial Belt of Low Pressure. This zone—although it is in the equatorial region—is not centered at the Equator. The equatorial low-pressure zone varies in width and is usually referred to as the **intertropical convergence zone,** or *ITCZ*. The ITCZ completely encircles the Earth, and its winds are generally light and variable with frequent calms. Since sailing ships were often becalmed in this area, the zone became known as the **doldrums.** The sea-level pressure throughout this belt is less than 29.9 inches of mercury (1012.5 millibars) during January; in fact, in equatorial South America, equatorial Africa, and northern Australia the pressures in January hover around 29.8 inches (1009 millibars). During July, the belt is almost

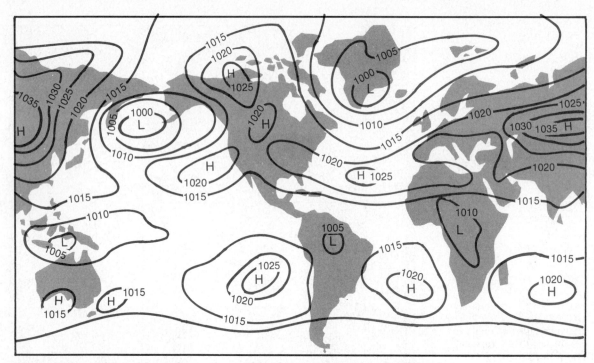

FIGURE 10–4 The isobars indicate the worldwide-pressure pattern (millibars) at the surface during January.

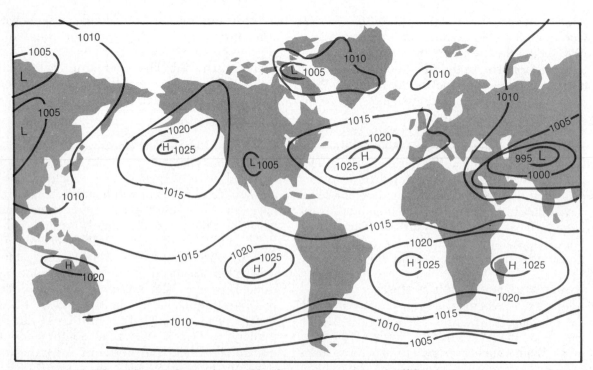

FIGURE 10–5 The isobars indicate the worldwide-pressure pattern (millibars) at the surface during July.

entirely north of the Equator, and low pressures extend into North America and Asia, with minimums of 29.7 (1005 millibars) in the southwestern United States and 29.4 (995 millibars) in northern India.

Subtropical High-Pressure Belts. These high-pressure zones are found in both hemispheres at about latitude 30°. Large land and water surfaces alternate to make the high-pressure zone in the Northern Hemisphere somewhat irregular. The average annual sea-level pressure in these zones is about 30.1 inches of mercury (1020 millibars). In July, in the Northern Hemisphere, high-pressure cells with pressures averaging 30.3 inches (1025 millibars) are found in the Azores area of the Atlantic and in the eastern Pacific. In the Northern Hemisphere during January, the highest pressures are found over the land areas of Asia, with pressures averaging 30.6 inches (1035 millibars); in fact, the highest pressure ever recorded—32 inches (1083.8 millibars)—was near Agata, Siberia (66°50′N, 93°40′E) on December 3, 1968.

Polar Low-Pressure Zones. In the Southern Hemisphere between latitudes 60° and 70°, there is a continuous belt of low pressure that lies over water and has sea-level pressures averaging 29.3 inches (992 millibars) during January. In the Northern Hemisphere, the picture is somewhat complicated because of large

This is Mitre Peak at latitude 44°38′S, longitude 167°50′E on the southwestern shore of South Island, New Zealand. The peak stands at 5560 feet (1695 meters) above sea level. Refer to Figure 10–4 and you will find that the sea-level pressure in this area during January runs around 1015 millibars, which makes it a high-pressure zone. (New Zealand Consulate)

We need a lot more information about Antarctica before we fully understand the flow and change of pressure patterns in the area. We must, for example, understand the impact of this active volcano, Mount Erebus, on the climatic elements of the region. Mount Erebus is 20 miles (32 kilometers) from McMurdo Station. (U.S. Navy)

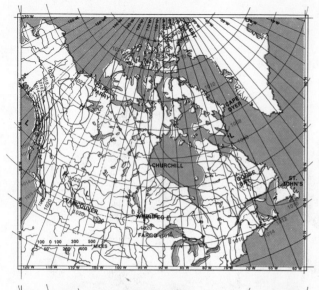

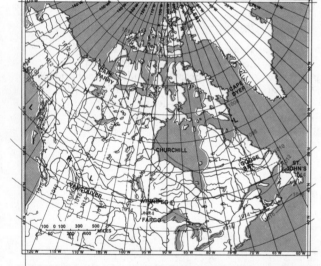

FIGURE 10–6 The isobars depict the sea-level pressure patterns in Canada and the conterminous U.S. during the months of January (upper maps A and C) and July (lower maps B and D).

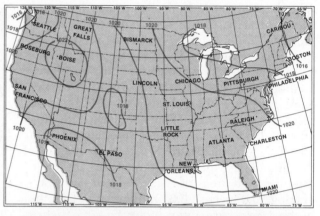

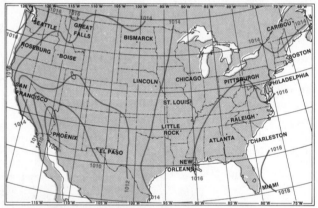

landmasses, but over the ocean areas at these same latitudes there are well-defined regions of low pressure. Low-pressure zones, for example, are found in the Greenland-Iceland area and in the Aleutian Island area where sea-level pressures average 29.5 inches (1000 millibars) during January; in the Greenland-Iceland region pressures can drop as low as 29.4 inches (995 millibars) during January.

High-Pressure Polar Caps. These are somewhat permanent areas of high pressure in the Arctic and Antarctic regions. The high-pressure polar zone in the Northern Hemisphere is well-

established; the center of the high-pressure zone extends from northern Greenland westward across the northern islands of Canada. The average sea-level pressure during January in this northern region is above 30 inches (1015 millibars). Pressures in the high-pressure zone of Antarctica are quite variable; for example, pressures in the range of 30.4 inches (1029 millibars) are regularly recorded, but the pressure can also drop as low as 27.5 inches (931 millibars).

U.S.–CANADIAN PATTERN

The average sea-level pressures for January and July are depicted in Figure 10–6. Note that pressures are lower in summer than in winter. From May through October, the pressure tends to be high off the northwest coast of the United States and the southwest coast of Canada because of the warm-season expansion of the semipermanent Pacific high close to 40°N latitude, which is shown on the world map in Figure 10–5. The high centered over southern Idaho in Figure 10–6C persists from November to February. The Bermuda-Azores high-pressure cell, which is shown in Figures 10–4 and 10–5, induces a zone of high pressure over the southeastern section of the United States from late September through December. Pressures over the deserts of the Southwest (U.S.) tend to be low throughout the year. The largest variations in sea-level pressure occur during the winter over the northern tier of states and provinces east of the Rockies.

THE CORIOLIS FORCE

As indicated on page 327, a difference in pressure, the pressure gradient, forces air toward a region of lower pressure. From this simple statement we might expect winds to travel in straight paths from one place to another. Once the fluid, called *air,* is set in motion, however, an additional force enters the picture; this new force causes winds in the Northern Hemisphere to move toward the right side of the expected path of travel, and causes winds in the Southern Hemisphere to move toward the left. This motion toward one side or another of the projected path is the result of the **Coriolis force,** which is produced as an effect of the rotating Earth.

The Earth's rotation can be described in terms of the instantaneous linear velocity of a point on its surface, which simply means the velocity of a point in the circular orbit that it follows. Two facts are important for an understanding of this concept: (1) The period of rotation is the same from the Equator to each pole. (2) The shape of the Earth causes the length of a parallel of latitude to decrease as you move from the Equator to the poles. Putting these two facts together, we find from Table 10–1 that the linear velocity of a point on the surface of the Earth depends on the parallel of latitude on which it is located. The linear velocity is at a maximum on

the Equator, but it diminishes as you move poleward. The linear velocity at the poles is zero.

The rotation of the Earth combined with the different linear speeds encountered on various parallels of latitude produces the Coriolis effect. The effect is easy to conceive when we imagine a simple analogy: If the Earth were a stationary object and we were to stand at the Equator and fire a rocket at the North Pole, the rocket would travel in a straight line to the target. But the Earth is not a stationary object; it is rotating in a west-to-east direction, and the velocity of a point on the Earth's surface is a function of latitude. For example, we find in Table 10–1 that the linear velocity of a point on the Equator is 1041 miles (1675 km) per hour; at a latitude of 40°, linear velocity falls to 798 miles (1284 km) per hour; and at a latitude of 90°, the eastward velocity of a point on the Earth's surface is zero. All of this means that if a rocket is fired toward the North Pole from the Equator, the vehicle also has, at the moment of launch, an eastward velocity of 1041 miles (1675 km) per hour. This eastward velocity is retained by the rocket while in flight toward the North Pole. However, since the eastward velocity of points on the Earth's surface decreases as you move from the Equator, the rocket will be "ahead" as it proceeds northward. The net effect is that the rocket's path curves to the right from the point of view of an observer on the Equator. If the rocket is fired instead toward the South Pole, the rocket will curve to the left of an observer on the Equator.

Thus, the Coriolis force is really a consequence of an unattached moving object conserving its angular momentum. Although this concept

TABLE 10–1 LINEAR VELOCITY ON VARIOUS PARALLELS OF LATITUDE

LATITUDE	MILES PER HOUR	KILOMETERS PER HOUR
90	0.0	0.0
80	181.3	291.7
70	356.9	574.4
60	521.5	839.3
50	670.1	1078.4
40	798.1	1284.3
30	901.8	1451.2
20	978.1	1574.0
10	1024.7	1649.1
0	1041.4	1675.9

can only be fully discussed in mathematical terms, a few summary remarks can be made in a qualitative way: The apparent deflection in the path of an object moving toward the Equator, in either hemisphere, results from the fact that it crosses increasingly larger circles of latitude as the Equator is approached. As an object crosses circles of greater circumference, its initial eastward velocity "lags behind" that of the Earth's rotational motion. When an object moves poleward, in either hemisphere, its initial eastward velocity "runs ahead" of the Earth's rotational motion as it crosses parallel circles of latitude that decrease in size. Thus, an unattached moving object, in effect, veers to the right of an observer in the Northern Hemisphere and to the left of an observer in the Southern Hemisphere.

FORCES GOVERNING WINDS

The actual wind that flows over a surface is a balance of three separate forces: the horizontal deflecting force (Coriolis force), the frictional force due to the surface, and the pressure gradient force. In Figure 10–7, an actual wind is depicted—its flow is from the west to the east.

The Coriolis force acts at a right angle to the wind and pulls to the right in the Northern Hemisphere. The frictional force, produced through contact with the surface, is opposite to the direction of the wind. The pressure gradient force is perpendicular to the isobars and acts in the direction from higher to lower pressure. As a result of the interplay of these three forces, the actual wind blows at an angle across the isobars, in a direction from higher to lower pressure.

The angle between actual wind direction and the isobar increases with increasing friction. This means you can expect the angle to be greater over land than over water. Over land, the angle between the isobars and the actual wind averages about 30°. Since there is generally less friction over water surfaces than over land, the angle averages about 15° between the actual wind and the isobars at sea.

As we move upward into the atmosphere from the Earth's surface, the frictional force diminishes. At a height of 3000 feet (900 m), surface friction is usually negligible, which means the wind blows parallel to the surface isobars.

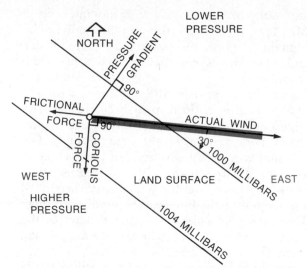

FIGURE 10–7 The windflow in this figure is shown cutting across a land-surface isobar at an angle of 30 degrees.

Thus, surface isobars indicate the wind direction at the 3000-foot (900 m) level. Wind blowing with steady speed and parallel to the isobars at upper levels is referred to as the **geostrophic wind.**

Over land and sea, then, as you rise into the atmosphere there is a normal variation of wind direction. A 30-degree angle between the actual wind and the isobars at the surface, over land, becomes a 20-degree angle at the 1000-foot (300-m) level, and a 10-degree angle at the 2000-foot (600-m) level. The 15-degree angle at sea level becomes a 10-degree angle at 1000 feet (300 m) and a 5-degree angle at 2000 feet (600 m).

Think through what you have just learned and examine Figure 10–7 again. Note that if you stand with your back to the wind, the low-pressure zone is to your left and the high-pressure zone is to your right in the Northern Hemisphere. The situation, of course, is reversed in the Southern Hemisphere. This law is known as the **Baric Wind Law;** it was formulated in 1857 by a Dutch meteorologist, Buys Ballot, and sometimes bears his name.

CYCLONIC AND ANTICYCLONIC FLOW

We now know that the motion of air is influenced by the Coriolis force, the pressure gradient force,

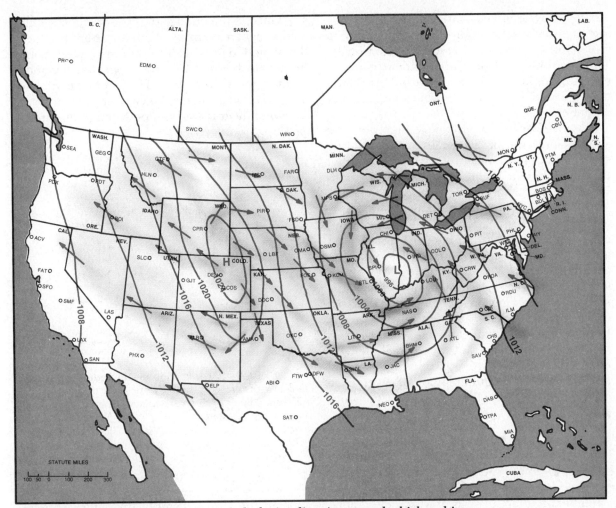

FIGURE 10–8   Surface winds move in a clockwise direction around a high and in a counterclockwise direction around a low-pressure area. (Cities are identified by airport code; see Appendix VIII.)

and the surface friction. In the Northern Hemisphere, these three forces cause surface winds over land to cross the isobars at an average angle of 30 degrees. In Figure 10–8, short, straight arrows have been drawn to show the direction of the wind as it blows at an angle of 30 degrees across the isobars, in a direction from the high-pressure zone to the low-pressure zone. From perusing this map, you will note that the surface winds move in a clockwise motion around the high-pressure area and in a counterclockwise direction around the low-pressure area. In the Southern Hemisphere, the situation is reversed.

A low-pressure system is known as a **cyclone,** and its counterclockwise wind movement is called a *cyclonic-wind flow*. Winds converge on the center of such a system. A high-pressure system, referred to as an **anticyclone,** has winds moving out of its center, that is, diverging is an anticyclonic, clockwise flow.

HIGH-LOW COUPLING

For a high-pressure area to be maintained, air must descend within it. Descending air, referred to as **subsiding air,** warms adiabatically. As air within a high-pressure system subsides and warms up, it imposes an area of fair weather within the system. The continuing existence of a low-pressure system requires the inflowing air to rise within it. Rising air cools adiabatically. Thus, within a low-pressure system, rising air cools and produces an area of turbulence often accompanied by the formation of clouds and precipitation.

Figure 10–9 shows the way high-pressure and low-pressure surface zones are coupled together in a convection system, which consists of two layers of convergence and two layers of divergence. Let's start at the left of the diagram: Within the high-pressure cell, air is subsiding; that is, air in the upper layers is being cooled and compressed, thus causing it to sink from higher altitudes to the Earth's surface where we encounter the first layer of divergence. The air diverging out of the surface-high converges on the surface-center of low pressure. Because the mass of air is conserved, the air must move vertically from the low's layer of convergence at the surface to the low's layer of divergence aloft. In turn, there is a corresponding upper-level movement of air from the region of divergence above the low toward the region of convergence above the high.

Thus, two motions are present when a high and a low are coupled together. There is a low-level motion of cool air toward the low-pressure region on the surface, and an upper-level motion in the opposite direction. This circular motion of air is termed a **convection current.** It is one way in which heat is transported through our atmosphere.

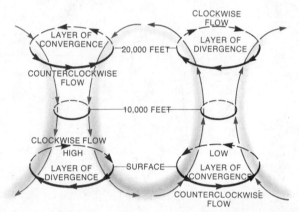

FIGURE 10–9 A convection system couples a high- and low-pressure system. There is an outflow of air from a layer of divergence and an inflow around a layer of convergence.

LOCAL WINDS

A number of different kinds of winds are generated as a result of specific local conditions: land-sea breezes, mountain and valley breezes, and chinooks. These winds, generally, do not have any large-scale significance. They are, however, rather important for the local areas in which they are generated.

LAND-SEA BREEZES

The rising Sun begins the period of insolation for each day. Land adjacent to a body of water heats up more rapidly than the water during the daytime for all seasons in the tropics and for the summer period in mid latitudes. Along the shores of the United States, the land becomes significantly warmer than the water by mid morning during a summer's day. Thus, parcels of

air over the land become warmer than similar parcels over the water; this causes a local low-pressure area to develop in the air over the land. The density difference causes colder, heavier air over the water to move toward the land and to replace the low-pressure parcel. A convection current is set in motion. The air moving over the surface of the water toward the land is called a **sea breeze.** Normally, a sea breeze can be detected by 9:00 A.M. local time. It increases in strength until late afternoon and penetrates inland for as much as 10 miles (16 km) in some locations. Sea breezes cool seashore resorts during the summer months.

At night, the situation reverses itself; the land cools by radiation and its temperature drops more rapidly than that of water. The land becomes colder than the adjoining water. We now find a local low-pressure cell developing over the water since the water gives up some of its stored heat to the air above it. The warming air over the water begins to rise. As soon as the pressure gradient reverses, the cooler higher-pressure air over the land begins to move toward the lower-pressure zone over the sea. This offshore night wind is called a **land breeze.** The nighttime flow is an exact reversal of the prevailing winds during the day.

The height to which a sea breeze extends varies; in the mid latitudes, for example, it may reach a height of 600 to 1600 feet (200 to 500 m) off large lakes; on the other hand, along a tropical sea coast, the upper surface of a sea breeze may be at a height of 6000 feet (1800 m). The land breeze that develops in the same locale is generally much shallower. Velocities also show great variation; in the mid latitudes, for example, average sea-breeze speeds of 10 miles (16 km) per hour prevail, but maximums can go above 25 miles (40 km) per hour. The strongest and fastest sea breezes generally develop when there is a large contrast between land and water temperatures—as, for example, along dry tropical coasts that are paralleled by cool ocean currents.

The difference in specific heat, or thermal capacity, between land and water produces a modification in climate for land areas near large bodies of water. Since the ocean does not lose the summer heat it has stored until after the land has cooled, autumn sea breezes help raise land temperatures slightly above those of inland areas. In summer, when the land is hottest, the cooler sea breezes help keep temperatures below those of inland areas; in fact, the temperature differential produced by a sea breeze may be as much as 20°F (11°C).

A sea breeze's effect on precipitation is somewhat less obvious than its effect on temperature. Along some coasts, however, a sea breeze seems to produce some convergence, which leads to the lifting of the air and increases shower activity. Florida, for example, is bathed by an easterly sea breeze from the Atlantic Ocean and a westerly sea breeze from the Gulf of Mexico. It is thought that the convergence of these two sea breezes causes increased levels of precipitation over Florida.

MOUNTAIN AND VALLEY BREEZES

As their names imply, these winds exist as the result of local circumstance—topography causes thermal differences which, in turn, create convectional circulation. There may be more than one component to the diurnal upslope and downslope winds that develop; however, for the purpose of this discussion, the upslope winds are referred to as **valley winds** and the downslope winds as **mountain winds.** A daytime upslope-valley wind develops when one wall of a valley—in the Northern Hemisphere it is usually the southern slope—is heated by the Sun to a greater degree than the shaded valley itself. Air begins to rise where the heating takes place. The pressure gradient runs from the valley floor to the slope. Cooler air from the valley floor moves up the slope to replace the rising warm air. Valley winds are usually weak or absent on northern slopes in the Northern Hemisphere. When upslope winds are flowing, cumulus clouds gener-

A good sailor understands and takes advantage of the pattern of winds that flow over a region. The bay, gulf, and ocean in the San Francisco area are good areas in which to sail. (Redwood Empire Assoc.)

ally form over the peaks and summits of mountains during afternoon; the afternoon cloudiness usually culminates in showers.

After the heating process has stopped, the slope cools by radiation and the circulation pattern reverses. The air on the mountainside in contact with the cooling slope cools more rapidly at night than the air in the valley. The draining of cooler, denser air down the slope into the valley, under the influence of gravity, creates the downslope-mountain breezes. Temperature and humidity are greatly influenced by these local diurnal winds in mountainous regions.

CHINOOKS

Another topographical effect results in the common **chinook**—warm, dry winds that blow at intervals down the eastern slopes of the Rocky Mountains in the United States and Canada and also along the slopes of the Swiss Alps. In the Alps these winds are known as **foehn winds.** They are produced as air is forced to rise over a mountain. The air is cooled adiabatically during the rise in elevation; precipitation of water from the air usually occurs on the way up the slopes. As the air descends the leeward side of the

mountain, it is warmed adiabatically. It is warmed more rapidly on the lee side than it is cooled on the windward side because the dry adiabatic lapse rate operating on the lee side is greater than the wet rate of the windward side. In fact, the final air temperature on the leeward side is usually higher than the initial temperature on the windward side. Areas on the leeward side of the Rockies are usually warmed several degrees by the arrival of a chinook.

The temperatures of the Rocky Mountain chinooks during the winter average about 45°F (7°C); however, they sometimes have highs of 70°F (21°C). Since the average daily temperature during January, for example, is just under 30°F (−1°C) in Denver, Colorado, and just above 20°F (−6°C) in Great Falls, Montana, the chinook temperatures appear very warm by contrast. Chinooks also have desertlike relative humidities in the range of 10 percent, which makes their warming effect quite noticeable. A dry chinook passing over a snowfield causes the snow to sublimate—change from solid to gas—and disappear as if by magic.

Thus, in the western sections of the North American plains, chinooks tend to make the win-

This is Rogers Pass in the mountains of British Columbia. The slopes to the right are in shade; those at the left are bathed in sunlight. An upslope-valley wind has developed on this autumn day and cumulus clouds have developed over the peaks and summits as a result. (National Film Board of Canada)

ters less severe by their warming effect. In addition, chinooks have a decided effect on the snow cover, making it much less persistent than it otherwise would be. The drying effects of these winds tend to be quite negative in human terms because chinooks deplete soil moisture and can seriously damage a spring crop that is in its formative stage.

GLOBAL SURFACE-WIND SYSTEMS

Since air moves from regions of high pressure to regions of low pressure, the general distribution of pressure in each hemisphere—as summarized in Figure 10–10—dictates three zones of prevailing surface winds that move from high-pressure belts to low-pressure belts: polar easterlies, prevailing westerlies, and the trade winds.

POLAR EASTERLIES

Cold air moves from the polar caps of high pressure to the polar low-pressure zones located at approximately 60°N and S. The deflection produced by the Coriolis force causes these winds to become easterlies. In the Northern Hemisphere, these **polar easterlies** flow mostly from the northeast toward the southwest. In the Southern Hemisphere, the polar easterlies flow primarily from the southeast toward the northwest.

Examine the wind rose in Figure 10–11. This rose depicts the percentage of time the wind blows from each compass direction during the month of June off the coast of Norway at latitude 66°N, longitude 7.5°E. The northeasterly component, which blows 30 percent of the time, is the largest. The collective component from N, NE, and E amounts to 62 percent. As the data indicate, winds blow mostly from a northeasterly direction, but not exclusively so.

PREVAILING WESTERLIES

Major air movements flow with the pressure gradient from the subtropical high-pressure belts 35°N, 30°S) to the polar low-pressure zones at approximately 60°. In each hemisphere the air moves poleward, but deflection causes the movement to become a largely westerly flow. In the Northern Hemisphere, the winds are usually from south of west and flow toward the northeast. In the Southern Hemisphere, the winds are slightly north of west and flow toward the southeast.

The wind rose in Figure 10–12 depicts the percentage of time the wind blows from different compass directions during the month of June at latitude 37.5°N, longitude 52.5°W, which is approximately 1300 nautical miles off the Virginia coast. The collective component of the wind from the S, SW, and W accounts for 64 percent of the flow-time. Thus, the data indicate that winds blow most of the time from a southwesterly direction, but not exclusively so.

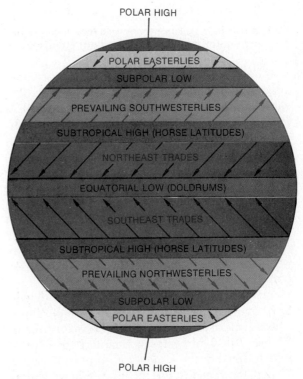

FIGURE 10–10 There are 3 zones of prevailing surface winds in each hemisphere. This schematic, idealized view of the prevailing-wind zones does not take into account the disrupting influence of large landmasses.

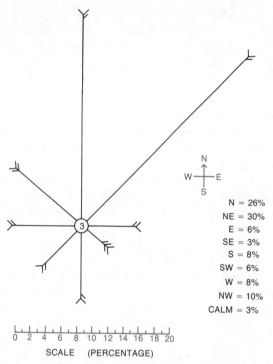

N = 26%
NE = 30%
E = 6%
SE = 3%
S = 8%
SW = 6%
W = 8%
NW = 10%
CALM = 3%

SCALE   (PERCENTAGE)

FIGURE 10–11   This is a wind rose for a location in the polar-easterly flow of the Northern Hemisphere.

The mid-latitude westerly flow is not composed exclusively of air supplied from the poleward flank of the subtropical high (Fig. 10–10). A lot of air from farther south—some of it genuinely tropical—"leaks" into the **prevailing-westerly** flow through the extensive longitudinal gaps between the individual high-pressure cells identified in Figures 10–4 and 10–5. These invasions, or leaks, occur primarily around the western ends of the oceanic high-pressure cells.

The poleward boundary of the mid-latitude westerlies fluctuates over time. There is a definite seasonal shift, but shifts occur over shorter periods, too. During the winter in the Northern Hemisphere, the westerlies often blow with gale force that approaches 50 miles (80 km) per hour. The Northern Hemisphere wind rose for June in Figure 10–12 indicates a Beaufort force of 4, which translates to a moderate breeze between 13 and 18 miles (21 and 29 km) per hour. Calms occur relatively infrequently as the 3-percent calm period noted in Figure 10–12 attests. In the Southern Hemisphere, which has few landmasses between the latitudes of 40° and 65°, gale-

force westerlies are common during summer and winter. In fact, these southern latitudes are referred to as the *roaring forties,* the *furious fifties,* and the *shrieking sixties* by mariners who sail through these seas. The winds around Cape Horn (latitude 56°S, longitude 67°16′W), for example, are known for their violence and the mountainous sea they produce.

TROPICAL EASTERLIES

These surface winds—sometimes referred to as **trade winds**—flow from the subtropical high-pressure belts toward the equatorial low-pressure belt. The trade winds have a large component from the east and are almost steady northeast winds in the Northern Hemisphere. In the Southern Hemisphere, the trade winds flow from the southeast toward the northwest. The trade winds from the two hemispheres converge to produce the intertropical convergence zone (ITCZ) near the Equator.

Figure 10–13 depicts the trade winds in June at latitude 12.5°N, longitude 42.5°W. A large

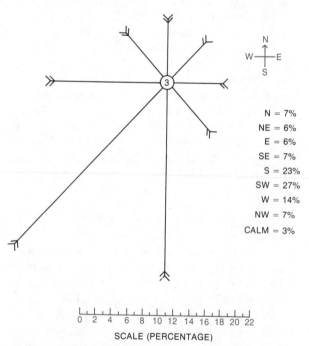

N = 7%
NE = 6%
E = 6%
SE = 7%
S = 23%
SW = 27%
W = 14%
NW = 7%
CALM = 3%

SCALE (PERCENTAGE)

FIGURE 10–12   This is a wind rose for a location in the prevailing-westerly flow of the Northern Hemisphere.

component from the east, which blows for 35 percent of the time, is shown, but the predominant wind flow is from the northeast for 63 percent of the time. The combination of easterly and northeasterly components accounts for 98 percent of the flow. The periods of calm, as depicted by this wind rose, are nonexistent.

The trade winds, you might say, are rather uniform in direction, especially when comparisons are made with the polar-easterly flow and the representative wind rose for the westerlies. A Beaufort force number of 5 is shown for the northeast component of the wind flow in Figure 10–13, which translates to a fresh breeze with wind speeds between 19 and 24 miles (31 and 39 km) per hour.

Generally, reference is made to the fair-weather trades, but weather and climate vary in different areas of the trade-wind zone. The poleward part of the zone is closest to the subtropical high-pressure belt, where subsidence and horizontal divergence are occurring; thus, the poleward part of the trade-wind zone has a lot of dry weather and sunshine. As we move equatorward in the zone, we find the air acquires large additions of moisture through evaporation. The ITCZ is on the equatorward side of the trade-wind zone; horizontal convergence and lifting occur on this side of the trades. The addition of moisture and the increasing convergence and lifting make the equatorward side of the trade-wind zone somewhat unstable; atmospheric disturbances develop readily along with convectional overturning, cloudiness, and precipitation.

The trade winds not only change character in a north-south direction, but they also change character in an east-west direction. The subtropical high-pressure cells (Figs. 10–4 and 10–5) from which the trades flow are most strongly developed on the eastern sides of the Atlantic and the Pacific oceans. Since it is in these high-pressure cells that subsidence is most pronounced, the trades are most stable and driest in these regions. Farther to the west—along the eastern margins of North America and Asia, for example—the high-pressure cells are weaker and the trades tend to be more buoyant and inclined to rise, which produces a greater incidence of cloudiness and precipitation.

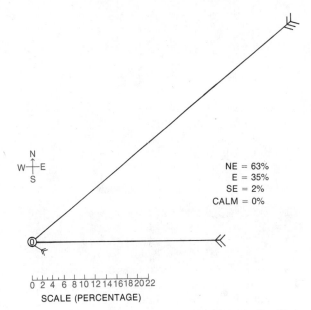

FIGURE 10–13 This is a typical wind rose for the trade winds of the Northern Hemisphere.

MODIFICATION AND SHIFTING

The modification and shifting of the idealized general pattern of prevailing winds described above occurs because (1) the inclination of the Earth's axis causes a seasonal north-south shifting of insolation and (2) the Earth's land-water surfaces cause the air above to develop contrasting temperature, moisture, and pressure characteristics.

For example, you know that during the course of one solar year, the Sun's vertical noon rays shift from their June position of 23°30′N to 23°30′S by the end of December. This is a total latitudinal shift of 47 degrees. The zone of maximum solar-energy input, however, has a latitudinal shift of as much as 70° degrees over the same period. Since pressure and wind belts are to an extent thermally induced, they, too, shift north and south with the "migration" of the Sun.

The north-south shifting of wind belts does not proceed on a uniform zonal basis. It varies in amount and rapidity according to the zone. Generally, however, it lags from 4 to 8 weeks behind the Sun's march. Over the sea, the latitudinal shift is fairly small and usually not much more than 10 degrees. Since there are wide variations

The wind beats the ocean's surface and drives the sea before it. A ship's captain studies the wind roses printed on the various quadrants of an ocean chart to anticipate the kind of sea that will be encountered when entering a certain sector. (U.S. Coast Guard)

in seasonal temperature over land, the north-south wind shift varies over a greater range than 10 degrees. In addition, the wind shift over land follows the Sun's latitudinal shift more closely than does the wind shift at sea.

The ocean wind roses depicted in Figures 10–11, 10–12, and 10–13 do have some order and pattern to them. The wind-flow patterns over land for these same latitudes are not as orderly. They are, in fact, much more chaotic because of irregularities in the elevation of the topography as well as strong seasonal and diurnal temperature contrasts.

THE TRICELLULAR THEORY

The classic **tricellular theory,** which is commonly used to explain the general circulation of the atmosphere, is based on heat convection principles and stems mainly from the studies of the eighteenth-century Englishman, George Hadley. In 1735, Hadley reasoned that air warmed in the tropics rises and flows poleward; air sinks as it cools and flows back toward the Equator along the surface. The basic premise of any heat-convection model is that wind systems are powered by a south-north overturning circulation—air rises in the hot latitudes and falls in the cold latitudes.

The tricellular theory's Earth-model has a uniform surface. Over half the area of the Earth lies between 30°S and 30°N. This is the zone that receives the greatest input of solar radiation, and intense heating in this zone causes air to rise. The tricellular theory assumes that the rising air from the Equator moves poleward after it reaches high altitudes.

According to the tricellular theory, the poleward moving air is chilled aloft and subsides at 30° latitude. Thus high-pressure zones are created by descending air at 30°N and S, the subtropical high-pressure belts. The descending air splits horizontally at the surface and moves equatorward as trade winds and poleward as prevailing westerlies. The Coriolis force turns these winds to the right in the Northern Hemisphere and to the left in the Southern Hemisphere. For example, the equatorward-moving surface wind in the Northern Hemisphere—which would be a north wind on a nonrotating Earth—becomes a northeast wind on a rotating Earth. The poleward-moving wind becomes a southwesterly in the Northern Hemisphere as a result of the Coriolis force. The surface branch moving toward the Equator, called the *trade winds,* completes the first Hadley cell of the three-cell motion.

The second cell has its beginnings at latitude 30° in the poleward-moving surface air, called the *prevailing westerlies.* The prevailing westerlies flow toward the polar easterlies, which originate at the cold poles and move equatorward along the surface. At about 60°, the prevailing westerlies and the polar easterlies converge horizontally and produce a zone of rising air. According to the tricellular theory, the ascending air splits when it reaches the upper levels of the atmosphere; one portion moves poleward, and the other moves equatorward. Thus, two additional cells of circulation are created to complete a three-cell model. Figure 10–14 depicts the three-cell circulation in a schematic way.

The problem with all of this, however, is that high-altitude investigation does not verify the upper flow which, according to the tricellular theory, should exist. The mid-latitude upper-air

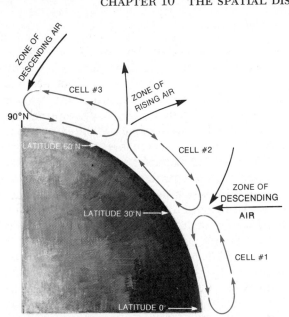

FIGURE 10–14 The three-cell circular model, the tricellular theory, was proposed by George Hadley in 1735.

wind, called the *jet stream*, for example, is a westerly flow aloft over a westerly flow at the surface.

THE EDDY THEORY

The modern account of global-wind circulation is called the **eddy theory.** It is not based on convection; rather, the eddy theory is based on the conservation of angular momentum, that is, the momentum resulting from rotation. The trade winds, for example, are flowing westward, which is opposite to the Earth's eastward rotation; thus, the friction between the trade-wind flow and the Earth's surface tends to slow the Earth's rotation. Angular momentum is transferred from the moving solid Earth to the trade winds as a result of the friction.

Since it is unlikely that the atmosphere causes any net change in the Earth's rotation rate, all the momentum taken from the Earth by the easterlies must be restored to it by the westerlies. This means that the momentum picked up by the prevailing easterlies in the tropical and polar regions must be transported by atmospheric circulation to the mid latitudes where the westerlies prevail.

In the zone of the prevailing westerlies, the

atmosphere rotates faster than the surface of the Earth, and the Earth draws angular momentum from the prevailing westerlies. The prevailing westerlies, under these circumstances, should be quickly slowed to nothing. They are not, however, because angular momentum is transferred poleward from the trade winds and equatorward from the polar easterlies to the prevailing westerlies. The transfer of angular momentum is accomplished through the mechanism of large-scale cyclones and anticyclones.

Until recently, most people thought that these eddies—cyclones and anticyclones—simply draw their energy of motion from the major circulations, making the transfer a one-way affair. But according to the modern concept, eddying masses of air in the atmosphere carry their own independent stores of energy and can contribute this energy to the general circulation. When a cold air mass moves into a warmer zone and subsides, for example, it converts potential energy into the kinetic energy of motion. The cold and warm air masses—coming into the mid latitudes from the north and south—contribute their energy of motion to the prevailing westerlies.

In the first instance, of course, the Sun sets the Earth's atmosphere in motion, but then the rotation of the Earth plays a double role in shaping the global pattern of circulation: (1) It is the Earth's rotation that breaks up the primary circulation into cyclones and anticyclones, and (2) the Earth's rotation channels the turbulent motion of the highs and lows into the prevailing west winds.

Thus, according to the eddy theory, there are three areas—trades, westerlies, and polar easterlies—that have imbalances in momentum. Eddies transfer momentum between these areas to restore the global equilibrium. The eddies are the familiar high- and low-pressure systems. These rotating systems migrate from one area to another. In the process, they transfer angular momentum and assist in maintaining a balance between the wind regions. The heaviest traffic in the transport of angular momentum occurs at 30 degrees latitude. In the upper levels of the atmosphere, wavelike motions aid in the transfer of momentum from one region to another.

THE UPPER-AIR SYSTEM

Broad belts of upper-troposphere westerly winds encircle each hemisphere in temperate latitudes. The overall westerly motion of these wind belts, however, includes northward and southward meanders—undulations which are sometimes called **Rossby waves.** The term **jet stream** was originally applied in a general way to all regions of strong wind that occur in the upper troposphere. Today, the term is applied in a restricted sense to the parts of these upper-troposphere currents where the winds are strongest and the typical structure is best developed. For example, according to current usage, a jet-stream current is a flattened narrow core that is thousands of miles in length, a hundred or so miles in width, and one or more miles in vertical thickness. Maximum winds in the center of such a core may reach 300 knots; however, 100- and 200-knot winds are more typical of the flow-speeds encountered in these streams of air.

There are two main jet-stream systems in the upper atmosphere (Fig. 10–15). In the northern latitudes, one jet stream seems to be associated with the polar front—an irregular shifting boundary where the relatively warm prevailing westerlies meet the cold polar easterlies—and is, therefore, called the **polar-front jet stream.** The core of this polar-front jet lies at the top of the troposphere, generally in the range from 40° to 60°N latitude. The velocities in this jet stream show a seasonal variation, being much higher in winter than in summer. Velocities of 200 knots over the United States in winter are common. The second jet stream system is referred to as the **subtropical jet stream;** it is normally located between 20° and 30°N. The polar-front jet and the subtropical jet flow from the west to the east between altitudes of 35,000 and 40,000 feet (10.7 to 12.2 km).

A third jet-stream system has recently been identified in lower latitudes at 13°N. It is called the **tropical-easterly jet stream.** The flow of air is reversed in this jet stream. It flows from east to west. The tropical-easterly jet stream occurs only in the summer season and seems to be limited to the Northern Hemisphere.

Jet-stream analysis is an important key to understanding surface weather conditions as well as the way in which the solid and gaseous spheres of the Earth are physically interrelated: On a global basis, for example, the west-to-east atmospheric motion increases during the winter of the Northern Hemisphere while the Earth's rotation slows. Specifically, the polar-front jet stream is intense in January and February, but it is weak in July; on the other hand, the speed of the Earth's rotation slows during the Northern Hemisphere's winter and increases during its summer. The eddy theory, of course, makes the relationship between these facts understandable. Since the total angular momentum of the Earth must remain constant, a change in the angular momentum of the Earth's gaseous sphere, the atmosphere, is met by an opposite and compensating change in the angular momentum of the Earth's solid sphere, the geosphere.

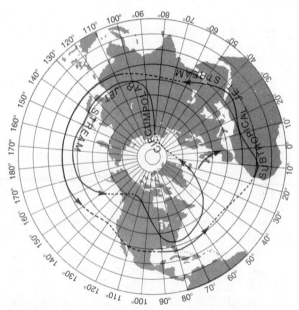

FIGURE 10–15 The polar front-jet stream and the subtropical jet stream flow from west to east at latitudes around 35,000 feet (10,700 meters). The range of the polar-front jet stream is largely from 40° to 60°N. The subtropical system is normally between 20° and 30°N.

MONSOON CIRCULATIONS

The term **monsoon,** as originally used in the literature of meteorology and climatology, means *seasonal wind.* Beyond this simple definition we find basic and widespread disagreement as to the cause of monsoons. There is, for example, some doubt as to whether monsoons are purely thermal systems. But, alas, most of the modern attempts to get at the causative factors of monsoons lack the simple clarity and consistency needed in an introductory text. Therefore, in this instance, the common definition of monsoon and the suggestion that it arises from a thermal origin, which is rooted in the differential heating of extensive land and water surfaces, will be used.

THE CLASSIC THEORY

In some way, then, a monsoon is a modification of the general planetary wind system. According to conventional wisdom, a monsoon circulation depends on the ability of land areas to cool quickly in winter and to warm just as quickly in summer. The temperature fluctuations from summer to winter produce pronounced changes in the pressure distribution, which causes seasonal modification and reversal in the surface-wind circulation. The enormous land area of Asia, for example, is subject to rapid fluctuations in its temperature as the seasons change. In some areas of Siberia, for example, temperatures plunge to −108°F (−78°C) during the winter.

During the summer, the continents heat very rapidly—especially over the deserts. As the air in contact with the land is heated, it becomes less dense and its surface pressure decreases. On the other hand, the sea with its great heat capacity remains relatively cool during the summer. As a consequence, the air in contact with the sea also remains relatively cool and dense. Continental interiors become centers of low pressure, and the pressure gradient moves air from over the sea toward the land. The summer monsoon, then, is a sea-to-land wind in which tropical maritime air normally brings large amounts of moisture and latent heat to a landmass.

The cloud-covered landmass of India, with the island of Ceylon off its southeast coast, stretches before us. The cloud-covered Arabian Sea lies to the west of India, at the left of the photo; the Bay of Bengal is to the east. India covers 1,262,274 square miles (3,269,139 square kilometers). A sea-to-land wind, the summer monsoon, is carrying moisture-laden air from the Arabian Sea toward the Indian landmass. The rather regular break in the cloud pattern as the wind approaches the coast is due to a warm ocean current, which moves along the coast and transfers enough heat upward to dissipate the clouds in the air immediately over the current. (NASA)

The interior of a large continent cools rapidly as summer fades into winter. The air overlying the cooling continent also cools, and its density and surface pressure increase. The ocean, on the other hand, cools slowly and as a consequence becomes relatively warmer than the continental interior. The winter pressure gradient runs from relatively cool high-pressure air over the land to the relatively warm low-pressure air over the sea; as a consequence, air is moved from the land toward the sea. The winter monsoon, then, is a wind of land origin that carries cold, dry continental air toward the sea.

Monsoons are common in many parts of the world where large continents border large oceans. They occur, to some extent, over the north coast of Australia where there is an onshore northwesterly flow of humid equatorial maritime air during the summer and offshore dry southeast trade winds during the winter. Monsoons also flow over the Iberian Peninsula and the Guinea coast of Africa. However, eastern and southern Asia together have the largest and most perfectly developed monsoon circulation.

ASIAN MONSOONS

The large Asiatic landmass naturally intensifies any weather effect that is associated with its continentality; monsoons fall into this category. The winter high-pressure area over Asia is shown in Figure 10–4; it is, as you can note, a very strong high, which is positioned somewhat east of the continent's center. The Asiatic summertime lows are shown in Figure 10–5; the deepest low on the Asian continent is situated over the hot lowlands of northern India-Pakistan.

THE SOUTHERN-ASIA MONSOONS

Summer monsoons are quite pronounced over the Arabian Sea, India, and the Bay of Bengal. The Indian monsoon is effectively separated from the northern Asiatic landmass by the highlands of the Himalaya-Tibet system, which bar the movement of air to the north (Fig. 10–16). Heating to the south of the mountains is intense during May and June. A low-pressure cell forms over the Indian landmass, and a continuous flow of air takes place from the subtropical high of the South Indian Ocean toward the low-pressure

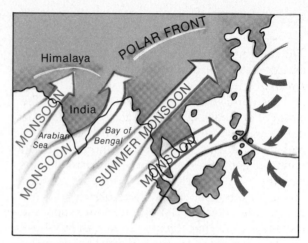

FIGURE 10–16 The moist ocean air of the southwest monsoon sweeps over the land surfaces of India, Burma, the Indochina Peninsula, and on into eastern Asia.

area over the land. The continuous flow of air is called the *southwest monsoon* (Fig. 10–16).

The summertime low (Fig. 10–5), which is thermal in origin, extends across India-Pakistan into Southeast Asia. The moist ocean air of the summer monsoon that sweeps over the land surfaces of India and Burma continues eastward over the Indochina Peninsula and then moves northward over much of eastern Asia (Fig. 10–16). The moist winds cool adiabatically as they are forced to higher elevations over the land. The rains that fall from the rapid and widespread condensation begin during May in Burma. The start of the Monsoon rain in India is usually early in June. The normal dates for the onset of the southwest monsoon over India are depicted in Figure 10–17. During the four months of the summer monsoon, as much as 400 inches (1016 cm) of rain may fall on the southern slopes of the Himalayas. The record for the greatest annual rainfall in India is more than 1000 inches (2540 cm).

On the average, the monsoon winds travel at about 20 knots while over the Arabian Sea. The winds entering the Bay of Bengal blow from the southeast as they move up the Ganges River Valley (Fig. 10–16). The summer monsoon blows with great steadiness over India. The northwestward spread of the monsoon rains over India is accompanied by a fall in daily maximum temperatures. In May, the average maximum

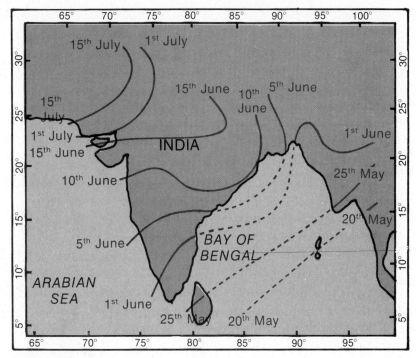

FIGURE 10–17 The onset of the southwest monsoon is an easily documented event because the weather changes abruptly to cool, humid, and rainy conditions. The isolines indicate the days for the onset of the southwest monsoon conditions throughout India.

temperatures in India are above 104°F (40°C), except in coastal regions. The monsoon depresses average maximum temperatures by as much as 18 Fahrenheit degrees (10C°) over central India.

India, with 600 million people to feed, is a country whose economy is dominated by agriculture. The monsoon rains that lash India every summer are crucial to the country's survival. If the onset of the summer monsoon is delayed for a considerable period, drought and famine can occur in India. In 1973 and 1974, the summer monsoons did not develop properly and India got very little rain. The drought caused great difficulty. After two bad years, the heavy monsoon rains of 1975 were a welcome relief. India produced a record crop of grain and increased the output of water-generated electricity as a result of the 1975 summer monsoon.

During the winter, relatively dry winds move from the Indian landmass toward the ocean. This winter monsoon is referred to as the *northeast monsoon*. It develops and is maintained because the ocean water is warmer than the land and a low-pressure system is set up over the sea.

THE EAST-ASIA MONSOON

The strongest winter monsoon in Asia has its beginning in a low-pressure cell that forms off the coast of eastern Asia (Fig. 10–4). The Himalayas also play a role in the way this monsoon develops because they form a southern barrier that does not extend east of Burma. During the winter when temperatures plunge in central and eastern Siberia, the cold air over Siberia produces the highest air pressures of anywhere in the world—an average of 30.6 inches (1036 millibars). The highest pressure seems to be centered in the Lake Baikal area of Siberia (53°N, 107°40'E). The steep pressure gradient that develops between the land and the sea drives northwest winds across China and the China Sea, through the Formosa Channel, and on to the south. The circulation mainly affects China, Taiwan, Korea, and Japan.

WEST-AFRICAN MONSOON

The **Sahel**—an Arabic name meaning *fringe* that describes the sub-Sahara region of Africa—is a monsoon-dependent area (Fig. 10–18). Lying roughly between 14° and 18°N latitude, the Sahel even in normal times rarely gets as much as 23 inches (58 cm) of rain a year in its southernmost parts. The rainfall tapers off to almost none as you move north toward the Sahara. The farmers of the Sahel live south of the 14-inch (35-cm) rainfall line and sow millet and sorghum at the start of the monsoon rains in June. They harvest these crops after the close of the rainy season in October. To the north of the 14-inch (35-cm) rainfall area, tamarisk and acacia trees, savannah grass, and thorny bush survive in temperatures that sometimes reach 122°F (50°C). Nomadic herders move north into this area during the rainy season. Their herds of camels, sheep, goats, cattle, and donkeys are constantly on the move in search of forage and water.

Beginning in 1968, the summertime monsoon rains that normally came to the Sahel began to fail. The southern farms got less than half the usual rain; the rangelands to the north got almost none. The desperate nomads began to move their herds southward. By 1973, the drought had caused a full-scale disaster. Famine took a heavy toll during the 6 years: 100,000 people and some 20 million head of livestock starved to death. Mauritania, Senegal, Mali, Upper Volta, Niger, and Chad were devastated.

Some meteorologists believe that high-altitude cooling altered the large-scale circulation pattern of the atmosphere and produced a change in the summer position of the jet-stream system—the great band of high-altitude winds that circule the Northern Hemisphere. This change in the jet-stream system probably blocked the summer monsoons and their life-giving rains! No one, of course, knows with any certainty the mechanism that is operating. Heavy monsoon clouds did return to the Sahel in June 1974. Monsoon rains fell until October and produced a good harvest.

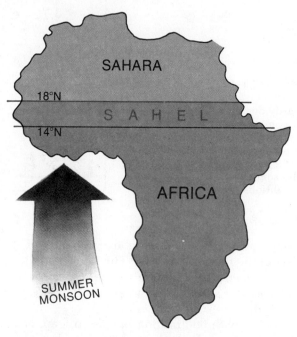

FIGURE 10–18 The sub-Sahara region of Africa, known as the *Sahel*, is a monsoon-dependent area.

THE LANGUAGE OF PHYSICAL GEOGRAPHY

A physical geographer uses a technical vocabulary to make explanations. Review your understanding of the vocabulary used to develop the concepts in this chapter. Among the important words and terms used are:

anticyclone
Baric Wind Law
chinook
constant-pressure chart
convection current
Coriolis force
currents
cyclone
depression
doldrums
eddy theory

foehn winds
geostrophic wind
high
high-pressure area
isobar
intertropical
    convergence zone
jet stream
land breeze
local breeze
low

low-pressure area
monsoon
mountain winds
polar easterlies
polar-front jet stream
pressure-contour chart
pressure gradient
prevailing westerlies
Rossby waves
Sahel
sea breeze

subsiding air
subtropical jet
  stream
trade winds
tricellular theory
tropical-easterly
  jet stream
valley winds
winds

SELF-TESTING QUIZ

1. Explain the two basic climatic functions served by wind.
2. What causes a parcel of air to move?
3. Describe the way in which the horizontal pressure is plotted and displayed.
4. How would you describe the direction of the most rapid change in pressure at sea level?
5. Compare and contrast the wind pattern around a high and a low.
6. How is the problem of representing the pressure field for upper-air conditions handled?
7. Explain the relationship that exists between the highs, lows, troughs, and ridges on the 500-millibar surface and those found at the Earth's surface.
8. Why are the high-pressure zones in the Northern Hemisphere somewhat irregular?
9. What is the net effect of the operation of the Coriolis force in the Northern and Southern hemispheres?
10. List the various forces that make an impact on wind flow.
11. Explain the coupling that exists between high-pressure and low-pressure zones at the surface.
12. Trace the development of a sea breeze throughout the day.
13. What conditions must exist for a valley breeze to develop?
14. In what way does topography condition the development of a chinook?
15. Explain the global pattern of surface winds.
16. Describe the pattern of upper-troposphere westerly winds known as the *jet stream*.
17. Identify some important monsoon-dependent areas.

IN REVIEW
SPATIAL DISTRIBUTION OF CLIMATIC ELEMENTS BY WIND

I. WIND AS A CLIMATIC CONTROL

  A. The Pressure Field
    1. The Sea-Level Pattern
    2. Highs and Lows
    3. The Upper-Air Pattern
    4. Worldwide Pressure Patterns
      a. Equatorial Belt of Low Pressure
      b. Subtropical High-Pressure Belts
      c. Polar Low-Pressure Zones
      d. High-Pressure Polar Caps
    5. U.S.-Canadian Pattern
  B. The Coriolis Force
  C. Forces Governing Winds
  D. Cyclonic and Anticyclonic Flow
  E. High-Low Coupling

II. LOCAL WINDS

  A. Land-Sea Breezes
  B. Mountain and Valley Breezes

  C. Chinooks

III. GLOBAL SURFACE-WIND SYSTEMS

  A. Polar Easterlies
  B. Prevailing Westerlies
  C. Tropical Easterlies
  D. Modification and Shifting
  E. The Tricellular Theory
  F. The Eddy Theory

IV. THE UPPER-AIR SYSTEM

V. MONSOON CIRCULATIONS

  A. The Classic Theory
  B. Asian Monsoons
    1. The Southern-Asia Monsoons
    2. The East-Asia Monsoon
  C. West-African Monsoon

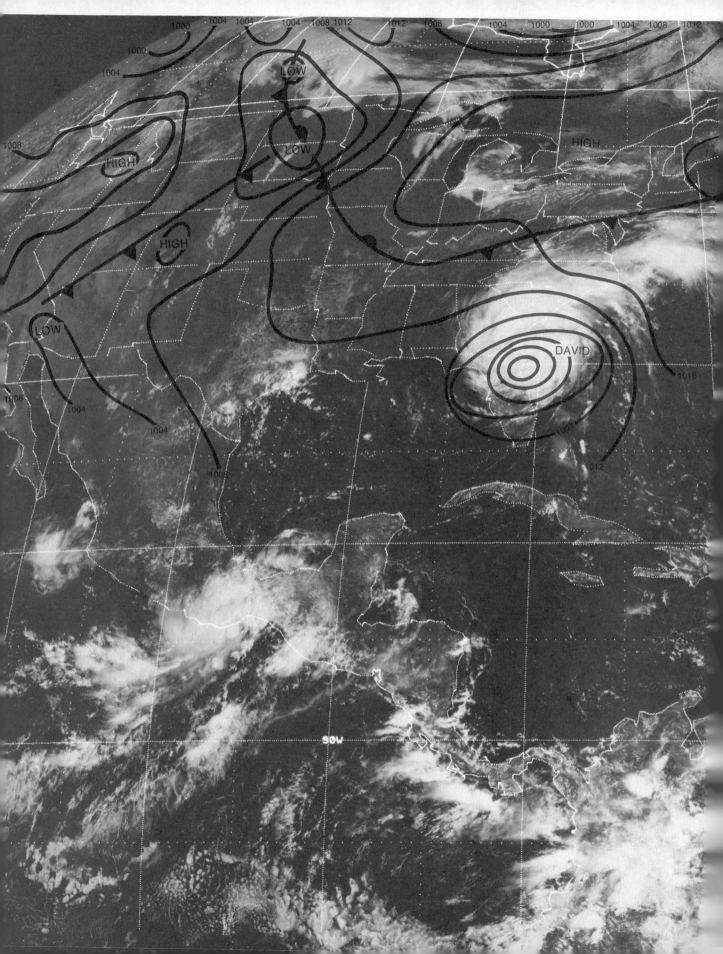

11 *Air-mass climatology*

In a very real sense, the prime causative climatic factor has to do with air masses since every locality derives its weather and ultimately its climate from the air masses that pass over. The general circulation transports a variety of air masses to you and away from you; the continents, mountains, oceans, lakes, and ice fields modify these air masses as they move from one locale to another across the Earth's surface.

High-pressure cells are the source regions for all air masses. The high-pressure areas of the subtropics are the spawning grounds for warm-moist maritime tropical air and hot-dry continental tropical air. The high-pressure areas in high latitudes produce cold continental arctic, cold continental polar, and cool-moist maritime polar air masses.

Climatic monotony prevails in locations that do not have a variety of air masses passing over.

On some tropical islands in the trade-wind belt, for example, temperature, humidity, cloudiness, rainfall, and wind hover in the same range day-after-day and month-after-month. Some low-latitude deserts suffer from climatic monotony also. On the other hand, in the prevailing-westerlies belt, locations are subjected to rapid and, at times, violent displacements of air masses. The borders of the continents in the westerlies are particularly affected. The same air mass rarely prevails for more than a few days.

The frequency with which various air masses move over a location in the westerlies belt changes with the seasons. Continental and maritime air masses and tropical and polar air masses clash in this belt. As a result, the seasonal amplitude of climatic elements in the prevailing westerlies is significant: the largest amplitudes occur in the interior of continents.

The pattern of air masses, pressure, and fronts is spread before us in this satellite view (September 4, 1979) as Hurricane David swept over the Florida-Georgia-South Carolina area. David was the most intense hurricane of this century in the eastern Caribbean area. Maximum strength of 150 knots and minimum pressure of 924 millibars were reached south of Puerto Rico on August 30. David was not a major hurricane when it struck the United States. The landfall pressure was 972 millibars just north of Palm Beach around midday on September 3. (NOAA)

AIR-MASS CHARACTERISTICS

The term **air mass** means a large homogeneous body of air that has approximately the same temperature and humidity characteristics in a horizontal direction. These two physical properties—temperature and water-vapor content—are, recall, the primary climatic elements. It is, however, the vertical distribution of temperature and moisture within the air mass that determines the actual or potential development of clouds and precipitation.

SOURCE REGIONS AND FRONTS

Some locations are more conducive to air-mass development than others; an area in which air-mass formation occurs is called a **source region.** Air-mass source regions are large regions with physically homogeneous surfaces where the wind movement is light and there is sufficient stagnation within the atmospheric circulation to allow time for the air to acquire the temperature and moisture properties of the underlying surface. Thus, the temperature and humidity characteristics of an air mass are determined directly by the nature of its source region. The arctic plains of North America, tropical oceans, and great deserts are some of the kinds of areas that can and do serve as good air-mass source regions.

The degree and depth to which an air mass achieves temperature and moisture equilibrium with the surface it overlies depends on the length of time it overlies the region and the initial temperature differential between the air mass and the underlying source-region surface. The principal processes that bring about equilibrium are radiation and vertical convection. When an air mass is initially colder than the surface it overlies, heat is transferred from the source-region surface to the lower layer of the air mass. This upward transfer creates ascending currents of warm air. Heat and any moisture present are carried aloft and the air mass is modified to considerable heights. On the other hand, if the air mass is warmer than the surface it overlies, it transfers heat to the surface below; since convection currents do not develop, the modification of the air mass occurs very slowly over a very shallow vertical range.

Areas of high pressure tend to be favorable regions for air-mass formation, but not exclusively so. On a worldwide basis there are four basic source regions: arctic, polar, tropical, and equatorial. Most source regions tend to bound the belt of prevailing westerlies. For example, there are source regions along the poleward boundary of the prevailing westerlies at about 60° latitude, which is in the vicinity of the subpolar low-pressure belt. Other source regions are clustered along the equatorward boundary of the prevailing westerlies in the high-pressure belt vicinity of the horse latitudes. The zone of separation between two differing air masses is called a **front.** The general location of air masses and fronts in the Northern Hemisphere is set out in Figures 11–1 and 11–2.

ARCTIC AIR MASSES

These air masses form over the large areas of ice and snow typically found near the poles in both hemispheres (Fig. 11–1). Wintertime—December to March in the Northern Hemisphere, June to September in the Southern Hemisphere—in the region of the poles is quite cold and without much insolation. Thus, arctic air masses tend to form quickly.

In general, during the winter, an arctic air mass is colder than any air mass from a different source region. Since there is some conduction of heat from the seawater below the frozen surface of the Arctic Ocean to the air above, the lower limit for air temperatures over the frozen Arctic Ocean is about −49°F (−45°C). Over snow-covered solid ground—in northern Siberia, for example—there is much less heat conduction from below the frozen surface; thus, the surface air temperature in such arctic air can plunge to −94°F (−70°C). The humidity in very cold arctic air is necessarily very low.

ARCTIC FRONTS

The zone of demarcation between arctic air and its more southerly neighbor, polar air, is referred to as an **arctic front** (Fig. 11–1).

FIGURE 11–1 Air mass source regions tend to bound the belt of the prevailing westerlies. A front is a zone of collision between two differing air masses. The color is used in this figure to designate the average location and extent of the fronts in the Northern Hemisphere during January.

In the Northern Hemisphere during January, the Atlantic arctic front extends from near the Icelandic low toward the Barents Sea, which situates it roughly along the boundary between the open Atlantic waters and the polar ice (Fig. 11–1). The temperature contrast on either side of this front is often considerable. West of the Icelandic low no arctic front exists. On its Pacific side, the arctic air of the Northern Hemisphere is hemmed in by the northern part of the Rocky Mountains. The southern end of the arctic front over North America bends east from the mountains in Canada toward the Great Lakes (Fig. 11–1).

In the Southern Hemisphere, arctic fronts form in the trough of low pressure around Antarctica. There are strong temperature gradients from Antarctica's ice and snow surface to the open ocean. Thus, we find a relatively persistent ring of zonal fronts around the continent of Antarctica even during the summer of the Southern Hemisphere.

POLAR AIR MASSES

Strictly speaking, the term *polar* should be subpolar since it designates high latitudes below the arctic source region. The source regions for **polar air masses** lie roughly between latitude 55° and 65° in both hemispheres. There are a number of source regions for polar air in the Northern Hemisphere: Siberia, the Gulf of Alaska, northern Canada, the North Pacific to the south of the Aleutian low, the northwestern Atlantic, and the North Atlantic just west of the British Isles (Fig. 11–1). On the average, polar air covers about 40 percent of the total surface of the Earth.

POLAR FRONTS

All lines of demarcation between polar and tropical air are called **polar fronts.** A comparison of Figures 11–1 and 11–2 shows that polar fronts are situated at lower latitudes during winter than during the summer. Three main polar fronts are depicted: the Pacific, Atlantic, and Asiatic.

During January in the Northern Hemisphere, the Pacific polar front extends primarily over water (Fig. 11–1). Trace its extent from the west coast of North America to the Philippines. The July position of the Pacific polar front is at much higher latitudes and is primarily over land (Fig.

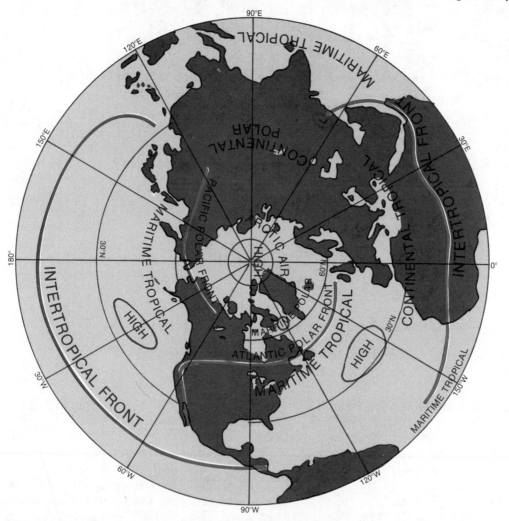

FIGURE 11–2 Areas of high pressure tend to be favorable regions for air-mass formation. The color is used in this figure to designate the average location and extent of the fronts—that is, zones of collision between differing air masses—in the Northern Hemisphere during July.

Stratocumulus clouds hang over these farmers working a soybean field near Culiacán (24°48′N,107°24′W) in the state of Sinaloa, along the mainland coastal strip that borders the Gulf of California. The area is west of the Sierra Madre Occidental in what is considered to be the Pacific Northwest of Mexico. Figure 11–2 shows that the Atlantic Polar Front is in the Gulf of California, just to the west of Culiacan, during July. The average July temperatures in this farming region are in the 80s (very close to 30°C). The area is generally dry, with 20 to 30 inches (50 to 76 centimeters) of yearly precipitation. River water flowing from the mountains to the east is used for irrigation on this farm. (World Bank)

11–2). During the summer, it extends from Canada's Northwest Territories, through Alaska, and on to the Asian landmass.

In Figure 11–1, the January position of the Atlantic polar front extends westward across the Atlantic from Europe to Florida. It moves across the Gulf to the highlands of Mexico and then bends north along the east side of the Rocky Mountains. Under normal conditions, polar air from Canada is not deep enough to cross the Rockies from east to west. The July position of the Atlantic polar front shown in Figure 11–2 has it running through the gap in the Canadian Rockies and then moving south to the east coast of Baja California.

The Asiatic polar front exists only during the winter of the Northern Hemisphere. It extends from the Mediterranean near Italy along the mountain ranges of Asia Minor, Iran, Afghanistan, and Tibet (Fig. 11–1). The eastern position of this front sweeps through the Burma area.

In the Southern Hemisphere, the polar fronts do not seem to have preferred positions over land. The northern ends of the polar fronts are found near the subtropical highs. They tend to be at lower latitudes, farther north, during the July winter of the Southern Hemisphere.

TROPICAL AIR MASSES

There is a chain of tropical air-mass source regions that encircles the Earth in the Northern Hemisphere, and there is another in the Southern Hemisphere. These source regions girdle the Earth in the vicinity of the Tropic of Cancer (latitude 23.5°N) and the Tropic of Capricorn (latitude 23.5°S). Tropical air masses form over five regions in the Northern Hemisphere: the southern area of the North Pacific, the southwestern United States (summer only), northern Mexico (summer only), the southern area of the North Atlantic, and the northern mainland of Africa.

The land areas of the subtropical latitudes of the Northern Hemisphere, from northern Africa in the west to China in the east, are sufficiently large to produce a **continental-type tropical air mass** (Figs. 11–1 and 11–2). The tropical source regions that are over land coincide with the great tropical deserts found at these latitudes. These tropical deserts result, in part, from the adiabatic warming of the air that is subsiding in the subtropical high pressure belts, or horse latitudes. It is, however, only over northern Africa that we have a year-round high within which the continental-type tropical air is produced on a 12-month basis.

LOW-LATITUDE FRONTS

In Figures 11–1 and 11–2, the **low latitude fronts** are represented as the intertropical front. This is, in reality, the ITCZ discussed on page 341. The trade winds from both hemispheres come together to form this front, which like all fronts is in a trough of low pressure. The weather phenomena associated with the ITCZ changes from day-to-day and also varies from one longitude section to another.

AIR-MASS CLASSIFICATION

There are three different systems by which air masses are classified: one is an absolute system and the other two are relative systems.

The absolute classification system is based on the temperature, humidity, and temperature lapse rate of the air mass. Since a relation exists between these properties and the location of the air-mass source region, the absolute classification is geographical. The study of source regions suggests an absolute geographical classification of three main types: arctic, polar, and tropical.

One of the relative classification systems describes the way a given air mass is transformed during its movement from one location to another. In other words, the classification is based on the effects produced as heat is exchanged with the ground. When the air mass transfers heat to the ground, it is called a **warm air mass.** If the surface transfers heat to the air mass, the air mass is referred to as a **cold air mass.**

Another relative classification system is based on the comparison of air masses on either side of a front. The air mass with the lower temperature is called *cold air,* and the other air mass is given the designation *warm air.* This system is useful, as we shall see on page 361, in naming fronts.

The air-mass classification system described below is a combination of the absolute and first relative system described above—it is based on geographic source regions and the subsequent modifications an air mass undergoes.

THE BASIC SYSTEM

Capital letters are used to identify the general latitude of the source regions: A for arctic, P for polar, and T for tropical. A lower case c for continental or m for maritime is prefixed to indicate whether the source region is over land or water. Thus, an identification of an air mass as cT means that the source region is over land in a tropical area.

An air mass is also identified as to its low-level temperature relative to the surface over which it is passing. An air mass is designated as either warm (w) or cold (k). A w identification means that it is warmer than the surface over which it is passing. A k designation indicates that the air mass is colder than the surface with which it is in contact. Thus, an identification of cTw means a continental air mass of tropical origin is warmer than the surface over which it is currently passing.

A w designation is used when the lower level of the air mass is transferring heat to the surface over which it is passing; in other words, the air mass is being cooled from below. The k identification means that heat is flowing from the surface below to the colder air mass. A k air mass is warmed from below. Warming from below leads to instability in the air mass; cooling from below favors stability.

TEMPERATURE-HUMIDITY PROPERTIES

As a rule, maritime air contains more moisture than continental air. During winter in a hemisphere, maritime air is generally milder than continental air because the sea over which the maritime air develops its characteristics is usually warmer than the continental surfaces. The situation reverses in the summer, however. Con-

Cool trade winds keep the climate on the islands of Hawaii pleasantly mild all year. Rainfall, however, varies from hundreds of inches (centimeters) a year on the mountaintops to less than 10 inches (25 centimeters) in the lowlands. The Iao Needle, standing above the clouds, is a prominent point on a winding knife-edge ridge on West Maui. This needle formation was carved from the surrounding lava by stream erosion fed by more than 300 inches (700 centimeters) of yearly precipitation. (Hawaii Visitors Bureau)

In both hemispheres, ice and snow surfaces cover the source regions for cA and mA air masses. These air masses tend to be cold with very little moisture content. The cP air masses tend to be cold, too. During the winter, a cP air mass may have a surface air temperature of 0°F (−18°C) and specific humidity in the range of 1 gram of water per kilogram of air. At an altitude of 16,000 feet (4900 m), the air temperature may be less than −40°F (−40°C), and the specific humidity may be in the range of 0.2 gram of water per kilogram of air.

The mP air masses are cold, but winter mP air masses are not subjected to the intense chilling that cP air masses receive. As a result, an mP air mass usually has a comparatively high surface air temperature—quite often in the range of 45°F (7°C). The specific humidity of an mP air mass is relatively high in comparison to the water content of cP air. At the surface, the specific humidity of mP air may be 4.4 grams of water per kilogram of air.

The cT air masses tend to have their source regions over desert areas. A desert has very little surface water and vegetation to add moisture to the air. Daytime surface temperatures in cT air during the summer may be well over 100°F (38°C). The mT air forms over sea surfaces at these same latitudes. The sea surface during the summer is quite a bit cooler than the adjoining land surfaces, and the surface air temperatures of mT air may be in the range of 75°F (24°C). The mT air picks up a lot of moisture over its source region, and during the summer it may have a specific humidity of 19 grams of water per kilogram of air, which will certainly make the air "feel" moist.

tinental surfaces generally warm faster than the sea; as a result, maritime air masses tend to be cooler than continental air masses during the summertime.

AIR-MASS MODIFICATION AND ASSOCIATED CLIMATIC ELEMENTS

Eventually the general circulation pulls an air mass from its source region and carries it to other locations. In North America, the Rocky Mountains serve as a barrier that prevents the unimpeded penetration of air masses from the Pacific into the heartland of the continent. This mountain barrier leaves the way open for arctic and continental polar air masses to sweep southward out of Canada into the United States. As an air mass pulls loose from its source region and moves over an area with different characteristics, the process of modification begins. The lower portion of the air mass undergoes a transition in temperature and humidity. The upper layers remain relatively unchanged unless the air mass becomes unstable.

cP MODIFICATION

Vertical movement produced by thermal instability is an important modifier of polar air masses. The addition or the removal of water vapor is another important way in which these air masses are modified. Canadian cP air masses are subject to continuous change as they move from their source region into the United States (Fig. 11–3). In the following paragraphs an attempt is made to describe some of these cP modifications.

During winter, a Canadian cP air mass overlies a frozen, snow-covered surface for a long time and becomes quite cold. Its moisture content is very low, too, and the air mass is characteristically free of any form of condensation. As it moves south, the air mass is designated as cPk because it is colder than the land over which it is traveling. If this cPk air is heated or moistened from the ground, under conditions of strong turbulence it will develop stratocumulus clouds, that is, limited convection types, with scattered light snow showers.

When the heating-moistening modification takes place as a result of Canadian cPk air moving over the Great Lakes region (Fig. 11–3), a strong cyclonic circulation develops. The relatively warm lake water transfers heat to the lower level of the cPk air. As the air in the lower level warms, it also picks up moisture from the lake surface. The warming-moistening air rises as vertical currents. As the moisture is carried aloft, adiabat-

ic cooling leads to condensation and the release of heat, which produces cumulonimbus cloud types. If active convection develops, precipitation in the form of a heavy snowfall may be produced locally and continue as the air moves south. The heavy snowfall in the Buffalo area during the winter of 1976–77 resulted from such a sequence.

When the cPk air mass passes over the Great Lakes, the modification actually causes the air mass to become mP in character. If this modified air moves to the southeast and crosses the Appalachian Mountains, it usually produces stratocumulus clouds and snowfall on the windward side of the mountains. Once across the mountains the air tends to be clear.

During the summer, the coolness and dryness of Canadian cP air masses are modified by the addition of heat and water as they move to lower latitudes. Cumulus clouds develop in summer cP air, but there is usually sufficient upper-level stability to prevent the formation of cumulonimbus clouds. If cumulonimbus clouds build in cP air during the summer, hail may fall when convection within the clouds is strong.

mP MODIFICATION

There are two source regions for mP air that enters the mainland United States (Fig. 11–3). The first source is the cold northwestern Atlantic off the coasts of Greenland, Labrador, and Newfoundland. The second source of mP air is from the area south of the Aleutian low in the North Pacific Ocean. It should be pointed out, however, that the mP air from the Aleutians started out originally as cP air from Siberia; as the Siberian cP moves across the Pacific, it is thoroughly modified, and it becomes mP air after spending time in the Aleutian area.

During the winter, cool, moist maritime polar air is strongly heated and moistened as it moves to lower latitudes over the ocean. If its air pressure is relatively high, warming usually produces stratocumulus clouds, which may yield scattered light showers. When pressures are relatively low and falling, this type of winter modification produces cumulonimbus clouds and moderate-to-heavy showers. The summertime ocean is relatively cool, and thus pressures are

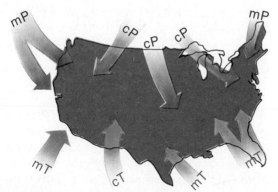

FIGURE 11–3 Tropical and polar air masses move into the conterminous U.S. and clash within its borders.

The five Great Lakes—Superior, Michigan, Huron, Erie, and Ontario—and Georgian Bay, above Lake Huron, are clearly visible in this satellite photo. The combined area of these lakes is 95,170 square miles (246,490 square kilometers). The relatively warm water of these lakes transfers a lot of heat to cPk air moving across the region. (NOAA)

relatively high as mP air moves to lower latitudes; mP modification by the addition of heat and moisture during the summer produces only stratocumulus clouds with occasional light drizzle.

During the summer, mP air moving toward lower latitudes characteristically develops stratocumulus. As summertime mP air moves toward the Washington, Oregon, and California coasts, it crosses the California Current, which is a zone of cold water that hugs the shoreline. The mP air is chilled at its interface with the sea, and low stratus and advection fog develop in the air. The air mass moves inland as mPk air. The mPk summer air that moves over coastal areas in the San Francisco region carries sea fog through the Golden Gate and over the hills. As the fog cascades down the eastern slopes of the hills, it evaporates because it is warmed by compression and contact with the hot, dry surface. The warming reduces the relative humidity of the air mass and dissipates the stratus clouds and fog. However, the mPk air usually continues to remain cooler than the surface over which it is traveling

as it moves inland. By the time mPk air crosses the Rocky Mountains, it is completely modified and indistinguishable from cP air.

During the winter, the mP air moving in toward the West Coast may develop cumulonimbus in addition to the stratocumulus. The wintertime air reaches the West Coast as mPk air, but as it moves inland it becomes mPw because the land surface at this time of year is not receiving much insolation. The air is cooled at its interface with the surface, and when it undergoes uplift, as a result of the topography, heavy rain results. San Francisco, for example, usually gets 15.5 inches (39.4 cm) of rainfall from December through March, which amounts to 72 percent of its annual precipitation of 21.5 inches (54.6 cm). This same mPw air produces heavy snowfalls on the western slopes of the mountains along the West Coast.

Along the East Coast, mP air occasionally moves inland from its source region (Fig. 11–3). Such a situation, however, occurs more frequently during the summer because the pressures over land are lower at this time of year, and,

Lubbock at 33°35′N,101°51′W forms the background in this photo of the Great Plains of Texas. The sky cover consists of altostratus and altocumulus clouds. The part of the Great Plains in which Lubbock is found lies along the border of Texas and New Mexico; it is called the *Llano Estacado*. This is a vast region where little rain falls. Irrigation and dry-farming methods are used in the region. (Interior Department)

thus, encourage the westward movement of the mP air. This eastern maritime polar air has relatively low humidity, and only thin stratocumulus develop as it moves toward lower latitudes. Summertime mP air that intrudes itself along the East Coast becomes mPk air over land; the warming of the air as it moves inland dissipates the clouds. Thus, along the East Coast, summer mP from the Atlantic often brings clear and cool weather to coastal states from New England to Virginia. If mP air penetrates the East Coast during the winter, it brings raw, northeast winds sometimes called *nor'easters*.

cT MODIFICATION

Continental tropical air masses originate in arid or desert regions principally during the summer. A cT air mass is usually strongly heated, and turbulence and intense convection develop to heights of 10,000 feet (3050 m) or more. Its moisture content is very low, however, and even intense convection does not normally bring the air to its condensation level; thus, the air is without significant cloud cover. The low moisture content of the air results in its rapid heating by day and rapid cooling by night. The temperature

curve for El Paso (page 291), with its large diurnal range, is a good example of the way such air behaves.

In the summer, cT air forms in the northern interior of Mexico over the Sonoran Desert region and in the arid Southwest of the United States, including the southern parts of Arizona and New Mexico and western Texas (Fig. 11–3). A cT air mass is usually confined to the region of origin and thus the extent of its effect is very limited. When cT air does move away from its source region, it mixes with other air masses and quickly loses its identity.

mT MODIFICATION

Maritime tropical air is warm and moist. It is our most important moisture-bearing and rain-producing air. In both hemispheres, as mT air moves poleward, its lower layer is cooled by the sea surface over which it passes. Thus mT air usually develops low stratus or fog, or even stratocumulus, depending on the wind velocity over the ocean.

The Gulf of Mexico, the Caribbean, the western Atlantic from the region of the Sargasso Sea, and the Pacific Ocean between Baja and Hawaii are the source regions for wintertime mT air masses that move into the United States (Fig. 11–3). As Pacific mT air moves toward California, it becomes mTw air. The cooling that takes place in its lower level as it moves northward causes decreasing visibility, fog, and drizzle, particularly at night. When this air is forced upward, as it moves inland and encounters mountains, there is often moderate-to-heavy precipitation.

In summer, it is not easy to distinguish between mT and mP air along the Pacific coast. For example, the maritime air at San Diego has surface temperatures similar to those at Seattle because cold, upwelled water along the coast tends

to cool and modify any northerly flow of mT air. Thus, one might say, practically no normal mT air moves toward the West Coast of the United States from the Pacific during the summertime. Atlantic Ocean and Gulf of Mexico mT air masses are numerous, however. The specific humidity and the temperature of Gulf and Atlantic mT air masses tend to be high, and any small uplift produces strong convection currents and frequent thundershowers. This summertime air becomes mTk as it moves over land. Radiation cooling of humid mT air at night causes stratus and stratocumulus clouds to develop. These low clouds dissipate in the morning and are usually replaced by cumulus types in the afternoon.

FRONTAL SYSTEMS AND ASSOCIATED CLIMATIC ELEMENTS

When two masses of air lie side by side—each horizontally uniform in density but with appreciable density differences between them at each elevation—the heavier air lies as a wedge under the lighter air. The sloping area of contact between the two air masses is called a **frontal surface.** Since air masses have a vertical dimension, the frontal surface starts at ground level—where it is called a *front*—and moves aloft. The frontal surface aloft is never vertical; it is always inclined. Air of differing physical characteristics intermingles along the frontal surface. Temperature, humidity, and wind direction are different on each side of the line of discontinuity.

The formation of a front is called **frontogenesis.** In the middle latitudes, fronts—which are generally moving—extend for hundreds and sometimes thousands of miles along the surface. The interplay and intermingling of air along fronts often results in dramatic weather changes. As the air masses become mixed and indistinguishable, the process is called **frontolysis,** which means the dissolving of a front.

FRONT SYMBOLS

Fronts at the surface are represented by a set of symbols: triangles and half-circles arranged

along or on either side of a line (Fig. 11–4). The symbol for the cold front represents a situation in which colder air—designated on the basis of the second relative classification system discussed on page 356—is replacing warmer air at the surface of the Earth. A warm front is one in which warmer air replaces colder air; and an occluded front results when a cold front overtakes a warm front. Occluded fronts are discussed in detail on pages 367–369; there are two types of occluded fronts—cold-front and warm-front occlusions—but one symbol is used for both. A stationary front is one in which there is no forward motion along a line of transition between two different air masses.

The three principal worldwide zones in which air masses clash are depicted in Figures 11–1 and 11–2. These geographic zones of air-mass collision are arctic, polar, and low latitude. There are, of course, variations in the specific locations of these zones from winter to summer; the locations of the fronts depicted in Figures 11–1 and 11–2 should be taken as being very general within broad regions. The arctic frontal zone marks the area of collision between arctic air and either **maritime polar** or **continental polar** air masses. The polar frontal zone marks the region of collision between cP and mP air masses, or between either cP or mP air masses and mT air masses. Over North America, this polar frontal zone is located within the conterminous United States during the winter half-year. During summer, the collisions can and do occur along the Canadian-United States border.

CHARACTERISTICS OF FRONTS

Before considering each of the four types of fronts—cold, warm, occluded, and stationary—in detail, we need to consider some general characteristics and facts that apply to almost all fronts:

■ First of all, fronts form at the outer boundaries of high-pressure cells. In Figure 11–5, 1028 isobar lines surround two high-pressure areas. The cold high is labeled H-1; the warm high is labeled H-2. The cold front is at the boundary between these high-pressure zones.

■ Another important fact is that fronts form only between air masses of different temperature and moisture conditions. The temperatures in the vicinity of H-1 (although not noted in the figure) are in the mid 40s, and those in the vicinity of H-2 are in the mid 80s. In Figure 11–5, the temperatures close to the front in each air mass are noted—65°F (18.3°C) at station A in the cold air and 76°F (24.4°C) at station B in the warm air.

■ In the Northern Hemisphere, the wind near the ground always shifts clockwise as a front passes over your position. For example, at station B in Figure 11–5, the wind is blowing from a southeasterly direction, and at location A the wind is blowing from a northwesterly direction. Visualize the front moving ahead and passing over station B. In other words, the whole figure, except locations A and B, should be shifted to the right. Such a shift would place station B along with station A in the cold air behind the front. After the front passes station B, the winds moving over this station would be the northwesterly winds that are presently blowing over station A. Thus, the wind shift is from SE to NW as the front passes, which is a shift in a clockwise direction.

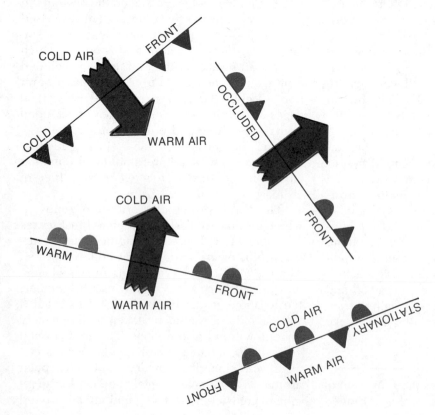

FIGURE 11–4 Fronts at the surface are represented by a set of symbols: triangles and half circles arranged along or on either side of a line.

■ Isobars at a front are always V-shaped. Between stations A and B (Fig. 11–5), you can note the way the 1012 isobar forms a V. The 1008 and the 1004 isobar also have a pronounced V-shape at the front.

■ The center of a cyclone, or low-pressure system, is found at one end of the front, that is, at one end of the line marking the boundary between adjacent cold and warm air masses. Pressure at the center of the low in Figure 11–5 is less than 992 millibars. The circulation of air around the low is counterclockwise.

■ Inspect Figure 11–5 again. Note that the front is found along a low-pressure trough. Moving toward the front from either high—H-1 or H-2—the pressure drops. Consequently, the barometric pressure always falls as a front approaches and rises after it passes.

COLD FRONTS

A **cold front** occurs at the boundary face along which a warm air mass is being displaced by a colder, heavier air mass. Figure 11–6, which de-

picts the frontal situation for the mainland of the United States one November day, shows two cold fronts: The first, a short front, stretches from Cuba to the northeast over the Bahamas for a distance of 1000 miles (1600 km). The second cold front is the result of a tongue of cP air pushing into the United States from Canada; it stretches more than 2500 miles (4000 km) from near Boise, Idaho, southeastward toward Kansas and then northward through Iowa and Minnesota into Canada.

On the map (Fig. 11–6) the temperature in the "warm" air over the eastern Bahamas is 74°F (23.3°C). The air temperature at Miami in the "cold" air mass is 70°F (21.1°C). The terms *cold* and *warm* are used in a relative sense; often, there is not a large temperature difference across the front.

If our reference is to the Cuba-Bahama cold front, the high-pressure air centered over Kentucky is considered the cold air mass. But if our reference is to the Idaho-Kansas-Minnesota cold front, the high-pressure air centered over Kentucky is considered the warm air mass; the cold, high-pressure center is over Helena, Montana, as the 1028 isobar indicates (Fig. 11–6). The air

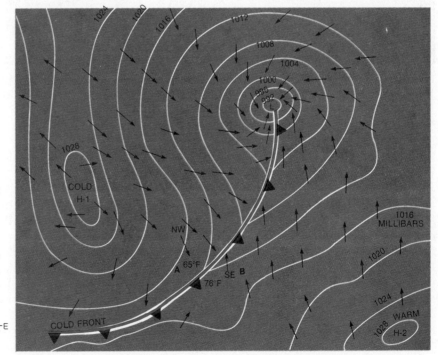

FIGURE 11–5 The cold front depicted in this figure is at the outer boundary of two high-pressure cells. The isobars at the front are V-shaped and the winds near the ground shift clockwise as the front passes.

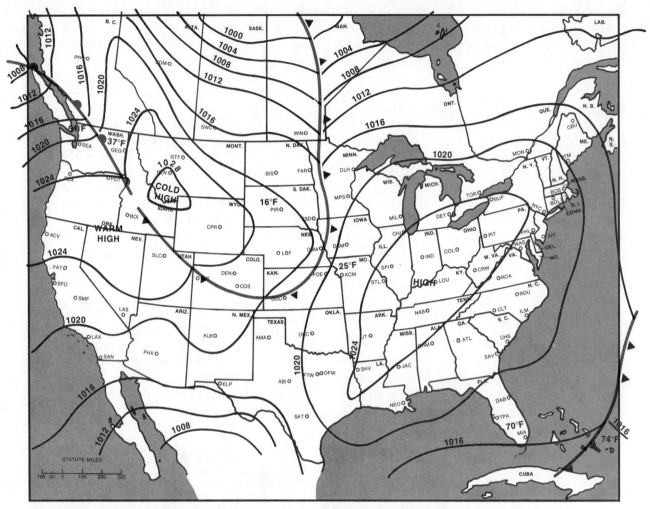

FIGURE 11–6 Two cold fronts and one warm front are depicted in this figure. The air behind the warm front is mP air from the Pacific. The air over the Dakotas, behind the cold front, is cP air from Canada. The air over Texas is mT air from the Gulf of Mexico (cities are identified by airport codes; see Appendix VIII).

temperature at Kansas City in the so-called warm air mass is 25°F (−4°C), and in the cold air mass at Pierre, South Dakota, the air temperature is 16°F (−9°C).

Along a cold front, the cold air, with its greater density, pushes under the warmer air. As indicated in Figure 11–7, cold fronts have steep leading edges aloft where the warm air is being pushed sharply upward. The frontal weather associated with a cold front lies in a narrow band, is often violently active, and depends on whether the warm air is thermodynamically warm, *w* (stable air), or thermodynamically cold, *k* (unstable air). Stable air is warmer than the surface it

overlies, which means the air is giving up heat to the colder surface; since its lower layer is cooling, the air resists vertical motion and is called *stable*. Unstable air—k air—is colder than the surface it overlies; the heat it receives from the warmer surface produces vertical motion and unstable conditions.

In Figure 11–7A, the warm air ahead of the cold front is stable. Under these conditions nimbostratus clouds form almost directly over the cold front's contact with the ground. Towering cumulus form over the nimbostratus, and altostratus and altocumulus also form as a result of the lifting of the warm air by the denser, colder air.

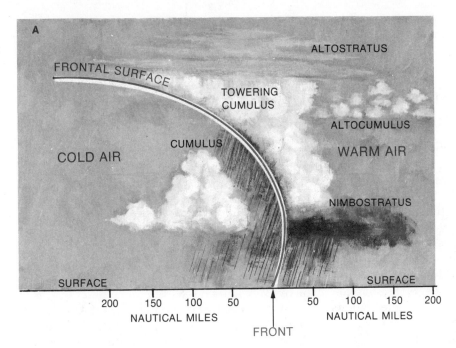

FIGURE 11-7A. The warm air ahead of the cold front is stable; it resists vertical motion. Towering cumulus are produced in the warm air along the frontal surface.

The rain from the nimbostratus and the towering cumulus falls slightly ahead of and well behind the front.

In Figure 11-7B, the warm air ahead of the cold front is unstable and very humid. The "warm" air is warm with respect to the cold air mass, but it is cold with respect to the ground; therefore, the "warm" air receives heat from the ground. The nimbostratus, in this case, form on both sides of the front. As the advancing cold front lifts the unstable air, cumulonimbus develop in addition to the altostratus and altocumulus clouds. There is a steady downpour from the nimbostratus cloud deck, and periodic

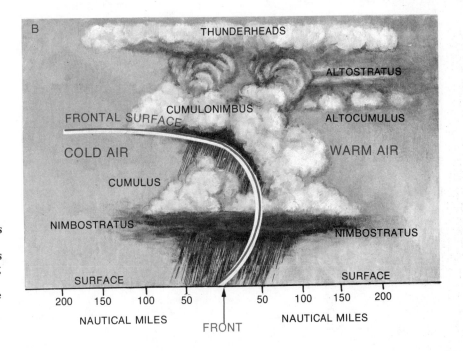

FIGURE 11-7B. The warm air ahead of the cold front is unstable; the heat being received from the land adds to the vertical motion being induced by the advancing cold air. Cumulonimbus are produced in the warm air along the frontal surface.

heavy showers and thunderstorms are produced by the cumulonimbus cloud formations. These thunderstorms are usually very narrow in width but hundreds of miles long. A narrow band of thunderstorms, called a *squall line*, may precede the front by 100 miles (160 km), although the front itself is only about 10 to 50 miles (16 to 80 km) in width.

The speed with which cold fronts move varies within rather wide limits. A cold front may, for example, travel as slowly as 10 knots or as fast as 50 knots. As a general rule, the speed of these fronts is considerably slower in summer than in winter.

The sequence of events as a cold front approaches is somewhat as follows: There is an increase in the winds from the south or southwest. Altocumulus clouds appear on the horizon to the northwest or west. The barometer begins to fall. The altocumulus clouds are seen for two to three hours prior to the arrival of the cold front. As the front approaches, the clouds lower and towering cumulus or cumulonimbus begin to dominate the sky. Rain begins to fall and increases in intensity as the front passes. The winds shift rapidly to the west or northwest with strong gusts, the barometer begins to rise, and the temperature drops with the passage of the front. Well behind the front the sky clears and the weather is good.

WARM FRONTS

Warm fronts are found along boundary faces where warm air is displacing a colder air mass. In Figure 11–6, a warm front extends for more than 800 miles (1290 km) from the Pacific Ocean across the northern part of Vancouver Island through the southeast corner of the state of Washington. The air temperature in the vicinity of Seattle—within the warm air mass—is 54°F (12°C). The warm air is mP air from the Pacific. On the other side of the warm front in the cold air mass near Spokane, the air is cP air from Canada with a temperature of 37°F (2.8°C). The lighter, warmer air—because its density is lower—rides up on top of the cold air mass along a gently sloping line similar to the frontal surfaces depicted in Figure 11–8.

When warm, stable air (Fig. 11–8A) is lifted up the slope, it cools and produces a sequence of stratus-type clouds: nimbostratus, altostratus, cirrostratus, and cirrus. The precipitation is heaviest under the nimbostratus clouds and decreases as the cloud bases become higher. Altostratus clouds are precipitation makers, but the high clouds above 20,000 feet (6100 m) are glaciated and do not drop precipitation. The rain that falls into the wedge of cold air adds to the moisture of the cold air, which can lead to the development of stratus clouds in the cold air

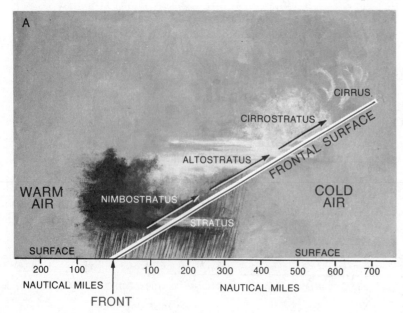

FIGURE 11–8A. The warm air behind the front is stable; since it is being cooled from below, the air resists vertical motion and a series of stratus clouds are produced.

mass. The high cirrus clouds can be seen hundreds of miles in advance of the warm front.

Unstable warm air (Fig. 11–8B) unleashes violent weather when it is lifted along the incline of a warm frontal surface. In addition to the stratus-type clouds that develop, turbulence and ascending air currents in the rising, unstable warm air produce cumulonimbus clouds in advance of the warm front. Heavy downpours cascade from the cumulonimbus clouds. If cirrocumulus—the mackerel-sky clouds—appear among the high clouds, the warm air aloft is very unstable. When the altocumulus and altostratus layers become very dense, rain or snow may fall from these middle-family clouds.

Warm fronts are relatively slow-moving; usual speeds range from 12 to 15 knots, and thus warm fronts tend to remain over a region for longer periods than cold fronts. Warm fronts cover a broad area; aloft they extend for hundreds of miles in advance of the front line at the ground. Typical cloud sequences starting with

the high cirrus can, at times, be seen as far as 1000 miles (1600 km) in advance of the front's contact with the surface. Cirrus clouds are followed in succession by cirrostratus, altostratus, nimbostratus, and stratus clouds. Pressure falls as a warm front advances. As the front passes, the sky clears, the temperature increases, and the barometer rises. The wind shifts in a clockwise direction in the Northern Hemisphere, usually from southeast to southwest.

OCCLUDED FRONTS

The word **occlusion** means to shut something in or out. An **occluded front** develops when a cold front overtakes a warm front and lifts the warm air mass behind the warm front completely off the ground. The warm air mass is literally caught in the middle between two cold air masses, and it is squeezed sharply upward. With respect to what happens when the fronts collide—either the warm front or the cold front is lifted off the

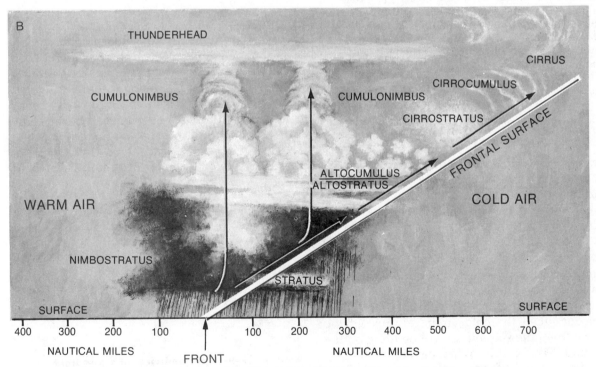

FIGURE 11–8B. The warm air is unstable; the additional lift produced by the flow of heat from the ground produces the cumulus-type clouds.

Vancouver Island is the largest island on the Pacific Coast of North America (Fig. 11–6). The city of Vancouver, which lies opposite Vancouver Island, covers about 45 square miles (117 square kilometers) on the southern shore of Burrard Inlet. The lone tree in this photo is Siwash Rock at the entrance to Burrard Inlet. The evergreen-covered slopes, with stratocumulus above, at the left of the inlet is the suburb of West Vancouver. The Fraser River empties its waters into Burrard Inlet. (National Film Board of Canada)

ground, depending on whether the cold air mass behind the cold front is colder or warmer than the cold air mass in advance of the warm front—the front that remains at the surface is the occluded front.

Figure 11–9 is a plan view that shows a cold front approaching a warm front prior to occlusion. A vertical cross section of this same situation is depicted in Figure 11–10A. If the cold air behind the cold front is colder—and thus heavier—than the cold air in advance of the warm front, the warm front will be lifted off the ground (Fig. 11–10B). When the cold front hugs the ground and lifts the warm air mass behind the warm front as well as the warm front, the

occlusion is called a **cold-front occlusion.** In Figure 11–10C, the cold air ahead of the warm front is colder—and thus heavier—than the cold air behind the cold front. The occlusion illustrated in Figure 11–10C is called a **warm-front occlusion** because the cold-front surface is lifted aloft.

Precipitation occurs on both sides of an occluded front. The cloud and weather sequence along an occluded front is that associated with warm and cold fronts, which have been described in the prior sections of this chapter. Occluded systems combine the cold front's narrow band of violent weather and the warm front's widespread precipitation. The plan view of the occlusion process is illustrated in Figure 11–11. Note the isobars are somewhat V-shaped at all the fronts. Along an occluded front, the warm air aloft—together with the two cold air masses in contact with one another on the ground—produces an area of cloudiness, precipitation, and localized thunderstorms.

STATIONARY FRONTS

A **stationary front** results when there is no forward motion along a line of transition between two different air masses. Under these conditions, a front can remain over an area for an extended period. For as long as the front persists, it may produce continuous rain, drizzle, and fog. The weather conditions along a stationary front are similar to those of a warm front. If the front becomes diffuse, the process of frontolysis is occurring.

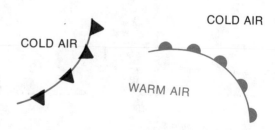

FIGURE 11–9 This plan view of a cold front approaching a warm front pictures the situation from the point of view of an observer aloft.

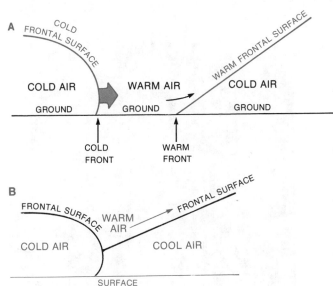

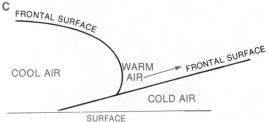

FIGURE 11–10   A. The vertical cross section below depicts the situation prior to occlusion. B. The cold front hugs the ground and produces a cold-front type occlusion. C. When the warm front maintains contact with the surface, the occlusion is referred to as a *warm-front occlusion*.

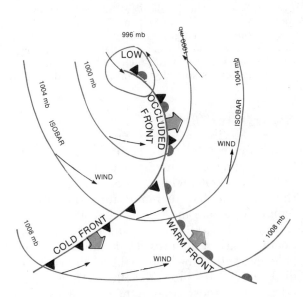

FIGURE 11–11   This plan view of an occlusion is the way in which the condition is depicted on a weather map.

ROUND-THE-WORLD REVIEW

Air masses play a significant role in the spatial distribution of the heat and moisture they carry from their source regions. In addition, the consistent and pervasive modification and interaction of air masses, which occur through exchanges of energy, leads to new and constantly evolving patterns of temperature and moisture within air masses. It is through the dynamics of air-mass formation and evolution that the normal and average ranges of the climatic elements are established in a region. We recognize the pattern of climatic elements thus produced as the region's climate and document the average pattern in climatological maps. The material that follows is a continent-by-continent review of the average climatic elements that prevail as a result of the pattern of air masses that flow and collide over each of the landmasses.

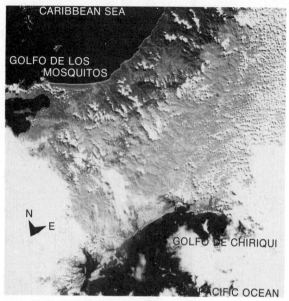

Both photos—the satellite view and the view of the 882-foot- (269-meter-) long container ship passing through the Gaillard Cut of the Panama Canal—show that wooded hills and low mountains cover most of Panama. The upper gulf is an extension of the Caribbean Sea; the lower cloud-covered gulf is an extension of the Pacific Ocean. Most of Panama has a hot climate with almost no seasonal changes in temperature. The Caribbean side of the country receives about 130 inches (330 centimeters) of rain per year; 68 inches (173 centimeters) falls annually on the Pacific side of Panama. (NOAA; Panama Canal Co.)

NORTH AMERICA

Along the coasts of northern Alaska, western Canada, and the northwestern part of the United States, moderate temperatures prevail during the summer. These moderate summertime temperatures are the result of the Pacific high shifting poleward; compare the position of the Pacific high in Figures 11–1 and 11–2. The poleward shift of the high off the western coast of the United States allows a southward flow of cool mP air along these areas of the coast. This same mP air produces a narrow strip of coastal area— from the Aleutian Peninsula to northern California west of the mountain crests—where annual precipitation is more than 40 inches (102 cm) and exceeds 100 inches (254 cm) on the coast of British Columbia.

These moderate mid-summer Pacific coastal temperatures are in sharp contrast to those prevailing in the interior of the continent east of the mountains. During the summer, mT-Gulf-and-Atlantic air masses are the dominant parcels of air over much of the United States east of the Rockies. There is a prevailing sea-to-land pressure gradient during the summer that allows mT air to penetrate deep into the continental area and often into Canada. The humid and hot summer air that characterizes so much of the mid-continental area is produced by mT air.

Because of the high humidity of this summer-time mT air, only a small amount of vertical lifting is needed to produce shower activity. Summer thunderstorms, even in the arid southwestern United States, appear to be related to westward thrusts of mT-Gulf air. In some instances, mT-Gulf air even reaches southern California, where it produces summer rainfall, which is rare for the region.

The mid-continental area of North America, which has no major highlands between the arctic plains and the Gulf of Mexico, has marked seasonal contrasts in temperature. The contrasts are produced by the different air masses that dominate the region during the year. The summer-time domination by mT air (Fig. 11–2) gives way to a wintertime domination by polar air (Fig. 11–1). The cP air, which enters the United States between the Rocky Mountains and the Great Lakes, can move rapidly southward because of the openness of the continent.

The low wintertime temperatures throughout the mid continent—to the subtropical latitudes of the east-central United States—are due to outbreaks of cP air. These polar outbreaks can even move across the Gulf and penetrate into Mexico and Central America. Of course, by the time the cP air moves as far south as Mexico, it has been modified considerably and is generally classified as cool mTk air, which produces heavy convective precipitation.

The prevailing westerly wind flow over the United States carries the mid-continental type of climate, with its marked temperature contrasts, eastward. This eastward movement of mid-continent conditions makes the region of maritime climate along the Atlantic Ocean very narrow. Nevertheless, the wintertime temperatures in the eastern coastal areas tend to be mild when contrasted with the severe conditions found in the mid continent. Intrusions of wintertime mP-Atlantic air into New England do, however, produce surface temperatures near freezing or below. And although cP air prevails in the central eastern United States during the colder months, mT air is associated with mild humid winter days along coastal areas in the South.

In the West Indian region, temperature conditions are subtropical. The climatic regions in Mexico and Central America are controlled by elevation, which greatly modifies the air masses flowing over these regions. The temperature conditions range from subtropical at low elevations to temperate at high elevations. Rainfall in the West Indies, southern Mexico, and Central America is generally abundant, but it is spotty and varies widely from the windward to the leeward sides of the mountains.

SOUTH AMERICA

The widest and largest portion of the South American landmass lies within the tropical zone from the Equator to 23°30′S (Fig. 11–12). The tropical portion of the continent, both north and

south of the Equator, is dominated by mT air masses, which make the region—except in the high Andes—warm throughout the year. The remaining southern portion, which spans the mid latitudes, is too narrow to develop the continental type of climate with hot summers and cold winters; it, in fact, has cool summers and mild winters.

The cool Humboldt Current moves northward to the Equator along the western coast of South America. The prevailing winds coupled with the cold ocean current exert a strong cooling influence over the western coastal regions from the southern tip of the continent to the Equator. Along the eastern coastline, the warm, southerly-moving Brazilian Current and the prevailing-wind system produce a warming effect from the Equator to the approximate latitude of 45°S, where the Falkland Current takes over and moves along the southern coast of Argentina.

There are four regions—the Amazon River Valley, the Guyana coastal lands, the northwestern coasts of Colombia and Ecuador, and a small portion of Chile south of Santiago—where rainfall averages more than 60 inches (150 cm) per year (Fig. 11–13). The sharply contrasted dry and wet seasons in the north are related to the trade winds. The Atlantic's southeast trade winds, for example, bring rains to the continent and drop the maximum amounts of rainfall in the extreme

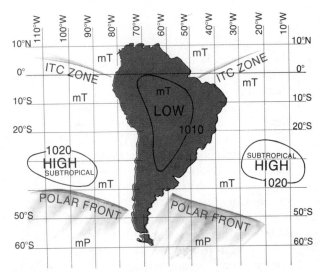

FIGURE 11–12 The South American continent is dominated by maritime tropical air masses.

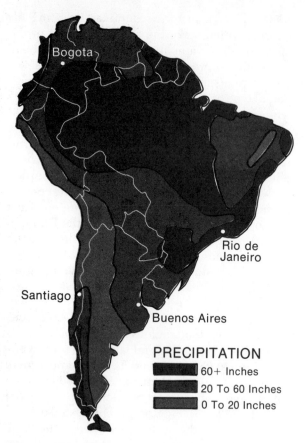

PRECIPITATION

■ 60+ Inches

■ 20 To 60 Inches

■ 0 To 20 Inches

FIGURE 11–13 Rain-bearing winds sweep into South America from the Atlantic and Pacific oceans. The tints used on the map show the pattern of precipitation that covers the continent.

west, where they ascend the Andean slopes. In the South, it is the Pacific's moist northwesterlies that keep Chile's central valley well watered in winter.

The desert areas on the west coast of South America, from the Equator to Santiago (Fig. 11–13), are due primarily to the cold Humboldt Current, which cools the surface air that blows onto shore. When the cool, moist ocean air flows inland, it absorbs heat from the land with a consequent decrease in its relative humidity. Dew points in this air are not reached until it has been lifted to high elevations in the Andes. In the prevailing northwesterly belt of southern Chile, the Andes wring moisture from the air and cast a great rain shadow over southern Argentina, which is in the lee of the mountains (Fig. 11–13).

EUROPE

Most of Europe is above latitude 40°N, which places it north of Philadelphia and in the same latitude range of Canada. There are no extensive north-south mountain systems, and the general east-west direction of European highlands allows westerly winds that blow off the Atlantic to warm the continent far inland during winter. One-third of Norway lies north of the Arctic Circle, but most of its coast remains free of ice throughout the year because of the warming effect of an ocean current, the North Atlantic Drift.

The weather and, thus, climate of western and central Europe is dominated throughout the year by mP air that moves in from the west (Figs. 11–1, 2). The conditions in the maritime west change gradually as you move eastward: the winters become longer and colder, and the summers become shorter and hotter. Glasgow (55°53′N, 4°15′W), in Scotland, for example, has average temperatures of 38°F (3°C) in January and 58°F (14°C) in July; Moscow (55°45′N, 37°35′E), at the same general latitude, has an average January temperature of 14°F (−10°C) and an average July temperature of 65°F (18°C).

The European precipitation pattern also exhibits a shift in a west-east direction. It, in fact, follows a diminishing trend as you move eastward, except in the elevated Alpine and Caucasus regions. Glasgow's annual precipitation, for example, is 38 inches (97 cm) and Moscow's is a mere 21 inches (53 cm) per year. In some areas of the maritime west, the precipitation exceeds 60 inches (150 cm) per year.

The west-to-east change in the spatial distribution of climatic elements is primarily controlled by the mP flow, but not exclusively so. Continental polar air masses have some input because the weather of Russia and central Europe is affected by cP air. The wintertime cP air that intrudes into central Europe (Fig. 11–1), however, originates in western Russia and has temperatures that are less severe than cP air from Siberia. The Russian summertime cP air is comparatively warm and humid at the surface and does not generally penetrate into central Europe.

The countries of southern Europe have mild winters and hot, dry summers. The mT air masses that intrude over the European mainland are cooler, drier, and more stable than United States mT-Gulf air. The mT air that moves into western Europe makes its chief impact in the cooler seasons and does not play a significant role.

Continental tropical air masses are found over Europe mainly during the summer. Since European cT air originates over the Sahara Desert, it is extremely hot and dry. Unmodified cT air from Africa is much too arid to produce significant rainfall.

ASIA

The vast landmass of Asia provides ample opportunity for continental conditions. A cold, high-pressure region develops northeast of the Himalayas during the winter (Fig. 11–1). The hot, low-pressure area, which develops during the summer, stretches west-to-east across northern India. The well-known monsoons described on pages 345–348 develop as a result of these temperature and pressure patterns. The summertime flow of air over southern, southeastern, and eastern Asia is dominated by the mT air of the monsoon.

The Asiatic cP air originates in the cold high centered over eastern Siberia (Fig. 11–1). Over its source regions, this cP air is very cold and very dry. When wintertime air from Siberia surges seaward, a number of modifications occur. As this cP air moves over extensive highlands and then descends to the plains of China, its temperature is increased. The arid winters of northern China (Mukden and Tientsin, Table 11–1), including Manchuria, are caused by this flow of dry cP air.

When the wintertime cP air surges from the mainland toward the islands of Japan, it crosses an open sea and picks up large amounts of heat and moisture, which cause the air to become unstable. As a result of these modifications, the cP air drops a lot of snow on the windward western slopes of the Japanese mountains. The wintertime precipitation in Korea and subtropical southern China (south of the Yangtze River, Shanghai, and Canton, Table 11–1) is also produced by sea-modified cP air.

During the wintertime, the Pacific polar front runs eastward from north of the Philippines in the vicinity of the northern section of Southeast

A gray, overcast sky hangs over Saint-Malo (48°39′N, 21°W) in the Golfe de Saint-Malo along the Brittany-Normandy coast of France. The westerly winds, that blow in from the Atlantic, give this coastal region a rainy climate with mild winters and cool summers. (French Embassy)

TABLE 11-1   CLIMATIC ELEMENTS FOR SELECTED STATIONS IN ASIA

STATION	LATI-TUDE	LONGI-TUDE	ELE-VATION (feet)	EXTREME TEMP. Max°F	EXTREME TEMP. Min°F	J	F	M	A	M	J	J	A	S	O	N	D	An-nual
China:																		
Canton	23°10'N	113°20'E	59	101	31	0.9	1.9	4.2	6.8	10.6	10.6	8.1	8.5	6.5	3.4	1.2	0.9	63.6
Chunking	29°30'N	106°33'E	855	111	28	0.7	6.8	1.5	3.8	5.7	7.1	5.6	4.7	5.8	4.3	1.9	0.8	42.9
Mukden	41°47'N	123°24'E	138	103	-28	0.2	0.2	0.7	1.2	2.6	3.8	7.0	6.3	2.9	1.7	0.9	0.4	28.2
Shanghai	31°12'N	121°26'E	16	104	10	1.9	2.4	3.3	3.6	7.0	7.0	5.8	5.5	5.2	2.9	2.1	1.5	45.0
Tientsin	39°10'N	117°10'E	13	109	-3	0.2	0.1	0.4	0.5	1.1	2.4	7.6	6.0	1.7	0.6	0.4	0.2	21.0
Korea:																		
Pusan	35°10'N	129°7'E	6	97	7	1.7	1.4	2.7	5.5	5.2	7.9	11.6	5.1	6.8	2.9	1.6	1.2	53.6
P'yang-yang	39°1'N	125°49'E	94	100	-19	0.6	0.4	1.0	1.8	2.6	3.0	9.3	9.0	4.4	1.8	1.6	0.8	36.4
Vietnam:																		
Hoh Chi Min City	10°49'N	106°39'E	33	104	57	0.6	0.1	0.5	1.7	8.7	13.0	12.4	10.6	13.2	10.6	4.5	2.2	78.1
Thailand:																		
Bangkok	13°44'N	100°30'E	53	104	50	0.2	1.1	1.1	2.3	5.2	6.0	6.9	9.2	14.0	9.9	1.8	0.1	57.8
Singapore	1°18'N	103°50'E	33	97	66	9.9	6.8	7.6	7.4	6.8	6.8	6.7	7.7	7.0	8.2	10.0	10.1	95.0
Japan:																		
Kushiro	43°2'N	144°12'E	315	87	-19	1.8	1.4	2.8	3.6	3.8	4.1	4.4	4.9	6.6	4.0	3.1	2.0	42.9
Nagasaki	32°44'N	129°53'E	436	98	22	2.8	3.3	4.9	7.3	6.7	12.3	10.1	6.9	9.8	4.5	3.7	3.2	75.5

Asia (Fig. 11–1). The air that flows over the two large peninsulas of Southeast Asia is dominated by the northeast trade winds. Wintertime precipitation is slight except along the windward eastern sides of these peninsulas.

AFRICA

This continent lies almost completely within the tropical zone (Fig. 11–14), which ranks it as one of the warmest of the continents. Large regions near the Equator are hot all year-round. Any great variations of temperature within an area are produced primarily by altitude.

The cool Benguela Current moves along Africa's western coast from its southern tip to the Equator. The warm tropical currents of the Indian Ocean move south along most of its eastern coast. These ocean currents—the cool flow along the southern portion of the western coast and the

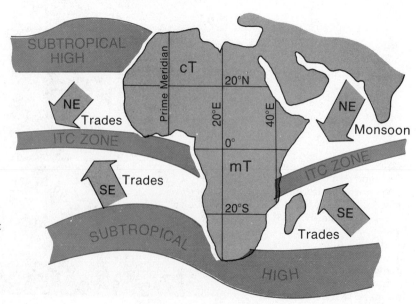

FIGURE 11–14 Africa lies almost completely within the tropical zone and it is completely within the belt of the northeasterly and southeasterly trade winds.

warm eastern-coast flow—create climatic conditions that closely parallel those produced by the same kind of flow around South America.

Figure 11–14 indicates that cT air dominates northern Africa and mT air is the dominant flow over the southern portion of the continent. In fact, this figure shows that Africa is completely within the belts of the northeasterly and southeasterly trade winds.

The Atlas Mountains in the northwestern portion of the continent are the only barrier in the path of the northeast trades. The orographic lifting of the trades by the Atlas Mountains produces some moderate rainfall in the region. Other than this small area of moderate rainfall—extending along the northwestern coast from Casablanca northeast to Algiers and on to Tunis (Table 11–2)—northern Africa is a very arid desert as shown by the data for the station at Sabhah, Libya (Table 11–2). Sahara-desert conditions extend from the Atlantic to the Red Sea and from the Mediterranean southward to the Sahel (described on page 348).

South of the Sahara, mT air dominates the landmass and precipitation increases as you move southward. In fact, the rainfall becomes abundant to heavy along the west coast from the Guinea coast (Conakry, Table 11–2) south, with one exception. The marked increase in precipitation extends eastward from the west coast to the Lake Victoria region and beyond, again, with one exception. The exception in the east is along the eastern portion of the middle region of the continent—along the coast of Somalia (Berbera, Table 11–2)—where precipitation decreases to less than 10 inches (25.4 cm). The other exception to this precipitation pattern is in the southwest along the arid coasts of Angola, Southwest Africa, and South Africa, where desert conditions prevail (Keetmanshoop, Table 11–2).

AUSTRALIA

During August, which is winter in the Southern Hemisphere, the subtropical high-pressure belt crosses the interior of Australia (Fig. 11–15A). The continent is dominated by cT air at this time, and all except the southern parts of the landmass are dry. In the southern summer—during February, for example—the high-pressure belt borders the southern fringe of the continent and conditions are still dry over the southern and western areas (Fig. 11–15B).

The highest levels of precipitation are found along the northern, eastern, and southern coasts from Darwin to Adelaide (Table 11–3). Moderate rainfall also occurs on the southwest coast in a small area around Perth. In the south, the wintertime precipitation is of the cyclonic type—it develops within a low associated with a weather front. The heavy summertime rains of the north are of monsoon origin; the monsoon flow develops in response to the low-pressure zone over the northern coast at this time (Fig. 11–15B). The precipitation that falls along the eastern coast develops, in part, from the orographic uplift of southeast trades by the coastal highlands. Along the outer border of the eastern precipitation strip, the rainfall averages more than 40 inches (100 cm); Brisbane (Table 11–3) is in this outer strip.

TABLE 11–2   CLIMATIC ELEMENTS FOR SELECTED STATIONS IN AFRICA

STATION	LATITUDE	LONGITUDE	ELEVATION (feet)	EXTREME TEMPERATURE Max°F	EXTREME TEMPERATURE Min°F	PRECIPITATION (Annual Inches)
Casablanca (Morocco)	33°35′N	7°39′W	164	110	31	15.9
Tunis (Tunisia)	36°47′N	10°12′E	217	118	30	16.5
Sabhah (Libya)	27°1′N	14°26′E	1457	120	24	0.3
Conakry (Guinea)	9°31′N	13°43′W	23	96	63	169.0
Berbera (Somalia)	10°26′N	45°2′E	45	117	58	2.0
Keetmanshoop (South West Africa)	26°36′S	18°8′E	3295	108	26	5.2

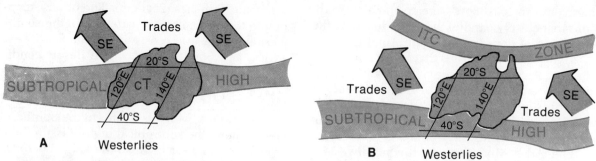

FIGURE 11-5 A, The subtropical belt crosses the interior of Australia during August. B, By February, the subtropical high has migrated south and is at the southern fringe of the continent during the Southern Hemisphere's summer.

The Tropic of Capricorn cuts across Australia. Thus, below-freezing temperatures are found in only a small part of the continent, primarily in the south at high elevations. Canberra is an example (Table 11-3). The lowest temperature ever recorded in Australia was −8°F (−22°C) in the mountains of New South Wales. In the arid interior, maximum temperatures are very high; the highest ever recorded was 127°F (53°C) at Cloncurry (Table 11-3).

TABLE 11-3 CLIMATIC ELEMENTS FOR SELECTED STATIONS IN AUSTRALIA

STATION	LATI- TUDE	LONGI- TUDE	ELEVATION (feet)	EXTREME TEMPERATURE Max°F	Min°F	PRECIPITATION (Annual Inches)
Adelaide	34°57′S	138°32′E	20	118	32	21.1
Alice Springs	23°48′S	133°53′E	1791	111	19	9.9
Brisbane	27°25′S	153°5′E	17	110	35	44.7
Broome	17°57′S	122°13′E	56	113	40	22.9
Canberra	35°18′S	149°11E	1886	109	14	23.0
Cloncurry	20°40′S	140°30′E	622	127	35	18.0
Melbourne	37°49′S	144°58′E	115	114	27	10.1

THE LANGUAGE OF PHYSICAL GEOGRAPHY

A physical geographer uses a technical vocabulary to make explanations. Review your understanding of the vocabulary used to develop the concepts in this chapter.
Among the important words and terms used are:

air mass	continental tropical	maritime polar	source region
arctic front	front	maritime tropical	stationary front
cold air mass	frontal surface	occluded front	tropical air masses
cold front	frontogenesis	occlusion	warm air mass
cold-front occlusion	frontolysis	polar air masses	warm front
continental polar	low-latitude fronts	polar front	warm-front occlusion

SELF-TESTING QUIZ

1. Under what conditions does climatic monotony prevail?
2. List the locations that are conducive to air-mass development.
3. What factor determines the degree and depth to which an air mass achieves temperature and moisture equilibrium with the surface it overlies?

4. Discuss the pertinent conditions under which arctic air masses develop.

5. Trace the location of the arctic front during January.

6. At what time of year are polar fronts situated at the lowest latitudes?

7. How is the ITCZ formed?

8. Discuss the various air-mass classification systems currently in use.

9. Compare the temperature and humidity characteristics of cT and mA air.

10. Describe the modifications that take place in cP air as it moves out of Canada over the Great Lakes.

11. What kind of air masses are involved in the production of the coastal fogs in the San Francisco region?

12. Discuss the various changes that can occur when air of differing physical characteristics intermingles along a frontal surface.

13. How are fronts represented on a surface map?

14. List some characteristics of fronts.

15. Describe the sequence of events that might occur as a cold front approaches your area.

16. Contrast a cold-front occlusion with that of a warm-front occlusion.

17. Why do warm temperatures prevail along the coasts of northern Alaska, western Canada, and the northwestern part of the United States during the summer?

18. How do you account for the fact that the mid-continent area of North America has marked seasonal contrasts in temperature?

19. What accounts for the sharply contrasted dry and wet seasons in the northern part of South America?

20. Why does most of Norway's coast remain free of ice throughout the year although one-third of Norway lies north of the Arctic Circle?

IN REVIEW
AIR-MASS CLIMATOLOGY

I. AIR-MASS CHARACTERISTICS
  A. Source Regions and Fronts
    1. Arctic Air Masses
    2. Arctic Fronts
    3. Polar Air Masses
    4. Polar Fronts
    5. Tropical Air Masses
    6. Low-Latitude Fronts
  B. Air-Mass Classification
    1. The Basic System
    2. Temperature-Humidity Properties

II. AIR-MASS MODIFICATION AND ASSOCIATED CLIMATIC ELEMENTS
  A. cP Modification
  B. mP Modification
  C. cT Modification
  D. mT Modification

III. FRONTAL SYSTEMS AND ASSOCIATED CLIMATIC ELEMENTS
  A. Front Symbols
  B. Characteristics of Fronts
  C. Cold Fronts
  D. Warm Fronts
  E. Occluded Fronts
  F. Stationary Fronts

IV. ROUND-THE-WORLD REVIEW
  A. North America
  B. South America
  C. Europe
  D. Asia
  E. Africa
  F. Australia

12 *Climate classification and analysis*

Scientific classification systems are developed and designed to bring order to large quantities of information. Once a classification scheme is understood, it should facilitate reference and communication. Generally, there are two broad approaches to climate classification: genetic and empirical.

Genetic classifications organize climates according to their causative factors. The Greeks, for example, recognized the importance of latitude. On the basis of this perception, the Earth was eventually divided into five climatic regions based on insolation: one tropical zone, two temperate zones, and two polar zones. The tropical zone is bounded by the northern and southern limit of the direct rays of the Sun—the Tropic of Cancer at latitude 23.5°N and the

Tropic of Capricorn at latitude 23.5°S. In the Northern Hemisphere, the line of demarcation between the temperate and the polar zones is the Arctic Circle at latitude 66.5°N. The Antarctic Circle at latitude 66.5°S separates the temperate and polar zones in the Southern Hemisphere. Differences in insolation obviously produce important temperature differences in these climatic zones, but this simple classification ignores a number of other causative factors such as precipitation, oceans and continents, mountain barriers, and altitude. Any one of these factors could be used as a basis for a genetic classification of climates.

The system of **empirical classification** does not focus on causative factors; rather, the emphasis is on observed characteristics. Vegeta-

We are at Suruga Bay, about 70 miles (110 kilometers) southwest of Tokyo, on the eastern coast of the island of Honshu. Mount Fuji, about 20 miles (30 kilometers) in the distance, seems to be close enough to touch. This area of Honshu has warm, humid summers and mild winters; it is classified as Cfa. Two ocean currents influence Japan's climate: the Japan Current and the Oyashio Current. The warm Japan Current, which starts northeast of the Philippines, flows northward along Japan's southern and eastern coasts and makes southern Japan warmer than places at the same latitude on the Asiatic mainland. The cold Oyashio Current flows south along the western coasts of Hokkaido and northern Honshu. (Japanese Govt.)

tion, animals, and soils are some of the characteristics that can form the basis of an empirical classification of climates. Natural vegetation, of course, integrates the effects of climate better than any other mechanism available to us, and it is the basis of many empirical classification schemes.

SYSTEMS OF CLASSIFICATION

Different combinations of climatic elements produce a large number of climatic types. The search for order in the world climatic pattern is an exercise in human ingenuity. Various researchers in the field have used the climatic elements singly and in various combinations to establish criteria for climatic types.

Outstanding among the modern attempts at climatic classification are those of W. P. Köppen (1900), C. Thornthwaite (1931), H. Flohn (1950), E. Kupfer (1954), B. Allisow (1956), L. Holdridge (1959), and M. Hendl (1960). Each of these systems possesses genuine merit. Almost all the schemes have subdivisions and boundaries partly based on the climatic elements of temperature and rainfall.

The classification system devised by Waldimir Peter Köppen is the most used and most popular worldwide system among physical geographers. The Köppen system is an empirical-genetic climatic scheme that relates climate to vegetation but also provides an objective numerical basis for defining climatic types in terms of climatic elements. The first Köppen system, devised in 1900, was based largely on vegetative zones; since the 1918 revision, the system has devoted greater attention to temperature, rainfall, and their seasonal characteristics.

The Thornthwaite system uses four factors to describe the climatic situation of a particular location: moisture index, index of thermal efficiency, the summer concentration of thermal efficiency, and an index expressing the seasonal variation of effective moisture. Primary emphasis in Thornthwaite's classification is assigned to the moisture index. As with Köppen's classification, each of Thornthwaite's four factors is expressed in terms of a letter. A location's complete climatic classification consists of four letters.

Flohn's scheme of climate classification involves eight zones: inner tropical, outer tropical, subtropical dry, subtropical winter-rain, moist temperate, boreal, subpolar, and high polar. For each of these climatic zones, Flohn indicates the corresponding pressure and wind belts as well as the prevailing winds in the extreme seasons. He relates each of the eight zones to a corresponding climatic type and indicates the typical vegetation forms found in each. Precipitation plays the major role in differentiating among Flohn's eight climatic zones.

THE KÖPPEN SYSTEM

Five major categories are included in the Köppen classification system (Table 12–1). In Figures 12–1 and 12–2, areas of each continent are classified according to Köppen's five major categories. Each major category is designated by a capital letter: Category A includes equatorial and tropical forest climates, which are hot during all seasons; the B category is reserved for dry climates of the arid zone; category C is used for warm, temperate, rainy climates of the mainly broadleafed forest zone; the D category designates colder temperate forest climates; and category E identifies treeless polar climates.

MAJOR CATEGORY BOUNDARIES

The boundaries of four of these categories are defined by temperature. The fifth category boundary is defined initially by precipitation. The boundaries are as follows:

TABLE 12-1   KÖPPEN'S MAIN CLIMATIC TYPES

CLIMATE CATEGORY	SYMBOL	DESCRIPTION
Tropical	Af	Tropical rain forest. Hot; rainy all seasons.
	Am	Tropical monsoon. Hot; seasonally excessive rainfall.
	Aw	Tropical savanna. Hot; seasonally dry—usually in winter.
Dry	BSh	Tropical steppe. Semiarid and hot.
	BSk	Mid-latitude steppe. Semiarid and cool.
	BWh	Tropical desert. Arid and hot.
	BWk	Mid-latitude desert. Arid and cool.
Warm Temperature	Cfa	Humid subtropical. Mild winter; moist all seasons; long, hot summer.
	Cfb	Marine. Mild winter; moist all seasons; short, cool summer.
	Cfc	Marine. Mild winter; moist all seasons; short, cool summer.
	Csa	Interior Mediterranean. Mild winter; dry, hot summer.
	Csb	Coastal Mediterranean. Mild winter; short, dry, hot summer.
	Cwa	Subtropical monsoon. Mild, dry winter; hot summer.
	Cwb	Tropical upland. Mild, dry winter; short, warm summer.
Snow	Dfa	Humid continental. Severe winter; long, hot summer.
	Dfb	Humid continental. Severe winter; moist all seasons; short, warm summer.
	Dfc	Subarctic. Severe winter; moist all seasons; short, cool summer.
	Dfd	Subarctic. Extremely cold winter; moist all seasons; short summer.
	Dwa	Humid continental. Severe, dry winter; long, hot summer.
	Dwb	Humid continental. Severe, dry winter; warm summer.
	Dwc	Subarctic. Severe, dry winter; short, cool summer.
	Dwd	Subarctic. Extremely cold, dry winter; short, cool summer.
Ice	ET	Tundra. Very short summer.
	EF	Perpetual ice and snow.

Category A: The average monthly temperature is above 64.4°F (18°C).

Category B: The average annual rainfall expressed as R and reported in centimeters is
(1) less than 2T* + 28 when summer is the rainiest season.
(2) less than 2T* + 14 when rainfall is fairly uniform throughout the year.
(3) less than 2T* when winter is the rainiest season.

Category C: The average temperature of the coldest month is in the range of 64.4°F (18°C) to 26.6°F (−3°C); the warmest month must be above 50°F (10°C).

Category D: The average temperature of the coldest month is below 26.6°F (−3°C); the warmest month must be above 50°F (10°C).

Category E: The average monthly temperature is below 50°F (10°C).

*T is the average temperature over the year expressed in degrees Celsius.

SUBGROUP BOUNDARIES

The five major categories of the Köppen system are classified into **climatic subgroups** according to more narrowly defined rainfall and temperature characteristics. The subgroups are applied to the continental areas in Figures 12–1 and 12–2 by the use of additional symbols. Except in the case of the ice climates, the second letter refers to the precipitation regime and the third letter to temperature characteristics.

A-CATEGORY SUBGROUPS

The letters f, m, and w, all used as second letters, refer to rainfall characteristics of the A category as follows:

f Subgroup: The precipitation in the driest month is more than 2.4 inches (60 mm).

m Subgroup: The driest month has less than 2.4 inches (60 mm), but the precipitation during the driest month is equal to or greater than 3.94−r/25, where r is the average annual precipitation in inches.

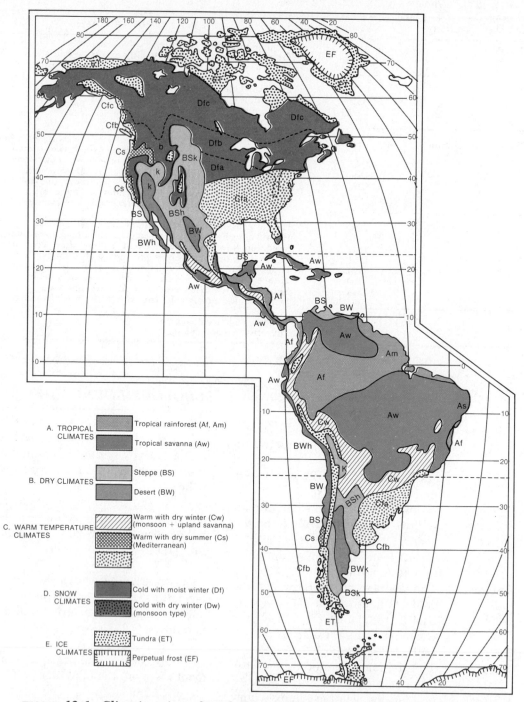

FIGURE 12–1 Climatic regions of North and South America according to the Köppen system (see Table 12–1 for an interpretation of the symbols used).

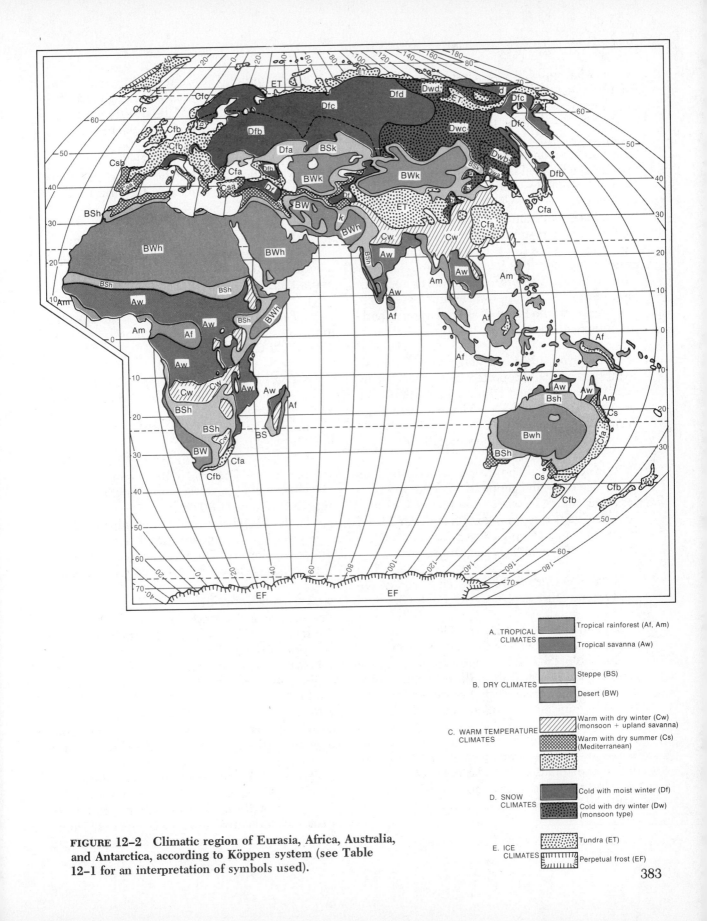

FIGURE 12–2 Climatic region of Eurasia, Africa, Australia, and Antarctica, according to Köppen system (see Table 12–1 for an interpretation of symbols used).

Legend:

A. TROPICAL CLIMATES
- Tropical rainforest (Af, Am)
- Tropical savanna (Aw)

B. DRY CLIMATES
- Steppe (BS)
- Desert (BW)

C. WARM TEMPERATURE CLIMATES
- Warm with dry winter (Cw) (monsoon + upland savanna)
- Warm with dry summer (Cs) (Mediterranean)

D. SNOW CLIMATES
- Cold with moist winter (Df)
- Cold with dry winter (Dw) (monsoon type)

E. ICE CLIMATES
- Tundra (ET)
- Perpetual frost (EF)

w Subgroup: The precipitation in the driest month is less than 3.94−r/25, where r is the average annual precipitation in inches.

Table 12–1 shows that an Af climatic type produces a tropical rain forest. The Am subgroup is a tropical-monsoon climate. And an Aw type produces a tropical savanna, that is, a vegetation that is largely grassland with a few trees.

B-CATEGORY SUBGROUPS

In the Köppen system the capital letters S and W apply only to the dry B climates and designate rainfall regimes. The lower case letters h and k are used as third letters to designate temperature boundaries, or **temperature subclasses.** The boundaries of the S and W as well as the h and k subgroups are as follows:

S Subgroup: The average yearly rainfall, expressed as R and reported in centimeters, exceeds T* + 14 in areas with summer rainfall. R is between T* + 14 and T* + 7 in areas where rain is spread over the year. R is between T + 7 and T in areas with mainly winter rainfall.

W Subgroup: The average yearly rainfall, expressed as R and reported in centimeters, is less than T* + 14 where summer is the rainy season. The yearly average of R is under T* + 7 where the rains are spread over the year, and less than T* where winter is the rain season.

h-Temperature Subclass: The average annual temperature is above 64.4°F (18°C).

k-Temperature Subclass: The average annual temperature is less than 64.4°F (18°C).

In Table 12–1, four combinations of the B category are listed: BSh, BSk, BWh, and BWk. The BS subgroups are listed as steppes and the BW subgroups are classified as deserts.

C-CATEGORY SUBGROUPS

Three lower case letters, *s*, *w*, and *f*, are used to define precipitation boundaries in the C category. Refinements in the temperature boundaries of this category are designated by the letters *a*, *b*, and *c*.

*T is the average temperature over the year expressed in degrees Celsius.

s Subgroup: The wettest month of the year, which occurs during the winter, has more than three times the rainfall of the driest month's average of less than 1.2 inches (3.0 cm).

w Subgroup: The wettest month of the year, which occurs during the summer, has more than ten times the rainfall of the driest winter month.

f Subgroup: This subgroup is moist throughout the year and does not meet the s or w conditions. The driest summer month receives more than 1.2 inches (3.0 cm).

a-Temperature Subclass: The average temperature of the warmest month is above 71.6°F (22°C).

b-Temperature Subclass: The average temperature of each of the four warmest months is 50°F (10°C) or above; the temperature of the warmest month is below 71.6°F (22°C).

c-Temperature Subclass: The average temperature of from one to three months is 50°F (10°C) or above; the temperature of the warmest month is below 71.6°F (22°C).

Table 12–1 lists seven subgroups of the C category. The Cs climates produce broadleafed evergreen forests. The Cw subgroup produces the evergreen forests of the mountain heights.

D-CATEGORY SUBGROUPS

The precipitation subgroups—s, w, and f—have the same boundaries as those of the C category described above. The a-, b-, and c-temperature subclasses also have the same boundaries described for those of the C category. A fourth temperature subclass, *d*, is added for the D category, however. The d-temperature-subclass designation is applied when the average temperature of the coldest month is below −36.4°F (−38°C).

Table 12–1 lists eight subgroups in the D category. The milder D climates produce deciduous, broadleafed forests. Needletree forests are associated with the colder D climates. The D climates are only found on the great continents in the Northern Hemisphere.

The small village of Alkali Lake (51°47′N, 122°14′W) is on the Fraser Plateau in British Columbia, Canada. Its climate is classified as Dfc in the Köppen system. (National Film Board of Canada)

E-CATEGORY SUBGROUPS

In the Köppen classification, the capital letters *T* and *F* are applied only to the E climates. The boundaries of the T and F subgroups are as follows:

T Subgroup: The average temperature of the warmest month is between 50° and 32°F (10° and 0°C).

F Subgroup: The average temperature of the warmest month is 32°F (0°C) and below.

ANALYSIS BY CLIMOGRAPH

A graphic presentation of climatic data—which is almost always limited to a plot of the simultaneous variations of two climatic elements, precipitation and temperature, through an annual cycle —is called a **climograph.** It is, in a sense, a condensed version of the main features of each climatic region.

The graph consists of 12 vertical columns. Each column represents one month; the labeling for the horizontal scale starts at the left with January and follows in sequence so that December is the column at the extreme right. This sequencing of columns has a special meaning for each hemisphere: In the Northern Hemisphere, the middle columns represent summer and the outside columns are a visual plot of winter conditions. In the Southern Hemisphere, the reverse is true, that is, middle columns represent winter and outside columns give the summer conditions.

PLOTTING THE DATA

A temperature scale is placed along the left margin of the graph. The average monthly temperatures for the region are plotted on the basis of this scale. A dot representing the average temperature for each month is placed at the appropriate level on the scale in the center of the month, and the dots are connected by a line.

A precipitation scale is placed along the right margin of the climograph. The average rainfall for each month is plotted by drawing a vertical bar. The height of the bar is controlled by the precipitation scale. The completed bars give an instant picture of the precipitation during any one month as well as distribution of precipitation throughout the year.

The climograph for Montreal (latitude 45°30′N, longitude 73°34′W) is shown in Figure 12–3. The elevation of the station at which the

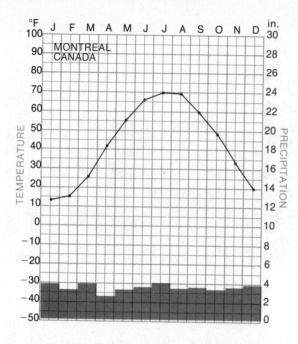

FIGURE 12–3 This climograph is for Montreal, Canada, which is classified as Dfb in the Köppen system. The precipitation scale for use with the bar graph is at the right. The temperature scale is at the left margin of the climograph. The dots connected by the lines represent the average temperature for each month.

data were collected is 187 feet (57 m) above sea level.

The average daily maximum temperature in Montreal during January is 21°F (−6°C), and the January minimum is 6°F (−14°C); the average temperature for January plotted on the climograph is 13.5°F (−10°C). This Canadian city in Quebec has average daily maximum and minimum temperatures during April of 50°F (10°C) and 33°F (0.56°C) respectively; April's average is 41.5°F (5.3°C). July's average daily maximum is 78°F (25.5°C) and minimum is 61°F (16°C); the July average is plotted as 69.6°F (20.8°C) on the climograph. The fall month of October, with an average maximum of 54°F (12.2°C) and an average minimum of 40°F (4.4°C), has a monthly average of 47°F (8.3°C).

ANALYZING DATA

Examine the Montreal Climograph in Figure 12–3 and evaluate it using the Köppen system. The average temperature of the coldest month, January, is 13.5°F (−10.3°C); the warmest month, July, has an average of 69.5°F (20.8°C). These temperature boundaries fit Category D of the Köppen system (page 381) in which the average

Fort Garry on the Winnipeg River in Manitoba is in an area that is classified as Dfb in the Köppen system. Its climatic conditions are similar to those of Montreal, which is east of this location in Quebec-Province. (National Film Board of Canada)

temperature of the coldest month is below 26.6°F (−3°) and the warmest month must be above 50°F (10°C).

The average yearly precipitation in Montreal is 40.8 inches (103.6 cm). The bar graphs on the climograph start with a January average of 3.8 inches (9.65 cm). All succeeding months, except April, are above the 3-inch (7.6-cm) mark (Fig. 12–3). Even April has a precipitation of 2.6 inches (6.6 cm). Montreal is, in fact, moist throughout the year and fits the f subgroup of the Köppen system. We have now placed Montreal as a Df climate in the Köppen system.

Let's take a closer look at the temperature boundaries for Montreal (Fig. 12–3). The four warmest months are June, July, August, and September; each of these months has an average temperature above 50°F (10°C) and the average of the warmest month is below 71.6°F (22°C). According to the Köppen system, this pattern places Montreal in the b-temperature subclass, which gives Montreal a Dfb climate.

A-CATEGORY CLIMOGRAPHS

Climatic data (temperature and precipitation) for Belém (Brazil), Rangoon (Burma), and Timbo (Guinea) are detailed in Table 12–2. Climographs can be constructed from these data; the climograph for Rangoon is shown as Figure 12–4. The average temperature for every month for each of these stations is above 64.4°F (18°C), which classifies these regions as tropical **A climates** according to the Köppen system.

The subgroup possibilities in this category are f, m, and w (Table 12–1). Review the precipitation data for each of these locations. Belém's driest month, November, has rainfall of 3.4 inches (8.6 cm), which places this station in sub-

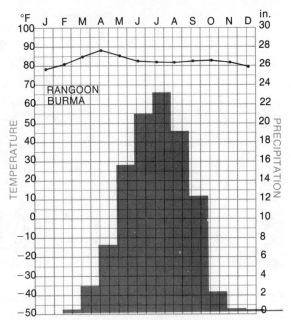

FIGURE 12–4 The climatic elements of temperature and precipitation are plotted in this climograph of Rangoon, Burma. Rangoon is classified as a tropical monsoon, Am climate in the Köppen system.

group f. Rangoon and Timbo, on the basis of their precipitation pattern, cannot qualify for subgroup f.

The driest months in both Rangoon and Timbo have less than 2.4 inches (6.1 cm) of precipitation. Reread the requirements for subgroups m and w on page 381. In order to discriminate among these subgroups, we must determine the value of $3.94 - r/25$ for each station.

The average annual precipitation, r, in Rangoon is 103 inches. Substitute 103 for r in $3.94 - r/25$. The value of $3.94 - 103/25$ is -0.18. Rangoon's driest month, January, has a rainfall of $+0.1$ inch; thus, the precipitation during Ran-

TABLE 12–2   AVERAGE MONTHLY TEMPERATURE AND PRECIPITATION DATA

STATION	LATI-TUDE	LONGI-TUDE		JAN	FEB	MAR	APR	MAY	JUN	JUL	AUG	SEP	OCT	NOV	DEC	CLI-MATE
Belém	1°27′S	48°29′W	°F	77.5	77.0	77.0	78.0	78.5	78.5	78.5	78.5	78.5	79.0	79.5	79	
(Brazil)			in	13.4	16.0	17.2	13.5	11.3	6.9	5.7	5.0	4.7	3.6	3.4	6.9	Af
Rangoon	16°47′N	96°10′E	°F	77	79.5	83.5	86.5	84.5	81	80.5	80.5	81	82	80.5	77.5	
(Burma)			in	0.1	0.2	0.3	2.0	12.1	18.9	22.9	20.8	15.5	7.1	2.7	0.4	Am
Timbo	10°38′N	11°50′W	°F	72	76	81	80	77	73	72	72	72	73	72	71	
(Guinea)			in	0.0	0.0	1.0	2.4	6.4	9.0	12.4	14.7	10.2	6.7	1.3	0.0	Aw

Gabon lies on the Equator and is classified as Af; its climate is hot and humid most of the year. Many regions receive as much as 100 inches (254 centimeters) of rainfall. The lumber being loaded on trucks has been cut from the surrounding tropical rain forest southwest of Makokou near the Ivindo River. (World Bank)

goon's driest month is greater than the value of 3.94 − r/25, which places Rangoon in subgroup m according to the definition on page 381.

The average annual precipitation, r, in Timbo is 64.1 inches. Substitute 64.1 for r in 3.94 − r/25. The value of 3.94 − 64.1/25 is +1.4. There are three dry months with no rainfall, that is, 0.0 inch, in Timbo. Since Timbo's driest month is less than the value of 3.94−r/25, this region is classified as subgroup w.

On the basis of climographs, then, we are able to discriminate among these stations. Belém is an Af climate, that is, a tropical rain forest according to the Köppen classification. The Am or tropical monsoon designation is applied to Rangoon; Timbo falls in the Aw category of tropical savanna (Table 12–1).

B-CATEGORY CLIMOGRAPHS

The data represented in Table 12–3 is for stations in dry, or arid, areas. The data for In Salah is plotted as a climograph in Figure 12–5. The total yearly precipitation for Denver is 14.47 inches (36.75 cm). Kayes in Mali has a total of 29.1 inches (73.9 cm) of rainfall, which is concentrated in the summer months and falls primarily as a result of the summer monsoon that sweeps over the Sahel of Africa. The station at In Salah, Algeria has a meager 0.6 inch (1.5 cm) of rainfall over the year. The total at Albuquerque is 8.24 inches (20.93 cm).

It simplifies matters in this category if we work in degrees Celsius and centimeters. Refer to page 384 for a definition of the precipitation parameters for this category. Denver has an average temperature over the year, T, of 10.1°C. The T for In Salah is 25.3°C. At Kayes, T is 29.4°C. Albuquerque has a T of 13.2°C.

Denver's annual precipitation of 36.75 centimeters is less than 2T+28, where T is 10.1°C. The annual precipitation at Kayes is 73.9 centimeters, which makes it less than 2T+28, where T is 29.4°C; the data in Table 12–3 show summer as the rainiest season and this range of less than 2T+28 conforms to our parameters on page 381. At In Salah the 1.5 centimeters of precipitation is less than 2T. Albuquerque, with a T of 13.2°C, has an annual precipitation of 20.93 centimeters, which is less than 2T.

The dry **B climates** have two subgroups, S and W, based on precipitation (page 384). Denver, with summer rainfall and an annual amount of 36.75 centimeters, which exceeds T+14, is in subgroup S. Kayes also has summer rainfall, which exceeds T+14, and is classified in subgroup S. The region of In Salah, with 1.5 centimeters of annual precipitation, which is less than T, falls in subgroup W. Albuquerque falls in subgroup W, too, with a summer rain season and an r of less than T+14.

The temperature subclasses, h and k, in the B category of the Köppen system are assigned on the basis of the average annual temperature

TABLE 12–3   MEAN MONTHLY TEMPERATURE AND PRECIPITATION DATA

STATION	LATI-TUDE	LONGI-TUDE		JAN	FEB	MAR	APR	MAY	JUN	JUL	AUG	SEP	OCT	NOV	DEC	CLI-MATE
Denver	39°43′N	105°1′W	°F	30.2	32.7	38.5	47.5	56.7	66.6	72.7	71.3	62.8	51.5	39.7	32.4	
(Colorado)			in	0.47	0.59	1.09	1.99	2.41	1.50	1.68	1.38	1.11	1.01	0.64	0.60	BSk
Kayes	14°27′N	11°26′W	°F	77	81	89	94	96	91	84	82	82	85	83	77	
(Mali)			in	0.0	0.0	0.0	0.0	0.6	3.9	8.3	8.3	5.6	1.9	0.3	0.2	BSh
In Salah	27°12′N	2°28′E	°F	56	61	68	77	84	95	99.5	96.5	91	80	66.5	58	
(Algeria)			in	0.1	0.1	<0.1	<0.1	<0.1	<0.1	0.0	0.1	<0.1	<0.1	0.2	0.1	BWh
Albuquerque	35°5′N	106°40′W	°F	34.6	39.5	46.2	54.9	63.8	73.2	77.1	75.1	68.4	56.8	43.9	35.1	
(New Mexico)			in	0.36	0.35	0.41	0.56	0.63	0.59	1.44	1.31	0.91	0.80	0.42	0.46	BWk

being above or below 64.4°F (18°C). Kayes (85°F) and In Salah (77°F) are in the h-temperature subclass; Denver (50°F) and Albuquerque (56°F) are in the k-temperature subclass.

Thus, on the basis of the data for these arid regions, we have worked out and assigned classifications according to the Köppen system. Denver, is BSk, a mid-latitude steppe. Kayes, Mali, is a tropical steppe, BSh, in the African Sahel. The Algerian station at In Salah (Fig. 12–5) is a tropical desert with the designation BWh in the

Köppen classification. Albuquerque, designated BWk, is a mid-latitude desert.

C-CATEGORY CLIMOGRAPHS

Generally, the **C climates**, a subtropical group of mild, temperate, rainy climates, occupy the equatorward side of the middle latitudes. The stations selected for Table 12–4 are in the 29° to 38° latitude band. Within this subtropical belt three types of humid climates are distinguished

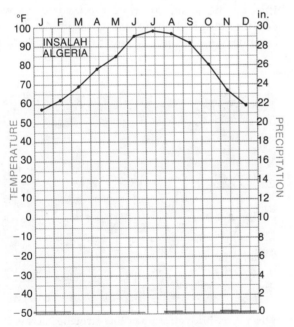

FIGURE 12–5   In Salah is a tropical desert with the designation BWh in the Köppen system. The climograph shows that the total annual precipitation in this desert region is 0.6 inch (1.5 centimeters).

California's Death Valley (36°30′N, 117°W) is lower, hotter, and drier than any other location in the United States. The average annual precipitation is around 2 inches (5 centimeters). Summer temperatures frequently exceed 120° (49°C); and, by any definition, Death Valley qualifies as a tropical desert (BWh). The burros are part of a wild-burro herd in Death Valley. (National Parks Service)

TABLE 12-4   MEAN MONTHLY TEMPERATURE AND PRECIPITATION DATA

STATION	LATITUDE	LONGITUDE		JAN	FEB	MAR	APR	MAY	JUN	JUL	AUG	SEP	OCT	NOV	DEC	CLIMATE
Charleston	32°48'N	79°57'W	°F	50	51.5	57.5	64.5	72.5	78.5	81	80.5	76.5	67.5	57.5	51	
(South Carolina)			in.	3.0	3.3	3.6	2.8	3.4	5.0	7.4	6.7	5.0	3.3	2.2	2.8	Cfa
Memphis	35°8'N	90°3'W	°F	41	44	52	62	70.5	78	81	80	74	63.5	51.5	43.5	
(Tennessee)			in	5.0	4.4	5.2	5.0	4.2	3.6	3.4	3.2	2.9	2.7	4.1	4.6	Cfa
New Orleans	29°58'N	90°7'W	°F	54	56.5	61.5	69	75.5	81	82.5	82	79	71	60.5	55.5	
(Louisiana)			in	4.5	4.6	5.2	4.4	4.6	4.6	6.9	5.8	5.5	2.9	3.7	4.9	Cfa
Melbourne	37°49'S	144°58'E	°F	67	67	65	59	54	50	49	51	54	57	61	64	
(Australia)			in	1.9	1.8	2.2	2.3	2.1	2.1	1.9	1.9	2.3	2.6	2.3	2.3	Cfb
Sacramento	38°35'N	121°30'W	°F	45.5	50.5	55	58.5	64	70	74	73	70	62.5	54	46.5	
(California)			in	3.8	2.8	2.8	1.5	0.8	0.1	<0.1	<0.1	0.3	0.8	1.9	3.8	Csa
San Francisco	37°48'N	122°24'W	°F	50	53	54.5	55.5	57	59	59	59.5	62	61	57	51.5	
(California)			in	4.7	3.6	3.0	1.5	0.6	0.2	0.0	0.0	0.3	1.0	2.4	4.3	Csb
Santa Monica	34°1'N	118°30'W	°F	53	53	55	58	60	63	66	66	65	62	58	55	
(California)			in	3.5	3.0	2.9	0.5	0.5	0.0	0.0	0.0	0.1	0.6	1.4	2.3	Csb
Chungking	29°39'N	106°34'E	°F	48	50	58	68	74	80	83	86	77	68	59	50	
(China)			in	0.7	0.9	1.3	4.0	5.3	6.7	5.3	4.4	5.8	4.6	2.0	0.9	Cwa

in the Köppen system by the subgroups s, w, and f (Table 12–1). Subgroup f is usually found on the eastern side of a continent and subgroup s is more often than not situated on the western side of a continent.

The Cf climates—Charleston, Memphis, New Orleans, and Melbourne in Table 12–4—have no distinct dry season, and the driest month of summer has more than 1.2 inches (3 cm) of precipitation. Charleston, Memphis, and New Orleans are in the a-temperature subclass, that is, Cfa, because the warmest month at each of these stations is above 71.6°F (22°C). Melbourne, Australia is in the b-temperature subclass (Cfb) because the temperature of its warmest month is below 71.6°F (22°C).

The Cs climates—represented by Sacramento, San Francisco, and Santa Monica—have rainfall peculiarities. The total annual amount of rainfall is fairly low, averaging roughly between 15 and 20 inches (38 and 51 cm); rain is concentrated in the winter months, while the summer is almost completely dry (Fig. 12–6). The year is broken into two distinct seasons on the basis of rainfall, and winter is the rainy season. Sacramento is Csa, interior Mediterranean, based on the fact that its warmest month is above 71.6°F (22°C) (Fig. 12–6). San Francisco and Santa Monica are classified as Csb, coastal Mediterranean, in the Köppen sys-

tem because the warmest month at each station is below 71.6°F (22°C).

Study Chungking's climograph (Fig. 12–7). The wettest month of the year occurs during the summer. Examine the boundaries for the s, w, and f subgroups on page 384. Chungking qualifies for the w subgroup on the basis of its wet summer and is, therefore, a Cw climate. The w boundary, according to the Köppen system, is

The white, bridgelike span that stretches across the Rance River Estuary on the Gulf of Saint-Malo, Brittany, France, is a facility that uses the rise and fall of the tides to produce electric power. This area of Brittany is a Cfb climatic zone. (French Embassy)

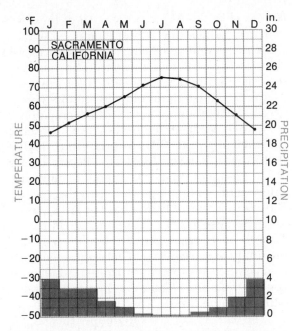

FIGURE 12–6 This climograph shows two distinct seasons on the basis of rainfall. Winter, obviously, is the rainy season. Sacramento is classified as interior Mediterranean, Csa, in the Köppen system.

temperature of the coldest month below 26.6°F (−3°C). The **D climates** tend to be found in the interior and leeward areas on the great landmasses. The average temperature of 50°F (10°C) for the warmest month, which these stations must have to qualify for this category in the Köppen system, coincides with the poleward limit of forests. These D climates are characterized by a snow cover of several months duration and frozen ground for part of the year.

Take special note of the precipitation pattern revealed by the D-category climographs in Figure 12–8. Two principal subdivisions can be recognized: the first, Df, with no dry season, and the second, Dw, with a dry season during the winter.

The Df subgroup is represented by Peoria, Halifax, and Tromso (Table 12–5). These stations are moist throughout the year, and even the driest month receives more than 1.2 inches (3.0 cm) of precipitation, as the climograph for

further qualified, and there should be 10 times as much rain in the wettest month of summer as in the driest month of winter. Chungking's driest month, January, has 0.7 inch. Ten times 0.7 inch is 7.0 inches; Chungking's wettest month, June, falls short by a mere 0.3 inch of this qualification. An alternative definition is often applied to the w subgroup—a station in the C category can qualify for the w subgroup if 70 percent or more of the average annual rainfall is received in the warmer 6 months. If you count Chungking's warmer 6 months as May through October, then 75 percent of its rainfall is received during this period and Chungking qualifies as Cw. Since the average temperature of its warmest month is above 71.6°F (22°C), Chungking is Cwa, subtropical monsoon, according to the Köppen system.

D-CATEGORY CLIMOGRAPHS

The stations selected in Table 12–5 represent the cold snow-forest climates with an average

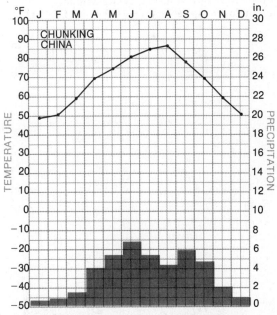

FIGURE 12–7 The wettest month of the year occurs in the summer according to Chungking's climograph. June gets almost 10 times more precipitation than January, which is the driest month according to the bar graph. Chungking is a Cwa climate in the Köppen system.

Maine, New Hampshire, and Vermont are in the Dfb climatic zone—that is, they have cold snow-forest climates. This photo captures a snowfall from stratocumulus clouds at New Hampshire's Mount Washington in the White Mountains. The air temperature at the time was 0°F (−17°C). (NOAA)

Halifax (Fig. 12–8A) indicates. All of these stations have a genuine winter combined with an authentic summer, which produces an annual climatic cycle of four seasons. The transition seasons—fall and spring—are brief, however. All the Df climates are cold with humid winters.

The temperature-subclass differentiation (a, b, and c) is made on the basis of the boundaries specified on page 384; Peoria (Dfa), Halifax (Dfb), and Tromso (Dfc). The Peoria data in Table 12–5 show that its temperature range from summer to winter is greater than the range at Halifax or Tromso. Peoria has the larger annual range because it is more continental; Halifax and Tromso have ranges that are modified by their proximity to ocean water. Temperature-subclass c is classified as subarctic in Table 12–1; Tromso, you can note from Table 12–5, is at the highest latitude of the three.

The data for the Dw subgroup—Inchon, Vladivostok, Irkutsk, and Verkhoyansk—show the wettest month of the year is during summer; the wettest month, in all cases, has more than 10 times the precipitation of the driest winter month. The Vladivostok climograph (Fig. 12–8B) is depicted as representative of the Dw

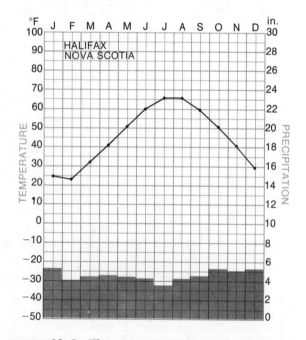

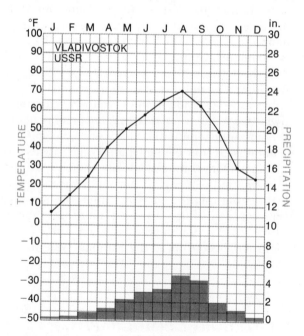

FIGURE 12–8 These are D-category climographs in the Köppen system. Two subdivisions are recognized on the basis of precipitation. Halifax is a Df climate with no dry season. Vladivostok is a Dw climate with a dry season during the winter.

TABLE 12–5   **MEAN MONTHLY TEMPERATURE AND PRECIPITATION DATA**

STATION	LATITUDE	LONGITUDE		JAN	FEB	MAR	APR	MAY	JUN	JUL	AUG	SEP	OCT	NOV	DEC	CLIMATE
Inchon	37°28′N	126°38′E	°F	26	29	38	50	59	68	75	77	68	58	43	30	
(Korea)			in	0.8	0.7	1.2	2.6	3.3	3.9	10.9	8.8	4.3	1.6	1.6	1.1	Dwa
Vladivostok	43°8′N	131°54′E	°F	6	15	25	40	50	57	65	70	62	48	30	24	
(USSR)			in	0.3	0.4	0.7	1.2	2.1	2.9	3.3	4.7	4.3	1.9	1.2	0.6	Dwb
Irkutsk	52°16′N	104°20′E	°F	−6	−1	13	31	44	56	60	58	46	31	11	−4	
(USSR)			in	0.5	0.4	0.3	0.6	1.3	2.2	3.1	2.8	1.7	0.7	0.6	0.6	Dwc
Verkhoyansk	67°35′N	133°27′E	°F	−58	−48	−26	4	32	54	56	49	35	4	−35	−54	
(USSR)			in	0.2	0.2	0.1	0.2	0.3	0.9	1.1	1.0	0.5	0.3	0.3	0.2	Dwd
Peoria	40°42′N	89°36′W	°F	24	28	40	51	62	71	75	73	65	53	39	28	
(Illinois)			in	1.8	2.0	2.7	3.8	3.9	3.8	3.8	3.2	3.8	2.4	2.4	2.0	Dfa
Halifax	44°39′N	63°36′W	°F	24	23	31	40	50	58	65	65	59	50	40	29	
(Nova Scotia)			in	5.2	4.0	4.3	4.5	4.4	4.3	3.6	4.1	4.6	5.1	5.0	5.2	Dfb
Tromso	69°40′N	18°58′E	°F	26	25	26	32	39	47	52	51	44	36	30	27	
(Norway)			in	4.3	4.4	3.1	2.3	1.9	2.2	2.2	2.8	4.8	4.6	4.4	3.8	Dfc

subgroup. Dw climates, then, are cold with dry winters. Only the leeward side of Asia can generate this subgroup.

Again, the temperature-subclass differentiation (a, b, and c) is made on the basis of the boundaries specified on page 384. The fourth subclass, d, is applied when the temperature of the coldest month is below −36.4°F (−38°C); Verkhoyansk qualifies as Dwd since its lowest temperature, during January, is −58°F (−50°C). Inchon is classified Dwa, Vladivostok is Dwb, and Irkutsk is Dwc.

The latitudinal spread represented by these Dw climates is 30°. The latitudinal spread of the Df subgroup in both North America and Europe is also about 30°. The c and d subclasses in the Dw subgroup are classified in Table 12–1 as subarctic.

E-CATEGORY CLIMOGRAPHS

Two climatic subgroups are recognized in the **E climates:** ET and EF. Temperatures throughout the year in this category are quite low. Of course, once the temperatures are below freezing and the ground is frozen, it makes little difference to plants how cold it gets. Thus, the temperature boundaries for the subgroups in this category are expressed in terms of the warmest month because the intensity and duration of the season of warmth is crucial to plants.

The ET subgroup has a brief growing season during which temperatures are between 50°F (10°C) and 32°F (0°C). Two examples of the climatic-element pattern for the ET subgroup are detailed in Table 12–6. The data for Point Barrow are plotted as a climograph in Figure

TABLE 12–6   **MEAN MONTHLY TEMPERATURE AND PRECIPITATION DATA**

STATION	LATITUDE	LONGITUDE		JAN	FEB	MAR	APR	MAY	JUN	JUL	AUG	SEP	OCT	NOV	DEC	CLIMATE
Point Barrow	71°18′N	156°47′W	°F	−15.5	−18.5	−15	−0.5	18.5	34	39.5	38.5	30.5	17	1	−10.5	
(Alaska)			in	0.2	0.2	0.1	0.1	0.1	0.4	0.8	0.9	0.6	0.5	0.2	0.2	ET
Angmagssalik	65°36′N	37°33′W	°F	21	18	22	27	36	42	46	44	39	32	27	22	
(Greenland)			in	2.1	2.5	3.0	1.9	2.0	1.4	1.6	2.5	3.0	3.6	3.8	2.3	ET
McMurdo Station	77°53′S	166°48′W	°F	23	14	8	−8	−13	−14	−15	−16	−14	−5	13	23	
(Antarctica)			in	0.5	0.7	0.4	0.4	0.4	0.3	0.2	0.3	0.4	0.2	0.2	0.3	ET

Caribou dot the horizon in this view of the Canadian tundra north of the Arctic Circle. This region of undulating, treeless plains is classified as ET in the Köppen system. (National Film Board of Canada)

12–9. In the Köppen system the ET subgroup is referred to as a **tundra climate.**

The ET climatic type is found along the arctic coasts of North America and Eurasia. It is generally found north of the Arctic Circle in undulating treeless plains called **tundra.** The tundra generally has a black muck soil and a permanently frozen subsoil. In eastern Canada and in eastern Siberia, the ET climate extends into lower latitudes in keeping with the continental influence. This climate does not occur in the Southern Hemisphere except in small areas of

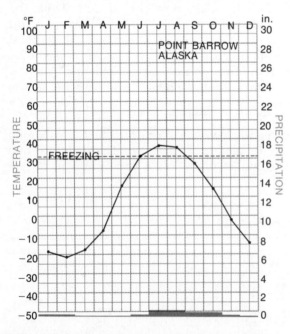

FIGURE 12–9 Point Barrow is classified as an ET climate in the Köppen system. An ET climate, as this climograph shows, has very few months in which temperatures rise above freezing.

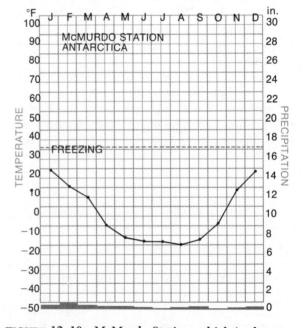

FIGURE 12–10 McMurdo Station, which is close to McMurdo Sound in Antarctica is in an area of perpetual ice and snow. The region carries a Köppen designation as an EF climate.

Antarctica and on a few islands in the Antarctic Ocean.

The climograph for Point Barrow (Fig. 12–9) shows a temperature range of 58°F (32°C), which is quite a range considering the fact that the station is located on the coast. The Greenland station—documented in Table 12–6—has a much narrower range. Most ET stations record less than 6 months with an average temperature above 32°F (0°C). The Angmagssalik data show 5 months with temperatures above 32°F (0°C) and the Point Barrow data show 3 months above 32°F (0°C).

The ET climate generally has less than 15 inches (38 cm) of annual precipitation, which falls largely during the warmer months. The Point Barrow pattern is in keeping with this general rule; a mere 4.3 inches (10.9 cm) of precipitation fall at this Alaskan station and 49 percent of it falls in the three warmest months of June, July, and August. If we include September's precipitation, when the temperature averages 30.5°F

(−0.8°C), we account for 63 percent of the annual amount. The annual precipitation at Angmagssalik, which amounts to 29.7 inches (75.4 cm), exceeds the average for ET climates because of the generally warm sea temperatures and a persistent upper-level trough of converging air streams in this area of Greenland.

The EF subgroup exists only over the permanent icecaps. EF climates—with mean monthly temperatures below 32°F (0°C), perpetual frost, and no vegetation—are found in Greenland and Antarctica. The data for the McMurdo Sound area are detailed in Table 12–6; a climograph constructed from these data (Fig. 12–10) shows a high of 23°F (−5°C) during January—summer in the Southern Hemisphere—and a low of −16°F (−26.6°C) during August, which is wintertime in the Southern Hemisphere. This coastal EF climate has a range of 39 Fahrenheit degrees (22C°); the ranges in the interior of Antarctica are much greater.

URBAN-ALTERED CLIMATE

In the last ten years many studies have gathered evidence that establishes a definite **urban effect** on climate. Today, urban-altered climate is recognized as a rather serious problem because the climatic anomaly that results is city-size or larger. And urbanization is increasing. In fact, some estimates indicate that by the turn of the century more than 80 percent of the people of industrialized countries will live in cities.

The longest studied of any aspect of **urban climate** is that of the urban heat island. The mean annual temperature excesses of urban over rural temperatures is in the range of 1.3 Fahrenheit degrees (0.72C°). The urban effect on precipitation is not as well-established because the complex coupling between surface conditions and cloud-level-rain processes is difficult to decipher.

INSOLATION AND SUNSHINE

The heavy load of gaseous and solid contaminants in urban air reduces the total annual insola-

tion that it is possible for a city to receive by 15 to 20 percent. The interception and subsequent reduction is most apparent during the winter when urban surfaces receive 30 percent less insolation than comparable surfaces in the surrounding countryside.

The duration of bright sunshine is also significantly reduced over urban areas. Urban smoke pollution is thought to be the factor that produces the decrease. Researchers have found that the central cities in some locales have less than one-half the sunshine of the surrounding suburbs during the winter. In the summer months, the difference between city and countryside sunshine seems to range around 10 percent.

HEAT-ISLAND EFFECT

The maximum daily difference between city and countryside temperatures ranges between 10 and 15 Fahrenheit degrees (5.5 and 8.3C°). This temperature differential, called a **heat-island ef-**

Winnipeg, Manitoba, lies on a wide flat plain at the junction of the Red River and its largest tributary, the Assiniboine River. The bridges crossing the Red River at the right connect Winnipeg with its suburbs. Winnipeg has the fourth largest metropolitan area in Canada. The prairie beyond the city can be seen at the top of the photo. There is a pronounced difference in temperature between Winnipeg and its surrounding countryside. (National Film Board of Canada)

fect, is found from the surface to heights of almost 1000 feet (300 m). Most cities, whether they are hilly like San Francisco or flat like Minneapolis, are **heat islands,** that is, they are enclosed in a mass of warm air. The heat-island effect occurs in large cities and in small cities.

The existence of the urban heat island is partially due to the physical and thermal properties of the urban fabric, that is, the city's steel, concrete, and asphalt. These materials conduct heat about three times faster than moist, sandy soil. City streets and buildings simply absorb more insolation in less time and store more heat than the open countryside does.

The urban-rural temperature discrepancy is also partly due to the heat derived from domestic and industrial space heating. In addition, tremendous amounts of heat are released to the urban environment through manufacturing processes and the operation of automobiles, buses, and trucks. Combustion during the winter on Manhattan Island, for example, releases 250 percent more heat than reaches the surface of the city through incoming solar radiation. Even during the summer there is no respite from cast-off heat because air conditioners pump hot air from buildings into the city's atmosphere.

THE WIND PATTERN

A definite pattern of circulation develops as a result of the heat-island effect. The warmth of a city's heat island produces a **suburb breeze**, which is illustrated in Figure 12–11. As warmer air rises over the center of the city, cooler air from the suburbs flows into the city. The particles of dust and smoke produced by activities within the city are carried aloft by this circulation and form a dust dome that extends from the city to the suburbs.

These thermally induced near-surface winds are a significant modification of the local circulation. The cycle of circulation shown in Figure 12–11 is strongest in the morning and at sunset, when the greatest temperature gradients exist between the city's center and the countryside. Toward midday, the temperature gradient between the countryside and the city is smallest, and the cycle of circulation is weakest.

Another major modification of wind flow is created by the city's many structures. Helmut Landsberg's studies indicate that the average annual wind speeds within the central city may be 20 to 30 percent less than in the surrounding countryside. The irregular skyline of a city (as shown in Fig. 12–12) has a braking effect on the wind that increases its turbulence. Under certain conditions, the local eddying, or turbulence, can increase gustiness and wind speeds at some sites within the city. We all know of situations where buildings channel the airflow down a particular street or around a corner at gusts and speeds that make it difficult to maintain a footing.

IMPACT ON PRECIPITATION

The precipitation that falls on cities is removed in rather distinct ways: Rainwater is channeled from housetops, building tops, sidewalks, and streets to sewers. The runoff is usually so rapid and efficient that very little moisture is available for evaporation. Snow, too, is cleared from sidewalks and streets, and when a significant

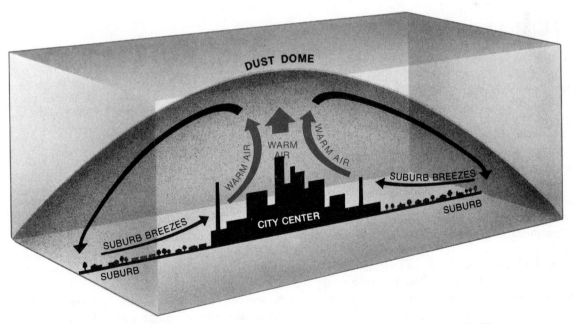

FIGURE 12–11 A definite cycle of circulation develops as a result of the heat-island effect. The surface flow toward the city is referred to as a *suburb breeze*. The flow aloft is away from the city.

FIGURE 12–12 The clustering of buildings and an irregular skyline interrupt the orderly flow of the air and produces turbulence in a city's airflow.

amount of snow falls in large cities, it is collected and carried away by trucks.

In the open countryside, on the other hand, most precipitation remains on the surface in natural runoff channels or sinks into the ground. Thus, more of the precipitation that falls in rural areas is available for evaporation. Since evapora-

tion requires heat energy, greater opportunity for the evaporation of precipitation produces a larger cooling effect in suburban and rural areas that don't have a city's efficient sewer system.

The lack of open water surfaces and transpiring vegetation in a city generally produces lower urban humidities than in the surrounding countryside. Comparative studies of relative humidity in a city and its surrounding countryside indicate that the annual average for humidity is about 6 percent less in the city. The contrast in relative humidity is greater during the summer (8 percent less for the city) than during the winter (2 percent less for the city).

Although urban humidities are lower, studies indicate that cities generally get 5 to 10 percent more precipitation than the surrounding countryside. This may seem to be a contradiction, but, remember, the pollutants in the city's atmosphere constitute a large quantity of condensation nuclei. In addition, there is an increased component of upward-flowing air over a city due to low-level turbulence and thermal updrafts. All these conditions conspire to produce precipitation from the moisture available in the air.

CLIMATE AND LANDFORMS

The same rock type produces a variety of landscapes in different settings. Normally jointed granite, for example, is found as forested narrow-crested ridges in the Blue Ridge Mountains. In the Southwest, the same type of jointed granite produces jumbles of rounded boulders. In the Sierras, jointed granite forms intricate angular shapes. In New Hampshire's White Mountains, jointed granite produces bumpy uneven slopes. The character of any landscape is, in fact, developed through an intricate interplay of rock type, geologic structure, tectonic conditions, and climate.

Although climate is only one of four factors, it is crucial in the processes of weathering and erosion that shape most landforms and, as such, its influence on geomorphic processes is significant. The nature of atmospheric stresses affecting rock material differs in kind and intensity from one climatic region to another. Frost-

heaving of the ground, which may dominate in a very cold region, for example, is insignificant in a hot, dry region where the transport of sediment by wind is the dominant atmospheric stress. These kinds of differences in geomorphic processes can and do lead to the development of distinctive landforms characteristic of particular climatic regions.

ICE-CLIMATE LANDFORMS

Köppen's ET climate is dominated by subfreezing temperatures throughout much of the year and the landscape takes on a highly distinctive appearance. These low-temperature regions are distinguished by a sparse cover of vegetation, surface soils that freeze and thaw seasonally, and permanently frozen subsoils. In an ET climate frost weathering is a dominant, active process that is capable of shattering rock at and below the surface.

Permanently frozen rock and soil is called **permafrost.** In northern Alaska, Canada, and Siberia, the permafrost reaches to depths of 2000 feet (600 m). It is only the top 3 feet (1 m) of soil that thaws each summer.

When the surface soil in an ET climate thaws, it becomes saturated with water that cannot move downward because of the permafrost barrier. The water lubricates the soil to such an extent that the soil slips and moves downslope in lobes (Fig. 12–13). This type of movement is known as **solifluction.** The long-term effect of solifluction is to smooth the landscape. Depressions and small valleys are gradually filled with slope deposits.

Solid rock masses that protrude above the surface in an ET climate are rapidly reduced to rubble by intense frost weathering. The rock rubble is incorporated into solifluction lobes and, thus, high areas are effectively lowered. Any broad summits that exist are usually covered by angular rock rubble. At the edges of large rock outcrops, there are accumulations of talus.

In ET lowlands composed of silty alluvial deposits, a patterned ground of ice-wedge polygons develops (Fig. 12–14). During the freezing process, water-soaked silt expands at first but, at prolonged low temperatures, it gradually begins to contract and cracks into a network of **polygonal fissures,** called **ice-wedge polygons.** The polygons are often 3 feet (1 m) across and 15 feet (5 m) deep.

Lenslike masses of ice that form in the ground heave up the overlying sod in a dome called a **pingo** (Fig. 12–14). These dome-shaped pingos develop in old lake beds or marshes that are filling with sediment and vegetation. The pingo develops when the encroachment of permafrost into the sediment traps a lens of unfrozen water, which subsequently freezes and domes up to produce the pingo. When a pingo is destroyed by thawing, it leaves a depression ringed by a rampart of sediment and soil. A large pingo may be more than 300 feet (100 m) across and 100 feet (30 m) high.

TROPICAL-CLIMATE LANDFORMS

Köppen's A-climate category produces two characteristic landscapes in low-latitude regions: vast lowlands covered with lush tropical rain forests, which occasionally extend up the slope of bordering mountains, and tropical savannas consisting of vast grasslands with a scattered growth of pine or other trees and bushes.

The almost daily rainfall in tropical rainforest climates generates large amounts of runoff, which produce some of the world's largest rivers—the Amazon, for example. Dense vegetation on tropical riverbanks seems to limit the ability of streams to cut laterally to the extent seen in mid latitudes. Waterfalls are common along the rivers, but deeply incised gorges are rare because thorough chemical decay of rock material prevents the streams from acquiring cutting tools in the form of large rock debris.

Soil wash and removal of material by solution are important tropical erosional processes in these moist climates. The sliding of water-saturated soil on steep slopes is conspicuous in

FIGURE 12–13   The soil within 3 feet (1 meter) of the surface in an ET climate zone is shown as thawed and is saturated with water in this figure. The water-saturated soil slips and moves downslope in lobes.

FIGURE 12–14 A patterned ground of ice-wedge polygons surrounds the lenslike mass of ice heaved up in a dome called a *pingo*.

areas where Af and Am climates reach into highlands. On horizontal surfaces the weathered mantle is generally thick, at times exceeding 300 feet (100 m). Chemical weathering is a dominant process in these A-type climates.

The tropical-savanna landscape of the Aw climate is found where crystalline rocks are dominant—Brazil, south of the Sahara in Africa, India, and two small areas of northern Australia (Figs. 12–1, 2). The word "savanna" refers to the plant association of drought-resistant tree species rising from a sea of grass. The geomorphic features in many savannas consist of escarpments, enormous rock domes, and very flat

plains into which shallow saucer-shaped valleys are cut by ephemeral streams.*

The peculiarities of granitic rock and an aggressive climate—in which chemical weathering decomposes rock below the surface as fast as or faster than surface erosion can wear down the land—produces the Aw landscape. Where granite is closely jointed, water penetrates it deeply, and the chemical weathering of the rock occurs, at times, 300 feet (100 m) or more below the surface. The enormous granitic rock domes

*During the dry season in an Aw climate, water tables sink far below the surface, which causes all but the largest stream channels to dry up.

400

that dot the surface of a tropical savanna have so few joints that water is excluded and chemical decay cannot begin. These domes, then, are cores of granite that resisted chemical decay below the surface and have been left behind as the surrounding material was carried away.

ARID-CLIMATE LANDFORMS

In arid regions the vegetation cover is limited and quite sparse. Generally, bare rock and accumulations of rock fragments dominate the landscape. The slope profiles in arid regions tend to exhibit a greater angularity than those of other regions because there is little or no soil to smooth over variations in rock type.

Since vegetation is sparse and soils tend to be compact in these dry regions, water infiltration into the soil is slow and runoff tends to be heavy during the occasional downpours. Ephemeral flows of surface water, then, are the dominant means by which the surface is eroded. A large water flow of short duration does not transport rock debris great distances, and, thus, depositional landforms are common and conspicuous in the landscapes of dry regions. Dry stream beds filled with debris, alluvial fans, talus accumulations, and sand dunes are some of the obvious depositional forms that attract attention in these landscapes.

When surface rocks in an arid region are granitic, the landscape generally consists of long, smooth ramps leading up to extremely rough boulder-covered slopes. The projecting relief forms are called **inselbergs,** or island mountains, because they rise abruptly above the ramplike surfaces. These steep-sided inselbergs are remnants of a long period of erosion during which slopes have retreated, that is, worn back, at a constant angle.

Vast areas of sand dunes, called **ergs,** are the dominant features in some arid regions. To produce an erg like those found in the deserts of North Africa and Saudi Arabia, several conditions must be met: An abundant source of sand must exist, and a system of delivery must operate to transport the sand from the source to the accumulation area. In those regions where the accumulations of sand are enormous and totally beyond the capacity of local desert weathering processes to supply, climatic change is a further condition that must be met to produce the erg. In other words, some erg formations require a period of humid-climate weathering and stream transport of the sand out of the source region to the zone of accumulation. This humid-climate period is then followed by a climatic change toward aridity when wind becomes the effective agent for reshaping the sand into dunes.

One of the largest present-day ergs is found in the deserts of Saudi Arabia. Wind is the effective agent of erosion and deposition that is constantly reshaping the Arabian erg. The Nebraska Sand Hills—covering an area of some 24,000 square miles (62,000 km²)—is an ancient erg that is presently grass-covered. This portion of west-

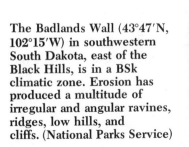

The Badlands Wall (43°47′N, 102°15′W) in southwestern South Dakota, east of the Black Hills, is in a BSk climatic zone. Erosion has produced a multitude of irregular and angular ravines, ridges, low hills, and cliffs. (National Parks Service)

Saudi Arabia's Rub al Khali *(left)* is a vast sand-dune-spotted desert which is classified as a BWh climate. Wind erosion has removed most of the grass cover in this area of the Nebraska Sand Hills *(right)*. (Aramco, *left;* USDA, *right*)

ern Nebraska was probably a desert as recently as 15,000 years ago.

HUMID MID-LATITUDE LANDFORMS

The humid mid-latitude regions are the source of many of our traditional and commonly accepted ideas concerning landform development. A wide range of geomorphic processes operate in these humid mid-latitude climatic regions.

The landscapes of the eastern United States and most of western Europe, for example, have developed in humid mid-latitude climates. The distribution of Köppen's Cf and Df climates in the eastern United States is indicated in Figure 12–1. Refer to Figure 12–2 for the climatic pattern of western Europe.

The cover of vegetation in these humid climates minimizes slope erosion by running water. There are, however, a large number of stream channels, and an intricate pattern of fluvial erosion—previously described in great detail in Chapter 7—that leaves its imprint on the relief in these climatic regions. The convex nature of the soil-covered hilltops and ridge crests is due to the process of soil creep, which results from the largely permeable soils of these humid regions.

THE LANGUAGE OF PHYSICAL GEOGRAPHY

A physical geographer uses a technical vocabulary to make explanations. Review your understanding of the vocabulary used to develop the concepts in this chapter. Among the important words and terms used are:

A climates
B climates
C climates
climatic subgroups
climograph
D climates
E climates
empirical classification
ergs

genetic classification
heat island
heat-island effect
icecap climate
ice-climate landforms
ice-wedge polygons
inselbergs
permafrost
pingo

polygonal fissures
solifluction
suburb breeze
temperature subclasses
tropical-climate landforms
tundra
tundra climate
urban climate
urban effect

SELF-TESTING QUIZ

1. Compare the various systems of climatic classification.
2. List, describe, and contrast the five major categories in the Köppen classification system.
3. How are subgroup boundaries established in the Köppen system?
4. Discuss the plotting and uses to which a climograph can be put.
5. Describe a typical A-category climograph in the Köppen system.
6. Compare and contrast A-category and B-category climographs in the Köppen system.
7. What is the latitudinal spread of Dw climates in North America?
8. Compare and contrast ET and EF climates.
9. What is meant by urban-altered climate?
10. Why is the urban effect on precipitation not well established?
11. What is the maximum daily difference between city and countryside temperature?
12. Compare the duration of bright sunshine in urban and countryside areas.
13. To what is the existence of the urban heat island due?
14. Describe the pattern of circulation that develops as a result of the heat-island effect.
15. Discuss the way climate produces a variety of landscapes from the same rock type in different settings.
16. What is the long-term effect of solifluction?
17. Under what kinds of conditions are pingos produced?
18. What kinds of rocks predominate in the tropical savanna landscapes of Brazil, India, and northern Australia?

IN REVIEW
CLIMATE CLASSIFICATION AND ANALYSIS

I. SYSTEMS OF CLASSIFICATION

II. THE KÖPPEN SYSTEM
- A. Major Category Boundaries
- B. Subgroup Boundaries
  - 1. A-Category Subgroups
    - a. f subgroup
    - b. m subgroup
    - c. w subgroup
  - 2. B-Category Subgroups
    - a. S subgroup
    - b. W subgroup
    - c. h-temperature subclass
    - d. k-temperature subclass
  - 3. C-Category Subgroups
    - a. s subgroup
    - b. w subgroup
    - c. f subgroup
    - d. a-temperature subclass
    - e. b-temperature subclass
    - f. c-temperature subclass
  - 4. D-Category Subgroups
  - 5. E-Category Subgroups
    - a. T subgroup
    - b. F subgroup

III. ANALYSIS BY CLIMOGRAPH
- A. Plotting the Data
- B. Analyzing the Data
- C. A-Category Climographs
- D. B-Category Climographs
- E. C-Category Climographs
- F. D-Category Climographs
- G. E-Category Climographs

IV. URBAN-ALTERED CLIMATE
- A. Insolation and Sunshine
- B. Heat-Island Effect
- C. The Wind Pattern
- D. Impact on Precipitation

V. CLIMATE AND LANDFORMS
- A. Ice-Climate Landforms
- B. Tropical-Climate Landforms
- C. Arid-Climate Landforms
- D. Humid Mid-Latitude Landforms

PART FOUR

SOILS AND VEGETATION OF THE EARTH

Soils are the product of a long and intricate formative process. The texture of soils, their depth, their composition, and their ultimate use by humanity are determined by inorganic and organic factors. Bedrock serves as the parent material of soils; weathering influences the structure of soils. Organic matter added to soils by plants and animals determines their usefulness to humankind.

Finally, in the last two chapters of Part 4, we come to the keystone of physical geography: plant geography. Natural vegetation offers the best indicator of the combined effects of surface features, soils, and climate in producing the landscape. Thus, we speak of the prairies of America or of the steppes of European Russia. The last chapter of this book examines present-day environments in terms of the vegetation found in various areas. In more than a few instances the present-day vegetation reflects the impact of humankind on nature.

A. Lichens are flowerless plants that grow on bare rock. The flat, green patches of these lichens attached to granitic rock in Australia are helping to form soil. (John Navarra)

B. Plants have moved onto a lava field in Hawaii, and the process of soil formation is well underway. (John Navarra)

C. Windswept Spruce Knobs is the highest point in West Virginia. (Department of Tourism, West Virginia)

D. Cypress swamp in South Carolina. (John Navarra)

E. Huge fern trees in an Australian forest. (John Navarra)

a.

b.   SPODOSOL

c.   ALFISOL

d.   ARIDISOL

e.   ENTISOL

f.   HISTOSOL

g.   INCEPTISOL

h.   MOLLISOL

i.   OXISOL

j.   ULTISOL

k.   VERTISOL

The soil order on Prince Edward Island, Canada (shown in photo A), according to the U.S. system of classification is spodosol (profile shown in B). The profiles of the other 9 orders in the U.S. classification system are shown in photos C through K (see pages 434–449 for descriptions).

13 *Soil as an essential resource*

Soil develops slowly from weathering rock. The loose surface deposits weathered from rock are classified as either **residual material**—formed from the underlying rock which is in place—or **transported material**—material which has been transported by flowing water, wind, glaciers, and gravity. Most soils in the United States are developing in transported parent material.

There are a variety of ways in which the term **soil** is used. Most engineers simply think of soil as being all the loose or unconsolidated material that lies on hard rock. The climatologist refines the definition somewhat and thinks of soil as a product of physical, chemical, and biological processes. Thus in the climatological sense, soil is what rock becomes through exposure to the atmosphere and weathering. The physical, chemical, and biological processes involved in its production make the soil distinct from the rock. The soil scientist places some additional restrictions on the definition by specifying that the detrital material must contain living matter and be capable of supporting plant life in order to qualify as soil.

In whatever way soil is defined, it is important to recognize that we, in one way or another, derive our foods from it. Soil is the source of 13 of the 16 or more **essential nutrients** that are of fundamental importance to plant growth. Of the commonly agreed upon essential plant nutrients, hydrogen, carbon, and oxygen are derived from air and water; phosphorus, potassium, sulfur, calcium, iron, magnesium, boron, manganese, copper, zinc, molybdenum, and chlorine can be derived naturally by plants from the soil; and nitrogen is contributed to plants by both air and soil. Clearly, then, plant dependence on the soil concentrations of these nutrients makes us ultimately dependent on soil.

SOIL FORMATION

Soil develops from the regolith through a continuation of the physical and chemical reactions described on pages 211–213 and by the addition of the biotic activity of plants and animals. Liv-

407

ing activity, from that of the earthworms to the microorganisms that decompose organic material, gradually changes the regolith into true soil.

For regolith to be classed as soil, it must have a well-developed profile. The several layers of altered rock material between the surface and the bedrock below are called the **soil profile.** The profiles of most soils have two or three well-defined layers that are called **soil horizons.**

SOIL HORIZONS

The study of the shape and nature of the profiles and horizons in soil is known as **soil morphology.** Most soil horizons are easily identifiable because they differ from other soil layers in color, texture, structure, salt content, and amount of organic material.

HORIZON SUBDIVISIONS

The major horizons in a soil profile are designated by the capital letters *O, A, B,* and *C*. The layer of consolidated rock beneath the soil is identified by the capital letter *R*. The major horizons and various subdivisions, which are designated by adding numerals to the letters, are identified in Table 13–1.

The horizons listed in Table 13–1 are of two major classes: organic horizons—identified by the letter *O*—and mineral horizons—identified by the capital letters *A* and *B*. The relative positions of the various horizons in the soil profile are depicted in Figure 13–1.

The O-horizons, consisting of loose leaves and organic debris, occupy the topmost position in the profile and are found on the surface of the soil as the top layer. The A-horizons are the layers of **topsoil.** Rain falls on the topsoil and moves down through the A-horizons on its way to becoming groundwater; thus, soluble material is dissolved or leached out of the topsoil. Nevertheless, the A1-horizon is a zone of accumulation for organic material. The B-horizons, called **subsoils,** are transition zones. The composition of the B-horizons depends on what is passed from the A-horizons above to the C-horizons below. For example, the B2-horizon is the layer in which clays leached from above are deposited. The C-horizon, at the bottom of the profile, is usually an active zone of chemical weathering; it consists of fragments of rocks that may have been recently dislodged from the bedrock below. The weathered material from the C-horizons is passed along to the B-horizon.

HORIZON CODE

The nature of a subdivision is defined more precisely by the use of a number of specially coded

TABLE 13–1   **HORIZONS IN A SOIL PROFILE**

HORIZON	SUBDIVISION	DESCRIPTION
Organic	O1	More than 20 to 30% organic matter content. Vegetative matter is identifiable, that is, leaves, stems, and so forth.
Organic	O2	Decomposition of organic matter is extensive. Source of organic matter is not identifiable.
Mineral	A1	The top mineral horizon. Darkened somewhat by organic materials.
Mineral	A2	A light-colored horizon. Found in high rainfall areas. Fine clays and organic material are leached out by percolation.
Mineral	A3	A transition zone that is more like A2 above than B2 below.
Mineral	B1	A transition zone that is more like B2 below than A2 above.
Mineral	B2	Small particles washed from horizons above accumulate. Generally higher in clay than horizons above; often higher than A1; always higher than A2.
Mineral	B3	A transition zone.
Regolith	C	Loose parent material without horizon development.
Parent	R	Underlying consolidated rock.

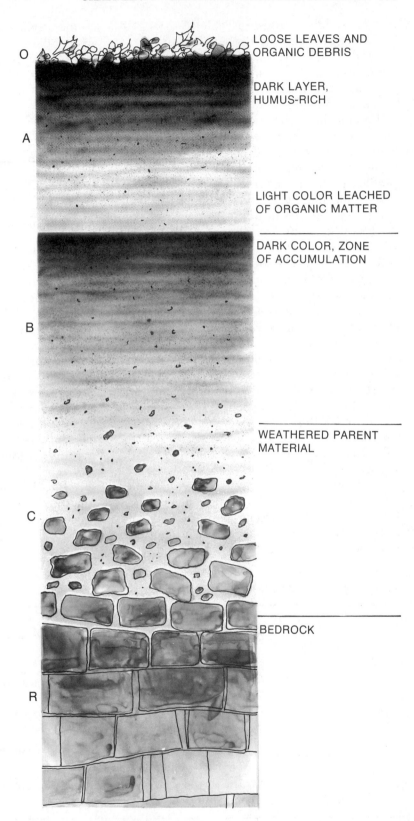

FIGURE 13-1 The soil horizons depicted in this figure are of two major classes: organic horizons (O) and mineral horizons (A & B). The C-horizon consists of fragments that have been recently dislodged from the bedrock (R) below.

lower-case letters. Refer to Table 13–2 for a few of these codes along with their meanings.

Within a soil profile, there may be any number or combination of horizons. The exact makeup of the profile depends on the soil's location and developmental history. For example, a soil profile on a comparatively young floodplain may have only a C-horizon or a thin A1-horizon over a C-horizon. In other words, some of the master horizons may be missing from certain soil profiles. In the case of a recent floodplain, where soil is forming on alluvial deposits—unconsolidated sediments deposited by a stream or river—the underlying transported sediment is the C-horizon. It takes time for the other horizons to develop. The soil on such a recent floodplain is so feebly developed that it simply does not show a B-horizon.

In other situations, the A-horizon of topsoil may have been removed by erosion. On the other hand, a mature grassland region may have a number of horizons including A1, A3, B1, B2t, B3, C, and R. A mountain soil under a dense forest canopy has a unique developmental history with a profile to match. Its horizons may include all of those listed in Table 13–1; in fact, its A2-horizon is often quite thick and the R-horizon may be 6 feet (almost 2 m) below the surface.

In humid regions, physical, chemical, and biotic soil-forming processes are much more intense in the surface than in the subsurface materials. In desert regions, chemical and biotic activity are less intense than in humid regions.

Generally, an A-horizon contains more organic matter than the B- or C-horizon. There are certain soils, however, in which conditions favor the accumulation of organic matter in the B-horizon.

FACTORS IN SOIL FORMATION

There are five factors that influence the rate and kind of soil development: (1) nature of the parent material, (2) topography, (3) climate, (4) biologic activity, and (5) time. All these factors interact to produce great variation in the rate of development as well as the kind of soil that is produced.

PARENT MATERIAL

The nature of the **parent material** determines the kind of regolith produced. Rocks—granite, sandstone, and limestone, for example—consisting of different mineral components will produce radically different kinds of regolith. Granite is slow to weather, but the transition from granite to soil produces an increase in bulk of almost 88 percent simply by the take-up of water, through a process called **hydration,** and the production of clays and silica. Some sandstone contains abundant, unaltered particles of the mineral called *mica*, which may weather by hydration and cause the rock to break down into its constituent grains; but the regolith produced from sandstone is predominantly quartz grains. The character of the regolith produced from a limestone is, of course, unique since calcite is the main mineral constituent of the rock.

TABLE 13–2   **SUBDIVISION CODE**

CODE	MEANING
ca	A depositional accumulation of calcium and magnesium carbonates.
f	Soil in horizon is frozen.
g	Gleying, which is a long-time condition of poor aeration, exists. Gleying is usually produced by excess water. Soil colors in the horizon are grey to pastel blues and greens.
h	There is an accumulation of deposited humus.
ir	There is an accumulation of deposited iron.
m	A strong cementation of many hardpans exists in the horizon.
p	The horizon is disturbed by plowing.
t	An accumulation of clay deposited from horizons above.
x	A brittle layer referred to as *fragipan*.

The northwestern corner of South Carolina rises into the peaks of the Blue Ridge Mountains. Caesar's Head, pictured here, is a huge rock outcrop in these mountains; it is an exposed igneous pluton. Soil, as you can see, has developed on Caesar's Head. A sufficient soil cover has developed to allow evergreen trees and other plants to establish themselves. From the top of Caesar's Head, you can see Georgia to the south and North Carolina to the north. (South Carolina Dept. of Parks)

THE CLIMATIC EFFECT

When soil is first formed, it is closely related to the rock from which it is derived. But other influences that form the character of the soil gradually become dominant. The single most important influence in soil formation is climate, and the most potent climatic element in the soil-forming process is precipitation. Climate influences soil formation through temperature, too. Weathering, erosion, and leaching—all important in the soil-genesis process—are directly affected and influenced by the temperature and precipitation patterns in a region.

The weathering of parent materials to make soils is continuous throughout each day of the year in hot and humid regions like Hawaii. The type of weathering in Hawaii, for example, is mostly chemical decomposition. In a dominantly cold region, such as subarctic Alaska, weathering is mostly through physical disintegration such as exfoliation and differential expansion.

The development of a soil profile is significantly affected through the leaching process, which is particularly sensitive to the precipitation pattern in a region. Precipitation patterns for the United States and Canada are detailed in Figures 9–20 and 9–23. Soils with calcium-rich upper horizons develop under subhumid to arid conditions where precipitation is less than 24 inches (61 cm) a year because the moisture supply is not sufficient to maintain a continuous downward flow.

With approximately 5 inches (13 cm) or less of annual precipitation, soluble salts, including lime, accumulate on the surface. When annual precipitation is in the range of 12 inches (30 cm), lime is leached downward but it accumulates at a depth of 6 to 20 inches (15 to 50 cm). At or above 24 inches (61 cm) of annual rainfall, most of the lime is leached from the upper layers and may be carried below 4 feet (1.2 m).

The intense weathering and leaching that results when the annual precipitation rises above 30 inches (76 cm) produces acidic soils. Many of the acidic soils have a lot of clay and iron in their B-horizons. Since intense leaching in these humid environments removes large quantities of soluble material—potassium, calcium, and magnesium, for example—from the soil, the acidic soils that develop under these conditions are relatively impoverished.

Climate also influences soil formation indirectly through its action on vegetation. Native vegetation (Fig. 13–2) exerts a tremendous influence on soil formation. A soil profile developed under a forest canopy—recall from page 410—has more well-devloped horizons than a profile developed under grass. Prairie grasses—under which we find deep, dark, uniform surface soils—develop in subhumid to semiarid cli-

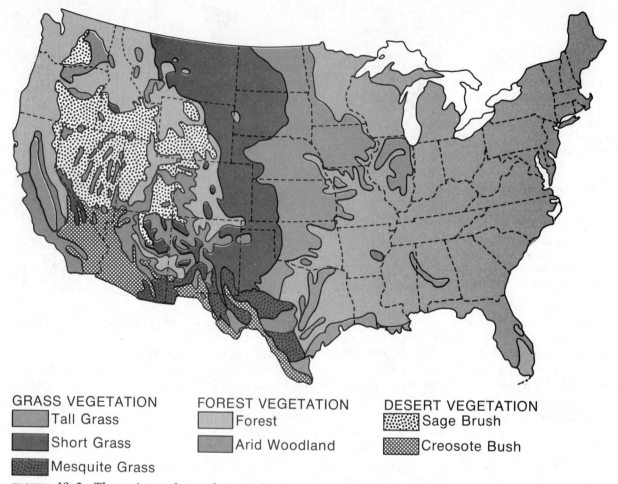

GRASS VEGETATION
█ Tall Grass
█ Short Grass
█ Mesquite Grass

FOREST VEGETATION
█ Forest
█ Arid Woodland

DESERT VEGETATION
▓ Sage Brush
▓ Creosote Bush

FIGURE 13–2 The various colors and tints indicate the pattern of native vegetation within the conterminous U.S. Native vegetation exerts an important influence on soil formation.

mates. Semiarid climatic conditions supply only enough moisture to encourage sparse and short plains grasses that do not afford adequate protection against wind and water erosion. The soil that develops under a cover of plains grass tends to be shallow and does not have a well-developed profile.

BIOLOGIC ACTIVITY

Plant and animal activity and the decomposition of their organic residues and wastes have a direct impact on the development of a soil. Organic matter and its distribution, soil acidity, and the compactness of the soil are characteristics influenced by the life forms in the soil.

Some of the first plants to move in and grow on weathering rock are the **lichens**, which are made up of an **alga** (green plant) and a **fungus** (nongreen plant) growing in symbiotic association. The fungus dissolves minerals out of the rock and passes them along with water to the green plant. The alga uses the minerals and water along with carbon dioxide derived from the air to produce sugars (food), which it shares with the fungus. During the process of living and surviving, lichens and the other plants that follow them in the plant succession of a region take up moisture and chemical substances and, in turn, add organic matter to the soils when they die.

Microorganisms help soil development by bringing about the decay of the organic matter added to the soil by plants and animals. During the process of slowly decomposing organic matter, the microorganisms form weak acids that dissolve minerals faster than water. The work of the microorganisms leaves a decomposed residue, called **humus,** in the soil. The humus is a brown or black complex material that forms an important part of the organic portion of the soil.

Burrowing animals—moles, gophers, earthworms, and ants—are also important in the formation of soils. These burrowing animals have different life styles but perform essentially the same function: they mix, overturn, and aerate the soil. The earthworm, for example, feeds on decaying plant matter but in the process of its movement through the soil in search of food, it tends to mix the soils somewhat. Soils with large numbers of burrowing animals have fewer horizons because of the consistent mixing within the profile that nullifies the downward movement produced by leaching.

Differences in soils that result primarily from differences in the vegetative cover are especially noticeable in areas called **tension zones,** where trees and grasses meet. Figure 13–2 shows that marked tension zones exist in Minnesota, Illinois, Oklahoma, and Texas. In Minnesota, a narrow tallgrass zone forms the western fringe of the state; the A-horizons under the Minnesota grass region are dark and deep and the B-horizons are thin or absent. Under the forest zone, which dominates most of Minnesota, the A-horizons are highly leached and there are well-developed B-horizons.

TOPOGRAPHY

The contour of the surface, called **topography,** is also referred to as **relief.** Topography, or surface relief, influences soil formation primarily through the water and temperature conditions associated with it. On a slope that is quite steep,

Lichens consist of irregularly shaped extensions called *thalli.* Lichens are often classified according to the type of thalli. This lichen is a common foliose rock species, *Parmelia conspersa.* The term *foliose* simply means that the thalli resemble leaves. The scale at the right center of the photo is in millimeters. (Mason Hale, Jr.).

however, gravity makes a direct impact on soil formation because gravity constantly moves the regolith downhill and fresh bedrock is continuously exposed to weathering. The zone of weathered rock under such conditions is extremely shallow; it is difficult for soil to form on a steep slope with an actively moving regolith.

Soils that do develop on hillsides typically have thinner A- and B-horizons as well as less organic matter than their contiguous counterparts in flatter terrain. The development of the A- and B-horizons is conditioned by the amount of water that moves down through the profile. Soils on sloped surfaces simply have less water passing through their profiles because the runoff rate is related to the slope of the terrain. Rapid runoff means there is less water available to run down through the profile and build the various horizons.

Soils that develop in landlocked depressions receive runoff from higher surrounding areas. If the conditions allow a lot of vegetation to develop, the soils that form will have large organic accumulations; peat or muck soils develop under such conditions. When the water that accumulates has a lot of dissolved salt, the depression may become a salt marsh.

TIME

Soil, like other things in nature, passes through a cycle of development. The soil's cycle includes (1) the weathering of bedrock, and (2) the processes that lead to the development of an orderly soil horizon from the regolith. Climate, vegetation, the character of the bedrock, and the topography play an important role in how fast a soil develops and evolves.

There are a number of conditions that can retard the development of a soil profile. Among the most prominent of these conditions are (1) low rainfall which slows the weathering process and places a limit on the amount of leaching that can occur; (2) low relative humidity that limits algal, fungal, and lichen growth; (3) poor aeration and slow water movement caused by a high percentage of clay in the profile; (4) very steep slopes; (5) temperatures that are cold enough to stop or slow chemical processes; (6) severe wind and water erosion that strips the soil and exposes new material; and (7) resistant parent materials, such as granite.

Horizons tend to develop rapidly under warm, humid, forested conditions where there is adequate water and large amounts of organic matter to decompose. In the tropics, a recognizable profile may begin to develop within 100 to 200 years. Under less favorable conditions in a temperate region, it may take more than 2000 years for a well-developed soil profile to emerge.

This is the Great White Throne, a huge mass of stone that rises nearly half a mile (1 kilometer) from the floor of Zion Canyon (37°10′N,113°W) in southwestern Utah. The Virgin River has cut through the sandstone of the region to produce Zion Canyon. It is difficult for soil to form on a steep slope, but (examine the photo carefully) the arrows point to groups of evergreen trees that have developed in soil pockets on a very steep slope on the face of the Great White Throne. (National Parks Service)

PHYSICAL PROPERTIES OF SOILS

Soils seem to exhibit unending differences to the casual observer, and it is often difficult to decide on the properties that need to be described to allow us to differentiate among them. Farmers, however, approach the task in a very practical way. From a farmer's point of view, it is important to decipher and understand the physical characteristics of a soil that affect its use. Since the availability of oxygen in the soils and the mobility of water into or through soils are key factors in the productive use of soils, a cue will be taken from the farmer and the discussion that follows will concentrate on certain physical properties—texture, structure, density, porosity, consistency, temperature, and color—that affect the use of a soil.

SOIL TEXTURE

Soils consist of particles of varying sizes. The soil-particle size groups, referred to as **soil separates,** are sands, silts, and clays. The soil separates and their diameter ranges are reported in Table 13–3. The limits for the various soil separates have been set by the United States Department of Agriculture.

A textural name is assigned to a soil based on the relative proportions of the soil separates found in it. Soils that consist primarily of clay are given the textural name **clay;** those with a preponderance of silt are called **silt;** and a soil with an overwhelming percentage of sand is given the textural-class name of **sand.** A soil that does not contain a dominant amount of any of the separates—a soil consisting of 40 percent sand, 40 percent silt, and 20 percent clay, for example—is called a **loam.**

After a laboratory analysis determines the percentage of sand, silt, and clay in a soil, the textural triangle (Fig. 13–3) is used to assign a textural name to the soil. Since a soil's textural classification is based only on mineral particles of less than 0.08 inch (2 mm), the total of the percentages for sand, silt, and clay always equals 100 percent. When we know the amount of any two separates—sand and silt, for example,—we automatically know the percentage of the third.

A soil whose analysis indicates it is composed of 60 percent sand and 30 percent clay is located at Point A in the triangle (Fig. 13–3). This soil falls in the sandy-clay-loam category. A mere 10 percent of its mineral particles consists of silt.

Table 13–4 lists the analyses of some soils and the textural class assigned to each on the basis of the textural triangle. The silty clay soil in Table 13–4, for example, shows an analysis of 50 percent clay and 45 percent silt. Find 50 percent on the clay scale. Run to the right along the 50 percent clay line until you intersect the 45-percent silt line. The point where these two lines intersect is in the silty-clay section of the triangle. The sand line that intersects with the other two at this point is the 5-percent sand line.

Mineral fragments found in soil that are larger than 0.08 inch (2 mm) fall in the **stone, cobble,** and **gravel** class. These larger fragments are considered part of the soil because they influence its use. When a soil contains a sufficient percentage of a large-fragment class, the name of the large-fragment class precedes the textural

TABLE 13–3   SOIL SEPARATE SIZES

SOIL SEPARATE	DIAMETER RANGE Inch	DIAMETER RANGE Millimeter
Sand		
very coarse	0.08–0.04	2.0–1.0
coarse	0.04–0.02	1.0–0.5
medium	0.02–0.01	0.5–0.25
fine	0.01–0.004	0.25–0.10
very fine	0.004–0.002	0.10–0.05
Silt	0.002–0.00008	0.05–0.002
Clay	less than 0.00008	less than 0.002

TABLE 13–4   REPRESENTATIVE SOILS

TEXTURAL CLASS	SOIL ANALYSIS (%) Sand	Silt	Clay
Clay	20	10	70
Silty clay	5	45	50
Silty loam	20	70	10
Sandy loam	65	25	10
Sandy clay	50	10	40
Clay loam	30	35	35
Silty clay loam	10	55	35
Silt	5	85	10
Loam	45	40	15
Loamy sand	80	15	5
Sand	90	5	5
Sandy clay loam	65	5	30

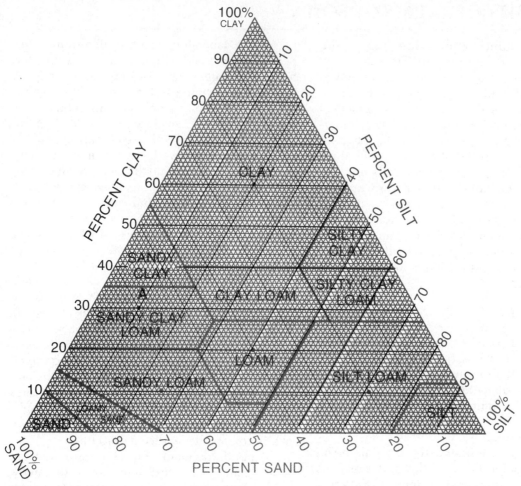

FIGURE 13–3   This is a graphic guide for the textural classification of soil. It is referred to as a textural triangle. Note that the (color) lines representing percentages of sand, clay, and silt converge at point A within the sandy-clay-loam area of the triangle. A soil with these percentages is classified as sandy clay loam.

name. For example, if 20 percent of a sandy loam's weight consists of gravel, the name of the soil becomes gravelly sandy loam. When the large fragment class makes up more than 50 percent of the soil's weight, the word *very* is added to the name; a sandy loam, which contains 60-percent-by-weight of gravel, is referred to as a *very gravelly sandy loam.*

SOIL STRUCTURE

The way in which the soil separates—sand, silt, and clay—are grouped together into stable col-

lections or aggregates is referred to as **soil structure.** Aggregates, which are secondary units or granules, consist of many soil particles bound together by substances that serve as cementing agents. Iron oxides, carbonates, clays, and silica are some of the materials that bind particles in soil into aggregates.

The natural aggregates that occur in soil are known as **peds.** Structural peds form in soil because of alternate swelling and shrinking. Wetting and freezing cause soils to swell; drying and thawing cause soils to shrink. The cracks that define and outline a ped develop along lines of

Both of these soil profiles have been developed under grass. The A-horizon in each profile is dark and deep. The Amarillo, Texas, profile *(left)* is a silty-clay-loam. The Minnesota profile *(right)* is a fine sandy-loam developed on a nearly level to gently rolling sandy outwash of the Mississippi River; it is an example of a somewhat excessively drained prairie soil. (USDA)

weakness in response to the stresses produced by shrinking.

The development of soil structure helps to protect soils from wind and water erosion. Structural peds are much more difficult to erode than discrete soil particles. Structure also modifies the effects of texture since it allows air, water, and plant roots to follow the structural lines of weakness between peds.

Based on the shape and arrangement of peds, four types of soil structure can be described: (1) **platy structure,** (2) prismatic and **columnar structure,** (3) angular blocky and subangular **blocky structure,** and (4) granular and **crumb structure.**

PLATY STRUCTURE
Peds, which consist of thin plates arranged in a horizontal position, exhibit a matted, flattened,

or compressed appearance as shown in Figure 13–4. Frost heaving, fluctuating water tables, compaction from above, and the loss of materials through translocation can cause a separation of soil into horizontal layers. This kind of structure can be readily observed where there is a thin layering of alluvium.

PRISMATIC AND COLUMNAR STRUCTURE
When the swelling-shrinking cycle in the soil takes place primarily in a vertical direction, the peds produced have a long vertical axis as shown in Figure 13–5. The columns of these peds are bounded by somewhat flattened sides. The horizontal dimensions of the peds are somewhat limited and considerably less than the vertical dimension. When the tops and bottoms of the columns are flattened, the peds are referred to as

FIGURE 13-4   This collection of thin plates of soil is referred to as a *ped*. This flattened horizontal structure is said to be *platy*. Soil structure does influence the infiltration rate of water. A number of water droplets are shown above the ped. The one drop of water below the ped indicates the rate of infiltration through a platy structure is slow.

FIGURE 13-6   Count the water drops above and below these blocklike peds. The rate of water infiltration through blocklike peds is moderate.

prismatic. If the tops of these peds are rounded, they are simply referred to as *columnar peds.*

BLOCKLIKE STRUCTURE

The peds in this type of structure resemble imperfect cubes (Fig. 13–6). They may have plane or curved surfaces. A blocklike ped with flattened faces and sharply angular vertices— sharply angular edges where the sides meet—is referred to as an *angular blocky structure.* If the ped has mixed, rounded and flattened faces with many rounded vertices, it is called a *subangular blocky structure.*

SPHEROID STRUCTURE

The peds in a **spheroid structure** are small, sand-sized, imperfect spheres as shown in Figure 13–7. When the peds are relatively nonporous the structure is termed **granular**; porous peds of this structure are given the name **crumb**. The mineral particles in these spheroidal peds are cemented by organic substances. The rounded edges and small size of the peds results from mixing by rodents, earthworms, frost action, and cultivation.

SOIL DENSITY

Density is defined as the mass of an object per unit volume. Water—using the English system—weighs 62.4 pounds per cubic foot, which gives it a density of 62.4 lb/ft³. Using metric units, we calculate water to weigh 1 gram per

FIGURE 13-5   Peds with a long vertical axis are referred to as being *columnar* or *prismatic.* The differentiation is made on the basis of whether the top of the ped is flattened or rounded. Prismatic peds are flattened. The rate of water infiltration through these peds is moderate as indicated by three drops below and five above.

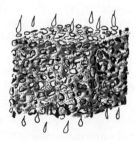

FIGURE 13-7   The peds shown in this figure are imperfect spheres. The fact that the number of water drops above and below this spheroid structure is the same indicates that the peds are arranged in a crumb structure; the rate of infiltration is rapid.

cubic centimeter and its density is reported as 1 g/cm³. In whatever system you make your calculations, the densities of the mineral components of soil are greater than the density of water; the densities of the organic components of soil are less than that of water.

Two different density measurements are commonly made for soils: particle density and bulk density. The measurement called *particle density* does not include the weight of water or the volume of the pore spaces in the soil; it is the density of the solid soil particles only. The density for a volume of soil as it exists naturally, called **bulk density,** includes the volume of the air spaces and the organic materials in a soil sample but not the soil moisture; bulk density is determined on a dry soil sample.

The measurement of bulk density is useful in a number of different ways. It can, for example, be used to calculate water storage capacity per soil volume or to evaluate soil to determine if its compaction is too great to allow root penetration. Aeration problems can also be identified through the assessment of bulk density. The average bulk density of cultivated loam is in the range of 68.6 to 87.4 pounds per cubic foot (1.1 to 1.4 g/cm³). Bulk densities for clay soils need to be below 87.4 pounds per cubic foot (1.4 g/cm³) to promote good plant growth. Sandy soils should have a bulk density below 99.9 pounds per cubic foot (1.6 g/cm³) if plants are to be grown successfully in them.

SOIL POROSITY

The **pore space** in a soil consists of that portion of the soil's volume not occupied by solids. We can formalize this obvious statement into an equation:

$$\text{Pore Space} = 100\% - \text{solid space}$$
$$\text{(as a \%)} \qquad\qquad \text{(as a \%)}$$

Reread the definitions of bulk density and particle density. Bulk density includes the volume of the pore space and is less than particle density. When you divide bulk density by particle density, you have the fraction of the soil's

volume occupied by solid space. We can recast the equation above to read:

$$\text{Pore Space} \atop \text{(as a \%)} = 100\% - \left(\frac{\text{bulk density}}{\text{particle density}} \times 100 \right).$$

Let's say that a clay loam has a bulk density of 1.19 g/cm³ and an average particle density of 2.65 g/cm³. If there were no pore spaces in the clay loam, a cubic centimeter of it would weigh 2.65 grams; but there are pore spaces in it, and, as a result, a cubic centimeter of the clay loam only weighs 1.19 grams. Use the formula to calculate the percentage of pore spaces in the clay loam:

$$\% \text{ Pore Space} = 100\% - \left(\frac{1.19}{2.65} \times 100 \right)$$

The percentage of the clay loam occupied by solid space $\left(\frac{1.19}{2.65} \times 100 \right)$ is equal to 44.9 percent. Subtract the solid space from 100 and you have the percentage of the pore space in the clay loam, which is 55.1 percent.

A sandy soil with a bulk density of 1.5 g/cm³ and an average particle density of 2.65 g/cm³ has 56.6 percent of its volume occupied by solid space. The percentage of pore space in this sandy soil is 43.4 percent.

As the examples above indicate, clay soils tend to contain more total pore space than sandy soils. Since soil particles have irregular shapes, the spaces or pores between them are irregular in size, shape, and direction. Sandy soils tend to be composed of large particles with large and continuous pores that encourage rapid water and air movements through the soil. Although the total number of pore spaces in a clay soil are numerous because of the small size of the clay particles, the individual pore spaces in a clay soil are very tiny and they transmit water very slowly.

The pore spaces in a soil are occupied at all times by either air or water. The relative amounts of air and water in a pore space fluctuates. During a downpour, for example, the water that moves into the soil tends to flush and drive air out of the pore space; but, as the soil water drains downward from the upper pore space, air gradually replaces the water.

Pore space is graded in Table 13–5 as coarse, medium, fine, and very fine. Gravity drains water from pores larger than 0.0012 inch (30 μ). In soils with pore spaces smaller than 0.0012 inch (30 μ), water has difficulty passing through; the small pore spaces tend to become clogged and the soil becomes waterlogged.

The very fine, individual pore spaces in clay soils tend to fill with water that blocks air movement. In fact, the air exchange in clay soils is often inadequate for proper plant root growth. To a growing plant, pore size is significant and relatively more important than the total percentage of pore space.

TABLE 13–5 PORE SPACE SIZE

PORE DESIGNATION	DIAMETER OF PORE	
	Inches	Micrometers*
Coarse	larger than 0.008	larger than 200
Medium	0.008 to 0.0008	200 to 20
Fine	0.0008 to 0.00008	20 to 2
Very fine	less than 0.00008	less than 2

*Micrometer (μ), formerly called a *micron*, is equivalent to one millionth of a meter.

SOIL CONSISTENCY

Large soil systems can rupture, undergo plastic flow, and creep when subjected to various conditions and stresses. The degree of cohesion or adhesion of a soil mass as well as its resistance to deformation or rupture are collectively referred to as *consistency*. This property is important to the engineer who must cope with placing structures and roadways in and on soil and to the farmer who must determine the best kinds of physical processes to employ in cultivating the soil. Soil consistency is measured at three moisture conditions: dry, moist, and wet.

AIR-DRY SAMPLE

Under field conditions, the consistency of an air-dried mass of soil is measured by its resistance to rupturing or fragmentation by hand.

The dry consistency is expressed in terms of hardness, ranging from loose to extremely hard: (1) *loose* is used to mean noncoherent; (2) *soft* is applied when the soil mass is very weakly coherent and fragile and it breaks to powder or individual grains under very slight pressure; (3) *slightly hard* means the air-dry soil is weakly resistant to pressure and is easily broken between

The degree of cohesion of the soil, along this section of the Seine River in France, as well as its resistance to rupture were important to the engineers who designed and built the bridge and roadways. Information about soil consistency is an ongoing concern for farmers in the area who are actively cultivating the soil. (French Embassy)

thumb and forefinger; (4) *hard* means the soil is moderately resistant to pressure and can be broken in the hands without difficulty, but it is barely breakable between thumb and forefinger; (5) *very hard* is applied when the dry soil is very resistant to pressure and can be broken in the hands only with difficulty; and (6) *extremely hard* means the sample is extremely resistant to pressure and cannot be broken in the hand.

MOIST SAMPLE

When a soil is at intermediate moisture—moist but not wet—the sample is tested for resistance to shearing forces. The test is accomplished by placing the moist soil between thumb and forefinger and using the thumb and forefinger to apply pressure and a shearing force to the sample.

The consistency of the moist soil is described as being loose, very friable, friable, firm, very firm, or extremely firm: (1) *loose* means noncoherent; (2) *very friable* applies when moist soil crushes under gentle pressure and coheres when pressed together; (3) *friable* is used to describe a sample that crushes easily under gentle to moderate pressure and coheres when pressed together; (4) *firm* indicates the moist sample crushes under moderate pressure between thumb and forefinger, and resistance is distinctly noticeable; (5) *very firm* means the moist soil sample crushes under strong pressure; and (6) *extremely* firm is reserved for a sample that cannot be crushed between thumb and forefinger.

WET-SOIL SAMPLE

With a mass of soil in the wet condition, its plasticity and its stickiness are measured. **Soil plasticity** is defined as the ability of a sample to change shape continuously under the influence of an applied stress and retain the impressed shape when the stress is removed. Stickiness involves the soil's ability to adhere to another object.

For the field evaluation of the plasticity of a wet-soil sample, roll the soil between thumb and forefinger and observe whether or not the sample can be formed and molded into a thin rod often referred to as a *wire*. The terms applied span the gamut from nonplastic to very plastic:

(1) *nonplastic* means no wire is formable; (2) *slightly plastic* is used when a wire is formable and the soil mass is easily deformable; (3) *plastic* indicates a wire is formable and moderate pressure is required for deformation of the soil mass; and (4) *very plastic* is reserved for a wet-soil sample in which a wire is formable but much pressure is required for deformation of the soil mass.

The field determination of stickiness is accomplished by pressing the wet-soil sample between thumb and forefinger and noting its adherence. Degrees of stickiness range from nonsticky to very sticky: (1) *nonsticky* is applied when practically no soil material adheres to thumb or forefinger after pressure is released; (2) *slightly sticky* means soil adheres to both thumb and forefinger but can easily be removed; (3) *sticky* is used when soil adheres to both thumb and finger, the adhered soil tends to stretch somewhat when thumb and forefinger are moved apart, and the soil is not easily removed from the digits; and (4) *very sticky* is reserved for a wet sample that adheres strongly and definitely and stretches when the digits are moved apart.

SOIL TEMPERATURE

The temperature of a soil undergoes definite changes at different depths, at different times throughout the day, and at different seasons. The temperature changes that result are produced by an interplay between a soil's thermal properties and the variations of insolation that reach the soil surface. For example, although latitude plays a role in the amount of insolation that reaches a surface, at the same latitude dark-colored soils capture a much higher percentage of insolation than do light-colored soils. Soils covered by vegetation and mulches are somewhat insulated by these covers and, thus, tend to be cooler than bare or fallow soils in the same vicinity. The energy that is absorbed by a soil surface is eventually disposed of through radiation, conduction, and convection; heat, for example, is passed to the air above by radiation and convection and to deeper soil layers by conduction.

The average monthly soil temperatures at four stations in South Carolina (Fig. 13–8) are recorded in Table 13–6. The temperatures were

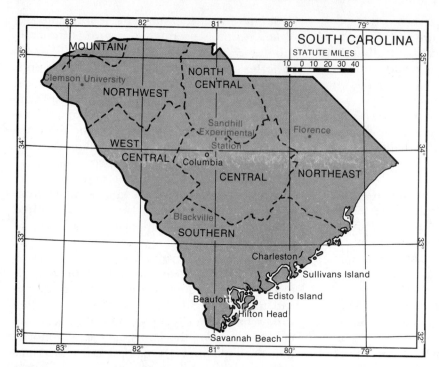

FIGURE 13–8 The soil temperatures at a depth of 4 inches (10 centimeters) at each of these stations in South Carolina are generally higher than the air temperature. The temperature range at the shallow-soil depth of 4 inches (10 centimeters) during any one month is rather large.

taken at a depth of 4 inches (10 cm) below the soil surface. When we compare the average monthly air temperature to the soil temperature at each of these stations, we find that soil temperatures are generally higher than the air temperatures. The monthly extremes of the soil temperatures are recorded in Table 13–7; the temperature range at this shallow depth of 4 inches (10 cm) is rather large.

In most regions of the world, the daily soil temperature seldom changes below a depth of 20 inches (51 cm). We can approximate the soil temperature at or just below the 20-inch (51-cm)

depth by adding 2°F to the average annual air temperature reported in degrees Fahrenheit. This means that the soil temperature at a depth of 22 inches (56 cm) in the Blackville area is about 66.3°F (19.1°C); simply add 2°F to the average annual air temperature (64.3°F) reported in Table 13–6.

Temperature is an influential factor in the development of soil characteristics since it controls the intensity and limits of chemical and biologic processes. For all practical purposes, chemical and biologic activity come to a standstill below 32°F (0°C). The germination of

TABLE 13–6 AVERAGE AIR AND SOIL* TEMPERATURE AT 4-INCH DEPTH IN °F

STATION		JAN	FEB	MAR	APR	MAY	JUN	JUL	AUG	SEP	OCT	NOV	DEC	ANNUAL
Blackville	air	37.5	47.5	61.0	67.2	73.4	79.4	82.2	78.8	76.1	62.1	59.0	47.8	64.3
	soil	40.6	48.5	60.2	71.1	77.8	82.9	87.3	84.2	81.8	67.7	63.4	51.4	68.0
Clemson	air	30.2	41.1	53.9	62.6	70.1	75.8	80.8	77.8	74.1	57.7	52.4	41.3	59.8
University	soil	36.0	41.7	53.6	64.7	76.5	85.3	89.1	84.8	79.4	67.5	60.5	50.6	65.8
Florence	air	33.5	44.6	57.5	66.1	71.8	77.4	83.1	79.3	76.3	59.9	56.3	45.1	62.6
	soil	37.2	44.6	56.0	66.8	73.7	77.8	83.3	78.7	76.4	62.8	57.3	48.0	63.5
Sandhill	air	33.4	44.0	56.5	66.1	70.8	76.1	82.5	78.8	75.7	60.0	56.5	45.0	62.1
	soil	35.7	45.5	57.9	70.4	77.9	83.5	89.4	84.6	82.4	65.8	59.0	47.6	66.6

*Soil temperature taken at a depth of 4 inches (10 centimeters).

seeds and root growth for most plants does not occur below temperatures of 41°F (5°C). Soil temperatures must be within the correct range to get maximum germination and growth of seeds: for corn, soil temperatures must be between 50 and 85°F (10–29°C); the range for potatoes is 60 to 70°F (16–21°C); wheat develops best at 40 to 50°F (4–10°C); and sorghums must have temperatures above 80°F (27°C).

SOIL COLOR

A number of soil characteristics are revealed by color. Whitish colors, for example, are common when salts or carbonate deposits exist in the soil. Spots of different color, called *mottles,* which are usually rust colored, indicate a soil that has periods of inadequate aeration. Very long periods of inadequate aeration, which develop in a waterlogged soil, may lead to bluish, grayish, and greenish subsoils. Within a small region, variations in color—darker colors—may indicate that a soil has more organic matter than an adjacent lighter soil; but, between contrasting climatic regions, color is not a good indicator of the organic content of the soil.

The determination and description of soil color is standardized. A soil's color is evaluated by the use of Munsell color charts. The charts are similar to books of color chips found in paint stores. The notation that defines a soil's color is divided into three parts: hue (dominant rainbow color), value (the relative lightness or darkness of the color), and chroma (purity or strength of the rainbow color).

This corn was grown near Lilongwe (13°59'S, 33°44'E) in Malawi, a small country in the eastern part of southern Africa. The African Rift Valley runs the length of the country from north to south. Lake Nyasa fills most of the valley. West of the lake, the land rises steeply to a plateau about 4000 feet (1200 meters) above sea level. Lilongwe is west of the lake; soil temperatures and air temperatures are conducive to the growth of corn. (World Bank)

TABLE 13-7   MONTHLY EXTREMES IN SOIL TEMPERATURES AT 4-INCH DEPTH IN °F

STATION		JAN	FEB	MAR	APR	MAY	JUN	JUL	AUG	SEP	OCT	NOV	DEC	ANNUAL
Blackville	high	51	67	74	84	88	91	98	96	92	82	74	64	98
	low	29	34	43	53	66	72	76	73	70	53	49	37	29
Clemson University	high	47	56	66	76	91	95	98	96	89	80	68	68	95
	low	27	31	37	51	59	74	78	73	71	60	43	32	27
Florence	high	52	66	70	80	84	93	96	91	86	76	70	60	96
	low	31	34	41	48	60	63	70	70	67	50	45	36	31
Sandhill	high	50	68	75	85	91	97	103	99	96	82	75	61	103
	low	30	30	40	50	62	70	77	72	68	51	42	35	30

SOIL MOISTURE

Water is held within a soil in both the liquid and vapor state since the air in all the pores is saturated with water vapor when liquid water is present. A number of concepts concerning **soil water** were detailed in Chapter 3 (pages 82–84) under the discussion of water in the zone of aeration. A differentiation was made between soil water and groundwater in Chapter 3. The material that follows in this section builds and expands on the concepts in Chapter 3.

Different portions of the water in soil are held at different suctions. The strengths of these suction forces are used as a convenient way of classifying soil moisture. The water films within a soil are identified as gravitational, available, field-capacity, and wilting-point water.

GRAVITATIONAL WATER

At any particular latitude, the force or pull of gravity is of constant magnitude and always acts in a downward direction. That portion of the soil water that will drain freely from the soil under the influence of gravity is called **gravitational water.** This portion of soil water—the gravita-

tional water—is held with suctions of less than one-third bar. In other words, gravitational water can be removed from the soil by the application of a pressure of one-third bar.

Gravitational water is available to a plant as it flows past the plant roots. In permeable soils, gravitational water is present only for short periods of time; thus, it cannot be depended on to satisfy a plant's water requirements.

Generally, the movement of water to or past a root system is not sufficient to meet the demands of the plant for water. Most plants satisfy their tremendous water needs by an extension of their root systems into fresh supplies of water held at low tension in untapped or recently filled soil pores. If a plant is to satisfy its water needs in this manner, the soil must have a structure that permits the rapid extension of a root system.

AVAILABLE WATER

The portion of the stored soil water that can be rapidly absorbed by a plant's root system is held with forces of suction between one-third and 15 bars. In other words, a plant is able to absorb

Upper Volta consists mainly of a huge plateau that slopes gently to the south. The Black, White, and Red Volta rivers have cut valleys through the plateau. The people in a village near Ouagadougou (12°22′N, 1°31′W) diked and flooded a lowland section for the growing of rice. The villagers are shown threshing and winnowing their harvest. The best crops of rice grow in river valleys, on deltas, and on coastal plains where the soil holds water well. Rice needs an average temperature of at least 75°F (24°C), but it grows well at temperatures between 70° and 100°F (21° and 38°C). (World Bank)

water held within these suction forces fast enough to sustain life. This rapidly absorbable water is referred to as **available water.**

Some species of plants can extract and absorb soil water that is held at stronger suctions than 15 bars. However, water held at greater than 15 bars is absorbed very slowly and in very small amounts. Generally, plants wilt when the only water present in the soil is held at 15 bars or more because water is lost through transpiration—through the leaves—at a faster rate than it can be replenished by absorption through the roots at these higher suctions.

WILTING PERCENTAGE

The wilting-point water, or wilting percentage, represents that portion of the soil water held at a suction of 15 bars or more. Moisture held at these tensions, as noted above, cannot be absorbed by plants rapidly enough to offset losses by transpiration.

Wilting occurs any time transpiration exceeds the ability of the plant to absorb water. On a hot, dry day, for example, a corn plant may transpire excessively and temporarily wilt when the available moisture is being held at 1 or 2 bars; a corn plant undergoing temporary wilting will recover at night when transpiration decreases and water absorption brings in a sufficient supply. Of course, if water is not available to the corn plant at suctions below 15 bars, the plant may wilt permanently.

FIELD CAPACITY

The **field capacity** of any soil layer represents the maximum amount of water it can hold against gravity when free drainage exists. A definite tension value cannot be assigned to this equilibrium point; however, some soil scientists consider the field capacity as the percentage of soil moisture held at greater than one-third bar.

Soil texture and organic matter content are important in determining the amount of water soils can hold. In Table 13–8, an average field capacity—measured in depth of water per unit depth of soil—is reported for various soil textures. Large amounts of clay and organic matter content increase the total water retention of a soil, but the large total surface area of clay and organic particles causes a large amount of the water retained to be held by adhesion at high suctions. Soils with small total amounts of clay hold more of the total water they retain at lower suctions. Loams, which are medium-textured soils, can hold large amounts of available water. Much of the water in sandy soils is held at low tensions, too.

The average field capacity of clay, as reported in Table 13–8, is a 4-inch depth of water per 1-foot depth of soil (33.3 cm of water per m of soil). The difference between the wilting point (2.6 inches per foot of clay soil) and field capacity (4 inches per foot of clay soil) is the available moisture (1.4 inches per foot of clay soil). The available moisture in clay amounts to 35 percent of its field capacity. On the other hand, the available moisture in medium sand amounts to 75 percent of its field capacity (Table 13–8).

TABLE 13–8 AVERAGE VALUES FOR THE WATER CONSTANTS OF VARIOUS SOIL TEXTURES

SOIL TEXTURE	FIELD CAPACITY (in/ft)	FIELD CAPACITY (cm/m)	AVAILABLE WATER (in/ft)	AVAILABLE WATER (cm/m)	WILTING POINT (in/ft)	WILTING POINT (cm/m)
Medium sand	1.2	10.0	0.9	7.5	0.3	2.5
Fine sand	1.5	12.5	1.1	9.2	0.4	3.3
Sandy loam	2.0	16.7	1.4	11.7	0.6	5.0
Fine sandy loam	2.6	21.7	1.8	15.0	0.8	6.7
Loam	3.2	26.7	2.0	16.7	1.2	10.0
Silt loam	3.5	29.2	2.1	17.5	1.4	11.7
Clay loam	3.8	31.7	2.0	16.7	1.8	15.0
Clay	4.0	33.3	1.4	11.7	2.6	21.7

SOILS AND PLANT NUTRITION

Although the higher plants contain at least 90 elements, only 16 (page 407) are considered to be essential for plant growth and reproduction. Of the 16 elements needed by plants, animals require all but boron; in addition, animals need sodium, iodine, selenium, and cobalt.

Plants absorb nutrients from the soil mostly in the form of ions, which are atoms that carry a positive or negative electric charge. Water containing dissolved nutrients is referred to as a soil solution. Nutrients move through the soil to a plant's roots where they are absorbed into the plant's cells by means of various mechanisms. In addition, a plant's roots grow into new areas of soil where they meet and intercept ions that are in solution.

SOIL NITROGEN

The most critical element in plant growth is nitrogen (N), which is an essential ingredient in plant proteins and chlorophyll. Plants absorb and use nitrogen in the form of the ammonium ion (NH_4^+) or the nitrate ion (NO_3^-). Unfortunately, only a small percentage of soil nitrogen occurs in these forms at any one time. The nitrate ion (NO_3^-) is highly soluble and easily leached from the soil. In addition, both the nitrate and

Rice is one of the world's most important food crops. About half the world's people eat rice as their chief food. Farmers in upland areas of the Philippines raise hill, or dry, rice, which gets enough water from the rainfall in the upland regions. These rice terraces at Banaue in northeastern Luzon were built 2000 years ago by the ancestors of the present-day farmers. The terraces are built as high as 4000 feet (1200 meters) above valley floors. (Philippine Government)

The dark-colored minerals in the lavas of Hawaii supply magnesium and many other essential minerals to its soils. Farming is an important enterprise in the Hanalei Valley on the island of Kauai in the Hawaiian chain. (USDA)

ammonium ions may be consumed by microorganisms or converted to gaseous nitrogen (N_2) or to gaseous ammonia (NH_3) and lost to the atmosphere.

A major portion of soil nitrogen is derived from nitrogen fixation. In the process of nitrogen fixation, microbial action converts gaseous nitrogen (N_2) taken from the atmosphere into forms of nitrogen that can be used by plants. Another source of soil nitrogen is the decomposition of organic matter, which contains approximately 5 percent by weight of nitrogen. The decomposition of soil organic matter releases ammonium ions (NH_4^+). Bacteria in the soil convert some of the ammonium ions into nitrate ions.

SOIL PHOSPHORUS

The second most critical plant nutrient in soil is phosphorus (P). This nutrient is important to the proper development of a plant cell's nucleus, which controls cell division and growth. Phosphorus tends to concentrate in plant cells that are dividing rapidly. It is found, for example, within the actively growing parts of roots and shoots.

The phosphates used by plants—other than those applied to soils as manufactured phosphate fertilizers—are derived primarily from the phosphates released during the decomposition of organic matter in soils. The phosphate ion used by plants is $H_2PO_4^-$. The total supply of phosphorus in most soils is low because phosphorus is not very soluble and the phosphate ion, $H_2PO_4^-$, tends to react in soils to form insoluble phosphates.

SOIL POTASSIUM

There are adequate amounts of potassium (K) in most soils since potassium is a constituent of the mineral orthoclase feldspar, which is a potassium-aluminum-silicate ($KAlSi_3O_8$). The mineral, however, is very slowly soluble, which means that only small amounts of potassium are available to a plant at any moment in time.

The potassium ion (K^+) helps to maintain cell permeability. It also aids in the movement of carbohydrates throughout the plant. In general, the potassium ion contributes to the well-being of the plant. For example, it maintains iron in a mobile condition within the plant and increases the plant's resistance to disease.

SOIL CALCIUM

Many calcium minerals are fairly soluble and, thus, there are great variations in the calcium (Ca) content of soil. Large quantities of moisture in the soils of humid regions, for example, means that maximum solution of calcium minerals occurs; but most soils are not generally deficient in this mineral because it is fairly prevalent. Of course, sandy soils, which contain no or only a few calcium minerals, and strongly acid soils may not have sufficient amounts of natural calcium for adequate plant growth.

Calcium is crucial to a plant's growth because it is an important constituent in the cell wall. But, in addition, calcium makes cells more selective in their absorption. Since plants are seldom deficient in calcium, the addition of lime (calcium carbonate, $CaCO_3$) to a soil is done to correct the acidity of soils and to improve nutrient absorption.

SOIL MAGNESIUM

Chlorophyll, the essential compound in the process of photosynthesis, contains one atom of magnesium (Mg) in each molecule. Thus, without magnesium, chlorophyll cannot be produced and the basic process of food making (photosynthesis) in green plants cannot occur. In addition, the presence of magnesium seems to be essential in the uptake of phosphorus.

Magnesium occurs in serpentine rocks and in many other dark-colored minerals. Dolomite is a limestone rock that consists of a mixture of calcium and magnesium carbonates ($CaMg(CO_3)_2$). Generally, the amounts of magnesium in soils seem to be adequate, but there are some soils along the Atlantic and Gulf coasts of the United States that are deficient in magnesium.

SOIL SULFUR

Sulfur (S) is required for the synthesis of certain vitamins and proteins in plants. Generally, the quantity of sulfur in soil is low. The major source of soil sulfur is the decomposition of organic matter.

SOIL MICRONUTRIENTS

The term *micronutrient* is applied to iron, manganese, zinc, copper, boron, chlorine, and molybdenum because plants require these elements in very small amounts. All of these elements, except chlorine, play essential roles as activators in numerous enzyme systems within plants. Most of the micronutrients are found in the common primary minerals or rocks from which soil is derived.

THE LANGUAGE OF PHYSICAL GEOGRAPHY

A physical geographer uses a technical vocabulary to make explanations. Review your understanding of the vocabulary used to develop the concepts in this chapter. Among the important words and terms used are:

alga	granular	relief	soil separates
available water	gravel	residual material	soil structure
blocky structure	gravitational water	sand	soil water
bulk density	humus	silt	spheroid structure
clay	hydration	soil	stone
cobble	lichens	soil consistency	subsoils
columnar structure	loam	soil density	tension zones
crumb	microorganisms	soil horizons	topography
crumb structure	parent material	soil morphology	topsoil
essential nutrients	peds	soil plasticity	transported material
field capacity	platy structure	soil porosity	wilting
fungus	pore space	soil profile	

——————————— SELF-TESTING QUIZ ———————————

1. Discuss the variety of ways in which the term *soil* is used.
2. How many of the essential nutrients that are of fundamental importance to plant growth are found in soil?
3. What is meant by a well-developed soil profile?
4. List the major horizons and subdivisions in a soil.
5. Describe the composition of the B-horizons.
6. Discuss the significance of a horizon code.
7. What causes some master horizons to be missing from certain soil profiles?
8. List and contrast the influence of the various factors that influence the rate and kind of soil formation.
9. What is the most potent climatic element in the soil-forming process?
10. How is humus produced?
11. Where does a significant tension zone exist?
12. How is a textural name assigned to a soil?
13. Why do structural peds form in soil?
14. Compare columnar and blocklike structures found in some soils.
15. Contrast the two different density measurements commonly made for soils.
16. Discuss the significance of the pore spaces in soil.
17. What effect does a vegetative cover have on soil temperature?
18. Why is temperature an influential factor in the development of soil characteristics?

IN REVIEW
——————————— SOIL AS AN ESSENTIAL RESOURCE ———————————

I. SOIL FORMATION

   A. Soil Horizons
      1. Horizon Subdivisions
      2. Horizon Code
   B. Factors in Soil Formation
      1. Parent Material
      2. The Climate Effect
      3. Biologic Activity
      4. Topography
      5. Time

II. PHYSICAL PROPERTIES OF SOILS

   A. Soil Texture
   B. Soil Structure
      1. Platy Structure
      2. Prismatic and Columnar Structure
      3. Blocklike Structure
      4. Spheroid Structure
   C. Soil Density
   D. Soil Porosity
   E. Soil Consistency

      1. Air-dry Sample
      2. Moist Sample
      3. Wet-Soil Sample
   F. Soil Temperature
   G. Soil Color

III. SOIL MOISTURE

   A. Gravitational Water
   B. Available Water
   C. Wilting Percentage
   D. Field Capacity

IV. SOILS AND PLANT NUTRITION

   A. Soil Nitrogen
   B. Soil Phosphorus
   C. Soil Potassium
   D. Soil Calcium
   E. Soil Magnesium
   F. Soil Sulfur
   G. Soil Micronutrients

14 *Soil classification and distribution*

Classification is an important form of scientific shorthand, and it is a way of bringing order to an otherwise complex array of information. The only way to sort out and make some sense of the myriad of soils and soil types found worldwide is to establish a system of soil classification. Over the years a number of classification systems have been developed in an attempt to relate parent materials, climate, and landforms to soil types and their distribution.

The traditional system that dominated soil classification during the first half of the twentieth century was organized into three orders: zonal, intrazonal, and azonal soils. The distinction between these orders was made on the basis of the soil's response to environmental conditions. The inadequacies in this system became apparent as new information accumulated and could not be integrated into the classification scheme.

During the 1950s, a new direction was set in soil classification. Soil scientists in the United States and other countries cooperated in developing a new, more comprehensive system. It took approximately 10 years for the Soil Survey Staff of the United States Soil Conservation Service to synthesize information and to prepare the new system. A summary of the classification scheme, known as the *Comprehensive Soil Classification System* (CSCS), was published in 1975. Since CSCS is the official system in the United States, it will be described in a very brief way to give you a general picture of modern practice in soil classification.[1]

---

[1] In the 1950s, the CSCS was judged to be too complicated and too tentative for Canadian needs. Thus, in 1955, Canadians developed their own system of soil classification. An introduction to the Canadian system can be found in Appendix IX. The Canadian system provides taxa for all known soils in Canada.

There are many steep slopes in the Truchas Peaks (35°58′N,105°39′W) area of New Mexico, north of Santa Fe. The peaks are 13,100 feet (3990 meters) above sea level. Elevation changes dramatically in short distances and the soils vary greatly within short distances and with changes in altitude. Boralfs and Argids (Table 14–1), which develop on gently sloping to steep slopes are found in the area. (New Mexico Travel)

THE UNITED STATES SYSTEM

In the CSCS, soils are classified on the basis of characteristics that are observed and measured, which means there is an effort to make the classification quantitative. Since soils often grade smoothly and gradually from one soil to another as an area is traversed, it is necessary to establish a minimum soil volume to represent the smallest distinctive division of the soil in a given region. This minimum volume, called a **pedon,** is a hexagonal column 40 inches (1 m) across that extends downward from the surface as deep as roots grow. A larger unit, referred to as a **polypedon,** is a unit of soil classification that consists of many or all of a particular soil family's pedons in a region.

The CSCS is based on six levels, or categories, of classification: order, suborder, great group, subgroup, family, and series.

SOIL ORDER

This is the most generalized of the six levels. In the CSCS, all soils are classified into one of the ten orders: Alfisols, Aridisols, Entisols, Histosols, Inceptisols, Mollisols, Oxisols, Spodosols, Ultisols, and Vertisols (Table 14–1).

The names for the orders have their origins primarily in Latin and Greek words. Histosol, for example, is derived from the Greek word *histos,* which means tissue. **Histosols** are organic soils. The Latin word *ultimus,* meaning last, is used as the root for the **Ultisol** soil order which is highly leached with an accumulation of clay in the B horizon. **Alfisol** is coined from pedalfer since it has a movement of aluminum (Al), ferric iron (Fe), and clay into the B horizon.

The Latin word *aridus* is translated into English as *dry;* the soil order that is dry for more than 6 months of the year in all horizons is appropriately named **Aridisol.** Soils without pedogenic horizons—without naturally developed horizons—may be thought of as recent soils and, thus, the *ent* of recent is used to form the order's name, **Entisol.** Weakly developed soils with incipient pedogenic horizons are given the name **Inceptisol,** which is based on the Latin word *inceptum* (beginning).

Mollisol is derived from the Latin word *mollis,* which means soft. **Mollisols** have organic-rich surface horizons that tend to feel soft. The French word *oxide* is used to derive the name **Oxisol,** which designates soils that are very highly oxidized throughout their profiles. The Greek word *spodos* translates as wood ash; the soil order with an ashy white to gray color in its A2 horizon is appropriately named **Spodosol.** The Latin word *verto,* which means turn, is used to derive the name **Vertisol,** which is an order of soils with a high content of clay that alternately shrinks and swells.

SOIL SUBORDER

The second level in the CSCS is designated as **suborder** (Table 14–1). The suborders within a soil order are differentiated largely on the basis of soil properties and horizons that result from differences in soil moisture and soil temperature. The suborders are distinctive to each order and are not interchangeable to other orders.

A suborder name is made up of a prefix and a root (Table 14–1). All suborder names contain a root—a portion of the order name—so that the soil order to which the soil belongs is always known. For example, the Ultisols are divided into five suborders, each of which has the root *ult* as part of its name: Aquult, Humult, Udult, Ustult, and Xerult. The root *ult* identifies the suborder as belonging to the **Ultisol** soil order. The portion of the suborder name that precedes the root is the prefix. The prefixes are usually based on Latin or Greek words that suggest properties of the soil. An Aquult, then, is a suborder of the Ultisols; its prefix *aqu* is derived from the Latin word *aqua,* which means water. The Aquults are soils that are overly wet.

The prefixes for some representative soil suborders are identified in Table 14–1. The root that identifies the soil order should be obvious from a perusal of the suborders listed under each order. Andept, for example, is a suborder of the Inceptisols; its prefix is *and;* its root is *ept.*

TABLE 14-1 SOME REPRESENTATIVE SUBORDERS (THIS TABLE AND ITS KEY ARE FOR USE WITH FIGURE 14-1 (UNITED STATES MAP.)

ORDER	SUBORDER	PREFIX	DERIVATION	CONNOTATION
Alfisol: soils with gray to brown surface horizons, medium to high base supply and subsurface horizons of clay accumulation				
A1	Aqualf	aqu	Latin, *aqua*, water	overly wet; seasonally saturated water
A2	Boralf	bor	Greek, *boreas*, northern	cool or cold areas
A3	Udalf	ud	Latin, *udus*, humid	wet all year; temperate or warm and moist
A4	Ustalf	ust	Latin, *ustus*, burnt	dry many months but moist in growing season
A5	Xeralf	xer	Greek, *xeros*, dry	dry in growing season for more than 60 consecutive days
Aridisol: soils with pedogenic horizons, low in organic matter, and dry more than 6 months of the year in all horizons.				
D1	Argid	arg	Latin, *argilla*, white clay	with horizon of clay accumulation
D2	Orthid	orth	Greek, *orthos*, true	typical profile but without horizon of clay accumulation
Entisol: soils without pedogenic horizons				
E1	Aquent	aqu	Latin, *aqua*, water	overly wet; seasonally saturated with water
E2	Orthent	orth	Greek, *orthos*, true	loamy or clayey textures
E3	Psamment	psamm	Greek, *psammos*, sand	sandy or loamy sand textures
Histosol: organic soils				
H1	Fibrist	fibr	Latin, *fibra*, fiber	largely undecomposed organic material
H2	Saprist	spr	Greek, *sapros*, rotten	unrecognizable fibers
Inceptisol: soils that are usually moist, with pedogenic horizons of alteration of parent material but not of accumulation				
I1	Andept	and	Japanese, *ando*, black soil	high volcanic ash content
I2	Aquept	aqu	Latin, *aqua*, water	overly wet; seasonally wet
I3	Ochrept	ochr	Greek, *ochros*, pale	light-colored surface
I4	Umbrept	umbr	Latin, *umbra*, shade	acidic dark-colored surface
Mollisol: soils with nearly black, organic-rich surface horizons				
M1	Aquoll	aqu	Latin, *aqua*, water	overly wet; seasonally saturated with water
M2	Boroll	bor	Greek, *boreos*, northern	cold areas; cool or cold areas
M3	Udoll	ud	Latin, *udus*, humid	seldom have drought; temperate or warm and moist
M4	Ustoll	ust	Latin, *ustus*, burnt	dry many months but moist growing season
M5	Xeroll	xer	Greek, *xeros*, dry	dry growing season
Oxisol: soils with pedogenic horizons that are mixtures principally of kaolin, hydrated oxides, and insoluble quartz sand, and are low in weatherable minerals.				
O1	Orthox	orth	Greek, *orthos*, true	hot, nearly always moist
O2	Ustox	ust	Latin, *ustus*, burnt	warm or hot, dry for long periods of time but moist more than 90 consecutive days in the year with a wet growing season
Spodosol: soils with accumulation of amorphous materials in subsurface horizons				
S1	Aquod	aqu	Latin, *aqua*, water	overly wet; seasonally saturated with water
S2	Orthod	orth	Greek, *orthos*, true	typical profile with subsurface accumulation of iron, aluminum, and organic matter
Ultisol: soils that are usually moist with a horizon of clay accumulation and a low base supply				
U1	Aquult	aqu	Latin, *aqua*, water	overly wet; seasonally saturated with water
U2	Humult	hum	Latin, *humus*, earth	high-humus surface
U3	Udult	ud	Latin, *udus*, humid	wet all year with a low organic content
U4	Xerult	xer	Greek, *xeros*, dry	dry during growing season
Vertisol: soils with high content of swelling clays and wide, deep cracks at some seasons				
V1	Udert	ud	Latin, *udus*, humid	wet all year; cracks open for short periods of less than 3 months per year
V2	Ustert	ust	Latin, *ustus*, burnt	dry many months, moist in growing season; cracks open and closed twice a year and remains open more than 3 months

GREAT GROUP

Soil **great groups** are subdivisions of suborders. Approximately 225 great groups have been identified worldwide. About 185 great groups are found in the United States.

Great groups are established by differentiating among soil horizons and soil features. For example, some soils have horizons that have accumulated clay, iron, and humus. Others have horizons that are hardened or cemented— pans—which interfere with water movement and root penetration. The soil features that are used in establishing great groups are the self-mixing properties—expansion and contraction—as well as soil temperature and major differences in the content of calcium, magnesium, sodium, potassium, and gypsum.

SUBGROUP

Each soil great group in the CSCS is divided into three kinds of subgroups: The first subgroup represents the central, called *typic*, segment of the soil group. The second subgroup has properties that integrade to other great groups. The third subgroup generally has properties that prevent its classification as typic or an integrade. About 1000 subgroups have been identified.

SURVEY OF SOIL ORDERS

In the sections that follow you will find a general discussion of each order. At the end of each discussion, reference is made to a series of soil maps. The map in Figure 14–1 shows the distribution of the soil orders and some prominent suborders throughout the United States. Soil maps for each of the continents are depicted in Figures 14–2 (North America), 14–3 (South America), 14–4 (Africa), 14–5 (Eurasia), and 14–6 (Australia).

ALFISOLS

A thumbnail sketch of this order and its important suborders is found in Table 14–1. These

FAMILY

The fifth level in the CSCS is the family. A soil family is a division within a subgroup. The distinction among soil families is made on the basis of soil properties important to the growth of plants or the behavior of soils when used for engineering purposes.

The soil properties used to differentiate among families include texture, mineral composition, acidity-alkalinity, soil temperature, the region's precipitation pattern, permeability, thickness of horizons, structure, and consistency. Approximately 5000 soil families have been identified.

SERIES

The sixth level in the CSCS is series. More than 10,500 soil series have been identified in the United States. A series—as a division within the soil family—has a narrower range of characteristics than the soil family.

The name of a series has no pedogenic, i.e., soil formation, significance. A series name represents a prominent geographic name of a river, town, or region near where the series was first recognized. Soil series are differentiated on the basis of observable and mappable soil characteristics such as color, texture, structure, consistency, and acidity-alkalinity as well as the number and arrangement of horizons in a soil pedon.

soils develop from high-lime parent materials under marginal-moisture forests or wetter areas where there is sufficient precipitation to move enough clay downward to form a clay-accumulation layer in the B horizon. The clay-accumulation horizon generally develops between 9 and 29 inches (23 and 74 cm) below the surface.

Most of the suborders have water available for plant growth for 3 or more warm-season months. When relief and climate are favorable, most Alfisols produce well when converted to cropland. Most of the suborders (Table 14–1) are leached of lime to at least 2 feet (0.61 m) deep and are slightly to moderately acid in the

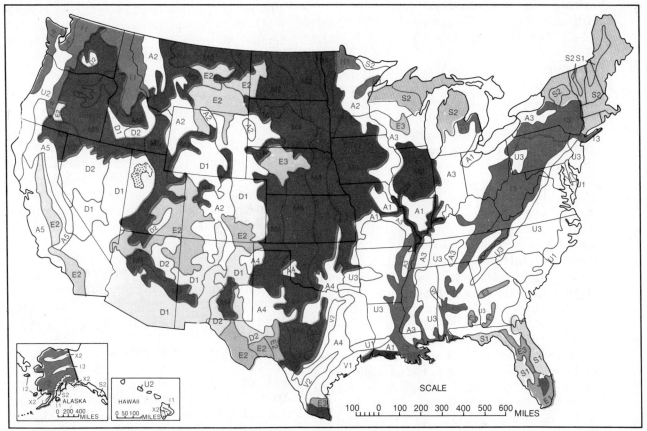

FIGURE 14–1   Reference should be made to Table 14–1 in deciphering the soil-pattern of the United States as indicated by the code within each unit.

surface horizon. The clayey accumulation in the subsurface horizons is not very favorable to plant growth, however.

UNITED STATES PATTERN

Large areas of Alfisols are found in the North Central states and the mountain states (Fig. 14–1). Suborders of the Alfisols are found on gently to moderately sloping land around Lake Ontario in New York. They are also found in Michigan and Ohio at the western edge of Lake Erie. Udalfs (A3 in Table 14–1) cover portions of Ohio, Kentucky, Indiana, and Michigan (Fig. 14–1). The Aqualfs (A1 in Table 14–1) are found on flat to gently sloping land that is seasonally saturated with water. The Aqualfs can be used for general crops if drained; if undrained, pasture and woodland develop on this kind of soil.

Boralfs (A2 in Table 14–1) are found in northern Minnesota around the western edge of

Lake Superior (Fig. 14–1). The soil is good for some small grain crops, but mostly woodland and pasture develop on it. The Boralf suborder is also found on steep, wooded slopes in Montana, Wyoming, Colorado, and a small section of western South Dakota.

Ustalfs (A4 in Table 14–1) are found in warm regions that are intermittently dry for long periods (Fig. 14–1). This suborder of the Alfisols is well-suited for range, small grain, and irrigated crops. Large pockets of Ustalfs (A4 in Table 14–1) are found in Texas. Xeralfs (A5 in Table 14–1) cover a significant area of California where it is warm and continuously dry in summer for long periods; the precipitation that does fall in the region is confined primarily to the winter period.

A perusal of Figure 14–1 shows a rather significant portion of the United States is covered by Alfisols. In fact, an accurate computation

WORLDWIDE DISTRIBUTION

Alfisols occur on all continents (Figs. 14–2, 14–3, 14–4, 14–5, and 14–6). They occupy 13.2 percent of the land area of the world. Humid and subhumid climates characterize the climate where Alfisols occur. Tallgrasses, savanna, and oak-hickory forests characterize the native vegetation that develops on Alfisols.

Significant areas of Alberta, Saskatchewan, and Manitoba in Canada are covered by Boralfs[2] (Fig. 14–2). One large pocket of Udalfs (A2 in Table 14–2) and two smaller pockets of Ustalfs (A3 in Table 14–2) are found in Mexico. In South America (Fig. 14–3), eastern Brazil contains a large area of Ustalfs and there is an elongate strip of Xeralfs (A4 in Table 14–2) along the coast of Chile from approximately 29°S to 36°S latitude. The Xeralfs of Chile are warm and continuously dry for long periods during the summer; they are, however, moist during the winter.

There are significant areas of Alfisols in western Europe (Fig. 14–5). Two important pockets of Alfisols are also found in China and there is one in Southeast Asia. A rather large portion of India is covered by Ustalfs (A3 in Table 14–2). A wide strip of Ustalfs spreads eastward across Africa (Fig. 14–4) from its western shore on the Atlantic; the strip is centered at about 10°S latitude. There are also important pockets of Alfisols on the eastern shore of Africa from the Equator south to the Cape of Good Hope. The western half of Madagascar is covered by Ustalfs (Fig. 14–4). Australia (Fig. 14–6) has four pockets of Alfisols; two of the southern pockets are Xeralfs (A4 in Table 14–2), which is to be expected with Australia's climatic pattern.

ARIDISOLS

The soils in this order develop in areas of low rainfall and from almost any kind of parent material. Since precipitation is minimal, there is an absence or a low incidence of leaching. Thus, in one or more horizons there is usually a concentration of calcium carbonate, gypsum, or soluble salts. There is a very brief summary of the order and its suborders in Table 14–1.

Aridisols have well-developed pedogenic horizons. Some of the profile features occurring in this order are: lime layers, salt or gypsum accumulations, low organic matter accumulation, and a calcareous profile. Some of the Aridisols have lime-cemented hardpans.

The natural vegetative cover of Aridisols is scattered desert shrubbery such as sagebrush, mesquite, creosote bush, and shortgrasses. The Aridisols, however, are among the most productive soils when they are irrigated and fertilized. Since they develop in dry climates, the Aridisols have large water requirements when they are brought under cultivation. The application of significant amounts of water during cul-

[2] The Boralfs are referred to as Gray Luvisols in the Canadian system of soil classification (see Appendix IX, Table A–1).

Tobacco is an important cash-crop in the agricultural economy of Malawi. This field of tobacco is being grown in an Alfisol—that is, in the suborder Ustalf. (World Bank)

FIGURE 14–B   (1 & 2) Two views of the dune area at Cape May Point, New Jersey. Beach grass, the vegetative cover, is apparent in both photographs. Soils of the area are not stable owing to the prevailing sea and wind conditions along the coast. In the photo on the right, note the thin layering of this very sandy soil in which there is practically no horizon development. The CSCS system classifies this soil as an Entisol (order), Psamments (sub-order), and Quartzipsamments (great group). The fence was erected to help protect against natural and man-made erosion. (Photos by Lloyd Black)

TABLE 14–3   SOILS OF THE WORLD*

A ALFISOLS	M MOLLISOLS	V VERTISOLS
A1 – BORALFS	M1 – ALBOLLS	V1 – UDERTS
A2 – UDALFS	M2 – BOROLLS	V2 – USTERTS
A3 – USTALFS	M3 – RENDOLLS	
A4 – XERALFS	M4 – UDOLLS	
	M5 – USTOLLS	SOILS IN AREAS WITH MOUNTAINS
	M6 – XEROLLS	
D ARIDISOLS		X1 – CRYIC (incl. Spodosols)
	O OXISOLS	X2 – CRYIC (incl. Alfisols)
D1 – UNDIFFER- ENTIATED		X3 – UDIC
D2 – ARGIDS	O1 – ORTHOX	X4 – USTIC
	O2 – USTOX	X5 – XERIC
		X6 – ARIDIC
E ENTISOLS		X7 – USTIC AND CRYIC
	S SPODOSOLS	X8 – ARIDIC AND CRYIC
E1 – AQUENTS		
E2 – ORTHENTS	S1 – UNDIFFER- ENTIATED	MISCELLANEOUS
E3 – PSAMMENTS	S2 – AQUODS	
	S3 – HUMODS	Z Z1 – ICEFIELDS
	S4 – ORTHODS	
H HISTOSOLS		Z Z2 – RUGGED MOUNTAINS
H1 – UNDIFFER- ENTIATED	U ULTISOLS	
		••••• SOUTHERN LIMIT OF CONTINUOUS PERMAFROST
	U1 – AQUULTS	
I INCEPTISOLS	U2 – HUMULTS	
	U3 – UDULTS	— — SOUTHERN LIMIT OF DISCONTINUOUS PERMAFROST
I1 – ANDEPTS	U4 – USTULTS	
I2 – AQUEPTS		
I3 – OCHREPTS		
I4 – TROPEPTS		
I5 – UMBREPTS		

*The key detailed above is to be used with the continent-wide maps identified as North America (Fig. 14–2), South America (Fig. 14–3), Africa (Fig. 14–4), Eurasia (Fig. 14–5), and Australia (Fig. 14–6).

shows that 13.5 percent of the total land area of the United States is covered by Alfisols.

tivation can—through the process of leaching—produce salt accumulations.

UNITED STATES PATTERN

Figure 14–1 shows that Aridisols are located primarily in the western Mountain and Pacific states. An accurate computation shows that the land area covered by Aridisols amounts to 11.6 percent of the total surface area of the United States. Climatically speaking, these regions are characterized by low rainfall. Scattered grasses and desert shrubs dominate the vegetative cover.

Argids (D1 in Table 14–1) dominate most of Nevada, the southeastern section of California, southern Arizona, and western New Mexico. More than 50 percent of Wyoming is also covered by Argids. The Orthids (D2 in Table 14–1) are found as significant pockets throughout the region. One very obvious isolated pocket is found in the middle of Washington. Another somewhat isolated Orthid pocket is found in eastern Colorado.

WORLDWIDE DISTRIBUTION

A continent-by-continent survey reveals that Aridisols cover more area on a worldwide basis than any other soil order. In fact, 18.8 percent of the total land surface of the Earth is covered by Aridisols. An examination of Figure 14–4 shows that most of North Africa—the Sahara Desert—is covered by Aridisols. Central Asia (Fig. 14–5) also seems to be dominated by Aridisols. Mexico (Fig. 14–2), southern South America (Fig. 14–3), southern and eastern Africa (Fig. 14–4), and Australia (Fig. 14–6) all have significant Aridisol covers.

ENTISOLS

The summary of this order in Table 14–1 indicates that these are soils of slight development without distinct pedogenic horizons. The absence of horizons does not mean that Entisols are simple, identical soils. Rather, they include among their number deep sands, stratified river-deposited clays, recent volcanic-ash deposits, and dry arid-lake beds. Some Entisols—floodplains and volcanic ash—make excellent agricultural soils.

The absence of distinct pedogenic horizons may be due to any one of a number of reasons. The simplest explanation, of course, is that the elapsed time from the deposit of a volcanic ash or river alluvium to the present has been insuf-

Cradled between the Sierra Nevada on the east and the Diablo Range on the west, California's fertile Aridisol-covered San Joaquin Valley is a farmer's paradise. Most of the valley gets less than 10 inches (25 centimeters) of rainfall, and irrigation is used extensively. Shallow layers of impermeable clay, however, act as obstacles to the downward movement of the irrigation water and large salt accumulations have begun to develop in the soil of the San Joaquin Valley. A massive salt buildup is in the foreground of this furrowed site south of Fresno. (California Dept. of Water Resources)

ficient to develop horizons. Another deterrent to horizon development exists when the Entisol is on a steep slope where the rate of surface erosion equals or exceeds the rate of soil-profile formation. Ecological conditions in very dry deserts and in permafrost regions also work to prevent horizon formations. And, then, there is the ultimate deterrent: some parent materials such as quartz and sand are simply too inert to develop horizons.

UNITED STATES PATTERN

The Entisols cover 8 percent of the surface area within the United States. An examination of Figure 14–1 indicates this order is widely dispersed. A review of the precipitation pattern for the United States shows that some Entisols are found in regions where rainfall is high and others are found in regions where rainfall is low. The vegetative cover seems to be varied. For example, some Entisols support forests and others have a

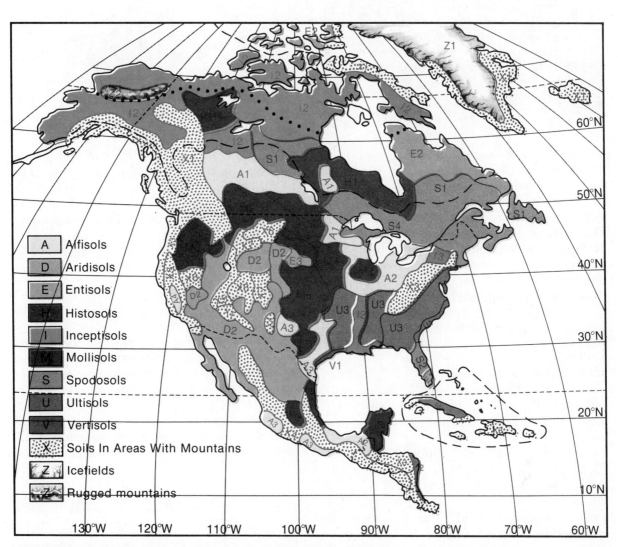

FIGURE 14–2 Reference should be made to Table 14–2 in deciphering the soil-pattern of North America as indicated by the code within each of the tinted areas.

The Negev *(left)*, which covers more than half of Israel from Berbsheba to the Gulf of Aqaba, is covered by Aridisols. Irrigation enables the Israelis to farm part of the Negev. The soils around this dry lake *(below)* in western Australia are Aridisols, too. (Govt. of Israel, *left;* Exxon, *below*)

cover of grass. Precipitation and vegetation are not diagnostic of this soil order.

The Entisols of southern Florida are Aquents (E1 in Table 14–1). The suborder found from central to northern Florida, in two pockets in Georgia, in southeastern Alabama, and in Nebraska, is Psamment (E3 in Table 14–1). Orthents (E2 in Table 14–1), are found in southern California, southwestern Texas, and Montana. A quick perusal of Figure 14–1 shows that this quick recital has not identified all the significant pockets of Entisols in the United States.

WORLDWIDE DISTRIBUTION

Entisols are found on all continents and at all latitudes (Figs. 14–2, 14–3, 14–4, 14–5, and 14–6). They occupy 8.3 percent of the total land surface of the world.

Entisols are found well within the Arctic Circle on the northern islands of Canada (Fig. 14–2). A shallow covering of Orthents (E2 in Table 14–2) is found in Canada's heavily glaciated Ungava Peninsula (60°N,74°W) as well as in the surrounding areas of Quebec and Labrador.* Psamments (E3 in Table 14–2) occur as large shifting dune-sand areas within the Sahara Des-

ert (Fig. 14–4). A good portion of the Arabian Peninsula is also covered by Psamments (Fig. 14–5). The Entisols of China are Aquents (E1 in Table 14–2), which are found on the floodplains and deltaic plains of the Yellow and Yangtze rivers (Fig. 14–5).

HISTOSOLS

The thumbnail sketch of this order in Table 14–1 indicates that Histosols are organic soils. The organic content of Histosols runs from a minimum of 12 percent to more than 20 percent. This substantial quantity of organic material gives these soils high water-retention capacities and low bulk densities. Sometimes, bulk densities are as low as 6.3 pounds per cubic foot (0.1 gram/cm³).

Histosols can form in any climate and at any latitude. Humid climates and the accumulation

---

*Some of the Entisols of Canada are underlain by permafrost and are referred to as *Cryosols* in the Canadian system of soil classification. Other "recent" soils found in Canada that are not underlain by permafrost are referred to as *Regosols* in the Canadian system (See Appendix IX, Table A–1.)

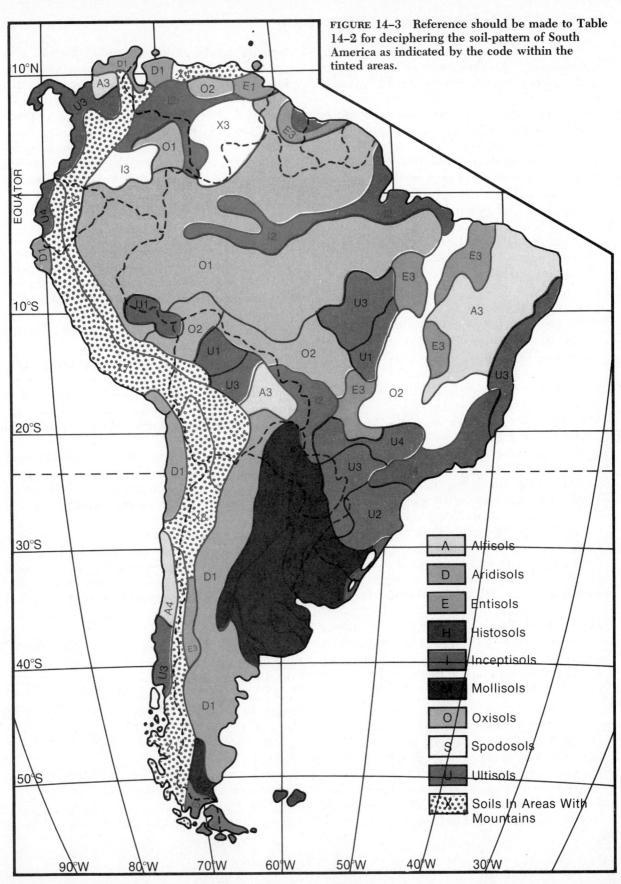

FIGURE 14–3 Reference should be made to Table 14–2 for deciphering the soil-pattern of South America as indicated by the code within the tinted areas.

A	Alfisols
D	Aridisols
E	Entisols
H	Histosols
I	Inceptisols
M	Mollisols
O	Oxisols
S	Spodosols
U	Ultisols
X	Soils In Areas With Mountains

441

of water in ponds and lakes are conducive to the accumulation of organic matter. The formation of Histosols is rather common in cold, wet areas such as Alaska, Canada, and Finland.

The classification of Histosols is based on physical and chemical properties. Among the physical properties considered are the kind of organic material the soil contains, the extent of decomposition of the organic matter, thickness of the soil, bulk density, water-holding capacity, temperature, and permeability. An analysis of the soil for carbonates, bog iron, sulfates, sulfides, acidity-alkalinity, and carbon-to-nitrogen ratio is used to establish the chemical parameters of Histosols.

UNITED STATES PATTERN
Histosols cover 0.5 percent of the land surface of the United States (Fig. 14–1). The Saprist suborder (H2 in Table 14–1)—consisting of decomposed muck—is found in Florida (in the Everglades), in Georgia (in the Okefenokee Swamp), and in the Mississippi Delta along the southern coast of Louisiana. Two prominent pockets of Fibrists (H1 in Table 14–1)—consisting of woody peats, largely undecomposed—are found in Minnesota.

Many acres of Histosols in the Everglades of Florida have been drained and used as farmland for a number of years. When drainage systems are established in these soils, a variety of problems develop. The soil organic matter, for example, decomposes rapidly and significantly reduces the soil volume; the compaction that results has caused the Histosol surface in some locales to subside by as much as 12 feet (3.7 m) over a 40-year period. In addition, a drained Histosol becomes a fire hazard because its rich, dry, organic matter ignites easily, and once a fire starts, it flourishes on the inexhaustible supply of fuel readily at hand in the soil.

WORLDWIDE DISTRIBUTION
Histosols are found on all continents but, more often than not, in areas too small to be shown on most soil maps. On a worldwide basis, Histosols cover about 0.9 percent of the total land surface.

Large accumulations of Histosols are found in Wales, Scotland, and Ireland (Fig. 14–5). The stagnant marshes and swamps of Ireland are especially conducive to the accumulation of organic matter and the development of Histosols.

There are two large accumulations of Histosols in Canada (Fig. 14–2). The smaller of the two is found in the Northwest Territories. And what is certainly the largest continuous body of Histosols* in the world makes up the second Canadian accumulation. It is found to the south of Hudson Bay and spreads across Manitoba and Ontario and into Quebec.

INCEPTISOLS

A brief summary of this order is found in Table 14–1. Inceptisols—for somewhat the same reasons cited for Entisols—are weakly developed soils. They do, however, have pedogenic horizons and are, therefore, more developed than Entisols.

The suborders that make up the Inceptisols are very different from one another. The Aquepts, for example, are poorly drained; the Andepts have a high-volcanic-ash content; and the Ochrepts are light-colored soils with weak B-horizons of clay.

The Inceptisols are found over a wide range of latitudes. They develop primarily on young geomorphic surfaces. Some Inceptisols make good agricultural soils; others do not.

UNITED STATES PATTERN
Inceptisols cover 18.2 percent of the land surface in the United States. A large area of Inceptisols (I3 in Table 14–1) runs northeastward from northern Alabama to New York (Fig. 14–1). These soils develop mostly under trees but sometimes under grasses.

Umbrepts (I4 in Table 14–1), with thick-colored surface horizons rich in organic matter, are found along the Washington and Oregon coasts (Fig. 14–1). Further inland, a strip of Andepts (I1 in Table 14–1) runs from the Canadian border southward through Washington and Oregon; another strip of this same suborder is found in Idaho.

In Figure 14–1, the Inceptisol strip that runs northward from the Histosol region of the Mis-

*Histosols correspond to the Organic order in the Canadian system of soil classification. (See Appendix IX, Table A–1.)

sissippi Delta belongs to the Aquept suborder (I2 in Table 14–1). This zone of Aquepts is actually on the floodplain of the Mississippi River. This suborder is developing on the accumulated alluvial sediments of the floodplain.

Andepts (I1 in Table 14–1) are found on the big island of Hawaii (Fig. 14–1). These Andepts commonly have a dark surface horizon and are rich in humus. This suborder forms chiefly from volcanic-ash deposits, which are quite plentiful on the island of Hawaii.

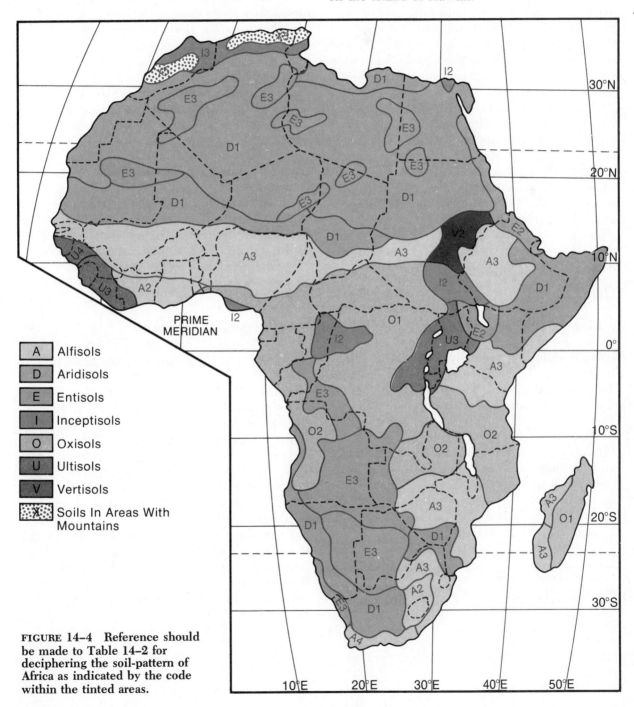

A	Alfisols
D	Aridisols
E	Entisols
I	Inceptisols
O	Oxisols
U	Ultisols
V	Vertisols
X	Soils In Areas With Mountains

FIGURE 14–4 Reference should be made to Table 14–2 for deciphering the soil-pattern of Africa as indicated by the code within the tinted areas.

Alaska—as shown in Figure 14-1—is almost completely covered with Inceptisols. Tundra climatic zones seem to be especially favorable locations for the development of Inceptisols.

WORLDWIDE DISTRIBUTION

About 8.9 percent of the land surface of the entire Earth is covered by Inceptisols. Accumulations of Aquepts (I2 in Table 14-2) spread across northern Canada (Fig. 14-2); a sizeable portion of these Canadian Aquepts are in the region of permafrost.*

Inceptisols tend to develop on recently accumulated alluvial sediments of floodplains, as noted in the discussion of the Mississippi floodplain on page 443. The extensive strip of Aquepts (I2 in Table 14-2) found in Brazil (Fig. 14-3) is actually on the floodplain of the Amazon River. The Ganges-Brahmaputra floodplain in northern India (Fig. 14-5) is also covered by Aquepts. The deltas of the Nile (Fig. 14-4) and the Mekong (Fig. 14-5) have a covering of Aquepts, too.

MOLLISOLS

A thumbnail sketch of this order and its suborders is found in Table 14-1. Mollisols are dark-colored soils that typically form under grasslands with a moderate to pronounced seasonal soil-water deficiency. The surface horizons of this order are saturated with bases—mostly calcium.

Five suborders are listed in Table 14-1 for this order. Two additional suborders, Albolls and Rendolls, are not listed. This rather large number of suborders is indicative of the diverse nature of these soils. The Mollisols are widely distributed and are found in cold, temperate, humid, and semiarid climates. Some suborders—Rendolls, for example,—form in par-

ent material high in calcium carbonate. Others, such as the Albolls, are well leached.

Xerolls are the driest of the Mollisols. Some of the drier Mollisols are used to grow dryland grains. Although the Udolls are quite moist, they are good agricultural soils and are very productive. Even though Mollisols are very fertile, they produce their optimum yields when fertilizer is used judiciously.

UNITED STATES PATTERN

The Mollisols are the most extensive of the soil orders in the United States. This order covers more than 25.1 percent of the land surface in the 50 states.

Mollisols dominate the Great Plains where they extend north into Canada and south to the Gulf of Mexico (Fig. 14-1). Extensive areas of Xerolls (M5 in Table 14-1) are found in Washington, Oregon, Idaho, northern California, and northern Nevada. A significant strip of Mollisols runs from Montana southward across eastern Idaho and central Utah. Another strip slashes across Arizona into the central portion of western New Mexico.

WORLDWIDE DISTRIBUTION

On a global basis, Mollisols cover 8.6 percent of the land surface. The largest continuous belt of Mollisols in the world is found on the Eurasian landmass (Fig. 14-5); the belt extends eastward from Romania across the steppes of Russia, through the grassy plateaus of Mongolia, and into China. Another large area of Mollisols is found covering the Pampas of Argentina and Uruguay (Fig. 14-3). Three areas of Mollisols are found in Mexico: in central Mexico (Ustolls), along Mexico's Gulf coastal plain (Ustolls), and in the Yucatan Peninsula where Rendolls are found (Fig. 14-2). An important area of mollisols in Canada represents a northward extension of the Great Plains; in Canada, the Borolls of the CSCS are referred to as *Chernozems* in the Canadian system. (See Appendix IX, Table A-1.)

---

*Aquepts underlain by permafrost are referred to as Cryosols in the Canadian system. Inceptisols found in the non-permafrost areas of Canada are generally referred to as *Brunisols* in the Canadian system. (See Appendix IX, Table A-1.)

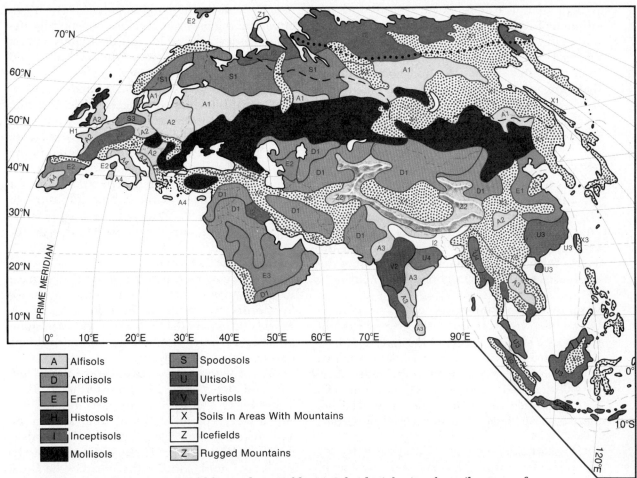

FIGURE 14–5   Reference should be made to Table 14–2 for deciphering the soil-pattern of Eurasia as indicated by the code within the tinted areas.

OXISOLS

This order is the most intensely weathered of all soils. They typically develop under hardwood forests in tropical and subtropical areas that are continuously hot and humid. High temperatures and ample precipitation are the key factors in the extreme weathering that characterizes the Oxisols.

Oxisols develop on old upland, medium-to-fine textured parent materials. The intense weathering converts most of the minerals, other than quartz, into crystalline kaolinitic-type clays and amorphous iron and aluminum oxides. Thus, weatherable minerals in Oxisols are absent or present in only very small amounts.

Oxisols are very low in nutrients. Cultivation of this soil order requires a well-balanced and carefully controlled program of fertilization. When phosphorus is simply added to Oxisols, for example, it combines readily with the iron in the soils and forms insoluble iron phosphates, which plants are not able to utilize. However, with a carefully devised program, adequate nitrogen, phosphorus, and potassium can be maintained in Oxisols so that bananas, sugarcane, coffee, rice, and pineapples can be grown productively.

A unique feature of Oxisols is the presence of plinthite, a noncemented soil material that is high in iron oxides. Some Oxisols that are stripped of vegetation, exposed, and eroded may, under the influence of repeated wetting and drying, harden due to the formation of ironstone from plinthite. In fact, the ironstone hardpans that form are so strong that they can be cut into bricks and used in the construction of buildings. Oxisols that are allowed to develop ironstone hardpans are impossible to cultivate.

UNITED STATES PATTERN

Oxisols occur only in Hawaii, Puerto Rico, and the Virgin Islands. The total area covered by Oxisols amounts to a mere 0.01 percent of the land surface of the United States. In fact, the Oxisol areas in Hawaii, Puerto Rico, and the Virgin Islands are too small to be shown at the scale used in the construction of our maps.

WORLDWIDE DISTRIBUTION

On a global basis, Oxisols cover 8.5 percent of the Earth's surface. Figures 14–3 and 14–4 show that extensive areas of tropical South America and Africa are covered by Oxisols. Both suborders Orthox (O1) and Ustox (O2) are found in South America and Africa. The eastern half of the island of Madagascar is covered by Orthox (O1).

Oxisols can be made highly productive for carbohydrate crops (bananas and sugarcane) and oil crops (coffee) since both of these crops consist primarily of carbon, hydrogen, and oxygen atoms, all of which are derived from air and water rather than from the soil. Another consideration in the productive use of Oxisols is the hours of daylight during a crop's growing season. Oxisols are found close to the Equator; daylight hours seldom exceed 12.5 hours over the range of the Oxisols. Some crops will not flourish with such a limited span of daylight. Corn production, for example, under a 12-hour-daylight span, is much less than in temperate regions where there are 14 to 15 hours of daylight during the growing season.

SPODOSOLS

This order has an acid sandy profile with an ashy white upper horizon over a dark-brown B horizon and yellowish subsoils. These soils are usually well leached and tend to be low in calcium and magnesium. Spodosols generally develop under coniferous and deciduous forests where temperatures run to the cool side and there is ample precipitation.

When Spodosols are brought under cultivation, a well-developed program of fertilization must be used to make them productive. Lime must also be added to these soils to decrease their acidity. Of course, the crop planted in the Spodosol determines the quantity of lime that needs to be used. Some acid Spodosols for example, are used without lime additions to grow blueberries. Strawberries, raspberries, potatoes, grains, and corn do well in these soils.

UNITED STATES PATTERN

Spodosols cover 4.8 percent of the land surface in the United States. Three large areas of Spodosols are found in Florida (Fig. 14–1). Florida's Aquods (S1 in Table 14–1) are quite productive; large tracts are used for citrus crops of oranges and grapefruits.

The Spodosols that cover most of the New England states (Fig. 14–1) are Orthods (S2 in Table 14–1). These New England Orthods have subsurface accumulations of iron, aluminum, and organic matter. A small pocket of Orthods is found in New Jersey. Three large areas of Orthods are found in the northern part of the Great Lakes states nestled around Lake Superior, Lake Michigan, and Lake Huron (Fig. 14–1).

WORLDWIDE DISTRIBUTION

On a global basis Spodosols occupy 4.3 percent of the world's land surface. Spodosols are found in Canada* from about 45°N to more than 60°N latitude (Fig. 14–2). Extensive areas of Spodosols stretch across Eurasia from west to east (Fig. 14–5). Eurasian Spodosols are found from 50°N to 70°N latitude.

ULTISOLS

A thumbnail sketch of this order and its suborders is found in Table 14–1. Ultisols develop

---

*The Spodosols of the CSCS are referred to as *Podzols* in the Canadian system. (See Appendix IX, Table A–1.)

Humults with very high organic matter content are found on these moderately sloping to steep slopes on the island of Kauai (22°N, 159°30'W) in the Hawaiian chain. (USDA)

under a cover of forest or savanna (grass-plus-forest) in a humid climate with tropical to sub-tropical temperatures and an average annual soil temperature of more than 47°F (8°C). The characteristic profile is one with a surface horizon that is dark with humus, a leached A2 layer, and a clay accumulation in the Bt horizon, which is moderately to strongly acid and, in a well-drained Ultisol, is characteristically red or yellowish-brown.

Pines mixed with hardwoods are the common natural cover on Ultisols. These soils have a potential for being productive when they are brought under cultivation because they are found in areas that are humid and free of frost for long periods. An Ultisol's nutrient reserve, however, is low, and successful productive cultivation requires the application of fertilizers and lime.

UNITED STATES PATTERN

Ultisols cover 12.8 percent of the land surface of our 50 states. Ultisols are found along our Atlantic coast from New Jersey to Georgia (Fig. 14–1). From Georgia, the Ultisols continue westward into Alabama and Mississippi. Ultisols are the primary soil order to the west of the Mississippi's floodplain in Louisiana, Arkansas, and Missouri. Ultisols are also found in the eastern portions of Texas and Oklahoma. A continuous strip of Ultisols stretches from northern California through Oregon and on into Washington. Except for the big island of Hawaii, Ultisols cover the Hawaiian Islands (Fig. 14–1), too.

The extreme eastern portions of the Delmarva Peninsula, Virginia, North Carolina, South Carolina, and Georgia are covered by Aquults (U1 in Table 14–1), as shown in Figure 14–1. Aquults are seasonally saturated with water, and the natural cover is woodland and pasture. When an Aquult is drained, it is quite suitable for truck crops.

Udults (U3 in Table 14–1) are by far the most prominent suborder as you move westward from the coast in Virginia, North Carolina, South Carolina, and Georgia (Fig. 14–1). This same suborder is found in Alabama, Mississippi, Tennessee, Kentucky, and a small section of southern Indiana. These soils have tradition-

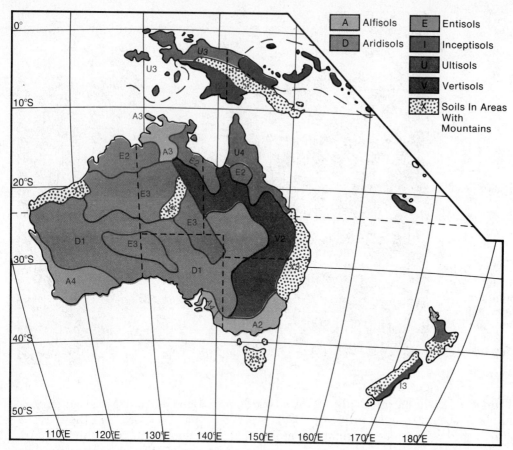

FIGURE 14–6   Reference should be made to Table 14–2 for deciphering the soil pattern of Australia and the nearby islands. The equivalent suborder for the codes within the tinted areas are identified in Table 14–2.

ally been used to produce feed crops, cotton, and tobacco.

WORLDWIDE DISTRIBUTION

On a global basis the Ultisols cover 5.6 percent of the world's land area. Strips of Ultisols are found along the western and eastern coasts of South America; isolated pockets of Ultisols can be found further inland (Fig. 14–3).

There are two patches of Ultisols on the western coast of Africa (Fig. 14–4). The Ustults (U4 in Table 14–2) are dry for more than 90 cumulative days in the year. The Udults (U3 in Table 14–2), which are located southeast of the other suborder, are never dry more than 90 cumulative days in the year. The other rather

substantial pocket of Ultisols in Africa is found along the northern and western shores of Lake Victoria; these Udults (U3 in Table 14–2) cover portions of Kenya, Uganda, and the Congo (Fig. 14–4).

A substantial area of southern China, as shown in Figure 14–5, is covered by Udults (U3 in Table 14–2). Formosa (Taiwan) is located to the east of the Chinese Udults; the western half of this island is also covered by Udults (U3). The Malay Peninsula and Borneo also contain Udults (U3). Significant patches of Ustults (U4) are found in Burma and India (Fig. 14–5). The continent of Australia also has a broad patch of Ustults (U4) along its northeast coast (**Fig. 14–6**).

VERTISOLS

A brief summary of this order and some of its suborders is found in Table 14–1. Vertisols typically form under tallgrasses or a savanna that is dominantly tallgrasses with scattered trees and shrubs. These soils develop in humid to semiarid climates from parent materials high in limestones, marls, or basic rocks. By definition they have more than 30 percent clay.

Vertisols expand and contract—swell and shrink—more than the soils of any other order because they have a high clay content. The clay makes these soils difficult to cultivate and unsuitable as a support-bed for roadways or buildings.

When these soils are dry, they have wide, deep cracks that may be more than 0.4 inch (1 cm) wide at a depth of 20 inches (51 cm). Loosened surface soil falls into the wide dry-weather cracks and, thus, this soil literally "swallows itself."

UNITED STATES PATTERN

Vertisols are found only at tropical and subtropical latitudes. They make up a mere 1.0 percent of the land surface of our 50 states (Fig. 14–1). Vertisols are found in Texas, Alabama, and Mississippi. The Alabama-Mississippi pocket of Vertisols consists of Uderts (V1 in Table 14–1). There are three rather narrow but substantial belts of Vertisols in Central and southeastern Texas. Two of the belts are Usterts (V2 in Table

14–1); the belt along the Gulf Coast of Texas is and Udert (V1 in Table 14–1).

WORLDWIDE DISTRIBUTION

On a global basis, Vertisols cover 1.8 percent of the land surface of the world. Three major accumulations of these soils, all at tropical or subtropical latitudes, are found outside the United States: an extensive belt of Usterts (V2 in Table 14–2) is found in eastern Australia (Fig. 14–6); a rather large Ustert (V2) pocket spreads across the Sudan (Fig. 14–4); and the third accumulation is found in India (Fig. 14–5). The Usterts (V2) of India have formed on the Deccan Plateau's weathered basalts, which were deposited as the India Plate drifted over a hot plume of rising magma during its journey north to its ultimate collision with and suture to the China Plate.

MOUNTAINS AND ICE FIELDS

Approximately one-fifth of the total land surface on the worldwide maps has been coded with the letters X (for soils of mountains) or Z (for ice fields). Various amounts of the ten soil orders are so intermixed in these areas that it is impossible at the scale used in our maps to differentiate and represent them adequately. The intermixing of soil orders is produced primarily by the effect of elevation, which causes temperature and soil-water regimes to undergo rapid change over short distances.

A CASE STUDY

Soil maps are prepared for most states, provinces, and counties; they are available in a variety of scales. The scale determines the detail that is shown and the differentiations that can be made within a given horizontal distance. Compare, for example, the area assigned to South Carolina in Figure 14–1, which is a general soil map of the United States, with the area assigned in Figure 14–7, which is a general soil map of South Carolina. The scale used for the soil map of the United States is 1:17,000,000; the scale used for the soil map of South Carolina is 1:750,000.

The soil map of the United States is useful in making comparisons among soils in different parts of the country. Such a map, however, does not give us very much detail about the soils within a state. On the general soil map of the United States (Fig. 14–1), South Carolina is represented as being covered by Ultisols, with Aquults along the coast and Udults inland. Ultisols are, in fact, the predominant soil order in South Carolina, but we also find at the scale of Figure 14–7 that there are five additional soil orders in South Carolina: Alfisols, Entisols, Inceptisols, Mollisols, and Spodosols.

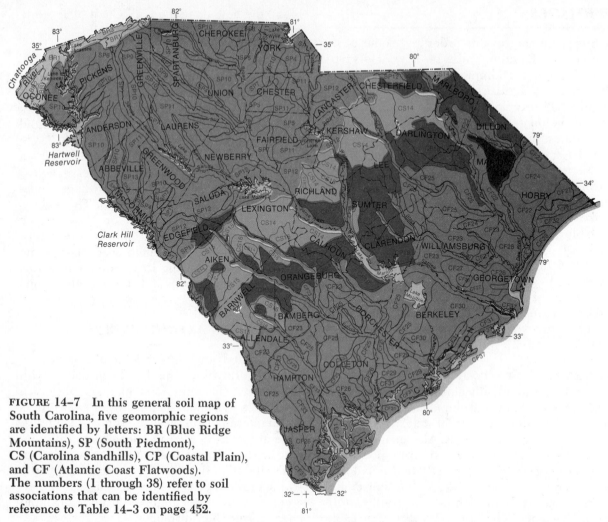

FIGURE 14–7 In this general soil map of South Carolina, five geomorphic regions are identified by letters: BR (Blue Ridge Mountains), SP (South Piedmont), CS (Carolina Sandhills), CP (Coastal Plain), and CF (Atlantic Coast Flatwoods). The numbers (1 through 38) refer to soil associations that can be identified by reference to Table 14–3 on page 452.

Let's examine the general soil map of South Carolina in the context of its geomorphic setting. South Carolina, remember, is only used as an example. You may select any other area and seek details about it through the study of its geomorphic setting and the soil maps that are available to you.

GEOMORPHIC REGIONS AND SETTING

Five broad regions can be identified in South Carolina: the Blue Ridge Mountains, the Southern Piedmont, the Carolina Sandhills, the Southern Coastal Plain, and the Atlantic Coast Flatwoods.

The Blue Ridge in South Carolina is a region of dissected rugged mountains with narrow valleys (BR areas in Fig. 14–7). The local relief within the region ranges from several hundred feet to a few thousand feet from valley floor to ridge crest. The vegetative cover is forest. There are a few scattered small farms in the valleys. The Blue Ridge Mountains occupy approximately 387,000 acres (155,000 ha) or 2 percent of the surface area of the state.

South Carolina's Piedmont is a region of gently rolling to hilly slopes with narrow stream valleys (SP areas in Fig. 14–7). Elevations within the Piedmont increase gradually from southeast to northwest. The local relief is mainly in tens of feet but ranges to several hundred feet. The Piedmont extends over an area of 6,786,000 acres (2,700,000 ha), which amounts to 35 percent of the state. Approximately two-thirds of South Carolina's Piedmont is forested. A significant amount of the remaining acreage is used to produce so-called cash crops.

The local relief in the Carolina Sandhills (CS areas in Fig. 14–7) is in tens of feet. The region consists of gently to strongly sloping uplands. More than 2,127,000 acres (850,000 ha) of land, about 11 percent of the state, lie within the Sandhills. Almost two-thirds of the region has a forest cover. Farming is a significant activity in the area, and a lot of the cleared land is used to produce crops.

The Southern Coastal Plain (CP areas in Fig. 14–7) covers 14 percent of South Carolina. It is a region of gentle slopes with local relief in tens of feet. The area of greatest dissection and most numerous slopes is found in the northwestern part of this region. There are almost 2,707,000 acres (1,080,000 ha) in the region; about one-half of this acreage is in forest. Significant areas of the cleared land are used for farming.

The region known as the Atlantic Coast Flatwoods is nearly level coastal plain dissected by broad valleys with meandering streams (CF areas in Fig. 14–7). The local relief is in the order of a few feet. The greatest range in the relief amounts to a mere 10 or 20 feet (3 or 6 m).

This region, like the others, is primarily forested; in fact, two-thirds of the area has a vegetative cover of forest. The 7,349,000 acres (2,940,000 ha) of land in the Flatwoods, which amounts to 38 percent of the state, makes the region the largest of the five.

SOIL ORDER AND SETTING

Thirty-eight soil associations are identified by code in Figure 14–7. Each soil-association area is assigned two letters and a number. The two letters are used to identify the geomorphic region: BR (Blue Ridge Mountains), SP (Southern Piedmont), CS (Carolina Sandhills), CP (Coastal Plain), CF (Coastal Flatwoods). Thus, the three soil-association areas found in the Blue Ridge Mountains are identified by map symbols BR-1, BR-2, and BR-3.

The major soil associations found within a coded area are identified in Table 14–3. In each of the three areas of the Blue Ridge Mountains, for example, Udults (U3), a suborder of the Ultisols, are found. In area BR-3, however, Inceptisols (I) are also present. These soils, which are moderately deep, form on rather steep ridges and side slopes. The depth to the bedrock of gneiss, granite, or schist runs between 3 and 5

In South Carolina, the Blue Ridge is a region of dissected rugged mountains with narrow valleys. The major soil association in this section of the Blue Ridge Mountains is Udults, a suborder of the Ultisols. (South Carolina Dept. of Parks)

TABLE 14–3   SOIL ASSOCIATION IN SOUTH CAROLINA

MAP CODE	MAJOR SOIL ASSOCIATIONS	PREVAILING SLOPE	UNDERLYING ROCK OR SEDIMENT
Blue Ridge			
BR-1	U3	steep ridges	gneiss, granite
BR-2	U3	steep, narrow ridges	schist, gneiss
BR-3	I, U3	steep, narrow ridges	gneiss, granite
Southern Piedmont			
SP-4	A	level to sloping broad ridges	diorite, gabbro, schist
SP-5	A	sloping to moderately steep ridges	diorite, schist
SP-6	U3	moderately steep ridges	gneiss, schist
SP-7	U3, A	moderately steep ridges	gneiss, schist
SP-8	U3	sloping to moderately steep ridges	schist, slate
SP-9	U3	sloping uplands	schist, gneiss
SP-10	U3	sloping uplands	gneiss, schist
SP-11	U3	sloping broad ridges	gneiss, granite, schist
SP-12	U3	sloping uplands	slate
SP-13	U3, A	sloping uplands	gneiss, schist, diorite
Carolina Sandhills			
CS-14	E, U3	sloping uplands	sandy and loamy sediments
CS-15	U3	sloping uplands	loamy sediments
CS-16	U3	sloping uplands	loamy sediments
Coastal Plain			
CP-17	U3, U1	level to sloping uplands	loamy sediments
CP-18	U3, U1	level to sloping uplands	loamy and clayey sediments
CP-19	U3	level to sloping uplands	clayey and loamy sediments
CP-20	U3	level to sloping uplands	clayey and loamy sediments
CP-21	U3, E	level to sloping uplands	loamy and clayey sediments
Atlantic-Coast Flatwoods			
CF-22	U1, U3	level uplands and depressions	clayey sediments
CF-23	U3, U1	level uplands	loamy sediments
CF-24	U3, E	level uplands	loamy and sandy sediments
CF-25	U1, U3	level uplands	loamy sediments
CF-26	A, U1	level uplands, depressions, low terraces	clayey sediments
CF-27	U3, U1	level uplands and depressions	clayey sediments
CF-28	U1	level upland, depressions	clayey sediments
CF-29	A, M	level, low terraces and flood-plains	clayey sediments
CF-30	U1	level upland, depressions	clayey sediments
CF-31	E, A, S	level uplands	loamy and sandy sediments
CF-32	S, I	level uplands, depressions	sandy sediments
CF-33	E, U3	level uplands	sandy and loamy sediments
CF-34	E, I	level floodplains	loamy sediments
CF-35	I	level floodplains, depressions	loamy sediments
CF-36	I, E	level floodplains	clayey sediments
CF-37	E	level tidal marshes	clayey sediments
CF-38	E	level to sloping dunes and beaches	sandy sediments

feet (1 to 1.5 m). The water table is usually more than 6 feet (1.8 m) below the surface.

The coded areas for the Southern Piedmont run from SP-4 through SP-13 (Fig. 14–7). The soil associations are predominantly Udults (U3), but Alfisol (A) is the major soil order found in SP-4 and SP-5 (Table 14–3). Alfisols are also part of the major association in SP-7 and SP-13 (Table 14–3). The soils in South Carolina's Piedmont are medium acid to moderately alkaline plastic soils. The slopes on which these soils form run from sloping uplands to moderately steep but rather broad ridges. The depth to the underlying bedrock is generally more than 5 feet (1.5 m), and the water table is found at depths below 6 feet (1.8 m).

Three code numbers—CS-14, CS-15, and CS-16—are sufficient to identify the soil association of the Carolina Sandhills (Fig. 14–7). Udults (U3) make up the dominant soil order in these sandy soils of dissected uplands (Table 14–3). Entisols—soils without pedogenic horizons—are found, however, as part of the major association in the rather extensive CS-14 areas. The materials from which the soils in the Sandhills have formed are presumably the underlying sandy and loamy sediments (Table 14–3), which are at depths of more than 10 feet (3 m) below the surface. The depth to the water table in this region runs from a mere 4 feet (1.2 m) to somewhat more than 6 feet (1.8 m).

Five codes—CP-17 through CP-21—identify the loamy and clayey coils of smooth uplands that are classified as belonging to the Southern Coastal Plain (Table 14–3). The soils in these areas are predominantly Ultisols (U3 and U1) that have formed from loamy and clayey sediments found at depths of more than 10 feet (3 m). We do, however, find Entisols in one area, CP-21, of Sumter County. The water table in these areas is generally less than 6 feet (1.8 m) below the surface, and in some few locations it is found less than 1 foot (0.3 m) from the surface.

Within the region of the Atlantic Coast Flatwoods, loamy and clayey soils of wet lowlands are found in coded areas CF-22 through CF-30 (Fig. 14–7). The Udults (U3) and Aquult (U1) suborders of the Ultisols dominate the soil associations, but Entisols (in CF-24), Alfisols (in CF-26 and CF-29), and Mollisols (CF-29) are also present (Table 14–3). The depth to the underlying sediments from which these soils pre-

sumably formed is greater than 10 feet (3 m) but the water table is, more often than not, less than 1 foot (0.3 m) from the surface.

Wet, sandy soils of broad ridges are identified by the codes CF-31, CF-32, and CF-33 (Fig. 14–7). Table 14–3 indicates that the major soil associations in these areas are made up of Entisols (E), Alfisols (A), Spodosols (S), Inceptisols (I), and Ultisols (U). These soils form from loamy and sandy sediments that are more than 10 feet (3 m) below the surface.

The soils of floodplains on the coastal plain of South Carolina are within the coded areas CF-34, CF-35, and CF-36 (Fig. 14–7). These floodplains—the Savannah's, Edisto's, Santee's, Congaree's, Wateree's, Pee Dee's, for example—are covered by soil associations that consist primarily of Inceptisols and Entisols (Table 14–3). The Inceptisols (I), recall, are moist soils. The water table in these areas is less than 1 foot (0.3 m) below the surface, but the loamy and clayey sediments from which these soils form are at depths of more than 10 feet (3 m).

The soils of marshes and dunes in South Carolina's coastal area are identified by the codes CF-37 and CF-38 (Fig. 14–7). The soil associations within these areas are dominantly Entisols (E) (Table 14–3), which are distinguished by their lack of pedogenic horizons.

SOIL SUITABILITY

Many states and provinces have developed soil- or land-suitability ratings for agriculture and

TABLE 14–4   **LAND CAPABILITY CLASSES**

Class	SUITABLE FOR CULTIVATION Management Needed	Class	NO CULTIVATION Use for Pasture, Hay, Woodland, or Wildlife
I	Requires good soil management practices only to obtain high level crop production.	V	No restrictions in use; poor drainage and possible boulders.
II	Moderate conservation practices necessary to maintain productivity; fertilizers must be added.	VI	Minor restrictions in use; must be improved for forage and pasture. May have steep slopes or shallow soil.
III	Some natural features that restrict use such as shallow soil, steep slopes, or shallow water tables; intensive conservation practices necessary to maintain productivity; fertilizers required.	VII	Severe restriction in use; suited for grazing or forestry.
IV	Restriction on crop choice; perennial vegetative cover must be maintained; suitable for cultivation infrequently in a limited number of years.	VIII	Suited only for wildlife and recreation; it may be land that is very steep sloped, rock lands, or swamps.

other purposes. Most systems for rating the suitability of a soil for a specific purpose are based on a number of criteria—topography, microclimate, wind exposure, and drainage, for example—in addition to soil properties.

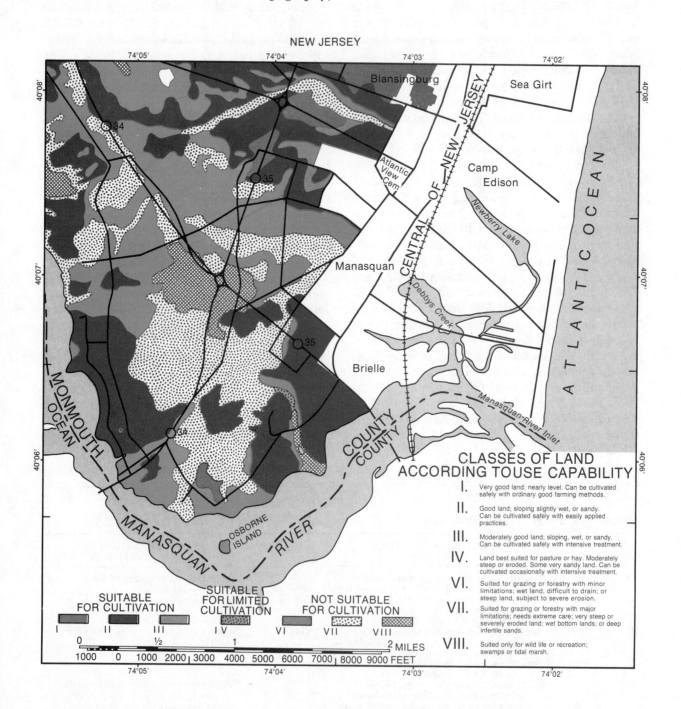

CLASSES OF LAND ACCORDING TO USE CAPABILITY

I. Very good land; nearly level. Can be cultivated safely with ordinary good farming methods.

II. Good land; sloping slightly wet, or sandy. Can be cultivated safely with easily applied practices.

III. Moderately good land; sloping, wet, or sandy. Can be cultivated safely with intensive treatment.

IV. Land best suited for pasture or hay. Moderately steep or eroded. Some very sandy land. Can be cultivated occasionally with intensive treatment.

VI. Suited for grazing or forestry with minor limitations; wet land, difficult to drain; or steep land, subject to severe erosion.

VII. Suited for grazing or forestry with major limitations; needs extreme care; very steep or severely eroded land; wet bottom lands; or deep infertile sands.

VIII. Suited only for wild life or recreation; swamps or tidal marsh.

FIGURE 14–8 This land area of Monmouth County, New Jersey, is classed and coded according to its suitability for an indicated use.

In some instances, a simple rating system of good, fair, and poor is applied to soils after a careful evaluation that is based on specific criteria. *Good* is used for those soils that are well-suited for the indicated use; they are highly productive with good management and have few or no soil properties that are problems. *Fair* is used for those soils that can be used satisfactorily for the indicated use, but good management and careful planning are required; they have some soil properties that are problems, but these problems can be overcome by practical means. *Poor* is used for those soils that are poorly suited for the indicated use; they have soil properties that produce problems, which can only be overcome at great cost.

Let's apply this simple rating system to a South Carolina soil association. The CF-22 soil-association area in South Carolina (Fig. 14–7), for example, is contiguous with the floodplain of the Pee Dee River. Ultisols (U1 and U3) make up the major soil association in this area of level uplands and depressions (Table 14–3). The soils develop from clayey sediments that are at depths of more than 10 feet (3 m). The water table, however, is at a depth of less than 1 foot (0.3 m). The risk of flooding is high since the water table is high and the area adjoins the floodplain of the Pee Dee. The shrink-swell potential of these clayey soils must be considered, too; these Ultisols contain plastic clays and undergo a moderate volume change on wetting and drying. The permeability of these soils is low. Taking all these facts into consideration, we can say this CF-22 area is well suited—has a rating of good—for row crops, pasture, and woodland; but it receives a rating of poor when its suitability for home sites, septic tank filter fields, or campsites is considered.

The United States Soil Conservation Service has developed one of the more easily understood and widely employed suitability ratings in which eight classes of land are recognized. Four classes represent land suited to cultivation; three additional classes are grazing or forestry land; and the eighth class is suited only to watershed, recreation, or wildlife support (Table 14–4).

The map in Figure 14–8 is a mosaic produced by aerial photography in the Manasquan-Brielle-Sea Girt area of New Jersey. Some of the land along the north shore of the Manasquan River is color-coded according to its capability. Classes I, II, and III are well represented in this area. A present survey of the area, however, shows that much of the land that was suitable for cultivation is now devoted to housing. Throughout the United States, we have lost and continue to lose a lot of good agricultural land to housing and other uses.

Farmers in Upper Volta near Ouagadougou (12°22′N, 1°31′W) are preparing their land for planting. This area is very close to the commonly accepted southern border of the Sahel; under normal conditions this farming area around Ouagadougou gets about 35 inches (89 centimeters) of rainfall. The soils in this area are Alfisols (Fig. 14–4). The International Development Association is attempting to strengthen the agricultural productivity in this area. (World Bank)

THE LANGUAGE OF PHYSICAL GEOGRAPHY

A physical geographer uses a technical vocabulary to make explanations. Review your understanding of the vocabulary used to develop the concepts in this chapter. Among the important words and terms used are:

Alfisol	Inceptisol	soil order
Aridisol	Mollisol	Spodosol
Entisol	Oxisol	suborder
great group	pedon	Ultisol
Histosol	polypedon	Vertisol

SELF-TESTING QUIZ

1. What is the basis of the comprehensive Soil Classification System?
2. What is the most generalized of the levels in the CSCS?
3. List the various soil orders in the CSCS.
4. Compare and contrast Alfisols and Mollisols.
5. How are the suborders within a soil order differentiated?
6. Explain the way in which great groups are established.
7. Discuss the concept of the subgroup in the CSCS.
8. List the soil properties used to differentiate among families in the CSCS.
9. What is the basis for the series name in the CSCS?
10. Give a thumbnail sketch of each soil order in the CSCS.
11. Describe the pattern of Alfisols in the United States.
12. What native vegetation develops on Alfisols?
13. Discuss some of the consequences of the application of significant amounts of water to Aridisols during cultivation.
14. Why are Entisols without distinct pedogenic horizons?
15. Which of the soil orders in the CSCS has a high level of organic material?
16. Discuss some of the problems that develop when Histosols are drained.
17. What is the predominant soil order in Alaska?
18. Which of the soil orders is the most intensely weathered?

IN REVIEW
SOIL CLASSIFICATION AND DISTRIBUTION

I. THE UNITED STATES SYSTEM

    A. Soil Order
    B. Soil Suborder
    C. Great Group
    D. Subgroup
    E. Family
    F. Series

II. SURVEY OF SOIL ORDERS

    A. Alfisols
      1. United States Pattern
      2. Worldwide Distribution
    B. Aridisols
      1. United States Pattern
      2. Worldwide Distribution
    C. Entisols
      1. United States Pattern
      2. Worldwide Distribution
    D. Histosols
      1. United States Pattern
      2. Worldwide Distribution
    E. Inceptisols
      1. United States Pattern
      2. Worldwide Distribution

    F. Mollisols
      1. United States Pattern
      2. Worldwide Distribution
    G. Oxisols
      1. United States Pattern
      2. Worldwide Distribution
    H. Spodosols
      1. United States Pattern
      2. Worldwide Distribution
    I. Ultisols
      1. United States Pattern
      2. Worldwide Distribution
    J. Vertisols
      1. United States Pattern
      2. Worldwide Distribution
    K. Mountains and Ice Fields

III. A CASE STUDY

    A. Geomorphic Regions and Setting
    B. Soil Order and Setting
    C. Soil Suitability

15 Spatial distribution of vegetation

Vegetation is a highly visible and important factor in the physical environment of our planet. Virtually the entire land surface of the Earth, with very few exceptions, supports some kind of vegetation. Grassy plains, forested slopes, chaparral, and tundra are some of the terms commonly employed to describe the nature of a region. Since our main concern, within the covers of this book, is with geographical phenomena—especially spatial relationships—we will examine the conditions of the environment that enable an assemblage of plants to grow in a particular locale and limit the area occupied by them.

MAIN HABITATS

At one time, the term *habitat* simply referred to the place in which an individual organism or community of organisms lived. Today, the term is taken to mean the *kind of place*, which includes the sum total of operating influences and effective conditions that are incidental to and characterize a particular area occupied by a particular community. A habitat, then, consists of various environmental factors that have any kind of influence on life within it. In fact, the life within a habitat interacts complicatedly with the environmental factors to modify them and to influence the kind of place the habitat is and becomes.

A lone cypress *(left)* clings to a point on the Monterey Peninsula of California. Huge evergreens *(bottom)* grow along the banks of the Bow River within sight of the rugged, eroded mass of 9390-foot (2860-meter) Mount Eisenhower in Alberta, Canada. A thick carpet of trees has moved to about the midpoint of the mountain's elevation. (USDA; National Film Board of Canada)

Water is used as a criterion for establishing a primary subdivision of habitats into terrestrial habitats and aquatic habitats. This initial step in categorizing habitats as recognizable entities should underscore the fact that there are almost endless variations within these primary subdivisions. With our main concern in mind, an enumeration of main habitat types will be attempted; we will use it as a basis for discussion of environmental factors, successions, and climaxes in the sections that follow.

AQUATIC HABITATS

Major differences in salinity, light, temperature, size and depth of the water body, tranquillity or shelter of the water from disturbance, content of dissolved substances, aeration, possibility of attachment, as well as seasonal or tidal changes are factors that lead to the existence of a whole series of aquatic habitats. In fact, on the basis of salinity alone, the aquatic subdivision is divided into two major subgroups: saline and freshwater habitats.

Freshwater habitats include those of lakes, tarns, and ponds, where the waters are relatively static, as well as those of rivers and streams, where the waters are on the move. Acidity and nutritive salt content are especially important factors in the development of freshwater communities. In fact, acidity and salt content levels are used to distinguish three types of freshwater habitats: (1) **oligotrophic,** with very little dissolved minerals, (2) **dystrophic,** with very little dissolved minerals, but acidic and rich in humus, and (3) **entrophic,** rich in combined nitrogen, phosphorus, and calcium, which typically support large communities of blue-green algae, a broad zone of rooted pondweeds, and a surrounding community of luxuriant reed-swamp.

TERRESTRIAL HABITATS

The general identity of vegetation with habitat and the ease with which vegetation is classified is used as the basis for differentiating among the **terrestrial habitats.** Four major terrestrial subgroups are identified: forest, savanna, grassland, and desert. When we examine each of these broad subgroups, we find we can identify specific habitat types among them; for example, among the forest-habitat types there are deciduous types as well as needleleaf types. A specific habitat type, sometimes called a **formation-class,** is a geographic unit of vegetation that seems to generalize a regional climatic pattern.

FOREST HABITATS

This terrestrial habitat receives annual amounts of precipitation that are characterized as being high. The precipitation may, however, be distributed unevenly throughout the year. Generally, there are no dry winds during the winter and relative humidity is high.

There is a typical vertical structure within a **forest** habitat. Trees, of course, are the dominant plants, which form the highest level in the habitat and tend to grow in a dense formation.

The canopy of this oak-hickory hardwood stand in a bottomland forest along the coastal plain of South Carolina is not very dense. Rhododendron are among the plants in the understory. (South Carolina Forestry Commission)

The crowns of the trees grow together to form a somewhat continuous cover called a **canopy.** But since the forest includes trees of different kinds and ages growing at different rates, individual trees, called **emergents,** often have crowns that rise well above those of their neighbors.

The height of a canopy's foliage varies from forest to forest. The canopy in some forests is quite low; in others, the canopy is several hundred feet above the forest floor. The foliage within a canopy may be contributed by plants other than trees; climbing vines, air plants, and parasitic plants also live within the canopy level.

Within a forest habitat, the level beneath the canopy is called the **understory.** This level includes trees that are completely submerged under the canopy. The trees of the understory consist of young members of the canopy species and other species that cannot tolerate open sunshine. There is generally more than one layer within the understory.

The third level within a forest habitat is referred to as the **ground cover.** Although the understory and the ground cover may both occupy the space immediately above the ground, there are distinct differences. The ground cover consists of shrubs, herbs, and creepers. The height of the ground cover can vary, again, depending on the kind of forest habitat.

The lowest level within the forest habitat is at or below ground level. It is sometimes known as the **subterranean strata.** This fourth level consists of roots and other underground plant parts, as well as bacteria, molds, fungi, and algae.

Within a forest, there will be one or several species that dominate the habitat. The other plants in the lower levels have adapted to living under the conditions influenced and maintained primarily by the dominant forms. Among the forest habitats we find **tropical rain forests, subtropical rain forests, monsoon forests, temperate rain forests, summergreen deciduous forests, needleleaf forests,** and **evergreen hardwood forests.**

SAVANNA HABITATS

This terrestrial habitat—consisting of grassland with scattered woody plants—is intermediate between a forest and a grassland. In some classification schemes, a **savanna** is classified simply as a disturbed grassland; but the vegetation and its pattern are so different that it deserves separate consideration.

Savannas develop under a climate of limited rainfall whose pattern is unevenly distributed with a definite dry period. Rainfall is, in fact, so low, on an annual basis, that it prevents the tree growth from forming any kind of continuous canopy. On a year-to-year basis, the rainfall regimen is somewhat unreliable: some years are much drier than others.

As a result of the limited rainfall, soils within a savanna are usually dry at the surface and water tends to accumulate farther down. The basic vegetative cover of a savanna consists of tall grasses, shrubs, bunch grasses, tufts of annuals, and even patches of lichens. Although trees found as part of the savanna are widely spaced and usually low, each produces—within its sphere of influence—a special microclimate.

Various types of savanna are found in tropical, temperate, and cold regions. Among the

A portion of the east side of the San Joaquin Valley (at about 35°30′N) between Bakersfield and Woody, California, is spread before us. The Greenhorn Mountains of the Sierras are beyond us to the east. This habitat consists of grassland with scattered woody plants. This grass-woodland is commonly referred to as a *savanna*. (USDA)

Hereford cattle are watering at the pond and grazing on this section of grassland-prairie about 40 miles (60 kilometers) east of Wichita in the Southeastern Plains of Kansas. This is gently rolling country used chiefly for grazing (USDA)

savanna habitats are those known as *savanna woodland, thornbush* and *scrub, tropical savanna, half-desert, heath,* and *cold woodland.*

GRASSLAND HABITATS

This terrestrial habitat is characterized by dry winds during the winter and by limited precipitation, which is unevenly distributed throughout the year. Grasses, herbs and shrubs are the dominant plants of **grasslands.** The few trees that are found in grasslands are limited to the stream edges and are referred to as **galleria forests.**

Although grasslands do not have an upper story of trees, they are multistoried because there is more than one level of herbs and shrubs. Some of the grasses and legumes, which are herbaceous plants, develop to considerable heights; 10-foot- (3-m-) high grass is not uncommon in the more humid grasslands. Bamboo is a member of the grass family; it is fast-growing, and some species reach heights of 100 feet (30 m).

The herbaceous plants of grasslands are perennials. These plants can tolerate the removal of their tips or ends. Thus, it is natural to find grazing animals in large numbers in grassland habitats.

Different types of grasslands are identified on the basis of secondary structural characteristics. A simple discrimination between tallgrass and **shortgrass** regions gives us the distinction between prairie and steppe. Grassy tundra with its lichens that form a mat under sedges and grasses is another type of grassland habitat.

DESERT HABITATS

The extreme aridity of these terrestrial habitats allows only a very dispersed and discontinuous plant cover to develop. The plants that do exist in these arid environments tend to clump. In some **desert** environments large plant forms tend to dominate; most of the world's deserts, however, are characterized by brush. Some desert areas are almost completely devoid of vegetation and are referred to as *barren.* In any event, survival in a desert is possible only for those plant species that are very resistant or highly specialized.

The unprotected and unbound surface of a desert is subject to wind erosion. Water erosion,

A lonely plant has established itself on this windblown, gypsum-sand dune in the White Sands region of southeastern New Mexico. (National Parks Service)

during the very infrequent downpours that produce flashfloods, is also a problem in some deserts. Desert soils are often rich in nutrients; these nutrients, however, tend to be unavailable to plants except after rainfalls.

Desert habitats are found at various latitudes from the tropics to the polar regions. Some deserts are hot; others are cold. Aridity is the factor that produces the desert environment. Among the types of deserts in the world are **tropical deserts, subtropical deserts, the arctic fell-field,** and **cold deserts.**

ENVIRONMENTAL FACTORS IN PLANT GEOGRAPHY

The term **environment,** in this context, is used to mean the aggregate or sum of all the external conditions that may act on an organism or community to influence its development or existence within a given region. Thus, the surrounding air, light, moisture, temperature, wind, terrain, soil, dissolved substances, and living organisms are all part or elements of the environment in any particular community.

Most environments, then, contain each of the elements noted above. It is simply variations in these environmental elements that characterize the different main terrestrial habitats such as forest, savanna, grassland, and desert. The differentiation among main-habitat vegetational types is in itself a consequence of existing variations in the environmental elements. Through long processes of evolution, which involve among other things the elimination of unsuitable features, different kinds of plants become adapted to different environments.

The list of environmental elements enumerated above may be grouped into four main factor-classes: climatic, geomorphic (topography and landforms), edaphic (soil), and biotic (living organisms). These environmental factor-classes are, without doubt, inextricably interwoven—they act and react together to produce a final effect. For example, topography influences the local climate that in turn affects the soil and eventually the kind of vegetative cover in a particular region.

THE ROLE OF CLIMATE

The different kinds of climate and their individual rhythms make an impact on a plant species in terms of its form, function, and range. The kind and intensity of the biological response that prescribes a vegetation type are related to different combinations of five climatic elements: light, temperature, precipitation, evaporating power, and wind.

LIGHT

In prior chapters, latitude is identified as a climatic factor along with distribution of continents, relief, barometric pressure, and marine currents. These climatic factors, as noted on page 283. bear on and affect the climatic elements; for example, with an increase in latitude, light,

The environment of Wharton Lakes, Alberta, is the aggregate of all conditions that act on the community to influence its development. Topography influences the local climate, that, in turn, affects the soil and eventually the kind of vegetative cover in the region. (National Film Board of Canada)

which is identified in this bioclimatological context as a climatic element, becomes strongly periodic. The length of daylight during the summer season increases rapidly with higher latitude and reaches a maximum in polar regions where the Sun may be above the horizon for 24 hours.

Although the climatic factors affect the climatic elements, it is the climatic elements that influence living things. Light, as you know, is essential for photosynthesis. It may also be important in reproductive processes. In addition, light governs such processes as molting in birds and flowering in plants.

Duration, time-distribution, and intensity all play a part in the biological impact made by light through its role as a climatic element. The effect of light on photosynthesis, for example, depends largely on intensity. For successful flowering, however, many plants require a relatively long daylight period, which limits them to high latitudes; those plant species that require short days for the flowering process to occur are limited to low latitudes. The annual rhythm of daylight also determines the timing of budding, fruiting, and leaf-shedding.

TEMPERATURE

This climatic element is vitally important to plants. Temperature governs the rate at which the chemical reactions that sustain biological processes occur. There are lower and upper temperatures beyond which a species cannot survive. Thus, we find that a specific plant species is adapted to minimum, optimum, and maximum temperatures for its life as a whole as well as for component physiological functions.

Since each species has an optimum temperature range for each of its various physiological functions, a distinction is made among them on the basis of temperature preference. Plant species that favor warm habitats are classified as **megatherms**. At the other end of the temperature spectrum are the plants that function well in cold habitats; the term **microtherm** is used to distinguish this cold-favoring group. Plants adapted to intermediate-temperature ranges—between those of the megatherms and microtherms—are classified as **mesotherms**.

Reelfoot Lake in northwestern Tennessee is in low, flat lands, called the *Mississippi Bottoms*, along the Mississippi River. Numerous swamps and ponds are found in the area. These bald cypresses are growing in the shallow-water area of Reelfoot Lake. Note the buttresses on the lower part of each tree trunk. Each buttress ends in a long-branching far-spreading root that gives stability to the standing tree. (State of Tennessee)

Temperature is affected by many factors, including cloud cover, relative humidity, marine currents, the albedo of soil surfaces, and the amount of water held by a soil. In addition, temperature varies markedly at different levels above the surface as well as at different times during the day. Average maximum and minimum temperatures over specified time periods of a day, month, and year are of much more value to the plant geographer than annual average temperatures. But, in order to develop a complete picture of biological response to this climatic element, temperatures must be recorded simultaneously at each different level of vegetation from the root-infested soil to the top of the canopy in the case of a forest.

PRECIPITATION

The availability of water, which is crucial to a plant because it enables the plant to carry out vital processes and grow, is largely determined by the precipitation falling in a region. The movement of air masses, the distribution of continents, and the relief within an area are factors that contribute to the amount, kind, and timing of precipitation.

A differentiation among terrestrial habitats is made on the basis of water in the soil. Three adjectives—**xeric** (dry), **hygric** (wet), and **mesic** (intermediate)—are used to indicate the degree to which the soil of a habitat is saturated with water; for example, a xeric soil or a xeric habitat is one that is essentially and prevailingly dry.

Hygric, xeric, and mesic are converted to prefixes and used with phyte—a combining word form that means *plant* and is used as the final element of the compound word—to identify those plants that grow in wet, dry, or intermediate-wet habitats. For example, those plants that grow in wet habitats such as a marsh are referred to as **hygrophytes**; plants that grow in dry places are **xerophytes**; and plants found growing in habitats that show neither an excess nor a deficiency of water are known as **mesophytes**. The term **hydrophyte** is reserved for plants that live more or less submerged in water.

EVAPORATING POWER

The relative humidity of the air—or more accurately, the saturation deficit, which takes temperature into account—establishes a crude approximation of the evaporating power of the air.

Low relative humidity, which means dry air, exerts a "pull" on the water economy of a plant and causes it to discharge water vapor into the air. Although water vapor is given off by any part of the plant exposed to the atmosphere, most of the loss occurs through the "breathing pores" of the leaf, which are called *stomata*. As relative humidity rises, the pull decreases and the plant discharges less water vapor into the air. A combination of high temperature, low humidity, and high wind exerts the greatest pull on the water economy of a plant.

In spite of the absolute necessity of water for growth and metabolism, most terrestrial plants are very inefficient in their use of water. A large proportion of the water absorbed is given off as water vapor through the process called **transpiration**. Many of the adaptations and changes in structure that have developed in plants through the evolutionary process are attempts to hold down the loss of water from aerial parts. The development of thick and impervious bark, waxy or hairy coverings, protection of the breathing pores or stomata, redesign of the leaf, and a reduction in the total surface are examples of plant adaptations designed to control its water economy and prevent transpirational losses.

WIND

This climatic element influences vegetation in a variety of ways, some direct and others indirect. On a regional basis, winds interacting with other factors affect the pattern of precipitation as well as the pattern of evaporation. Wind also makes an impact on the regional temperature pattern through the opposite processes of evaporation and precipitation. In a very direct and restricted sense, wind influences and increases a plant's transpirational loss of water when it brings unsaturated air in contact with leaves and young shoots.

A constant, long-term regime of wind makes a mechanical impact when it carries particles of soil, snow, and ice that cause the abrasion of vegetation and the loss of branches and leaves on the windward side of a plant. A physiological impact occurs when increased transpiration caused by a constant pattern of dry winds reduces the turgor of a plant's organs and limits growth. In fact, in strong dry winds—chinooks on the lee side of mountains that raise air temperatures as much as 30 to 40 Fahrenheit degrees (17 to 22 C°) in a very short time, for example—young parts of plants may shrivel and die in a few hours.

The stunted, flagged appearance of the trees in this open area of the Monongahela National Forest has been produced by the prevailing westerly winds. We are facing south; the westerly winds, coming from the right, abraded and caused the loss of branches on the right side of each tree. (State of West Virginia)

THE GEOMORPHIC FACTOR

Topography and landforms influence plant growth somewhat indirectly through their influence on other environmental conditions. Strong topographical relief, for example, makes an impact on the climatic elements and tends to produce a distinctive local climate which, in turn, supports a distinctive vegetative cover. Temperature, as you know, decreases with increasing elevation in accordance with the adiabatic lapse rates. In addition, in the Northern Hemisphere, north-facing slopes tend to be cooler and wetter than south-facing slopes at the same altitude because the south-facing slope receives greater amounts of insolation.

Differences in vegetation that are due to topography are generally correlated with moisture. The windward slopes of most mountains tend to be wetter than the leeward slopes. For example, there is a rain forest along the Hoh River on the Olympic Peninsula in Washington. The rain forest is on the windward side of the Olympic Mountains (Fig. 15–1). Dense vegetation characterizes the area, and the precipitation exceeds 100 inches (254 cm) per year. Directly across the mountain from this rain-forest area—on the leeward side of the mountain—constant irrigation is needed to grow crops because precipitation is only about 10 to 20 inches (25 to 50 cm) per year. The dry area on the leeward side is in the vicinity of Sequim. Sequim and the Hoh River Rain Forest are separated by slightly more than 30 miles (48 km).

In humid regions, valley floors tend to be wetter than the surrounding slopes because surface runoff causes water to converge there. In fact, the water table may be quite close to the valley's surface in a humid region. When the water table in a valley is at ground level, a bog, marsh, or swamp may develop with distinctive communities of hygrophytes.

In an arid region with strong topographic relief, the plains and valleys are often unproductive and sparsely covered by vegetation. The slopes and the mountainside, however, may have rich forest cover. The differences in vegetative cover in an arid region are, of course, due largely to differences in the local climate; but the differences would not exist if it were not for topography, which triggers the release of moisture at higher elevations.

There are, of course, many graduations in the steepness of slopes. Some are gentle inclines; others rise almost perpendicularly. Erosion and gravity, which constantly combine to wash and move the regolith to lower elevations as talus and other kinds of deposits, make their greatest impact on very steep slopes. The rate of surface runoff, which influences the development of soil

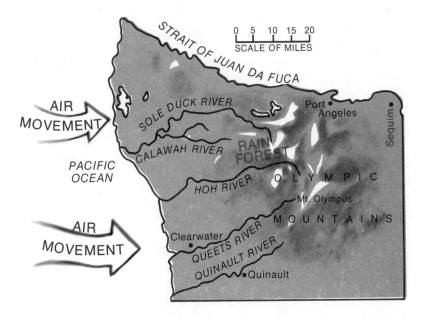

FIGURE 15–1 More than 100 inches (250 centimeters) of precipitation falls along the Hoh River on the windward side of the Olympic Mts. The rain forest that occupies the area is populated by a dense cover of tall trees. At Sequim on the leeward side of the Olympics, there is less than 20 inches (150 centimeters) of precipitation.

The rain forest on the Hoh River in the Olympic Peninsula of Washington is the result of good soil and exceptionally heavy rainfall that encourages the growth of Douglas fir, Sitka spruce, western hemlock, and western cedar. Mosses drape the branches and tree trunks in this photo. (National Parks Service)

and the water available to plants, is almost directly proportional to slope steepness. The fact that a steep south-facing slope in the Northern Hemisphere receives midday insolation more or less perpendicularly while a steep north-facing slope receives only oblique and weak morning and evening insolation means incident light and heat are related to and affected by steepness. Since the stability of a slope's regolith, the availability of water, as well as the incident light and heat are directly affected by steepness, it is not surprising to find many different and contrasting habitats along contrasting sloped surfaces at the same altitude and in the same general locale.

THE EDAPHIC FACTORS

The role played by soil in the development of vegetation is quite complex. It has been recognized for a long time, for example, that differences in soil are often largely responsible for differences in vegetation within the same climatic region. But the condition of the soil is, in turn, affected by and related to its vegetative cover, which supplies organic matter to the soil. Thus, soil development and the development of a veg-

etative cover are intimately interwoven. The factors in soil that exert the greatest influence in the development of vegetation are its mineral fragments, soil water, soil atmosphere, and organic matter.

MINERAL FRAGMENTS

The fragments from which a soil develops may be derived from the underlying bedrock or may be imported and bear no relation to the local bedrock. The soil chemistry, which is an important determining factor in plant geography, results and evolves from the composition of these mineral fragments. Limestone fragments, for example, tend to develop into alkaline soils; soils that develop from some volcanic rocks are acidic. And in the final analysis, it is the basicity or acidity of soil that determines the presence or absence of particular species in a habitat—where all other conditions are favorable—since some species are acid-loving while others require alkaline conditions.

SOIL WATER

Water is the main constituent by weight in most plants. And the soil is the chief source of a plant's water. Large quantities of water must be absorbed from the soil to cover a plant's transpirational and other needs. The soil's content of available water is often the chief contributing factor responsible for the differences between local plant communities.

SOIL ATMOSPHERE

The interstices between soil particles are occupied by air and water. The atmosphere within a soil tends to contain a slightly lower proportion of oxygen and a much higher proportion of carbon dioxide than ordinary air. An adequate supply of oxygen in the soil atmosphere is necessary for the respiration of the underground parts of higher plants. Waterlogged soils tend to be deficient in oxygen; this deficiency makes them unsuitable for most plants.

ORGANIC MATTER

All true soils contain organic matter, which is usually broken down to humus. The amount of organic matter in a soil can vary from very little to almost 100 percent of dry weight in some soils such as peat. Earthworms, fungi, and bacteria have a role to play in the breakdown of organic matter and the production of humus in the soil. Generally, when humus is added to a soil, it improves the soil's quality; for example, the addition of humus to a heavy clay soil lightens the soil. The addition of humus to light sandy soil gives a new dimension to the soil's consistency and water-holding capacity. Humus is a source of essential nutrients for plants, but it is also the main source of acidity in soils—and acidity affects plant geography.

THE BIOTIC FACTORS

A great variety of living organisms affect plant communities in numerous and diverse ways. Plants, for example, through competition, parasitism, and symbiosis limit or enhance the growth of other plants. The activities of animals—the spread of seeds by birds, the role of bees in pollination, and the bark-eating activity of the beaver, for example—also produce constructive as well as destructive impacts within plant communities.

The living organisms within soil, including primarily microscopic flora and fauna whose lives are usually centered on the soil's humus content, are a good example of the beneficial effects of biotic activity on vegetation and plant geography. These organisms make various essential food substances, including nitrogen, available to the higher plants. The role of the earthworm in disintegrating organic matter and aerating the soil is also an important part of the sustained biotic activity that is immediately beneficial and eventually modifies the environment and makes an impact on the vegetative cover of the area.

This is a typical mountain meadow in the Helena National Forest of Montana. The meadow is in the foothills of the Rocky Mountains on the east side of the continental divide, on a southerly exposure, at an elevation of 6000 feet (1800 meters). The trees on the hillside, in the background, are Douglas fir. (Phil Schlamp)

The caterpillar stage of the gypsy moth does great damage to the foliage of trees in New England, New York, New Jersey, and Pennsylvania. Caterpillars hatch in spring and reach their full size of 3 inches (7.5 centimeters) by midsummer. (USDA)

The activity of parasitic lower plants is a biotic influence that has devastating effects on plant distributions. A well-known example, which is often cited, involves the once luxuriant chestnut forests of eastern North America; these magnificent trees were parasitized and destroyed by a fungus that arrived from Asia in the 1890s. The so-called Dutch-elm disease, which reached North America from Europe in the 1930s, is another case of parasitism; over a period of 50 years, the populations of feathery, fan-shaped American elms has been significantly reduced.

Small herbivores—plant-eating animals—are another potent biotic factor. Snails, slugs, locusts, and the larval stages (caterpillars) of butterflies and moths can cause considerable damage and devastation to whole tracts of vegetation. The gypsy moth is a prime example of a plant-eating pest; the caterpillar, brought into Massachusetts for scientific experiments in 1869, eats the foliage of nearly all forest, shade, and fruit trees. In spite of the attempts to limit the gypsy moth, it has spread over much of New England, New York, New Jersey, and Pennsylvania causing enormous damage to trees.

The European rabbits brought to Australia, New Zealand, and South America had few natural enemies in these regions. As a result, they proliferated in these new environments and became a constant threat to plant life. The rabbits damage fruit trees by gnawing the bark and kill berry bushes by eating the sprouts. Some authorities indicate that rabbits can convert heaths or even forests into grasslands by gnawing and thus destroying tree seedlings, which prevents the the regeneration of the dominant trees.

The single most important biotic factor at work in the world today is not the rabbit but human activity. People burn forests and clear the

The tent homes (left) of the Great Basin tent caterpillars are attached to the limbs of these aspen trees. Great stands of aspen have been defoliated in southern Colorado and northern New Mexico. The hillside of quaking aspen (bottom), which stretches from the left of the photo to the foreground, was completely defoliated by tent caterpillars; the photo was taken at Elk Mountain in New Mexico's Santa Fe National Forest. (U.S. Forest Service)

land for agriculture, plow prairie grasslands and plant domesticated plants, allow overgrazing by domesticated livestock, clear-cut forests for lumber and pulp, spread concrete and macadam where there was once natural cover, develop huge strip mines that leave gaping scars in the landscape, and build reservoirs that flood huge valleys. All these activities produce great transformations in the natural vegetation of the Earth. The variety, breadth, and impact of such human activity on the distribution of vegetation in the world is overwhelming and irreversible. But like it or not, we must live with the Earth we have shaped!

COMMUNITY, SUCCESSION, AND CLIMAX

The vegetation that presently inhabits the Earth is the product of a long period of evolution. Each species has undergone constant modification in response to environmental pressure since it first came into being. And over the aeons, plant species have not lived in isolation; specific groups of species have existed and are existing together. An interacting population of various kinds of plants that occupy a common area is called a **biotic community.**

When biotic communities develop in a natural, undisturbed setting, they tend to change in a particular direction, usually from less complex communities of small plants to more complex communities dominated by larger plants. The change is continuous and seems to progress through recognizable stages. The term **succession** is applied to the sequence of changes that occurs in a biotic community as it develops on a given site.

The entire sequence of communities that is characteristic of a site is called a *sere.* Theoretically, at any site there is a specified and orderly sequence leading to an ultimately stable community. The "final" community that caps a long, undisturbed succession and is in a state of dynamic equilibrium is referred to as the *climax.*

PRIMARY SUCCESSION

A primary succession—also known as a **primary sere,** or **prisere**—begins on a bare substrate that has not previously supported life. A newly exposed sand dune or a recent lava flow are examples of bare substrates. As soon as life establishes itself on a bare substrate, an orderly sequence of developmental stages, which eventually lead to a climax, is initiated. The developmental stages are known as **seral stages.**

The first step, then, in a primary sere is the production of the bare area. Any one of a number of processes and events can produce a bare substrate. The process of emergence along a coast, for example, exposes an essentially new region to colonization. Glacial recession as well as fluvial erosion and deposition are additional events that develop and ready new substrates for the process of primary succession. The action of the wind in producing new dunes and the role of volcanic eruptions in producing lava flows also serve as examples of ways in which bare substrates are produced.

The seral stages of a primary sere are initiated by plant migration, that is, the initial colonization of the bare substrate by some species. From this point on, the seral stages involve a number of somewhat simultaneous and continuous developmental events. For example, the successful colonization of the area by a single species leads naturally to the development of colonies in which two or more species are successfully established on the substrate. Competition—involving the struggle for survival among the colonists—is another part of the ongoing seral stages; the struggle involves competition for space, light, water, and nutrients. Invasion of the substrate by new species from adjacent areas constantly adds to the number of colonists as the succession progresses. Gradually the number and variety of colonists begin to have an impact on the habitat; and, the colonists by their own activity produce changes in the conditions of the habitat. Changes in the microenvironment—especially the microclimate—produced by the succession of evolving biotic communities makes conditions suitable for new species to invade the habitat and establish themselves as part of the succession.

These floating-leaf colonists in this pond located near Athens, Georgia, are part of the succession that will convert this pond to a swamp habitat. (USDA)

The productivity of a primary sere is usually limited by the lack of some necessary ingredient in the environment. The process of succession, however, tends to increase the biologically important chemicals as the accumulation of different kinds of organic materials, including humus, increases. As the productivity of the biotic community increases, new levels of stability are constantly achieved until climax is attained. The competition within a climax community is generally so intense that further invasion is not very probable unless the community is drastically disturbed.

There are a number of types of primary seres. Those initiated in freshwater are called **hydroseres.** A primary succession beginning in saline water is known as a **halosere.** On damp surfaces such as alluvial mud, the term **mesosere** is used to designate and call attention to the fact that a primary succession is in progress. The general designation **xerosere** is used to identify primary seres initiated on dry materials; the specific term **lithosere** is used for a dry-material succession on bare rock; and **psammosere** is reserved for a primary succession on dry sand. Investigations of primary successions on sand dunes and lava flows indicate that, in some instances, from the

initiation to the climax more than 1000 years are involved.

A TYPICAL HYDROSERE

Let's assume that a new freshwater lake is created in a temperate region by one of the processes previously described on pages 124–128. An ideal succession, which starts as the lake bottom is colonized by aquatic vascular plants, can be visualized. The next step in the succession is initiated by the invasion of mosses into the shallow water of the lake; this stage is followed by the accumulation of dense mats of partially decayed vegetation that collect silt and humus.

Gradually, the bed of the lake is built up by the accumulation of decaying mats. When the water depth and other conditions are just right, the developing habitat is invaded by floating-leaf types such as water lilies, which send their long, stout leaf and flower stalks up from the mud bottom of the clear, shallow water. Pondweed, which grows in especially calm water, is another of the floating-leaf types that is usually part of the succession at this stage. The floating-leaf colonists tend to shade out the submerged plants. The long stalks of the water lilies trap silt; when they die, their coarse bodies fall to the bottom

Reed is the term applied to several distinct species of large water-loving grasses. The common water reed occurs along the margins of lakes, marshes, and placid streams. These giant reeds are a stage in the succession of a hydrosere. (USDA)

extreme exposure, the general lack of water, and the difficulty in retrieving nutrients from the slick surface of a rock. The activity of lichens prepares the way for hardy terrestrial mosses by producing soil particles, which accumulate in crevasses and depressions.

Mosses usually invade the developing community as spores. A single moss plant has tiny leaves growing around a stem. The rootlets, called *rhizoids,* look like hairs and grow from the bottom of the stem. Mosses are soil makers; their small rootlets work on the rock and produce an accumulation of fine soil.

The soil accumulations produced by the mosses make excellent nests for invading herbs, especially xerophytic annuals. The annuals are typically followed by herbaceous perennials. In time, taller woody plants enter the succession, overtop the herbs, and eventually change conditions sufficiently so the herbs can no longer compete in the succession. In due course, some kind of forest climax develops.

and build the lake bed to new heights until the water is shallow enough for swamp plants to enter the community.

A reed swamp—with plants only partially submerged—develops as the next stage in the succession. The cycle of plant growth in this new biotic community builds up the bed quickly and makes conditions suitable for fully terrestrial plants to invade the swamp. The terrestrial colonists eventually overwhelm the reed-swamp plants and the community gradually evolves into a sedge-meadow habitat.

The soil surface in this new community continues to build, and gradually shrubs and ultimately trees invade the habitat and cause it to evolve into a hygrophytic woodland. Alder, poplar, and willow trees gradually shade out the lower types and modify the habitat still further in preparation for the climax forest, which requires drier soil conditions.

A TYPICAL LITHOSERE

This primary sere is initiated on bare rock. It generally falls to the lichens to overcome the

This lush growth of ferns and mosses is on the floor of Allegheny National Forest in Pennsylvania. (National Parks Service)

PINE BARRENS

The Pine Barrens of central and southern New Jersey is a broad expanse of relatively level land covering approximately one and one-third million acres on the coastal plain between the piedmont and the tidal strip, roughly 80 miles (49.7 km) long and 30 miles (18.6 km) wide (Fig. 15–A). Though the largest single wilderness area east of the Mississippi River, it is less than 35 miles from New York at its northern boundary and 20 miles from Philadelphia at its western edge. The white sandy soil of the Barrens is largely infertile. With the exception of areas of cranberry and blueberry cultivation and some isolated truck farms, the region is not suited for commercial agriculture. The natural forest covering most of the region is predominantly pitch pine (Fig. 15–B), also known locally as "bull" pine, with scattered stands of shortleaf pine. Oak trees of several species are common where the pines have been removed, and oak becomes the climax type where firmly established. Oak and pine

FIGURE 15–B   A pitch-pine forest. (Photo by Lloyd Black)

FIGURE 15–A   The Pine Barrens cover a large area in New Jersey.

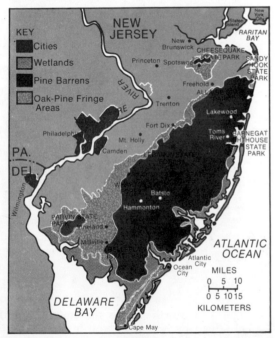

are mixed in the fringe areas surrounding the Pine Barrens and in those localities within the pinelands that have been settled. Figure 15–C, taken from the fire tower at Batsto, shows such a mixed forest that soon becomes mostly pine as one moves away from this pocket of civilization. In the swamps and along the streams, southern white cedar and swamp magnolia are quite common, with sour gum, red maple, and gray birch on somewhat drier soils nearby.

The greatest natural resource of the Pine Barrens lies under it—water. This water is extremely pure, and the Pine Barrens aquifer is so large that, in volume, it is equivalent to a lake 75 feet deep and covering a thousand square miles. So little of this water is used that it can be said to be untapped.

Evidence of this great water resource be-

FIGURE 15–C   A mixed forest in the Pine Barrens. (Photo by Lloyd Black)

low the Pine Barrens can be found in the numerous streams and small rivers that flow through the region, diminishing little in volume even in prolonged times of drought (Fig. 15–D). Though most major river and

FIGURE 15–D   A stream in the Pine Barrens. (Photo by Lloyd Black)

stream systems in the United States are in some way polluted, the small rivers and streams of the Pine Barrens are potable. The pinelands have their own divide. Some waterways, such as the Rancocas Creek and its branches, flow west to the Delaware River. Most flow southeast or east directly into the Atlantic Ocean. Examples are Great Egg Harbor at the southeast; Westecunck, Oyster, and Cedar Creeks and the Toms, Metedeconk, and Manasquan Rivers at the east and north; and the Mullica River and its tributaries (the Wading, Bass, and Batsto Rivers and the Nescochague and Landing Creeks) easterly. Since there are no through-flowing streams in the pines, outside pollution is almost nonexistent.

In addition to the rivers and streams that cross the Pine Barrens, large acreages of swamp land exist. Over the years many of the cut-over white cedar swamps have been converted into cranberry bogs. Figure 15–E shows an earthen dike separating two such bogs. When the berries are ripe, the bogs will be flooded; the berries are then knocked off the vines where they float to the surface for ease of harvesting. The bogs are also flooded throughout the winter to protect the vines from wind damage. Note the beehives, used for pollination purposes. The pine region still produces almost one third of the commercial cranberries in the United States. The only

FIGURE 15–E These cranberry bogs in the Pine Barrens are separated by an earthen dike. (Photo by Lloyd Black)

other major commercial enterprise in the region is blueberry growing. As with the cranberries, the commercial blueberry plants were developed from wild strains native to the Pine Barrens.

Currently, a major controversy has erupted over land use in the Pine Barrens involving environmentalists, developers, the State of New Jersey, and the Federal Government. Though most people agree that the still wild central area of the Pine Barrens must be pre-served, opinions vary widely about the thousands of acres of surrounding pine forest that act as a buffer between the heart of the Barrens and relatively close-by urban areas. Figure 15–F, taken from the Medford fire tower, shows recent inroads of civilization into these peripheral areas: homes and high-voltage towers. A moratorium on most construction in the Pine Barrens was instituted in 1979, though this is being vigorously challenged.

FIGURE 15–F Civilization has made inroads into the Pine Barrens. (Photo by Lloyd Black)

SECONDARY SUCCESSION

A secondary succession—also known as a **secondary sere,** or **subsere**—is the sequential development of a community after a disturbance. In other words, it is a new beginning after the plant succession in progress has been stopped or destroyed. The soil of the disturbed habitat must, however, remain in order for the new succession to be designated as secondary. If, in addition to the destruction of the plant succession, the soil is removed and a bare substrate is exposed, the succession is, by definition, a primary sere.

A biotic community is subject to a variety of events that can alter or destroy it. Fires, floods, hurricanes, parasites, and humans are potential threats to the ongoing succession of most communities. Of course, a disturbance may not necessarily be bad for the community as a whole. Some species of plants benefit from the disturbance because it destroys the competition. Others, in fact, depend on a disturbance for survival; some pines and cypress, for example, have seed cones that require fire to burst the cone and free the seed.

Secondary successions are often seen on farmland that has been abandoned for long periods of time. In the eastern United States, approximately 150 to 200 years are required for the succession in an abandoned field to reach its forest-community climax. Within the first 2 years an abandoned field develops into a grassland habitat with crabgrass, horseweed, and aster as the primary colonists. During the next 10 years the community is invaded by shrubs and then trees. At the end of 25 years, a pine-forest habitat is well on its way to being established. The pines develop into the dominant types over the following 75 years. The hardwood understory that develops in the pine forest gradually takes over the succession and, then, an oak-hickory forest climax begins to develop.

CLIMAX VEGETATION

Competition is the key to understanding plant succession. The final outcome of the competition results in the climax population, which theoretically is best fitted to take advantage of the conditions brought about by past reactions. Fluctuations in the climax community tend to be minor in the absence of any forceful change. The climax is really a mature state of vegetation living in a state of more-or-less dynamic equilibrium with the local environment. The climax community is usually lacking in successional stages and it generally persists—barring any disaster—as long as the climate remains unchanged.

This is a stage in the development of a secondary sere in a South Carolina forest. Two pond-pine trees in the left foreground and one pond-pine in the middle background are surrounded by broom-sedge growing on a marginally wet site with a history of frequent fires. (South Carolina Dept. of Forestry)

THE LANGUAGE OF PHYSICAL GEOGRAPHY

A physical geographer uses a technical vocabulary to make explanations. Review your understanding of the vocabulary used to develop the concepts in this chapter. Among the important words and terms used are:

aquatic habitats
arctic fell-field
biotic community
canopy
cold desert
desert
dystrophic
emergents
entrophic
environment
evergreen hard-
   wood forest
forest
formation-class
galleria forest

grasslands
ground cover
habitat
halosere
hydrophyte
hydroseres
hygric
hygrophytes
lithosere
megatherms
mesic
mesophytes
mesosere
mesotherm

microtherm
monsoon forests
needleleaf forest
oligotrophic
primary sere
prisere
psammosere
savanna
secondary sere
seral stages
shortgrass
subsere
subterranean strata
subtropical deserts

subtropical rain forests
succession
summergreen
   deciduous forest
tallgrass
temperate rain forests
terrestrial habitats
transpiration
tropical deserts
tropical rain forests
understory
xeric
xerophytes
xerosere

SELF-TESTING QUIZ

1. Discuss the factors that interact to produce a habitat.
2. What is used as a criterion for establishing a primary subdivision of habitats?
3. List and compare three types of freshwater habitat that are distinguished by acidity and salt content.
4. What are the four major terrestrial habitats?
5. Discuss the generalized precipitation pattern that characterizes forest habitats.
6. Identify the typical vertical structure within a forest habitat.
7. List the various types of forest habitats found across the Earth.
8. Discuss the pattern of rainfall found within a savanna.
9. In what sense are grasslands multistoried?
10. How would you characterize most of the world's desert vegetation?
11. Identify and compare the various desert types.
12. Explain the concept of factor classes.
13. What are the five climatic elements that interact to prescribe a vegetation type?
14. How does the time-distribution and intensity of light make an impact on vegetation?
15. Why is the evaporating power of the air crucial to plant growth and development?
16. How does wind make an impact on the regional distribution of vegetation?
17. Explain differences in vegetation that are due to topography.
18. What essential conditions must exist for a primary succession to develop?

IN REVIEW
SPATIAL DISTRIBUTION OF VEGETATION

I. MAIN HABITATS

  A. Aquatic Habitats

  B. Terrestrial Habitats

    1. Forest Habitats

    2. Savanna Habitats

    3. Grassland Habitats

    4. Desert Habitats

II. ENVIRONMENTAL FACTORS IN PLANT GEOGRAPHY

  A. The Role of Climate

    1. Light

    2. Temperature

    3. Precipitation

    4. Evaporating Power

    5. Wind

  B. The Geomorphic Factor

  C. The Edaphic Factors

    1. Mineral Fragments

    2. Soil Water

    3. Soil Atmosphere

    4. Organic Matter

  D. The Biotic Factors

III. COMMUNITY, SUCCESSION, AND CLIMAX

  A. Primary Succession

    1. A Typical Hydrosere

    2. A Typical Lithosere

  B. Secondary Succession

  C. Climax Vegetation

16 *Present-day environments*

One of the purposes of physical geography is to identify and study the unique features of a region that set it apart and make it different from other regions. The information in prior chapters indicates each environmental region has a unique vegetative cover that has evolved naturally through a succession of stages. The term **natural vegetation** is applied to the plant cover that develops primarily in response to the climatic conditions of the region.

The interwoven nature and relationship of climate and vegetation is clearly evident in the fact that Köppen used vegetation as his most important criterion in establishing climatic boundaries. Climate is also a potent determining factor in the intensity of various geomorphic processes as well as in the evolution of soil characteristics and profiles. Thus, it would seem that the best way to approach the study of the physical and biotic aspects as present-day environments is through a climatic framework.

Climate, as you know, is a composite idea; it is a generalization and distillation of the diverse weather conditions that prevail from day-to-day throughout the year. A climatic framework is built from the atmospheric conditions that affect life: light, temperature, precipitation, evaporating power, and wind. These components are interdependent; various combinations of them produce the characteristic climates of different parts of the world. For our immediate purpose in this chapter, however, it is convenient to work with the three classic climatic divisions—tropical, temperate, and polar—which are primarily temperature zones.

This is a view across a pond, called *Ocean Pond,* in the Osceola National Forest outside Lake City in northern Florida. Bald cypress trees dominate the area. (U.S. Forest Service)

TROPICAL ENVIRONMENTS

The climates of tropical lands are warm because the influx of solar energy is high throughout the year. In Figure 16–1, the noon altitude of the Sun is plotted for three latitudes within the tropics. The intensity of solar radiation, as you know, is related to the angle of the incoming rays.

At the Equator, the noontime solar altitude varies in a symmetrical pattern from 66.5° to 90° over the year (Fig. 16–1A). The variation at latitude 10°N is not symmetrical: the noon Sun stands at 90° above the horizon on April 16 and August 28; on June 22, it is around 76° at noon; and, by December 22, the Sun stands at its lowest noon position of approximately 56° (Fig. 16–1B). As we move to higher latitudes within the tropics, the asymmetrical pattern is even more evident. At latitude 20°N, a noon elevation of approximately 46° is recorded on December 22 (Fig. 16–1C).

The patterns plotted in Figure 16–1 are reflected in Table 16–1. At the Equator, for example, there is a symmetrical input of energy for each half of the year (Table 16–1). The energy input on a half-year basis, as recorded in Table 16–1, is, however, asymmetrical as you move to higher latitudes.

The data in Table 16–1 indicate that within the tropics there is not a lot of variation between the total energy received at various latitudes. Compare, for example, the total input at the Equator and the total received at latitude 20°. The input at latitude 20° is 94.5 percent of that received at the Equator on an annual basis. Use Table 16–1 to make a comparison between "summer" insolation and "winter" insolation within the tropics. Note that there is no significant periodic energy flux apparent in the data. At latitude 10°, the winter insolation is equivalent to 86.4 percent of that received during the summer period; at latitude 20°, the comparison is 73.9 percent. In a real sense, the tropical environments are winterless.

The primary periodic energy flux within the tropics is the diurnal flux—the variation in temperature from day to night is greater than the variation from season to season. San Juan, Puerto Rico (latitude 18°28′N, longitude 66°7′W) is within the tropics and its temperature data tend to substantiate this generalization (Table 16–2). The variation in San Juan's temperature over a day averages about 14.2 Fahrenheit degrees (7.9C°). The average monthly temperatures do not vary very much from San Juan's annual mean of 78°F (25.5°C). In fact, January and February are the months with the greatest variation from the annual mean, a mere 3.6 Fahrenheit degrees (20C°).

Other data that support this generalization can be cited. For example, Port-of-Spain, Trinidad (10°39′N, 61°31′W) reports a mean temperature of 79.9°F (26.6°C) for May, which is its warmest month, and a mean of 75.9°F (24.4°C) for its coldest month, February. The variation from season to season is a mere 4 Fahrenheit degrees (2.2C°), but the daily varia-

TABLE 16–1 RADIATION RECEIVED AT TOP OF ATMOSPHERE

LATITUDE (degrees)	SUMMER HALF-YEAR (cal/cm²)	WINTER HALF-YEAR (cal/cm²)	TOTAL
0	155,760	155,760	311,520
10	164,850	142,400	307,250
20	169,330	125,110	294,440
30	169,220	104,570	273,790
40	164,620	81,510	246,130
50	156,030	56,980	213,010
60	144,610	32,610	177,220
70	134,540	13,040	147,580
80	130,480	3,140	133,620
90	129,300	0	129,300

TABLE 16–2 AVERAGE DAILY MAXIMUM AND MINIMUM TEMPERATURES (°F) FOR SAN JUAN, PUERTO RICO

	JAN	FEB	MAR	APR	MAY	JUN	JUL	AUG	SEP	OCT	NOV	DEC	ANNUAL
Daily Maximum	81.3	81.8	83.1	84.0	85.8	87.1	87.1	87.8	87.8	87.1	85.0	82.7	85.0
Daily Minimum	67.4	67.0	67.5	69.2	71.5	72.9	73.7	74.0	73.2	72.8	71.4	69.6	70.9
Diurnal Range	13.9	14.8	15.6	14.8	14.3	14.2	13.4	13.8	14.6	14.3	13.6	13.1	14.2
Monthly Average	74.4	74.4	75.3	76.6	78.7	80.0	80.4	80.9	80.5	80.0	78.2	76.2	78.0

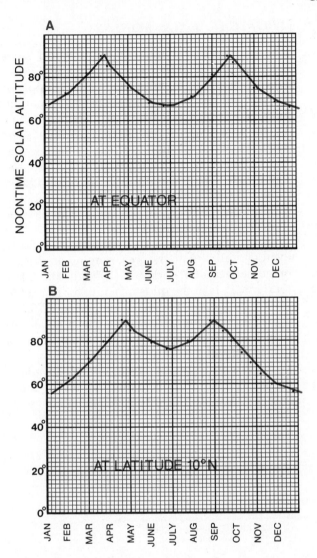

FIGURE 16–1   The noontime altitude of the Sun is plotted for three latitudes within the tropics.

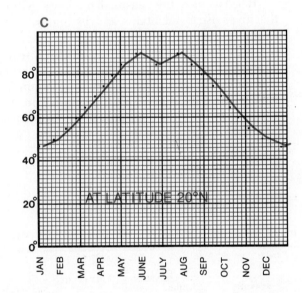

tion between maximum and minimum is much more. An average daily variation of 18 Fahrenheit degrees (10C°) occurs in January; an average spread of 21 Fahrenheit degrees (11.7C°) is found between the daily maximum and the daily minimum during April. The average daily variation at Port-of-Spain is 17 Fahrenheit degrees (9.4C°) in July; during October the average difference between the daily maximum and minimum is 18 Fahrenheit degrees (10C°). Belém, Brazil (1°27′S, 48°29′W), which is practically on the Equator, has an annual mean of 79.5°F (26.4°C) with a "seasonal" variation of less than 0.5 Fahrenheit degree (0.3C°); the average daily variation between the maximum and minimum

temperature is more than 16 Fahrenheit degrees (8.9C°), however.

The length of the alternating periods of light and dark is another aspect of tropical environments that needs to be placed in perspective. The **photoperiod** within the tropics is restricted to a very narrow range: Daylight, which is the interval between sunrise and sunset, varies at the Equator by no more than 1 minute from the September 21 duration of 12 hours 6 minutes. At latitude 10°N, the shortest daylight period of 11 hours 33 minutes occurs in December and the longest, 12 hours 43 minutes, occurs in late June, which amounts to a variation of 1 hour 10 minutes. The duration of daylight at latitude 20°N varies from 10 hours 55 minutes (December 21) to 13 hours 21 minutes (June 21), a difference of 2 hours 26 minutes. These very restricted variations in the length of day have a profound effect on vegetation.

Precipitation is another important element in tropical environments because the availability of water controls the productivity of the system. A large part of the tropics is desert with very restricted amounts of available water; only a small portion of the tropics has a year-round rainy season. On balance, however, most tropical environments have a marked "seasonal" regime of rainfall. Variations in precipitation patterns are documented in Table 16–3 for selected locations within the tropics.

It is, in fact, the seasonal moisture pattern that gives rise to, maintains, and distinguishes the major tropical environments—the rain forests, the savannas, and the deserts. The rain forest is characterized by a high frequency of precipitation in all months. The savanna environments have a rainy season and a dry season. The deserts, of course, are plagued by perennial drought.

TROPICAL RAIN FOREST

The climate that dominates this tropical environment places no limits on the growth of vegetation because it provides an abundant supply of energy and moisture during every month of the year. The **tropical rain forest**—consisting primarily of broadleaf evergreen trees, which shed some leaves throughout the year—is the vegetative response to the climatic regime. The large evergreen leaves maintain photosynthesis and growth throughout the year and also permit high rates of transpiration, which prevent the leaves from heating excessively.

Tropical rain forests are associated with fairly even and high temperatures, ranging on the average between 68° and 86°F (20° and 30°C). Belém, Brazil, is in the midst of a rain forest (Table 16–3); its annual mean temperature is 79.5°F (26.4°C). Andagoya, Colombia, has a mean annual temperature of 82°F (27.8°C) and is in a rain-forest environment, too (Table 16–3). The average monthly temperatures in these locations do not vary from their annual means by more than 1 Fahrenheit degree (0.5C°).

TABLE 16–3   **PRECIPITATION PATTERNS WITHIN THE TROPICS**

STATION	LOCATION (latitude, longitude)	ELEVATION (feet)	AVERAGE PRECIPITATION Inches												
			Jan	Feb	Mar	Apr	May	Jun	Jul	Aug	Sep	Oct	Nov	Dec	ANNUAL
Belém, Brazil	1°27′S,48°29′W	42	12.5	14.1	14.1	12.6	10.2	6.7	5.9	4.4	3.5	3.3	2.6	6.1	96.0
Quixeramobim, Brazil	5°12′S,39°18′W	653	0.7	5.0	6.6	5.0	7.0	1.7	0.7	0.6	0.4	0.6	0.7	0.6	29.6
Andagoya, Colombia	5°6′N,76°40′W	197	25.0	21.4	19.5	26.1	25.5	25.8	23.3	25.3	24.6	22.7	22.4	19.5	281.1
Cartagena, Colombia	10°28′N,75°30′W	39	0.4	0.0	0.4	0.9	3.4	3.4	3.0	0.6	0.5	10.8	8.9	4.5	36.8
Cuenca, Equador	2°53′S,78°39′W	8301	2.0	1.8	3.2	4.3	4.3	1.7	0.9	1.1	1.6	3.1	1.8	2.5	28.3
Georgetown, Guyana	6°50′N,58°12′W	6	8.0	4.5	6.9	5.5	11.4	11.9	10.0	6.9	3.2	3.0	6.1	11.3	88.7
Monrovia, Liberia	6°18′N,10°48′W	75	0.2	0.1	4.4	11.7	13.4	36.1	24.2	18.6	29.9	25.2	8.2	2.9	174.9
Lagos, Nigeria	6°27′N,3°24′E	10	1.1	1.8	4.0	5.9	10.6	18.1	11.0	2.5	5.5	8.1	2.7	1.0	72.3
Berbera, Somalia	10°26′N,45°2′E	45	0.3	0.1	0.2	0.5	0.3	—	—	0.1	—	0.1	0.2	0.2	2.0
Khartoum, Sudan	15°37′N,32°33′E	1279	—	—	—	—	0.1	0.3	2.1	2.8	0.7	0.2	—	—	6.2
Bombay, India	19°6′N,72°51′E	27	0.1	0.1	0.1	—	0.7	19.1	24.3	13.4	10.4	2.5	0.5	0.1	71.2
Broome, Australia	17°57′S,122°13′E	56	6.3	5.8	3.9	1.2	0.6	0.9	0.2	0.1	—	—	0.6	3.5	22.9
Darwin, Australia	12°28′S,130°50′E	104	15.2	12.3	10.0	3.8	0.6	0.1	—	0.1	0.5	2.0	4.7	9.4	58.7
Calcutta, India	22°52′N,88°20′E	21	0.4	1.2	1.4	1.7	5.5	11.7	12.8	12.9	9.9	4.5	0.8	0.2	63.0
Lusaka, Zambia	15°25′S,28°17′E	4191	9.1	7.5	5.6	0.7	0.1	—	—	—	0.4	3.6	5.9	32.9
Iringa, Tanzania	7°47′S,35°42′E	5330	6.8	5.1	7.1	3.5	0.5	—	—	—	0.1	0.2	1.5	4.5	29.3
Kaduna, Nigeria	10°35′N,6°26′E	2113	—	0.1	0.5	2.5	5.9	7.1	8.5	11.9	10.6	2.9	0.1	—	50.1
Niamey, Niger	13°31′N,2°6′E	709	—	—	0.2	0.3	1.3	3.2	5.2	7.4	3.7	0.5	—	—	21.6
Mogadisho, Somalia	2°1′N,45°20′E	39	—	—	—	2.3	2.3	3.8	2.5	1.9	1.0	0.9	1.6	0.5	16.9
Al Fashir, Sudan	13°38′N,25°21′E	2395	—	—	—	—	0.3	0.7	4.5	5.3	1.2	0.2	—	—	12.2
Port Sudan, Sudan	19°37′N,37°13′E	18	0.2	0.1	—	—	—	—	0.3	0.1	—	0.4	1.7	0.9	3.7
Cufra, Libya	24°20′N,23°15′E	1276	—	—	—	—	—	—	—	—	—	—	—	—	0.0

Another feature of rain-forest climates is the high frequency of precipitation year-round. In some areas, rain falls almost every afternoon and night throughout the year; in others, there are one or two dry seasons, which consist of months with less than 4 inches (10 cm) of rain. On an annual basis, a rain forest generally has more than 80 inches (200 cm) of precipitation. Perusal of Table 16–3 indicates that both Belém and Andagoya meet the minimum-precipitation requirement. Belém has a total of 96 inches (244 cm) for the year. Andagoya receives a somewhat evenly distributed 281 inches (714 cm) of precipitation over the year. Georgetown, Guyana, with a fairly even temperature of 80°F (26.7°C) throughout the year and an annual precipitation of 88.7 inches (225 cm) (Table 16–3) also qualifies as a tropical rain-forest environment and it falls well within the broadleaf-evergreen forest zone in Figure 16–2.

LOCATION AND EXTENT

Tropical rain forest regions are restricted to low elevations—usually below 3300 feet (1000 m)—since temperatures at higher elevations tend to fall below the limits cited earlier. The combinations of climatic elements that encourage the development of tropical rain forests have given rise to three distinct formations: the American, African, and Indo-Malaysian tropical rain forests.

The American tropical rain forests have their most extensive development in the Amazon Basin (Fig. 16–2). These broadleaf evergreen forests also extend southward from the Amazon Basin along river valleys and on to the Brazilian plateau of Mato Grosso. A southwestward extension is found to the north and to the east of Lake Titicaca in Bolivia; a narrow belt moves southward along the sub-Andean region of Bolivia into northern Argentina (Fig. 16–2). Tropical rain forests also move northward along the foothills of the Andes in Peru, Ecuador, and Colombia. Large portions of Venezuela and almost all of the Guianas are covered by these forests. The rain forests that occupy the lowlands of northern Ecuador and coastal Colombia extend into eastern Panama, eastern Nicaragua, eastern

The pond *(left)* is fed by a small waterfall deep beneath a canopy of trees in Puerto Rico's tropical rain forest, El Yunque, which is 25 miles (40 kilometers) east of San Juan. This stream *(right)* is moving through dense growth in the rain forest at Dalrymple Heights in the Eungella Range, west of Mackay, Queensland, Australia. (Puerto Rico Tourism; Australian Government)

FIGURE 16–2 The present-day natural vegetation pattern for South America can be deciphered using the key below. (Color areas are for visual contrast only and do not represent specific vegetation patterns.)

Alpine Vegetation
Deciduous "Beech" Forest
Evergreen Mixed Forest
Pampas And Other Grasslands
Patagonian Semidesert Scrub
Chilean Sclerophyllous Scrub
Tropical Rain Forest
Cactus Scrub
Desert
Tropical Semi-Evergreen Forest
Thorn Forest
Broadleaf Tree Savanna (Campos)
Thorn Tree–Desert Grass Savanna
Tropical Montane Forest

Honduras, and northern Guatemala to occupy most of the southern Yucatan (Fig. 16–3). The eastern parts of Cuba, Jamaica, and northern Puerto Rico (Fig. 16–3) are also covered by rain forests.

Large areas of the African rain-forest formation are still found in the Congo Basin (1°N, 18°E) (Fig. 16–4). The former extent of these rain forests, however, was much greater; vast areas have been disturbed and cleared in Nigeria, Ghana, and other West African countries. Human interference with the tropical rain forest of eastern Madagascar (Fig. 16–4) has done irreparable damage to the formation.

The Indo-Malaysian rain forests (Fig. 16–5) are widely distributed and extend north of the Tropic of Cancer in the Indian state of Assam (26°N, 92°E) and south of the Tropic of Capricorn along the coast of Queensland in Australia (Fig. 16–6). The westernmost segment of the Indo-Malaysian formation is in the Western Ghats, a nearly continuous chain of low mountain ranges

on the western shore of India that extends from the Tapti River in the north to Cape Comorin at the southern tip of the peninsula (Fig. 16–5). The easternmost extension of this rain-forest formation is found on the Fiji Islands in the southern Pacific (18°S, 178°E). Most of Malaya (4°N, 102°E), Sumatra (5°S, 102°E), Borneo (1°N, 114°E), and New Guinea (5°S, 140°E) are covered by tropical rain forests. Extensive areas of Bangladesh, Burma, Vietnam, and the Philippines (Fig. 16–5) are also covered by rain forests, although vast areas in all of these regions have been cleared and modified.

PRESENT STATUS

Rain forests were, at one time, much more widespread than they are today. In fact, at the present time, very few undisturbed tropical rain forests exist; most are subject to a variety of disturbances that range from the clearing of the land for food production and industrial crops to exploitation of large areas for timber and news-

Gabon, on the west coast of Africa, is heavily forested. Rainfall in many regions is above 100 inches (254 centimeters). Roads have been constructed to give access to logging sites in Gabon's rain forests. This convoy of trucks carrying logs to a processing center is in the vicinity of Ndjolé (0°11′S,10°45′E). The vegetation in the area is thick and dense. (World Bank)

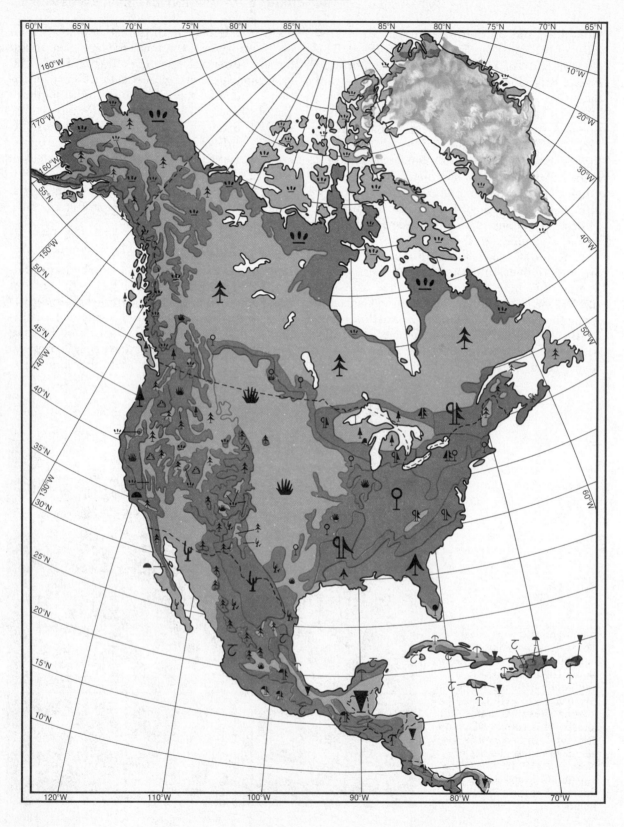

print. Thousands of acres of rain forest in the Amazon Basin have been cleared to provide new roads, charcoal for steel mills, and cattle pastures. One estimate indicates that 36 percent of these forests in South America and 63 percent in India have already been destroyed. Those in the Philippines and Indonesia are under similar attack.

Long-term agricultural projects in rain-forest regions have not been notably successful. Present attempts to farm these areas seem doomed to failure because the soil in which a rain forest develops is remarkably poor in nitrogen, phosphorus, and other elements necessary for plant growth. The abundant downward movement of water in a rain-forest region creates a leaching process that removes calcium, magnesium, and silica from the soil. The leaching process, operating over millions of years, has left tropical rainforest soils with little more than an accumulation of aluminum and iron oxides and hydroxides. It is difficult to imagine a worse environment for agricultural crops than these soils stripped of their cover and exposed to the direct rays of the tropical sun. Lush rain forests manage to thrive in such an environment because the trees are able to maintain an almost closed nutrient cycle; but, when the trees are cleared, the cycle is broken.

The trees within a rain forest send their roots out along the ground's surface and just below it to form a tight, incredibly thick root mass. When leaves and dead wood fall to the ground, fungi absorb the **nutrients** released from the decaying plant matter very efficiently and rapidly. Less than one-tenth of 1 percent of the nutrients released through the rapid and constant process of decay leach below the root mass into the soil. The lichens and mosses growing above the root mass in these forests also scavenge nutrients from the rainwater and decrease the supply of nutrients carried to the soil. The recycling of nutrients is so extremely efficient that streams flowing through undisturbed rain forests carry water whose composition is, for all practical purposes, that of pure rainwater.

The collecting and conserving of nutrient elements is only one of the ways in which rain forests affect the environment. Another, which is certainly of great significance, includes the effects these forests, in common with all other vegetation, have on the oxygen and carbon-dioxide levels of the atmosphere. These vast forested areas also have an immediate and profound effect on evaporation, condensation, and the hydrologic cycle generally. The possible long-range climatic effect of tropical deforestation is under investigation. Preliminary findings indicate the removal of tropical rain forests will lead to cooling in the middle and upper tropical troposphere, a lowering of the tropical tropopause, and a warming of the lower tropical stratosphere. A rather ugly question remains: Will the destruction of rain forests produce an ultimate impact that pushes us closer to an irreversible worldwide environmental deterioration?

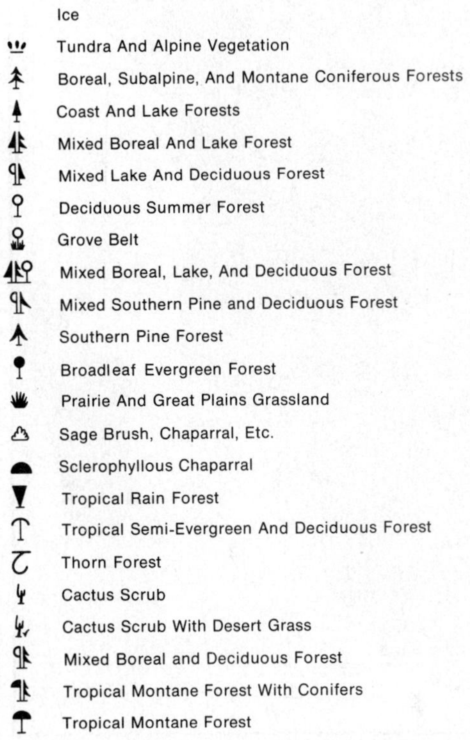
Ice
Tundra And Alpine Vegetation
Boreal, Subalpine, And Montane Coniferous Forests
Coast And Lake Forests
Mixed Boreal And Lake Forest
Mixed Lake And Deciduous Forest
Deciduous Summer Forest
Grove Belt
Mixed Boreal, Lake, And Deciduous Forest
Mixed Southern Pine and Deciduous Forest
Southern Pine Forest
Broadleaf Evergreen Forest
Prairie And Great Plains Grassland
Sage Brush, Chaparral, Etc.
Sclerophyllous Chaparral
Tropical Rain Forest
Tropical Semi-Evergreen And Deciduous Forest
Thorn Forest
Cactus Scrub
Cactus Scrub With Desert Grass
Mixed Boreal and Deciduous Forest
Tropical Montane Forest With Conifers
Tropical Montane Forest

FIGURE 16–3 The present-day natural vegetation pattern for North America can be deciphered by using the key. (Color areas are for visual contrast only and do not represent specific vegetation patterns.)

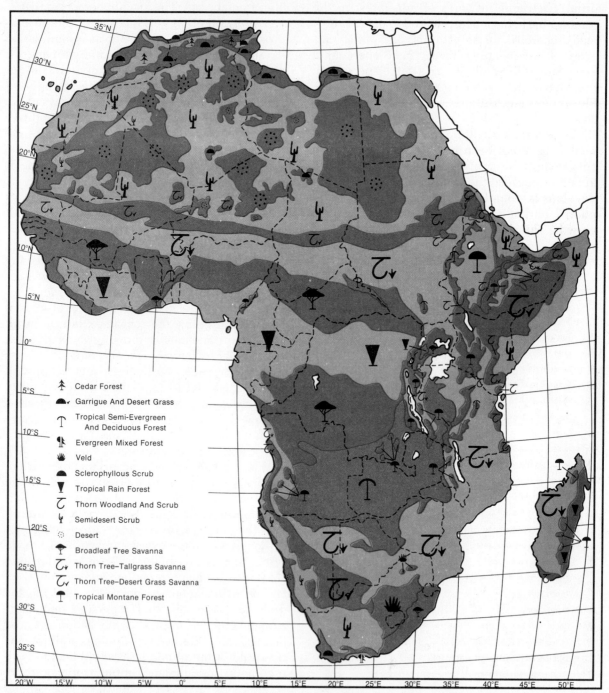

FIGURE 16–4   The present-day natural vegetation pattern for Africa can be deciphered using the key. (Color areas are for visual contrast only and do not represent specific vegetation patterns.)

TROPICAL SEASONAL FORESTS

Marked dry seasons occur in some areas of the otherwise humid tropics. The vegetation that develops in these regions is dependent on seasonal rainfall and is able to survive the one or sometimes two pronounced dry seasons of several months duration. When a tropical habitat with a pattern of dry seasons is primarily timbered, it is classified as one of three possibilities: monsoon forest (also known as semievergreen seasonal forest), deciduous seasonal forest, or thorn woodland.

The vegetation in these mixed tropical forests and shrublands is adapted to the moisture deficiency of the dry season. The three main groupings and the order in which they are listed above reflect decreasing availability of water. Many of the trees in these formations are deciduous and they shed their leaves for a month or two when soil moisture is low. The ground layer of vegetation in these formations tends to be more varied than in a rain forest because more light penetrates to the ground level.

MONSOON FOREST

These forests develop in regions with a pronounced wet and dry season. The wet season is one with abundant rainfall, but the dry season is one with a pronounced drought that may last from 3 to 6 months. The total amount of rainfall is usually less than in tropical rain forests; it commonly ranges between 40 and 80 inches (100 and 200 cm).

The **monsoon** or **semievergreen seasonal forest** is composed predominantly of evergreen trees, but more than 20 to 30 percent of the forest population is regularly deciduous. Even some of the evergreen species have the facility for losing their leaves in an abnormally dry season. The nutrient cycle in a monsoon forest is also very different from that in a tropical rain forest; in a monsoon forest, the ground is covered by freshly fallen leaves during the dry season and the litter does not begin to decay until the wet season.

Monsoon and allied forests are found in areas of true monsoons. India, recall, has a well-developed monsoon; both Bombay and Calcutta (Table 16–3) are in areas that support monsoon forests (Fig. 16–5). Darwin, Australia (Table 16–3), is also in a monsoon region and its precipitation pattern supports this kind of forest (Fig. 16–6).

These monsoon-dependent regions are marked by daily and seasonal changes in temperature (Table 16–4) and strong winds. Bombay's coolest month, January, has an average temperature of 75°F (23.3°C); its warmest month, May, has an average of 86.2°F (30.1°C). The average seasonal temperature difference in Bombay amounts to 11.2 Fahrenheit degrees (6.2C°). Calcutta has a larger seasonal difference of 20.5 Fahrenheit degrees (11.4C°) between its warmest average of 88°F (31.1°C), which occurs in May, and its coolest average of 67.5°F (19.7°C), which occurs in January.

Generally, we can say that the vegetation in a monsoon forest is not as luxuriant as that in a tropical rain forest; it varies in appearance, however, from an impoverished form of rain forest down to a savanna woodland. At times, the distinctions among the various tropical environments are somewhat blurred and it is difficult to categorize regions on the basis of specific parameters. The western coast of Africa, in the Nigeria-Liberia area, which is swept by a summer monsoon, is a case in point: Lagos, Nigeria (Table 16–3), with 4 consecutive months of precipitation below 4 inches (10 cm), is slightly below the water-parameter requirement to support a rain forest, *but* it does (Fig. 16–4). Monrovia, Liberia (Table 16–3) has a much larger annual precipitation, but it has 2 months of severe drought and presently supports vegetation that is best described as a broadleaf savanna or savanna woodland (Fig. 16–4).

TABLE 16–4 **DAILY TEMPERATURE IN MONSOON-DEPENDENT REGIONS** (°F)

| | JANUARY | | APRIL | | JULY | | OCTOBER | | EXTREME | |
	Max	Min	Max	Min	Max	Min	Max	Min	Max	Min
Darwin (Australia)	90	77	92	76	87	67	93	77	105	55
Bombay (India)	88	62	93	74	88	75	93	73	110	46
Calcutta (India)	80	55	97	66	90	79	89	74	111	44

DECIDUOUS SEASONAL FOREST

This kind of forest is frequently found intermixed with a semievergreen forest. But its maximum development occurs in areas that have a more protracted dry season than that normally associated with the semievergreen seasonal forest. In South America, for example, the tropical **deciduous seasonal forest** is found where there is a season of 5 consecutive dry months with less than 4 inches (10 cm) of precipitation and 2 of these dry months normally have less than 1 inch (2.5 cm); thus, the plants in this habitat must be adapted to withstand drought.

In the African formations, no clear distinction can really be made between a deciduous seasonal forest and a **savanna woodland,** which is a grassy forest. Lusaka, Zambia (Table 16–3) is surrounded by a deciduous seasonal forest in which you can see through the trees up to a distance of 0.5 mile (0.8 km) (Fig. 16–4). Note from Table 16–3 that Lusaka has 7 consecutive months with less than 1 inch (2.5 cm) of precipitation and for 5 months the drought is almost absolute. On the average, the habitats that support a deciduous seasonal forest get between 30 and 40 inches (76 and 100 cm) of precipitation.

THORN WOODLANDS

No abrupt change in vegetation type occurs where deciduous seasonal forests are flanked by even drier regions. As the dry season becomes longer and the mean annual rainfall decreases, the dominant deciduous trees decrease in size and take on a more gnarled and spreading appearance. Eventually, when such a habitat becomes dominated by low bushy trees and tall shrubs—many of which carry long spines or thorns— it is referred to as a *thorn forest.*

Iringa, Tanzania, Table 16–3, is in the midst of a **thorn woodland** (Fig. 16–4). There are 8 months in which the precipitation falling in the region is less than 4 inches (10 cm); in 6 of these months, the precipitation is below 1 inch (2.54 cm). The precipitation patterns for Quixeramobim, Brazil, and Cartagena, Colombia, are detailed in Table 16–3; both these towns are in extensive thorn-forest regions (Fig. 16–2). The patterns of precipitation for Quixeramobim and Cartagena are similar to that of Iringa.

TROPICAL SAVANNA

A tropical grassland in which trees or tall bushes occur in open formation is referred to as a *savanna.* The amount of tree cover or bush cover varies enormously, however. Some savannas are open woodlands with a ground cover of grass; others are simply grasslands with a few isolated trees or bushes.

Generally, you might say that a savanna has a parklike or orchardlike appearance. The trees, which are spread at various intervals, are usually stunted and gnarled, but sometimes they are quite lofty. The grasses in some savannas may be less than 3 feet (1 m) high; in others, the grasses

This is an area near Khartoum in the Sudan. The water encourages a denser thorn-tree growth in this particular location, which, if looked at in isolation, might be interpreted as a thorn woodland. However, the presence of grass in great abundance throughout the area indicates that we can classify the locale as a thorn tree-grass savanna. (Sudan Government)

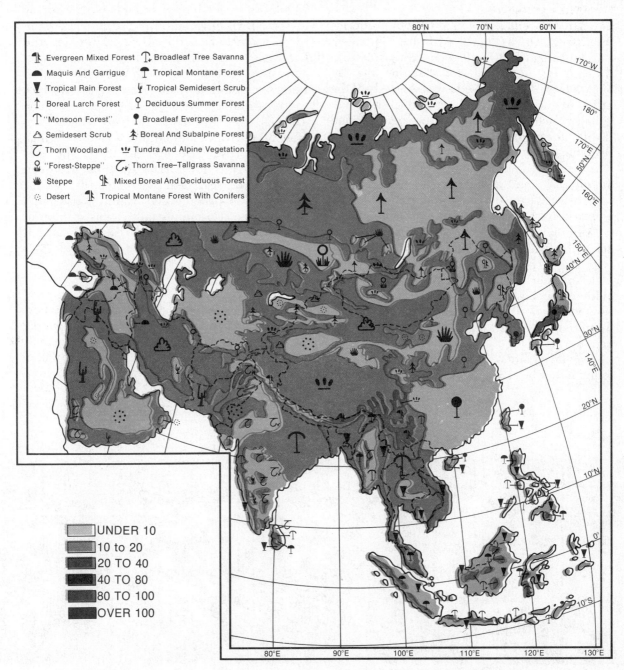

FIGURE 16–5   The present-day natural vegetation pattern of Asia can be deciphered by reference to the key. (Color areas are for visual contrast only and do not represent specific vegetation patterns.)

may exceed heights of 15 feet (4.5 m) and form impenetrable thickets.

LOCATION AND EXTENT

Extensive savannas are found in South America (Fig. 16–2), Africa (Fig. 16–4), India (Fig. 16–5), and Australia (Fig. 16–6). The savanna area in Africa between the rain-forest region and the Sahara Desert can be used as a case study of the different savanna types. Three distinct savannas, which stretch from west to east, can be identified in this African region (Fig. 16–4): a broadleaf tree savanna, known in Africa as the *Guinea zone;* a thorn tree–tallgrass savanna, referred to as the *Sudan zone;* and a third, somewhat narrower band of thorn tree–desert grass savanna, which is well within the Sahel, called the **Sahel zone.**

Monrovia and Kaduna, listed in Table 16–3, are in the Guinea zone. Throughout the Guinea zone the larger groves of trees contain many evergreens in addition to the predominantly deciduous species. The Sudan zone is represented by Niamey in Table 16–3; Khartoum, with its thorn tree–desert grass savanna, falls within the Sahel zone.

The three different kinds of savanna are also found in southern Africa (Fig. 16–4). A broadleaf tree savanna starts at the southern flank of the tropical rain forest and stretches from the southwestern part of the Congo to northern Angola. The thorn tree–tallgrass savanna is widespread in southern Kenya, eastern Tanzania, Mozambique, southern Zimbabwe-Rhodesia, and Botswana of South Africa; it also covers most of western Madagascar. South West Africa has an extensive area of thorn tree-desert grass savanna.

In Australia, a great belt of **tallgrass** savanna extends from southeastern Queensland into New South Wales and northern Victoria, which is well beyond the tropics (Fig. 16–6). The trees in this tallgrass savanna of Australia are evergreen eucalyptus. The most arid type of savanna, the thorn tree–desert grass savanna, is also found in Australia; it occupies large areas of interior Queensland and the Northern Territory, as well as extensive areas of Western Australia.

ORIGINS OF TROPICAL SAVANNA

Generally, a savanna is found in a region where there is a clearly defined wet and dry season. But the average annual precipitation is extremely varied and cannot be used as a criterion to distinguish a "savanna climate" (Monrovia, Kaduna,

Most Ethiopians are farmers who raise crops on the cool windswept highlands in much the same way their ancestors did in biblical times. This grain harvest is occurring near Mojo (8°38′N,39°7′E). Surrounding elevations are about 12,000 feet (3600 meters). The natural vegetation of much of Ethiopia's highlands is forest. In drier regions, on the higher parts and in deforested areas, a kind of savanna has developed. The trees beyond these people are in a thorn tree-tallgrass savanna setting. (World Bank)

This is typical of the semidesert-scrub region around Bikaner (28°N,73°E) in India. The native servan grass, in the foreground, is dwarfed, wiry, and exists in isolated bunches. There are acacia and castor saplings in the background that have been planted. (UNESCO)

Niamey, and Khartoum in Table 16–3, for example). Studies of the influence of climate, landscape, and soil on savanna formation tend to report that there is no basic correlation. Both grass savanna and deciduous forest are well adapted to a regime of alternating wet and dry seasons as well as various types of landscapes and soil conditions. The actual presence of a savanna or a deciduous forest at any particular site seems to depend on other factors in the total environment that favor one formation over the other.

Fire and human activity seem to be the two factors that are quite sufficient to reduce a forest to some kind of savanna. At first glance, it may seem inconceivable that vast savannas owe their existence to this combination. Human societies, however, do influence the development and spread of vegetation through the destruction of competing types. The extreme sharpness of the present boundaries between savanna and forest, especially in the Congo and West Africa, lends support to the conjecture that the forest canopy was opened by persistent cutting, cultivation, and fire. Recurrent burning, which results from either natural events or human activity, gives the herbaceous savanna vegetation an advantage over woody forest vegetation and aids savanna spread.

SEMIDESERTS AND DESERTS

On its dry side, a tropical thorn forest gives way to **semidesert** scrub. These arid bushlands are often included among deserts; but, in this discussion, semidesert scrub is considered to be a transitional habitat between true deserts and savannas or thorn woodlands. For example, Al Fashir, which is listed in Table 16–3, has a semidesert scrub area to its north (Fig. 16–4); semidesert scrub is transitional between the thorn tree–desert grass savanna of Al Fashir to its south and the Libyan Desert to its north (Fig. 16–4).

In the tropics, a semidesert-scrub habitat is generally found on stony hills, in open, rolling country, and on sandy areas exposed to the full impact of insolation. Three of the locations listed in Table 16–3 are in semidesert-scrub regions: Mogadisho, in southern Somalia, borders on the Indian Ocean; it has an annual rainfall of 7 to 20 inches (18 to 50 cm), which is on the high side of the climatic range normally found in these warm-region scrubs. Berbera, on the Gulf of Aden in northern Somalia, and Port Sudan, on the Red Sea, are both well-within semidesert-scrub regions; however, the annual precipitation for these stations is lower than the range generally reported as being needed to support a semidesert scrub. The seasonal distribution of the precipitation and its reliability are probably much more important than the precipitation range that is often specified.

The bushes in these semidesert regions are often of fair size. Grasses when present are dwarfed, wiry, and reduced to isolated bunches. Herbs tend to be leathery and fleshy-leafed. Cacti in the Americas and cactuslike plants on other continents are frequently a characteristic feature of the usually widely spaced vegetation of a semidesert region. Large-leafed agaves,

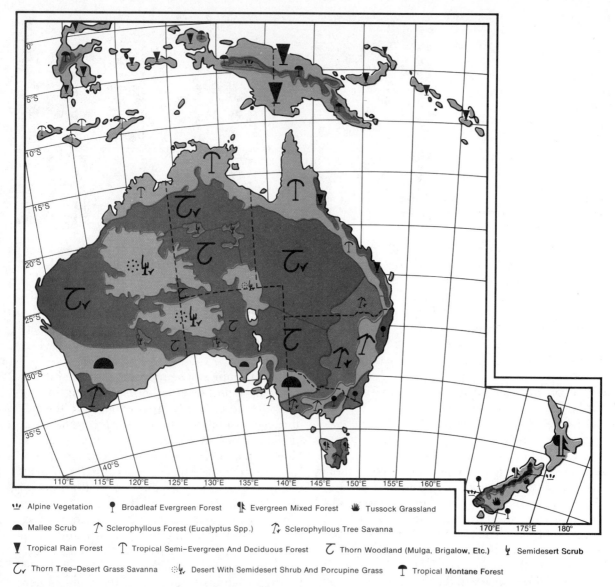

Alpine Vegetation ● Broadleaf Evergreen Forest Evergreen Mixed Forest Tussock Grassland

Mallee Scrub Sclerophyllous Forest (Eucalyptus Spp.) Sclerophyllous Tree Savanna

Tropical Rain Forest Tropical Semi–Evergreen And Deciduous Forest Thorn Woodland (Mulga, Brigalow, Etc.) Semidesert Scrub

Thorn Tree–Desert Grass Savanna Desert With Semidesert Shrub And Porcupine Grass Tropical Montane Forest

FIGURE 16–6 **The present-day natural vegetation pattern of Australasia can be deciphered by reference to the key. (Color areas are for visual contrast only and do not represent specific vegetation patterns.)**

aloes, and yuccas are also characteristic inhabitants of these regions, as are smaller succulents.

In a strict sense, the term **desert** should be applied exclusively to areas that are literally lifeless. True deserts, according to this definition, are quite rare. Some of the great sandy deserts of the Sahara (Fig. 16–4), the Rub-al-Khali of Saudi Arabia (20°N, 51°E) (Fig. 16–5), and parts of the Atacama Desert of the northern Chile (22°30′S,

69°15′W) (Fig. 16–2) are almost lifeless over large areas, however.

Cufra, listed in Table 16–3, sits in the Libyan Desert. For all practical purposes this region does not receive any precipitation; it is a true desert. Average daily maximum temperatures during July hover around 101°F (38°C); the daily minimum averages 75°F (24°C). Surprisingly, January temperatures in this barren environment are somewhat moderate; the daily maximum averages 69°F (20.5°C), and the average minimum is 43°F (6°C). The highest temperature ever recorded at this station is 122°F (50°C), and the lowest recorded temperature is 26°F (−3°C).

There are other rainless deserts in addition to the Central Sahara of North Africa represented by Cufra. The Rub-al-Khali and the Atacama Desert mentioned above are also rainless. The Takla Makan in Central Asia (39°N, 83°E in Fig. 16–5) is another rainless desert. It does not even rain once a year in these barren wastes.

DESERTIFICATION

The geologic evidence indicates that much of the Sahara was well treed during parts of the Pleistocene epoch. And the historical evidence of the last 3000 years tells us that the mountain slopes and plains of Lebanon, Syria, the Egyptian seaboard, and Tunisia were covered with rich vegetation. The celebrated cedars of Lebanon, for example, are legendary. Ancient Rome imported timber, grain, and other produce from the region. The destruction of forests and herbaceous vegetation by human activity as well as erosion by wind and water have transformed these areas into semideserts and deserts. In other words, the natural environment of Saudi Arabia and North Africa has been replaced by human-induced environments and landscapes.

The process, called **desertification,** by which desiccated, barren, desertlike conditions are created by natural changes as well as through the mismanagement of a semiarid zone has been observed and documented for a long time. In this century, the spread of the Sahara has been measured most precisely in the Sudan. A recent study of the Sudan's desert margins, for example, indicates that the desert boundary has shifted south by as much as 62 miles (100 km) in a 17-year period.

Desertification, of course, is not limited to the Saharan fringes. It is a major problem in other parts of the world, too. Many of the Western Hemisphere's drier regions and India's arid zones, as well as mismanaged rangelands in Australia are being impacted. Botswana in southern Africa has a genuine problem of desertification. Kenya, Tanzania, and Ethiopia also have vast semiarid regions that have been seriously damaged by overgrazing. Throughout most of the 1970s, the rangelands of Iraq and Syria have had triple the number of grazing animals they can safely support. Many farmlands in the Middle East presently show signs of ecological decline, and cultivation has been pushed onto lands with extremely low and unpredictable rainfall.

Where desert edges are moving outward, the process seldom involves the steady encroachment of a tide of sand along a uniform front. Rather, climatic fluctuations and land-use patterns interact to extend desertlike conditions irregularly over the susceptible lands. There is an intricate web of factors that controls the degradation and change in a habitat.

In 1974, J. Otterman of Tel-Aviv University proposed that desertification in regions of marginal rainfall may be due to an increase in surface albedo caused by the removal of vegetation by overgrazing. He suggested that when high-albedo soils are stripped of cover, the resultant increase in albedo causes lower surface temperatures, which in turn reduce the heat input to the lower atmosphere, decrease its temperature lapse rate, and hence somewhat reduce convective activity leading to rainfall. Otterman's data and proposals were disputed and challenged by R. Jackson and S. Idso in 1975. Many uncertainties and questions remain to be resolved. A lot of work needs to be done to identify the climatological mechanism of desertification. The human suffering caused by desertification, especially in the Sahel, gives a sense of urgency to this work.

MID-LATITUDE ENVIRONMENTS

The mid-latitude region in both hemispheres extends from latitude 23°30′ to latitude 66°30′. Two major centers of subsiding air serve as source regions for air masses that flow over the mid latitudes: the subtropical highs and the polar highs.

Each of these source regions has a unique temperature pattern that differs markedly from

The Middle Atlas Mountains above Fez (34°5′N,4°57′W) are covered with forests. Evergreen oak is an important tree of the subhumid and humid regions of the Middle Atlas. The Atlantic cedar is the typical tree above 5000 feet (1500 meters). There are also covers of maquis and garique—that is, scrublands characterized by small evergreen shrubs and low trees—in the Fez area. (Moroccan Government)

the other region. This fact causes the air masses emanating from these source regions to have quite different characteristics. The zone of collision or convergence of these dissimilar air masses occur between the latitude bands of 50° and 60°. Turbulent and stormy conditions are produced by energy exchanges between colliding air masses. Energy exchanges between an air mass and the surface it overrides also introduces temperature, moisture, and other changes into mid-latitude habitats.

The precipitation that falls into mid-latitude environments is quite varied in kind and amount. It runs the gamut from drizzle to ice pellets. Some is produced through convectional processes induced within a single air mass; most occurs as a result of energy exchanges between two dissimilar mid-latitude air masses; and a small amount of mid-latitude precipitation is introduced by hurricanes that are spawned in the tropics and migrate into the region. On balance, in terms of precipitation patterns, mid-latitude environments are not unlike those of the tropics—there are zones with a high frequency of precipitation throughout the year; some mid-latitude habitats have a pronounced wet and dry season, and others have only occasional precipitation.

At latitude 25°, the annual input of radiation at the top of the atmosphere is 285,100 calories per square centimeter. An examination of Table 16–1 indicates there is a progressive decrease in the total annual input as latitude increases. The input of radiation at latitude 30° is 96 percent of that received on an annual basis at latitude 25°; at 40° the input has fallen to 86.3 percent; the input stands at 74.7 percent at 50°; and at latitude 60°, the input of radiation is a mere 62.2 percent of the input received at 25°.

In addition, Table 16–1 shows a very asymmetrical pattern in the radiation received throughout the mid latitudes. The largest energy input occurs during the summer half of the year. Note, too, that the difference between the summer-half input and the winter-half input increases with increasing latitude. The winter input at latitude 30°, for example, is 73.9 percent of the summer-half input, but an increase of 10° to latitude 40° drops the winter-half radiation to 61.8 percent of the warm-period input; at latitude 60°, a comparison of winter and summer radiation places the low-Sun input at a mere 36.5 percent of the summer-half total.

Another important aspect of mid-latitude environments is the variation in the photoperiod over the year. Table 16–5 documents the dura-

TABLE 16–5   DURATION OF DAYLIGHT AT LATITUDE 40°N

DAY OF MONTH	JAN hr min	FEB hr min	MAR hr min	APR hr min	MAY hr min	JUN hr min	JUL hr min	AUG hr min	SEPT hr min	OCT hr min	NOV hr min	DEC hr min
1	9 23	10 10	11 18	12 39	13 54	14 49	14 58	14 16	13 05	11 47	10 29	9 33
5	9 27	10 19	11 28	12 50	14 02	14 53	14 55	14 08	12 55	11 36	10 20	9 29
9	9 31	10 28	11 38	13 00	14 11	14 57	14 52	14 00	12 44	11 26	10 11	9 25
13	9 36	10 37	11 50	13 10	14 19	15 00	14 47	13 51	12 34	11 16	10 03	9 22
17	9 42	10 47	12 00	13 20	14 27	15 00	14 42	13 41	12 24	11 06	9 55	9 20
21	9 49	10 58	12 11	13 30	14 34	15 01	14 36	13 32	12 13	10 55	9 48	9 20
25	9 56	11 07	12 21	13 40	14 40	15 01	14 29	13 22	12 03	10 46	9 42	9 20
29	10 03	11 18	12 32	13 49	14 45	14 59	14 22	13 13	11 52	10 37	9 36	9 22

tion of daylight at latitude 40°N on 8 days for each of the 12 months. The shortest daylight period occurs around December 21. There is a somewhat steady increase day-by-day to June 21. The variation in the duration of daylight from December to June amounts to 5 hours 41 minutes.

The inequality in the day and night periods increases with latitude. For example, at latitude 30°, the shortest daylight period is 10 hours 12 minutes and the longest is 14 hours 5 minutes, which is a variation of 3 hours 53 minutes; the duration of daylight at latitude 60° goes from a low of 5 hours 52 minutes to a maximum of 18 hours 53 minutes, which is a difference of 13 hours 1 minute.

Within the mid latitudes, the inequality in the day and night periods is coupled with the variation in solar intensity in a way that compounds seasonal differences—during the summer months, solar intensity and the duration of daylight are both at their peaks. In addition to seasonal differences at a specific latitude, this combination also causes the input of radiation during the summer season, at most mid latitudes, to exceed the summer half-year input at the Equator. Refer to Table 16–1 and note the summer radiation input from 30° to 50°; it exceeds the equatorial input. These large summer-period inputs have a profound effect on the development of mid-latitude habitats.

MID-LATITUDE FORESTS

The habitats of the middle latitudes support both broadleaf deciduous forests and needleleaf evergreen forests. These forest habitats develop where the climate provides an adequate supply of moisture. And although the forests are dominated by tree growth, the forest habitats support other kinds of vegetation, too.

Needleleaf evergreen trees, with their tapering symmetrical shapes, are commonly called **conifers.** They belong to the **gymnosperms,** one of the two classes of flowering plants. Broadleaf deciduous trees belong to the **angiosperms,** the other class of flowering plants. The gymnosperms are more primitive and ancient than the angiosperms. Although the conifers have adapted themselves successfully to a remarkable

White-barked aspen trees frame the Brazos Peak in northern New Mexico. (New Mexico Travel)

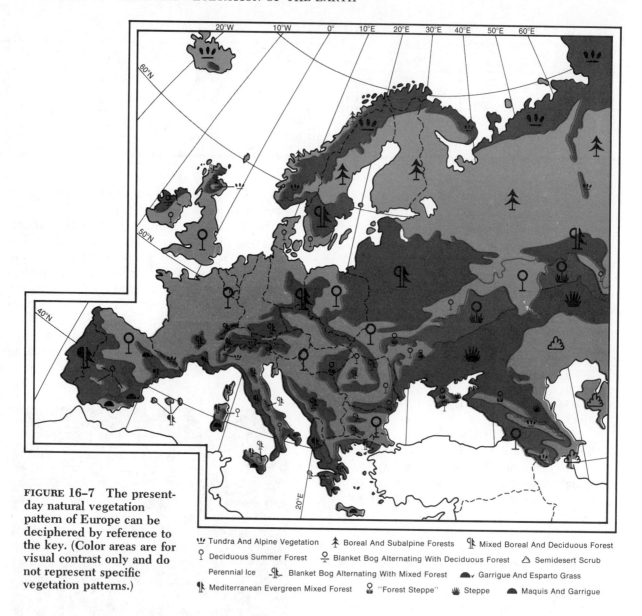

FIGURE 16–7   The present-day natural vegetation pattern of Europe can be deciphered by reference to the key. (Color areas are for visual contrast only and do not represent specific vegetation patterns.)

🌿 Tundra And Alpine Vegetation   🌲 Boreal And Subalpine Forests   🌲 Mixed Boreal And Deciduous Forest
🍥 Deciduous Summer Forest   🍥 Blanket Bog Alternating With Deciduous Forest   ⛅ Semidesert Scrub
   Perennial Ice   🌲 Blanket Bog Alternating With Mixed Forest   🦔 Garrigue And Esparto Grass
🌲 Mediterranean Evergreen Mixed Forest   🌿 "Forest Steppe"   🌾 Steppe   🌑 Maquis And Garrigue

variety of climates, on balance they have been unable to compete successfully with the broadleaf angiosperms during the post-glacial period—the last 15,000 years of the **Holocene.**

DECIDUOUS SUMMER FORESTS
These summer-green forests—dominated by broadleaf trees that lose their leaves during the winter period—are the main climax formation over much of eastern North America (Fig. 16–3), eastern Asia (Fig. 16–5), and Europe (Fig. 16–7).

In all the regions where these forests occur, there is a cool to severe winter with some precipitation all year-round and an annual total precipitation of 20 to 60 inches (50 to 152 cm) or more. Nashville, London, Bordeaux, and Tientsin are in regions with deciduous summer forests; note the average daily maximum and minimum temperatures during January and the precipitation pattern at each of these stations as detailed in Table 16–6.

From the physiological point of view, the cold winter tends to be a "dry" period because the low temperatures hinder absorption of water by the roots. The leafless condition of deciduous trees during the winter is a definite advantage since it is chiefly from the leaves that water is lost through transpiration. It is primarily the transpirational loss that must be replaced by absorption from the soil. If active transpiration continued during the winter when the water loss cannot be replenished by absorption, injury and death would result.

In a deciduous summer forest, the buds open and the leaves of the dominants expand quickly with the advent of suitable temperatures. April is normally the beginning of the green-up period, spring, in the Northern Hemisphere. The April data for Nashville, London, Bordeaux, and Tientsin show an increase in maximum and minimum temperatures, with the temperatures beginning to move into the growing-season range (Table 16–6).

The foliage in a deciduous summer forest develops quite early in the season. Flowering also tends to be completed early, which gives ample time for the development and ripening of the fruit. In fact, in many deciduous types, the flowers open before the leaves expand. Small perennial herbs that are part of a lower stratum in a deciduous forest send up flowering shoots and leaves very early in spring to take advantage of the brief period before the leaves of the dominants expand and cut off most of the sunlight.

The marked differences between the January and July temperature data for each station in Table 16–6 result from the asymmetrical radiation input at these latitudes (page 498). These seasonal inequalities in radiation are reflected in the strikingly different appearances of these forests during winter and summer. The spring and autumn temperature ranges for these stations are quite similar, but they occupy different positions in the radiation cycle; the April data are on the upside of the cycle and the October data are on the downside (Table 16–6). The deciduous cycle maintains a synchronous relationship with the radiation cycle by having the plants shed their leaves during autumn.

Prior to the eighteenth century, a great block of deciduous forest extended from the Appalachians to beyond the Mississippi. Today most of this forest has disappeared, due primarily to human intervention. In this case, however, there is a rationale for human intervention because the conditions that encourage the growth of these forests in North America, as well as in the rest of the world, are conditions favorable to the development of agriculture and grazing.

The American deciduous-summer formation has a number of common species in it. Oak and hickory are almost universal, while maple and beech occur throughout the eastern part of the formation. Birch and ash are particularly common in the subsere on burned-over areas. Although a number of species tend to coexist, three distinct sets of associations—two main trees frequently found growing close to one another—can be recognized in the American formation: a beech-maple association, chestnut-oak association, and an oak-hickory association.

The beech-maple association predominates in the deciduous forests of southern Michigan,

TABLE 16–6   PRECIPITATION AND AVERAGE DAILY MAXIMA AND MINIMA AT SELECTED MID-LATITUDE LOCATIONS

STATION	LOCATION (latitude, longitude)	ELEVATION (feet)		JAN	FEB	MAR	APR	MAY	JUN	JUL	AUG	SEP	OCT	NOV	DEC	ANNUAL
Nashville (Tennessee)	36°7'N,86°41'W	590	Precip. (in)	5.5	4.5	5.2	3.7	3.7	3.3	3.7	2.9	2.9	2.3	3.3	4.2	45.2
			Temp. (°F)	49			71			91			74			
				31			48			70			49			
London (England)	51°29'N,00°00'	149	Precip. (in)	2.0	1.5	1.4	1.8	1.8	1.6	2.0	2.2	1.8	2.3	2.5	2.0	22.9
			Temp. (°F)	44			56			73			58			
				35			40			55			44			
Bordeaux (France)	44°50'N,00°43'W	157	Precip. (in)	2.7	2.8	2.9	2.6	2.5	2.3	2.0	1.9	2.2	3.0	3.9	3.9	32.7
			Temp. (°F)	48			63			80			66			
				35			44			58			47			
Tientsin (China)	39°10'N,117°10'E	13	Precip. (in)	0.2	0.1	0.4	0.5	1.1	2.4	7.6	6.0	1.7	0.6	0.4	0.2	21.0
			Temp. (°F)	33			68			90			68			
				16			45			73			48			

Ohio, Kentucky, West Virginia, Pennsylvania, New Jersey, New York, Connecticut, Rhode Island, and Massachusetts. Of course, oak, sweet gum, chestnut, and tulip trees are also found interspersed among the predominantly beech-maple associations. Hemlock, a stately evergreen tree, is also found scattered among the deciduous trees of these forests; the hemlocks are generally found on north-facing slopes and in deep, shady valleys.

Prior to 1900, the beech-maple associations gave way in Tennessee and in the mountainous parts of North Carolina, South Carolina, and Georgia to a chestnut-oak association. During the first quarter of the twentieth century, however, a blight almost eliminated the chestnut tree from the American formation. Today, the oaks along with sweet gum, hickory, and tulip trees are the common dominants in these areas.

Significant areas of deciduous forest are still found west of the Mississippi River (Fig. 16–3). Oak-hickory associations are the dominants in these forests of eastern Texas, southeastern Oklahoma, Arkansas, and southern Missouri. The extensive oak-hickory forests found on the slopes of the Ouachitas and Ozarks can be traced

northward as a narrow fringe around the edge of the prairie (Fig. 16–3).

CONIFEROUS FORESTS

The vast majority of conifers are evergreens; the deciduous larches are the exception. Forests of coniferous trees are widely distributed in the mid latitudes of the Northern Hemisphere. In fact, purely coniferous forests are confined entirely to the Northern Hemisphere. Limited areas of evergreen mixed forest are of local importance in the Southern Hemisphere in parts of Chile and New Zealand.

In the Northern Hemisphere, coniferous forests occupy areas with a remarkably wide range of climatic conditions. The trees in these forests solve the problem of surviving the unfavorable cold period of having narrow, needlelike leaves with characteristics that reduce transpiration to very modest rates. Species of pine, spruce, and fir are particularly common and widespread in these forests.

Boreal Forest.   The most northern of the coniferous habitats is called a **boreal forest.** In North America, the boreal-forest region extends

The Ouachita Mountains stretch from eastern Oklahoma to central Arkansas. There are extensive oak-hickory forests on the slopes of the Ouachitas around Hot Springs, Arkansas. (Hot Springs Chamber of Commerce)

A pack train is passing a small lake in a coniferous forest in British Columbia. (National Film Board of Canada)

west-to-east in a 4000-mile (6400-km) band from Alaska to Newfoundland (Fig. 16–3). The North American boreal is located generally between

latitude 45°N and the Arctic Circle. Small areas, however, extend slightly beyond 68°N. In the western United States, the boreal forest merges with subalpine forests and moves southward to 35°N and beyond along the elevated ridges of the Sierra Nevada and Rocky Mountains. A similar southward movement of mixed boreal forests occurs along the elevated ridges of the Appalachians in the eastern United States (Fig. 16–3).

On the Eurasian landmass, the boreal forests extend from 45°N to well above the Arctic Circle. The boreal forest in Siberia, which extends to 72°50′N, has a winter climate—represented by the Yakutsk data in Table 16–7—that is so cold that a tree that carries any leaves at all finds it difficult to survive. Thus, in this northernmost forest area the dominants are larches (Fig. 16–5). The larch is a needleleaf tree that belongs to the pine family; unlike the other members of its family, the larch sheds its needles every fall and has a definitely deciduous habit, which gives it an advantage over the evergreens in this harsh setting.

The stations in Table 16–7 are all located in boreal-forest regions. Okhotsk and Yakutsk are in the boreal larch forest of Siberia. The European boreal is represented by Kuusamo, Finland. The Canadian data for Fort Simpson and St. John's are typical of climatic conditions in the North American boreal.

Reindeer are the cattle of Lapland. This setting is above the Arctic Circle in Finland. The boreal forest of the region can be seen in the background. There are many tall, slender larches in this forest. (B. Möller)

TABLE 16–7   **PRECIPITATION AND AVERAGE DAILY MAXIMA AND MINIMA AT SELECTED MID-LATITUDE STATIONS**

STATION	LOCATION (latitude, longitude)	ELEVATION (feet)		JAN	FEB	MAR	APR	MAY	JUN	JUL	AUG	SEP	OCT	NOV	DEC	ANNUAL
Okhotsk (Siberia)	59°21′N, 143°17′E	18	Precip. (in)	0.1	0.1	0.2	0.4	0.9	1.6	2.2	2.6	2.4	1.0	0.2	0.1	11.8
			Temp (°F)	−6			29			57			33			
				−17			10			48			21			
Yakutsk (Siberia)	62°1′N, 129°43′E	535	Precip. (in)	0.3	0.2	0.1	0.3	0.4	1.1	1.6	1.3	1.1	0.5	0.4	0.3	7.6
			Temp (°F)	−45			27			73			23			
				−53			6			54			11			
Kuusamo (Finland)	65°57′N, 29°12′E	843	Precip. (in)	1.1	1.1	1.1	1.1	1.4	2.3	2.8	3.0	2.1	2.1	1.6	1.1	20.8
			Temp (°F)	17			35			68			36			
				2			18			50			27			
Fort Simpson (N.W.T., Canada)	61°45′N, 121°14′W	554	Precip. (in)	0.7	0.7	0.5	0.7	1.4	1.5	2.0	1.5	1.3	1.1	0.9	0.8	13.1
			Temp (°F)	−10			38			74			36			
				−27			14			50			21			
St. John's (Newfoundland)	47°32′N, 52°44′W	211	Precip. (in)	5.3	5.1	4.6	3.8	3.9	3.1	3.1	4.0	3.7	4.8	5.7	6.0	53.1
			Temp (°F)	30			41			69			53			
				18			29			51			40			

The climate in boreal regions is generally cold, with short, cool summers and very cold winters. These areas characteristically have a snow cover throughout the winter. And although the summer is short and cool, the mean temperature of the warmest month—as detailed in Table 16–7—is above 50°F (10°C).

The total annual precipitation at most stations in boreal regions is said to range between 10 and 40 inches (25 and 100 cm). In Table 16–7, we have two exceptions to the quoted norm, Yakutsk and St. John's. Boreal climates are relatively moist because evaporation rates, even in summer, are never very high and, with fairly regular precipitation throughout the year, it does not take much absolute humidity at these low temperatures to produce a high relative humidity. Very rarely, if ever, does the surface soil beneath a boreal forest dry out.

Subalpine Forests. Mention has been made of the coniferous forests of the Sierras, Rockies, and Appalachians. Coniferous communities that inhabit mountain regions are referred to as **subalpine forests.** These forests dominated by conifers extend well beyond the southern limit of the boreal region, which is generally set at 45°N. Since air temperature decreases with an increase in altitude, the conifers find that the climatic conditions in the mountains of the middle latitudes resemble those of the boreal region.

The general north-south alignment of mountain ranges in North America forms the basis for the free migration of species from the boreal to the subalpine zones. In fact, some species of conifer in subalpine forests are identical with those found in the boreal regions. There is an almost continuous belt of subalpine forest extending the whole length of the Sierras, from Canada to southern California (Fig. 16–3). Another belt extends along the Rockies as far south as New Mexico.

The elevation at which you find the subalpine forest increases as you move to lower latitudes. For example, in the Rockies, to the north of Missoula, Montana, at latitude 47°N, subalpine forests lie between 3000 and 7000 feet (900 and 2000 m); farther south, in the Rockies of northern New Mexico, just beyond Santa Fe at 36°N, subalpine forests develop at elevations between 8000 and 12,000 feet (2400 and 3700 m).

Although there are similarities between mountain climates and the climates of boreal regions, there are also many differences. Compare the Missoula and Santa Fe data in Table 16–8 with the Fort Simpson data in Table 16–7. In addition to the temperature data found in Table 16–8, you need to know that Missoula has an annual precipitation of 12.9 inches (32.8 cm) and Santa Fe reports 13.8 inches (35 cm). The amounts as well as the patterns of precipitation are quite similar at the three stations. Note that

This small lake is at about 12,000 feet (3650 meters) in the Sangre de Cristo Mountains of northern New Mexico. It is about 8 miles (13 kilometers) north of Cowles in a region of subalpine forests. (New Mexico Travel)

Missoula is about 16 latitude degrees south of Fort Simpson, and Santa Fe is just about 11 degrees south of Missoula, but Santa Fe is at a higher elevation than either Missoula or Fort Simpson.

Late spring and early autumn frosts occur in the mountains around Missoula and Santa Fe when the same seasons are frost-free in the nearby lowlands. The relatively higher elevations of these stations compared with the nearby lowlands produces a decrease in their mean daily temperatures throughout the year. The frost-free period in these subalpine forests is often no longer than in parts of the boreal.

Although the spring and autumn nights in the mountains surrounding Missoula and Santa Fe may be frosty, the daytime temperatures at these seasons are significantly higher than in the boreal-forest region surrounding Fort Simpson.

These temperature differentials are due to an important difference between the subalpine-forest habitats in the United States and the boreal of Canada; that is, the midday elevation of the Sun is greater throughout the year in the subalpine locations and this produces a greater noontime-radiation intensity, which makes a basic contribution to higher maximum temperatures. On the other hand, a comparison of the duration of daylight in the boreal of Canada and the subalpine of the United States shows the boreal with a relatively greater span of daylight during the summer and a relatively shorter span during the winter. Thus, in some important ways, the conifers in the subalpine environments of the United States and those in the boreal of Canada are coping with very different environmental regimes.

TABLE 16–8   AVERAGE DAILY MINIMUM AND MAXIMUM TEMPERATURES (°F)

STATION	LOCATION (latitude, longitude)	ELEVATION (feet)		JAN	APR	JUL	OCT
Missoula	46°55′N,114°5′W	3190	Max	28	57	85	58
(Montana)			Min	10	31	49	30
Santa Fe	35°41′N,105°56′W	6950	Max	40	60	81	63
(New Mexico)			Min	19	34	56	38

Lake Forests. An extensive area that spreads from Minnesota, across Michigan, northern Pennsylvania, southern Ontario, and into northern New England is shown as an essentially coniferous-forest region in Figure 16–3. White pine, red pine, and eastern hemlock are the dominants in these forests. Since the habitats are centered around the Great Lakes, these tree communities are referred to as **lake forests.**

The precipitation over the region ranges from 23 to 45 inches (58 to 114 cm); there is a considerable range in the temperatures over the year. Data for three locations in the lake-forest region are reported in Table 16–9. The temperature patterns and the precipitation patterns at these stations are conducive to conifer development.

There is some debate as to whether or not the lake-forest conifers are the climax for the region. Some students of these lake-forest communities believe they are simply transitional between the boreal coniferous and southern deciduous forests. In any event, the dominant trees in these forests, which often attained heights of 200 feet (60 m), were prized as timber, and the deforestation that occurred during the last decade of the nineteenth and the first decade of the twentieth centuries swept away most of these trees.

The Pacific-Coast Forests. A band of coniferous forests stretches from southern Alaska down the coastal lowlands to central California. These same forests also extend inland across the basins and plateaus of British Columbia, Washington, Oregon, and Idaho (Fig. 16–3). Some of the largest and tallest trees in the

Shannon Falls slices through towering spruce along the shores of Howe Sound (49°22′N,123°18′W), which is north of Vancouver, British Columbia. (National Film Board of Canada)

world—coastal redwoods, sequoias, and Douglas fir—are found in these forests.

The species found in the lake forests have their counterparts in the Pacific-Coast forests.

TABLE 16–9 PRECIPITATION AND TEMPERATURE DATA

STATION	LOCATION (latitude, longitude)	ELEVATION (feet)		JAN	FEB	MAR	APR	MAY	JUN	JUL	AUG	SEP	OCT	NOV	DEC	ANNUAL
Port Arthur (Ontario)	48°22′N,89°19′W	644	Precip. (in)	0.9	0.8	1.0	1.5	2.1	2.8	3.6	2.8	3.4	2.5	1.5	0.9	23.8
			Temp. (°F)	17			44			74			50			
				−4			26			52			34			
Duluth (Minnesota)	46°47′N,92°6′W	1428	Precip. (in)	1.2	1.0	1.6	2.4	3.3	4.3	3.5	3.8	2.9	2.2	1.8	1.2	29.2
			Temp. (°F)	18			47			77			54			
				−6			27			54			35			
Houghton Lake (Michigan)	44°18′N,84°45′W	1149	Precip. (in)	1.5	1.3	1.7	2.3	2.9	3.1	2.9	2.7	2.9	3.0	2.4	1.7	28.4
			Temp. (°F)	28			52			80			58			
				11			29			54			36			

TABLE 16–10 PRECIPITATION AND TEMPERATURE DATA

STATION	LOCATION (latitude, longitude)		JAN	FEB	MAR	APR	MAY	JUN	JUL	AUG	SEP	OCT	NOV	DEC	ANNUAL
Eureka (California)	40°47′N,124°9′W	Precip. (in)	7.1	6.5	5.2	3.3	1.8	0.7	0.1	0.2	1.0	2.3	5.2	6.3	39.7
		Temp. (°F)	53			56			60			60			
			41			44			51			48			
Sitka (Alaska)	57°5′N,135°15′W	Precip. (in)	7.8	6.7	6.1	5.5	4.2	3.3	4.3	7.2	10.4	12.8	10.2	9.1	87.4
		Temp. (°F)	38			48			61			52			
			27			34			49			40			
Vancouver (British Columbia)	49°17′N,123°5′W	Precip. (in)	8.6	5.8	5.0	3.3	2.8	2.5	1.2	1.7	3.6	5.8	8.3	8.8	57.4
		Temp. (°F)	41			58			74			57			
			32			40			54			44			

For example, the eastern white pine is represented by the western white pine, the counterpart of the eastern red pine is the ponderosa pine, and the western lodgepole pine represents the Jack pine. The counterpart of the eastern hemlock is the western hemlock, and the tamarack or eastern larch is represented by the western larch. These pairs of species are very closely related, and, although the lake forests of the mid continent and the coniferous forests along our western coast are now separated, they were, at one time in the recent past, interconnected.

The data in Table 16–10 indicate the present climate in the coastal lowlands of Alaska around Sitka is cool throughout the year and the annual precipitation exceeds 80 inches (200 cm). The forests around Sitka are dominated by a spruce, called the *Sitka spruce*. The relative abundance of Sitka spruce decreases as you move toward southern British Columbia. In the vicinity of Vancouver, western cedar and western hemlock are the dominant species. Further to the east, in the drier parts of Washington, the Douglas fir is the dominant species. From Oregon south, the dominance of the cedar and hemlock is challenged by the gigantic coastal redwoods. The redwoods are the dominants around Eureka, California. The data in Table 16–10 show Eureka as an area of marked summer drought. The redwoods, it is believed, are able to survive at this southern end of their range because of the relatively high humidity and frequent fogs that roll into the area during the dry season.

Montane Forest. Another type of coniferous forest is the **montane forest.** It is found on the middle slopes of mountains just below the subalpine zone. In the United States, from the Rockies westward, the dominants in the montane forest are the ponderosa pine, the Douglas fir, and the white fir. The montane forest is usually interposed between the subalpine forest and the

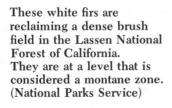

These white firs are reclaiming a dense brush field in the Lassen National Forest of California. They are at a level that is considered a montane zone. (National Parks Service)

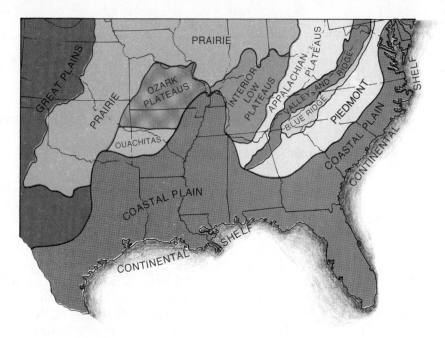

FIGURE 16–8 The portion of the coastal plain shown in this figure stretches from southern New Jersey to Texas. The present-day natural vegetation cover over most of this portion of the coastal plain is southern-pine forest and mixed southern-pine and deciduous forest.

"coast" forest; but in the drier mountains of the Southwest, the montane forest gives way on its downslope side to woodland and scrub communities.

The Pine Barrens. There is another large area of coniferous forest that covers much of the coastal plain, which is shown with some adjacent physical divisions in Figure 16–8. The map in Figure 16–3 depicts the relative extent of the southern-pine forest as well as the mixed southern-pine and deciduous forest. These forests extend from New Jersey southward through central Florida and then westward to Texas. Locally these forests extend upward on to the edge of the Piedmont. Note, at the scale used in Figure 16–3, that New Jersey, the Delmarva Peninsula, and Virginia fall within the mixed southern-pine and deciduous-forest region even though almost pure stands of a single species of pine are of frequent occurrence in these areas.

There are at least ten southern-pine species found in these coniferous forests of the coastal plain. Four—loblolly, slash, longleaf, and shortleaf—make up 90 percent of the trees, however. The natural range and volume-distribution of each of these four major southern pines has been mapped for easy reference in Figure 16–9. The shortleaf pine is the only one of the four with a natural range that includes New Jersey. The Delmarva Peninsula is the northern range of the loblolly pine. Longleaf pines spread into North Carolina; their natural range as indicated by the dashed line and areas shown in color includes a small portion of Virginia. The slash pine has the most restricted range of the four.

Since these southern-pine communities are almost entirely in areas of either immature sandy soils or low-lying marshy soils, most authorities believe that the pine barrens are not climatic climax communities. In fact, enclaves of broadleaf forest—dominated by oak and hickory—are found on better-drained, richer soils within the pine-forest region. The pine forests on sandy areas are, therefore, regarded as a late stage in the xerosere; on marshy and swampy lands, the pine-cypress communities are thought of as a late stage in the hydrosere. Thus, given sufficient time, the pine forests on well-drained land should ultimately give way to broadleaf species.

The mean annual precipitation as well as the normal daily temperatures for the southern-pine region indicate the climate is quite suitable for the typical broadleaf trees of the North American deciduous summer forest. The fact that the true climax has not been achieved within suitable soil areas is due to human intervention and the frequent burning of the forest. In fact, this pine-

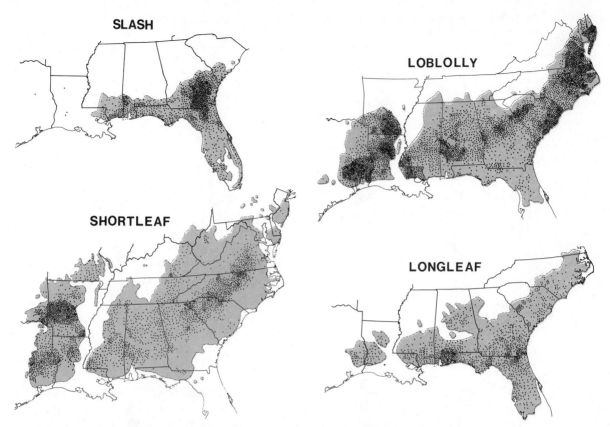

FIGURE 16–9 The distributions of four major southern-pine species—loblolly, slash, shortleaf, and longleaf—are shown in this figure. The dashed lines at the perimeter of the area shown in color indicate the natural range of the species. The black dots represent the volume of the species found in the area. Each dot stands for an average of 5 million cubic feet.

forest region has been disturbed extensively and persistently for centuries because the Indian tribes in these areas were cultivators long before the arrival of the Europeans. Today, there is a large-scale exploitation of the wood of these southern-pine species for timber and paper manufacture.

MID-LATITUDE GRASSLANDS

In tropical regions, grasslands typically take the form of savannas, with widely spaced trees or tall shrubs. The mid-latitude grasslands are quite different; they are usually without trees or bushes except along rivers or streams. Characteristically, the rooting of grasses is shallow and the underground parts tend to form a matted turf that holds and checks the penetration of rainwater to deeper layers. Grasslands are found in the middle latitudes of South America (Fig. 16–2), North America (Fig. 16–3), Africa (Fig. 16–4), and Asia (Fig. 16–5).

The popular conception of North America is of a humid, forested east giving way to a dry grassy interior. There are grasslands in the interior, and they do spread from Canada southward into Mexico (Fig. 16–3). Kansas City, Omaha, Bismarck, and Regina are located in the

TABLE 16–11 TEMPERATURE AND PRECIPITATION DATA IN MID-LATITUDE GRASSLANDS FOR SOME SELECTED STATIONS

STATION	LOCATION (latitude, longitude)	ELEVATION (feet)		JAN	FEB	MAR	APR	MAY	JUN	JUL	AUG	SEP	OCT	NOV	DEC	ANNUAL
Kansas City (Missouri)	39°7′N,94°36′W	742	Precip. (in)	1.4	1.2	2.5	3.6	4.4	4.6	3.2	3.8	3.3	2.9	1.8	1.5	34.2
			Temp. (°F)	40			66			92			72			
				23			46			71			49			
Omaha (Nebraska)	41°16′N,95°57′W	977	Precip. (in)	0.8	1.0	1.5	2.6	3.5	4.5	3.4	4.0	2.6	1.7	1.3	0.8	27.7
			Temp. (°F)	32			62			90			68			
				13			41			67			44			
Bismarck North Dakota)	46°46′N,100°45′W	1647	Precip. (in)	0.4	0.4	0.8	1.2	2.0	3.4	2.2	1.7	1.2	0.9	0.6	0.4	15.2
			Temp. (°F)	20			55			86			59			
				0			32			58			34			
Regina (Saskat-chewan, Canada)	50°26′N,104°40′W	1884	Precip. (in)	0.5	0.3	0.7	0.7	1.8	3.3	2.4	1.8	1.3	0.9	0.6	0.4	14.7
			Temp. (°F)	10			50			79			52			
				-11			26			51			27			
Buenos Aires (Argentina)	34°35′S,58°29′W	89	Precip. (in)	3.1	2.8	4.3	3.5	3.0	2.4	2.2	2.4	3.1	3.4	3.3	3.9	37.4
			Temp. (°F)	85			72			57			69			
				63			53			42			50			
Odessa (Ukraine, USSR)	46°29′N,30°44′E	214	Precip. (in)	1.0	0.7	0.7	1.1	1.1	1.9	1.6	1.4	1.1	1.4	1.1	1.1	14.2
			Temp. (°F)	28			52			79			57			
				22			41			65			47			

grasslands of North America. The annual precipitation data for these cities, as detailed in Table 16–11, show that rainfall decreases as we move from Kansas City to Omaha, to Bismarck, and on to Regina. The latitudes and longitudes reported in Table 16–11 indicate this sequence of travel takes us from east to west with a northward trend.

The grasses in the North American grassland fall into tall and short categories in general correspondence with the rainfall. **Tallgrasses,** for example, develop around Kansas City and Omaha. Bismarck and Regina, with precipitation in the 15-inch (38-cm) range, are shortgrass areas. The total amount of rainfall seems to control the tallgrass and shortgrass regions in other parts of

This is a tallgrass prairie in the Flint Hills of Riley County, Kansas. These slopes with a narrow valley are typical of those found in the Flint Hills. (USDA)

the world, too. Odessa (Table 16–11), for example, on the Russian steppes, with 14 inches (36 cm) of precipitation, is a shortgrass area; but Buenos Aires, on the South American pampas, with 37 inches (94 cm) of precipitation, is a tallgrass area.

At each of the North American grassland stations, there is a strong seasonal trend in the precipitation pattern, that is, spring and early summer are the periods of greatest rainfall and, thus, winter is much drier than summer (Table 16–11). The precipitation data and the temperature data for these stations do not, however, reveal anything that will inherently hold back the advance of trees. With this fact in mind, some authorities suggest that the grasslands of North America are not climatic climax vegetation at all; they believe that deflecting factors—primarily, numerous destructive fires prior to the arrival of European settlers—kept trees from occupying the region and encouraged the development of grass communities.

The extent of the tallgrass prairie in the Midwest prior to 1800 is shown by the color-shaded areas in Figure 16–10. This vast sea of tallgrass covered about 400,000 square miles (1.04 million km²). The westward movement of pioneers brought farms, towns, and roads to the region. The steel plow, which was introduced in 1837, turned the tallgrass prairie into the corn belt. The tallgrass prairie finally succumbed to this onslaught of cultivation and civilization; today, the only extensive area of tallgrass prairie is in the largely unplowed Flint Hills of eastern Kansas.

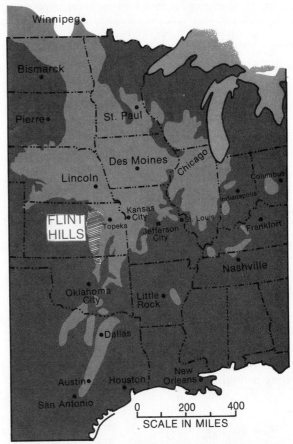

FIGURE 16–10   The color-shaded areas indicate the locations and extent of the tallgrass prairie region prior to 1800. The color-hatched region in eastern Kansas known as *Flint Hills* is the only extensive area of tallgrass prairie that survives today. Other remnants of tallgrass prairie that exist today are no more than pockets—most of which are less than 200 acres in extent.

SEMIDESERT SCRUB AND WOODLANDS

The unforested areas of the middle latitudes are not all dominated by grasses and herbaceous vegetation. Areas roughly equal to those of the grasslands are occupied by a vegetation composed mainly of woody plants. These scrublands occupy regions that are drier than those dominated by the grasses. In North America, for example, note the areas in Figure 16–3 that are designated as being covered by sagebrush and chaparral.

The woody vegetation of the scrublands varies from place to place in its structure and general appearance. In some places these habitats are referred to as woodlands because they are composed of small trees that form an open canopy. In other places shrubs form a dense cover over the ground; elsewhere the main shrubs of a scrubland habitat are discontinuous with lower shrubs and grasses filling the intervening space. Finally, some of the habitats in this category are places where large areas of completely barren ground intervene between sparse clumps of vegetation.

These scrub and woodland communities are of two distinct kinds and are found in rather

TABLE 16–12   TEMPERATURE AND PRECIPITATION DATA

STATION	LOCATION (latitude, longitude)	ELEVATION (feet)		JAN	FEB	MAR	APR	MAY	JUN	JUL	AUG	SEP	OCT	NOV	DEC	ANNUAL
Astrakhan (USSR)	46°21'N,48°2'E	45	Precip. (in)	0.5	0.5	0.4	0.6	0.6	0.7	0.5	0.4	0.6	0.4	0.6	0.6	6.4
			Temp. (°F)	23 / 14			57 / 40			85 / 69			56 / 40			
Reno (Nevada)	39°30'N,119°47'W	4404	Precip. (in)	1.2	1.0	0.7	0.5	0.5	0.4	0.3	0.2	0.2	0.5	0.6	1.1	7.2
			Temp. (°F)	45 / 16			65 / 31			89 / 46			69 / 29			
Santa Cruz (Argentina)	50°1'S,68°32'W	39	Precip. (in)	0.6	0.3	0.3	0.6	0.4	0.5	0.4	0.5	0.3	0.3	0.4	0.7	5.3
			Temp. (°F)	70 / 48 / 39			57 / 39			41 / 28			58 / 39			
Los Angeles (California)	33°56'N,118°23'W	97	Precip. (in)	2.7	2.9	1.8	1.1	0.1	0.1	—	—	0.2	0.4	1.1	2.4	12.8
			Temp. (°F)	64 / 45			67 / 82			76 / 62			73 / 57			
Valpariso (Chile)	33°1'S,71°38'W	135	Precip. (in)	0.1	—	0.3	0.6	4.1	5.9	3.9	2.9	1.3	0.4	0.2	0.2	19.9
			Temp. (°F)	72 / 56			67 / 52			60 / 47			65 / 50			
Capetown (South Africa)	33°54'S,18°32'E	56	Precip. (in)	0.6	0.3	0.7	1.9	3.1	3.3	3.5	2.6	1.7	1.2	0.7	0.4	20.0
			Temp. (°F)	78 / 60			72 / 53			63 / 45			70 / 52			

different climatic zones. The first kind of community—represented by Astrakhan, Reno, and Santa Cruz in Table 16–12—are semidesert scrub habitats that develop in semiarid areas where drought is persistent throughout the year; precipitation in these areas, as detailed in Table 16–12, is unreliable and light. Winter frosts and even protracted cold periods are part of the climatic pattern. The second kind of scrub and woodland community—represented by Los

Cactus scrub (left) is the cover in the Cordoba (31°25'S,64°10'W) area of Argentina. Cacti can be seen in the foreground of this open woodland (bottom) of oak and juniper in the Chiricahua Mountains (31°50'N,109°15'W) of southeastern Arizona. (Argentina Embassy; U.S. Forest Service)

Capetown lies at the foot of Table Mountain on the southwest coast of Africa. The elevation at the top of Table Mountain is 3563 feet (1086 meters). Table 16–12 indicates that Capetown has a pronounced summer drought. The vegetation on Table Mountain is sclerophyllous scrub. (Matter Welch)

Angeles, Vapariso, and Capetown in Table 16–12—is a habitat dominated by sclerophyllous shrubs, that is, shrubs characterized by thick, hard foliage of predominantly evergreen habit. Each of these stations—Los Angeles, Valpariso, and Capetown—is on the western side of a continent in an area marked by pronounced summer drought but where the winters are relatively mild, wet, and frost-free. The sclerophyllous shrubs, which dominate these frost-free habitats, are able to withstand the summer drought and make full use of the relatively short periods of favorable growth.

The sclerophyllous shrub vegetation in southern California is commonly referred to as *coastal chaparral*; only one common dominant in this formation, a species of dwarf oak, is deciduous. Between the Sierras and the Rockies, the semidesert scrub habitats are dominated by sagebrush; it is extremely xerophytic and does well in regions with 5 to 10 inches (13 to 25 cm) of mean annual precipitation. In the driest areas of southeast California, Lower California, southwest Arizona, southern New Mexico, and the Mexican Plateau, the sagebrush gives way to even more xerophytic formations—creosote bush and cacti, for example—in which individual plants are widely spaced.

There are relatively few regions in the middle latitudes where precipitation is so infrequent that complete deserts exist. Parts of Death Valley (Fig. 16–3) (36°18′N,116°25′W) as well as some sections of the U.S.S.R.'s Turkmenistan (Fig. 16–5) (43°N,60°E), Mongolia's Gobi (Fig. 16–5) (43°N,106°E), and China's Takla Makan (Fig. 16–5) (39°N,83°E) may be true deserts in which no plant life exists. In these mid-latitude deserts, however, it is difficult to say whether it is drought or the nature of the surface that prevents plants from developing. Some of these regions are maintained as completely bare rock, while others are characterized by persistently shifting sand dunes that destroy any invading plants.

POLAR AND HIGH-ALTITUDE ENVIRONMENTS

The environments that lie north of the Arctic Circle (66°30′N) and south of the Antarctic Circle (66°30′S) are generally referred to as **polar environments.** Except for the intrusion of the boreal forests, instances of which were cited on page 503, polar environments are beyond the limit of tree growth. The vegetative formation of polar lands is referred to as **tundra.**

In very high mountains there is an elevation beyond which trees do not grow. The elevation limit for arboreal growth, that is, the tree limit, is called **timberline.** The high-altitude environments, as defined and described here, are those above timberline. The plants in high-altitude environments are referred to as **alpine vegetation.**

THE ALPINE ZONE

The fact that temperatures decrease with increasing elevation in mountainous regions has been discussed in relation to subalpine forests on page 504. Precipitation, wind, and radiation are also affected by altitude. Where elevations are high enough for combinations of wind and temperature to produce conditions that no longer support tree growth, alpine vegetation— a combination of lichens, mosses, and flowering plants—is found.

The elevation at which the subalpine forest gives way to alpine vegetation rises gradually with decreasing latitude. The lower limit of the alpine zone is about 4000 feet (1200 m) in southern Alaska; but, at lower latitudes in southern British Columbia, the alpine zone begins around 7000 feet (2100 m). Further south, in the Sierra Nevada of California, the alpine zone's lower limit is at 10,000 feet (3000 m); and by the time we reach the mountains of Mexico, the lower limit of the alpine zone rises to at least 15,000 feet (4500 m).

At any given latitude, a combination of factors may place the alpine zone at either higher or lower elevations. Generally speaking, at a given latitude, there is a relationship between the mass of the mountain and the elevation at which tim-

The Middle and South Sisters, two of the tallest peaks in Oregon's Cascade Mountains, are shown under a heavy February snowfall. The peaks stand at over 10,000 feet (3000 meters). Heavy forests of fir and pine cover the lower slopes to the right. The tree growth thins out as you move to higher elevations at the left. (State of Oregon)

berline is found: the larger the mountain mass, the higher the timberline. The amount of precipitation is also of some significance; on the western side of the Cascades, for example, the timberline is higher and the snowline is lower than at the equivalent latitude in the Rockies, primarily because of differences in the amount of precipitation falling in the two regions. Larger amounts of precipitation falling in the Cascades produce a much more luxuriant growth and cause the lower boundary of the alpine community to be displaced upward.

The elevations at which timberline is found in seven different mountain ranges are detailed in Table 16–13. These are, of course, average elevations. For example, although the timberline elevation is reported as 11,810 feet (3600 m) for the Rockies, it is less than this in the northern part of the range and more than 12,000 feet (3660 m) in the southern Rockies of New Mexico.

Generally, alpine vegetation is exposed to relatively high light intensities with consequent daytime warming throughout the year. Of course, orientation of the slope with respect to the Sun is an important factor in determining the amount and intensity of radiation the alpine community receives. In the Northern Hemisphere, the intensity of radiation on a south-facing slope is significantly higher than the intensity on a north-facing slope in the same vicinity.

Except in well-sheltered locations, alpine vegetation must withstand very high wind speeds. The orientation of slopes with respect to prevailing winds is another important factor in the development of an alpine community since windward slopes receive more precipitation than leeward slopes. Data seem to indicate that the precipitation falling on a windward slope is often five times greater than the amount delivered to the leeward slope.

The upper limit of alpine vegetation seems to be around 19,500 feet (5900 m). At such high elevations in the Himalayas, lichens are the primary plants that persist but even their growth is generally poor. Above these very high altitudes, you encounter the so-called *cold deserts* of the middle and tropical latitudes, which are often snow- or ice-bound.

ANTARCTIC ENVIRONMENTS

The Antarctic continent is a generally circular landmass that, with the exception of the northward flare of the Antarctic Peninsula, lies almost wholly within the Antarctic Circle (Fig. 16–11). This continent has been isolated geographically from the other continents for more than 20 million years. More than 95 percent of its landmass lies buried beneath a vast ice sheet. Extensive ice shelves, which are hinged to coastal glaciers, extend seaward over the continental shelf.

The Transantarctic Mountains divide Antarctica into two regions: Greater and Lesser Antarctica (Fig. 16–11). Coal beds, fossil plants of the Permian, and amphibian fossils of the Triassic have been found in the rock structures of Greater Antarctica. Lesser Antarctica is an ice-covered archipelago of mountainous islands.

Except for two flowering plants that live on the Antarctic Peninsula, plant life consists of algae, mosses, liverworts, and lichens. These plants are found on ice-free land or exposed rock. Bare rock during the long days of January absorbs considerable radiation, which creates a thin, summerlike microclimate on the rock even when the ambient air temperature is below the

TABLE 16–13 ELEVATIONS AT WHICH TIMBERLINE IS FOUND

MOUNTAIN	MAXIMUM ALTITUDE OF RANGE		TIMBERLINE	
	Feet	Meters	Feet	Meters
1. Swiss High Alps	15,216	4638	7090	2160
2. White Mountains (U.S.)	6,283	1915	4590	1400
3. Central Pyrenees (France)	11,168	3404	7050	2150
4. Sierra Nevada (Spain)	11,420	3481	6890	2100
5. Himalayas (India)	27,887	8500	13,120	4000
6. Southern Andes (Chile)	12,795	3900	5900	1800
7. Rocky Mountains (U.S.)	14,272	4350	11,810	3600

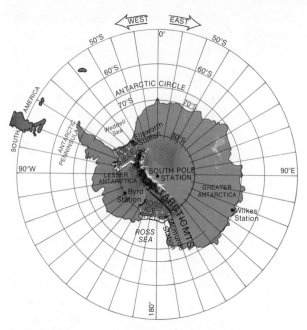

FIGURE 16–11 Antarctica is a generally circular landmass, which lies almost wholly within the Antarctic Circle. The Transantarctica Mountains divide the continent into two regions: Greater and Lesser Antarctica. Meteorological and other data are collected at five stations identified on the map.

freezing point of water. The lichens, which consist of algae and fungi living in **symbiotic association,** are widely distributed. The mosses and liverworts are found primarily in the upper Antarctic Peninsula.

The scarcity and simplicity of Antarctic plant life is due in large part to the severity of the climate. The January temperatures—the summer-period temperatures—at all stations in Table 16–14, except the maximum at the Wilkes Station, are below freezing. Precipitation over most of the continent, as shown by the data in Table 16–14, is very low; and, of course, precipitation is unavailable for plant growth until it melts. The lack of available water and the drying winds restrict plant growth during the brief summer period.

In recent years, Australia, Chile, France, Poland, Russia, South Africa, and the United States have sent expeditions to Antarctica. We tend to think of this continent as an international scientific laboratory, but it is now being recognized as a source of mineral and biological wealth. Russia, Japan, Poland, Chile, and West Germany, for example, are harvesting a small crustacean, the krill, from the sea surrounding Antarctica. The future harvest of the krill could reach 200 million

These mountains, which are about 5000 feet (1500 meters) above sea level, were discovered in an uncharted area of Queen Maud Land on Greater Antarctica between 20° and 30°E. The mountains are an extension of the Shackleton Range. (U.S. Navy)

TABLE 16–14 TEMPERATURE AND PRECIPITATION DATA FOR ANTARCTIC STATIONS

STATION	LOCATION (latitude, longitude)	ELEVATION (feet)		JAN	FEB	MAR	APR	MAY	JUN	JUL	AUG	SEP	OCT	NOV	DEC	ANNUAL
Byrd	80°1'S,119°32'W	5095	Precip. (in)	0.4	0.4	0.2	0.3	0.4	0.5	0.7	0.7	0.3	0.7	—	0.3	4.9
			Temp. (°F)	10			−11			−25			−15			
				−2			−30			−45			−33			
Ellsworth	77°44'S,41°7'W	139	Precip. (in)	0.3	0.2	0.3	0.6	0.2	0.2	0.2	0.2	0.3	0.4	0.5	0.2	3.6
			Temp. (°F)	22			−10			−21			−2			
				12			−25			−35			−15			
McMurdo	77°53'S,166°48'E	8	Precip. (in)	0.5	0.7	0.4	0.4	0.4	0.3	0.2	0.3	0.4	0.2	0.2	0.3	4.3
			Temp. (°F)	30			−1			−9			2			
				21			−13			−24			−12			
Wilkes	66°16'S,110°31'E	31	Precip. (in)	0.5	0.4	1.7	1.1	1.4	1.2	1.3	0.8	1.5	1.2	0.8	0.3	12.2
			Temp. (°F)	34			17			8			16			
				28			9			−3			6			
South Pole	90°S	9186	Precip. (in)	—	0.1	—	—	—	—	—	—	—	—	—	—	0.1
			Temp. (°F)	−16			−66			−67			−55			
				−23			−79			−81			−64			

tons and rival the soybean as a source of protein for humans and livestock. Because of the critical role of the krill in the food chain of these southern waters, there is some concern about unregulated harvesting.

There is also a great deal of interest in the mineral and hydrocarbon reserves of the continent and the surrounding seabed. And although certain articles of The Antarctic Treaty, signed in 1959, were designed to preserve the living resources of the continent, the treaty is silent on mineral reserves. Commercial activities on Antarctica, in the surrounding ocean, and on its seabed could increase in the very near future with momentous ecological implications!

THE ARCTIC TUNDRA

Close to but not exactly at the Arctic Circle, the trees of the boreal forest give way to low shrubs, grasses, flowering herbs, mosses, and lichens. This association of plants is known as *tundra*. The circumarctic tundra is regarded as a single formation type.

In Figure 16–12, a red line is drawn through all locations in Canada with an average July temperature of 50°F (10°C). Compare this 50°F (10°C) isotherm with the northern limit of Canada's boreal forest as detailed in Figure 16–3. This 50°F (10°C) line closely follows the tree line— the line north of which trees do not grow. Trees cannot endure unless the average temperature of the growing season exceeds 50°F (10°C) for a period of 2 or 3 months.

The arctic region—basically a sea surrounded by land—includes the Arctic Ocean, thousands of islands, and the northern parts of the continents of Europe, Asia, and North America. Arctic tundra is found on land areas that are north of the 50°F (10°C) summer isotherm, that is, the average position of the isotherm for the months of July, August, and September (Fig. 16–13).

The 50°F (10°C) summer isotherm, as shown in Figure 16–13, passes south of Greenland. In Canada, the isotherm runs from Goose Bay in Labrador, through Churchill on Hudson Bay, to Cape Parry along the Beaufort Sea. The 50°F (10°C) summer isotherm parallels the southern edge of the Brooks Range in Alaska; then, it swings south along the Bering Sea to the Aleutian Islands. The line turns northward again from the tip of the Aleutians to the Gulf of Anadyr in Siberia, where it turns to follow the Arctic Circle. In western Russia, the 50°F (10°C) summer isotherm swerves northward and barely touches the northern coast of Norway. Then it turns southward. In the vicinity of Iceland, the isotherm runs south of all but the southern coast of Iceland.

The winters in the Arctic are dark and cold. Direct solar radiation is absent or very restricted for several months during the winter. However, total darkness seldom exists because reflection

FIGURE 16–12 The 50°F isotherm during July defines the northern limit of the boreal forests in Canada. The plant associations north of this 50°F isotherm are referred to as *tundra.*

and refraction produce a twilight until the Sun is 18° below the horizon. The duration of daylight during the summer reaches a peak of 24 hours, with more than 20 hours of daylight during several summer months. The tundra turns into a carpet of flowers and grows at a furious pace during these long periods of daylight.

The pattern of temperature and precipitation for four stations in the tundra region are detailed in Table 16–15. The precipitation in most arctic regions is less than 10 inches (25 cm) per year.

The average July temperature at all the stations in Table 16–15 is below 50°F (10°C); only Vardo achieves a daily maximum above 50°F (10°C). Although insolation persists for many hours during each day of July, temperatures do not get very high because solar intensity is low; even on the southern margin of the Arctic, the noontime Sun rarely reaches more than 45° above the horizon. At most Arctic locations, the period between spring and fall frosts is less than 50 days; in addition, the subsoil is permanently frozen.

TABLE 16-15   TEMPERATURE AND PRECIPITATION DATA FOR TUNDRA ENVIRONMENTS

STATION	LOCATION (latitude, longitude)	ELEVATION (feet)		JAN	FEB	MAR	APR	MAY	JUN	JUL	AUG	SEP	OCT	NOV	DEC	ANNUAL
Alert, N.W.T., (Canada)	82°31'N,62°28'W	95	Precip. (in)	0.2	0.3	0.3	0.3	0.5	0.6	0.5	1.1	1.0	0.9	0.2	0.4	6.3
			Temp. (°F)	−19			−8			44			2			
				−29			−18			36			−7			
Thule (Greenland)	76°31'N,68°44'W	251	Precip. (in)	0.4	0.3	0.2	0.2	0.3	0.2	0.7	0.6	0.6	0.7	0.5	0.2	4.9
			Temp. (°F)	−4			10			46			19			
				−17			−7			38			8			
Barrow (Alaska)	71°18'N,156°47'W	31	Precip. (in)	0.2	0.2	0.1	0.1	0.1	0.4	0.8	0.9	0.6	0.5	0.2	0.2	4.3
			Temp. (°F)	−9			7			45			21			
				−23			−7			33			12			
Vardo (Norway)	70°22'N,31°6'E	43	Precip. (in)	2.5	2.5	2.3	1.5	1.3	1.3	1.5	1.7	1.9	2.5	2.1	2.4	23.5
			Temp. (°F)	27			34			53			38			
				19			26			44			32			

LIFE IN THE TUNDRA

Under the conditions that prevailed prior to the twentieth century, the tundra was grazed by herds of wild animals. Reindeer,* the most prominent of the grazers, fed primarily on the lichens in the tundra vegetation. Herds of reindeer have been domesticated and kept on the Siberian and European tundra for a long time. The Lapps of northern Norway, Sweden, and Finland, for example, have been reindeer herders for more than a thousand years. Today, however, no more than one-fifth of the Lapps herd reindeers, but those that do continue to use the same areas over which the animals ranged in their natural state.

The Eskimos of North America were hunters right up to the twentieth century. Canadian and Alaskan traders introduced firearms into the Eskimo culture during the nineteenth century. The use of firearms by the Eskimos caused the wild reindeer herds to become seriously depleted by 1900. During the next 30 years, it became increasingly difficult for the Eskimos to supply their needs from the dwindling reindeer herds, and Eskimo communities were at the verge of starvation.

In an attempt to aid the starving Eskimos, more than 2000 reindeer were introduced to the tundra via Alaska in 1935. The Eskimos returned to the hunt, but rangers were used to enforce some degree of conservation to ensure that the reindeer herds would flourish. In fact, the herds

*Caribou is the French-Canadian name for the wild reindeer of North America.

FIGURE 16-13   The 50°F (10°C) summer isotherm is a line of demarcation between boreal forests and the plant association known as *tundra*. Tundra is found in areas to the north of the isotherm.

Svalbard is a group of islands in the Arctic Ocean between longitude 10° and 35°E and latitude 74° and 81°N that form the Spitzbergen archipelago. The islands lie about midway between the north cape of Norway and the North Pole. The vegetation consists mostly of lichens and mosses—that is, tundra. Two reindeer can be seen moving across the tundra. (O. Salvigsen)

did very well and their numbers increased over a period of 20 years to the point where they began to seriously overgraze the lichens.

Tundra lichens grow and regenerate very slowly. It takes from 10 to 50 years for a really luxuriant growth of lichens to develop. In those areas where the reindeer population became too large and too concentrated, lichens were almost eliminated by 1960. Cotton grass, which was formerly a co-dominant with the lichen, was given an advantage through the decrease in the lichen cover. As a result, today there are some areas of tundra that are almost pure cotton grass. The purpose for an action—whether good or bad according to our scale of values—does not preserve a habitat when the equilibrium of the natural balance is disturbed.

OIL AND THE TUNDRA

The discovery of oil in the Prudhoe Bay field on Alaska's north slope has increased the level of human intrusion into the tundra. Since other options have not been developed to supply our insatiable need for energy, Alaska's oil seems to be indispensable for our immediate survival. As a result, we are faced with a dilemma: To get to Alaska's oil we must create roads, employ oversurface vehicles, and gain access to land for exploration and development. Although tundra surfaces can be crossed with very little harm done to them during the winter when they are frozen under a cushioning layer of snow, it is quite another story during the summer. The vegetative mat of the tundra is exposed during the summer. The removal or compression of the exposed mat by tracked vehicles or even by successive passes of vehicles with low-pressure tires produces a harmful effect. When the mat is compressed, its insulating quality is reduced and the underlying permafrost is in danger of melting, which can lead to the development of new water channels, erosion, and siltation.

Roads over the tundra must be built of gravel so that the permafrost is disturbed as little as possible. But even the removal of gravel to build a road involves intrusion and change in some other portion of the habitat. The source of gravel in these regions is the bed of a stream. Removal of large quantities of gravel changes a stream's course, its pattern of erosion, and its pattern of siltation. Even road dust raised by vehicles produces an impact when it settles on snow and causes an early melt during the spring.

With greater intrusion of people into the tundra, the disposal of wastes becomes an immediate problem, too. Liquids will not sink into the frozen ground and solid waste will not decompose. In addition, the flaring of gas at oil sites introduces a certain amount of atmospheric pollution; and, of course, there is the ever-present possibility of oil spills and leakage from pipelines.

The tundra seems to be particularly susceptible to the pressure of human intrusion. We are, nevertheless, intruding and the pace of this intrusion seems to be quickening. What kinds of safeguards *should* we apply to preserve this habitat? What kinds of safeguards *will* we apply?

This is an elevated section of the Alaskan pipeline. The swath cut through this environment for the pipeline and its service road constitutes a significant intrusion into the area. Leaks from the pipeline are a serious danger to this habitat. (Wide World Photos)

THE LANGUAGE OF PHYSICAL GEOGRAPHY

A physical geographer uses a technical vocabulary to make explanations. Review your understanding of the vocabulary used to develop the concepts in this chapter. Among the important words and terms used are:

alpine vegetation	mid-latitude	semievergreen seasonal forest
angiosperms	environments	subalpine forest
boreal forest	monsoon forest	Sudan zone
climax formation	montane forest	symbiotic association
conifers	natural vegetation	southern-pine forest
deciduous seasonal forest	nutrients	tallgrass
desert	photoperiod	thorn woodland
desertification	pine barrens	timberline
energy flux	polar environments	tropical rain forest
Guinea zone	radiation cycle	tropical savanna
gymnosperms	Sahel zone	tropical seasonal forests
Holocene	savanna woodland	tundra
lake forests	semidesert	

SELF-TESTING QUIZ

1. Discuss the interwoven nature of climate and vegetation.
2. Compare and contrast the solar energy input at various latitudes.
3. What is the primary periodic energy flux within the tropics?
4. What distinguishes and maintains the major tropical environments?
5. Describe the temperature pattern associated with tropical rain forests.
6. Where do you find the most extensive development of the American tropical rain forest?
7. How has human interference damaged rain forests?
8. Discuss the recycling of nutrients within a rain forest.
9. Why is the ground layer in a tropical seasonal forest more varied than in a rain forest?
10. Describe the precipitation pattern associated with a monsoon forest.
11. How do you characterize a thorn woodland?
12. Explain the origin of a tropical savanna.
13. Identify some areas in which desertification is occurring.
14. Describe the variation in the photoperiod over a year in a mid-latitude environment of your choosing.
15. Why does the cold winter tend to be a "dry" period with respect to the vegetation in a mid-latitude forest?
16. Identify, compare, and contrast the distinct sets of associations that occur in the American deciduous-summer formation.
17. Trace the extent of the boreal forest on the Eurasian landmass.
18. Identify the four southern-pine species that make up 90 percent of the trees in the southern-pine forests.
19. Trace the elevations at which subalpine forest gives way to alpine vegetation as you move to lower latitudes.

IN REVIEW
PRESENT-DAY ENVIRONMENTS

I. TROPICAL ENVIRONMENTS

  A. Tropical Rain Forest
    1. Location and Extent
    2. Present Status
  B. Tropical Seasonal Forest
    1. Monsoon Forest
    2. Deciduous Seasonal Forest
    3. Thorn Woodlands
  C. Tropical Savanna
    1. Location and Extent
    2. Origins of Tropical Savanna
  D. Semideserts and Deserts
    1. Desertification

II. MID-LATITUDE ENVIRONMENTS

  A. Mid-Latitude Forests
    1. Deciduous Summer Forests

    2. Coniferous Forests
      a. Boreal Forest
      b. Subalpine Forests
      c. Lake Forests
      d. The Pacific Coast Forests
      e. Montane Forest
      f. The Pine Barrens
  B. Mid-Latitude Grasslands
  C. Semidesert Scrub and Woodlands

III. POLAR AND HIGH-ALTITUDE ENVIRON-MENTS

  A. The Alpine Zone
  B. Antarctic Environments
  C. The Arctic Tundra
    1. Life in the Tundra
    2. Oil in the Tundra

Conversion Factors for
UNITS OF LENGTH
AND AREA

1 inch	=	2.54 centimeters
1 yard	=	0.9144 meter
1 mile	=	1.609344 kilometers
1 kilometer	=	0.62137 mile
1 meter	=	3.2808 feet

1 fathom	=	6 feet
1 statute mile	=	5280 feet
1 nautical mile	=	6076 feet
1 nautical mile	=	1.15 statute miles

1 micrometer = 0.0001 centimeter
  (previously called   also written as 1×10^{-4} centimeter
  a *micron*)

1 micrometer = 0.000001 meter
             also written as 1×10^{-6} meter

1 micrometer = 0.000039 inch

1 angstrom (Å) = 0.00000001 centimeter
             also written as 1×10^{-8} centimeter

1 square inch (in²)	=	6.45 square centimeters (cm²)
1 square meter (m²)	=	10.76 square feet (ft²)
1 square mile (mi²)	=	2.59 square kilometers (km²)
1 acre	=	43.560 square feet (ft²)
	=	4047 square meters (m²)
1 square foot (ft²)	=	144 square inches (in²)
	=	0.093 square meters (m²)

APPENDIX II

TEMPERATURE SCALES

When converting Kelvin degrees to Celsius, subtract 273°, the difference between the two scales. When converting from Celsius to Kelvin, add the 273° to the C reading.

Example

35°C = 35° + 273° or 308°K

Example

400°K = 400° − 273° or 127°C

When converting Fahrenheit to Celsius, remember that there is a 32° difference in the freezing points. Therefore, it is necessary to subtract 32 from the Fahrenheit reading. There is also a difference in the size of degree points on each scale. Between 32°F and 212°F there is a difference of 180°. Between 0°C and 100°C there are only 100° of difference. Each Fahrenheit degree is 5/9 of a Celsius degree. After subtracting 32 and multiplying by 5/9, the reading is converted to Centigrade.

Example

95°F = ?C (95° − 32°)5/9 = 35°C

To convert Celsius to Fahrenheit we reverse the previous operations. The multiplier is 9/5 and the 32° is added.

Example

80°C = ?F 80° × 9/5 + 32° = 176°F

Conversion Factors for
UNITS OF MASS

1 ounce (oz)	=	28.3495 grams (g)
1 g	=	0.035274 oz
1 kilogram (kg)	=	2.20462 pounds (lb)
	=	1000. g
1 lb	=	453.5923 g
1 metric ton	=	1000. kg
	=	1.10231 short tons
	=	0.9832107 long ton
1 short ton	=	2000. lb
1 long ton	=	2240. lb

APPENDIX IV

Conversion Factors for UNITS OF VOLUME

1 liter (ℓ)	=	1000 milliliters (ml)
	=	61.0255 cubic inches (in^3)
	=	33.815 U.S. fluid ounces (fl oz)
	=	1.0567 U.S. quarts (qt)
	=	0.264179 U.S. gallon (gal)
1 cubic inch (in^3)	=	16.3871 cubic centimeters (cm^3)
1 cubic meter (m^3)	=	35.3147 cubic feet (ft^3)
1 cubic foot (ft^3)	=	1728 in^3
	=	7.48052 U.S. gal
1 U.S. fl oz	=	29.5727 ml
1 cubic mile (mi^3)	=	4.168 cubic kilometers (km^3)

Conversion Factors for
UNITS OF PRESSURE

1 bar	= 1000 millibars (mb)
	= 1019.7 g/cm²
	= 1.0197 kg/cm²
1 mb	= 1.0197 g/cm²
	= 0.0295315 in mercury (Hg) at 45° latitude
	= 0.0145 pounds per square inch (psi)
1 kg/cm²	= 980.665 mb
	= 14.2233 lb/in²
1 atmosphere	= 1013.2 mb
	= 1.03318 kg/cm²
	= 760. mm/Hg
	= 14.695 psi
	= 29.92 in Hg
1 in Hg	= 33.8622 mb at 45° latitude
1 mm Hg	= 1.3332 mb
1 psi	= 68.9476 mb

MAPPING DATA

A map or chart, unlike a photograph, does not reveal all the detail that can be seen in a landscape or seascape; rather, it provides a more comprehensive and selective picture. The effectiveness of a map or chart as a means of communication is directly affected by its design. Map design involves the manner of portraying and arranging all the visual components of a map. The visual impact of a map—and, thus, the quality and effectiveness of its communication—is built by line, shape, pattern, and color. Lettering also plays a significant role in the development of an effective map or chart; preprinted letters make the typographer's entire stock available to the cartographer.

There are many different kinds of maps and charts, but we can classify all of them as belonging to one of two categories: (1) general reference or (2) thematic. A general reference map or chart provides general information such as the location of continents, oceans, countries, rivers, lakes, mountains, and cities. The map in Figure A–1, for example, is a general reference map because it provides information on the location and extent of the major drainage basins in South Carolina. Thematic maps or charts are specialized presentations of some particular feature such as temperature, rainfall, topography, pressure, elevation, and so on. Figure A–2 is a thematic map because it is a specialized presentation of the average dates when the first killing frost occurs in South Carolina.

The most familiar general reference map in use is the automobile road map, which is a transportation map. Transportation maps used by airplane pilots and by people operating boats are usually called *charts*.

DESIGN AND LANGUAGE

The amount of information you can learn from a map or chart depends on your ability to read it. An understanding of scale, direction, and the symbolism used is crucial to proper interpretation.

SCALE AND DISTANCE

All maps and charts are reductions of the regions they depict. This means there is an obligation to indicate how much reduction has occurred in the process of depicting the region. The term *scale*, in this context, is simply the index of reduction and indicates how much of the Earth's surface is represented by a given measurement on the map or chart. Scale can be expressed in one of three different ways: (1) graphic scale, (2) words and figures, and (3) representative fraction.

GRAPHIC SCALE

On many maps and charts, scale is shown graphically by means of a straight line called a *bar scale,* on which distances have been marked. The bar scale may be marked off in miles, feet, kilometers, or meters—whatever is most convenient. The map in

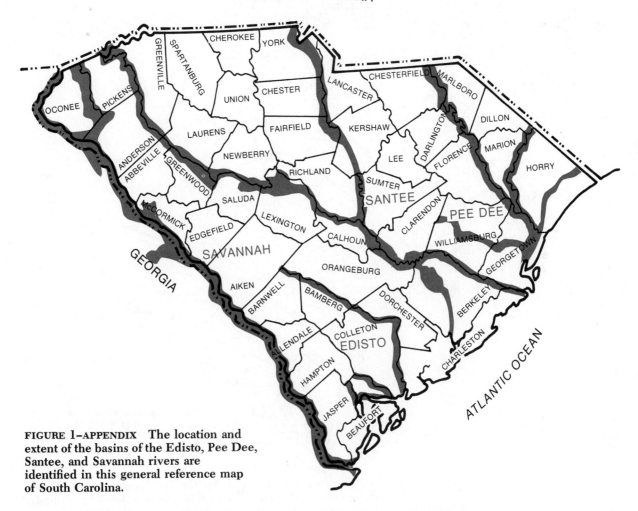

FIGURE 1–APPENDIX The location and extent of the basins of the Edisto, Pee Dee, Santee, and Savannah rivers are identified in this general reference map of South Carolina.

Figure A–2 includes a bar scale; the straight-line distance along the South Carolina-Georgia border, for example, from the November 10 to the November 15 frost line is about 40 miles (64 km).

Because no map or chart is perfect, the scale cannot be used with the same degree of reliability all over the map, especially if the map represents a large area. Generally, a scale is most accurate when it is used near the center and less accurate toward the margins of the map or chart.

Some maps and charts indicate the distortions that exist and the way in which the scale changes; for example, the scale on a surface-weather map distributed by the National Oceanic and Atmospheric Administration (Fig. A–3) indicates the variations that occur at different latitudes. The bar scale is smallest at high latitudes; its length increases as you move to lower latitudes. Note the length of each bar scale, that is, at latitudes 60°N, 40°N, and 20°N.

STATED SCALE

The scale of a map or chart is often expressed in words and figures as the number of miles (kilometers) that is the equivalent of 1 inch (1 centimeter). A stated scale might

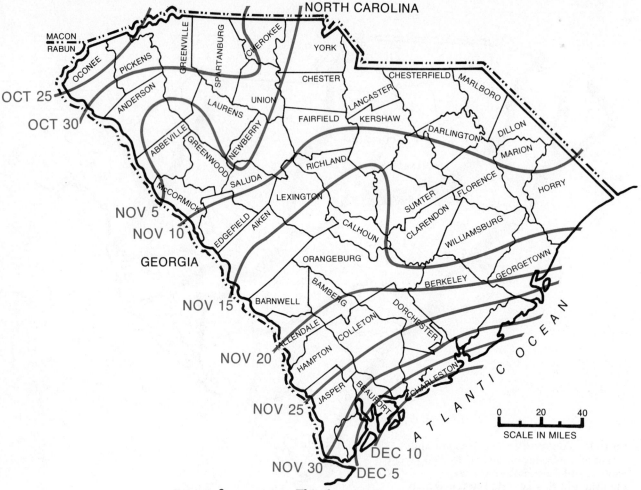

FIGURE 2–APPENDIX This thematic map indicates the first counties to experience a killing frost in South Carolina are Oconee, Pickens, and Greenville. The killing-frost line that runs through these counties carries the date October 25. Beaufort County in the southern tip of the state, along the coast, does not usually experience a killing frost until late in November.

appear as 1 inch = 16 miles (1 centimeter = 10 kilometers). In other words, 1 inch on the map equals 16 miles when measured on the Earth's surface; 1 centimeter on the map spans 10 kilometers of the Earth's surface.

REPRESENTATIVE FRACTION

One of the most common and versatile methods of expressing scale is by a fraction called the *representative fraction* (R.F.). A representative fraction of 1/62,500 means a length of 1 unit on the map represents 62,500 of the same units on the Earth's surface. The advantage of this method is that the scale is expressed by the R.F. regardless of the system used, that is, English or metric. If the R.F. is 1:62,500, for example, 1 inch on the

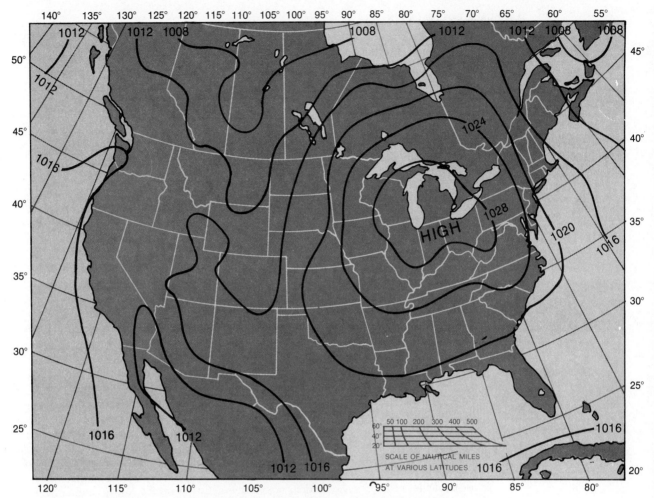

FIGURE 3–APPENDIX   This thematic map presents the pressure pattern over the U.S. and a portion of Canada. The high-pressure zone is centered over the Great Lakes and is ringed by an isobar of 1028 millibars. The scale for this map has distances marked in nautical miles. A special bar scale is designated for each latitude.

map represents 62,500 inches of the Earth's surface; alternatively, 1 centimeter on the map represents 62,500 centimeters on the Earth.

You can very easily convert 1 inch of map scale to feet or miles. Let's use the R.F. of 1:62,500 again; divide 62,500 by 12 (the number of inches in 1 foot) and you get 5208, which is the map-scale equivalent in feet. In Table A–1 you will find this means 1 inch on the map is the equivalent of 5208 feet on the Earth's surface. Divide 5208 by 5280 (the number of feet in 1 mile) and you get 0.986, which is the map-scale equivalent in miles; in other words, 1 inch on the map is equal to 0.986 miles (Table A–1).

Refer to Table A–1 and note that as you move from a scale of 1:2500 to a scale of 1:20,000, ground detail becomes simplified since 1 map-unit represents increasingly larger slices of the Earth's surface. Topographic maps most commonly used by physical geographers are in the scale range from 1:20,000 to 1:316,800.

TABLE A–1 **MAP-SCALE EQUIVALENTS**

SCALE	1 MAP-INCH REPRESENTS		1 MAP-CENTIMETER REPRESENTS	
	Earth-feet	Earth-miles	Earth-meters	Earth-kilometers
1:2,500	208	0.039	25	0.025
1:5,000	416	0.079	50	0.050
1:10,000	833	0.158	100	0.100
1:15,000	1,250	0.237	150	0.150
1:20,000	1,667	0.316	200	0.200
1:24,000	2,000	0.379	240	0.240
1:31,680	3,640	0.500	316.8	0.317
1:50,000	4,167	0.789	500	0.500
1:62,500	5,208	0.986	625	0.625
1:63,360	5,280	1.000	633.6	0.634
1:100,000	8,333	1.578	1000	1.000
1:125,000	10,417	1.973	1250	1.250
1:250,000	20,833	3.946	2500	2.500
1:316,800	26,400	5.000	3168	3.168

LOCATION AND DIRECTION

A network of accurately spaced north-south and east-west lines, called a *grid system*, is placed on a map or chart for finding and describing location. To be generally useful a grid system must be worked out from the geographic poles of the Earth.

MAGNETIC DECLINATION

The magnetic poles do not coincide with the geographic poles. The North Magnetic Pole, for example, lies at a point (77°N,100°W) in Prince Phillip Basin north of Bathurst Island in northern Canada. The South Magnetic Pole lies near the coast of Antarctica at about latitude 68°, longitude 143°E. Neither pole remains fixed in position; rather, each moves about irregularly within a limited area, often as much as 1° in 20 years.

The magnetic compass needle points in a direction toward the North Magnetic Pole, referred to as *MN* on maps and charts. Since the magnetic poles and geographic poles do not coincide, a compass needle does not point exactly toward the North Geographic Pole, or true north, referred to as *TN*. The difference between MN and TN is known as *magnetic declination*, which is expressed in degrees east or west of true north. This angle of declination can vary from zero to a full 180 degrees.

Lines of equal declination, called *isogonic lines,* are depicted on thematic maps known as isogonic maps. The isogonic map of the world (Fig. A–4) shows the lines are somewhat irregular. The irregularity and variations in the isogonic lines are brought about by different kinds of rock masses.

Three lines of zero declination are depicted on the isogonic map in Figure A–4. It is only along these zero lines that the magnetic needle points to true north. Elsewhere an angle to the right or left is formed by lines drawn to MN and TN. The amount of declination east or west of true north can be read from the map for any region.

Many maps and charts indicate the direction of true north by means of a star-tipped arrow or the symbol TN (Fig. A–5). The direction of the North Magnetic Pole is depicted by means of a half-headed arrow and the symbol MN. Across the 48 contiguous states of the United States, magnetic declination varies from 0° to as much as 25° east or west.

GRID SYSTEMS

The principal grid system used in specifying location on the Earth's surface is by latitude and longitude. The prime meridian in the latitude-longitude grid system is the one on which the Greenwich Observatory in England is located.

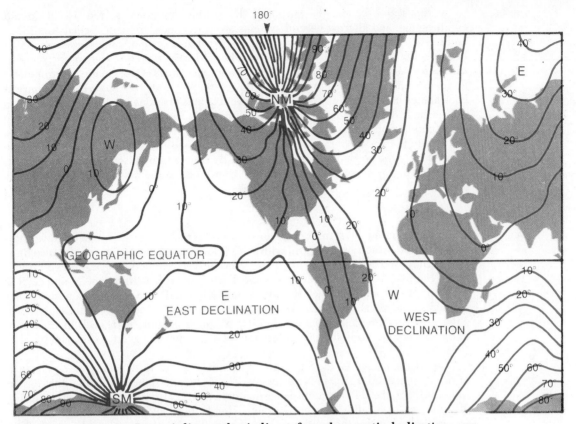

FIGURE 4–APPENDIX   Isogonic lines—that is, lines of equal magnetic declination—are indicated on this thematic map. Three lines of zero declination, along which the magnetic needle points to true north, are found on this map. Elsewhere the magnetic needle points either to the east or to the west of true north.

FIGURE 5–APPENDIX   The star-tipped line indicates the direction of true north (TN). The half-headed arrow points toward the North Magnetic pole (MN). The angle between the two is 7°, which means the magnetic declination is 7°W since the magnetic needle is pointing west of true north.

At times and for specific purposes, the latitude-longitude system is considered cumbersome, and alternative grid systems are constructed to meet special conditions and needs. The first step in constructing a special grid system is to choose a map that becomes the standard for the system. A square grid is overlaid on the map, and numerical coordinates are assigned to the grid's reference lines. Coordinates are generally given as units of distance from a selected reference point. In Figure A–6, numerical coordinates have been assigned to the reference lines of a square grid. Point A is at grid reference 427654; the grid coordinates are given as one number, which consists of an even number of digits—six in this case. The first half of the number, 427, is the coordinate to the east of the reference line; the second half of the number, 654, is the coordinate to the north of the reference line. By convention, the coordinate to the east, called *easting,* is specified first; then, the coordinate to the north, called *northing,* is specified.

UTM GRID

The grid system most frequently encountered is the Universal Transverse Mercator (UTM) grid in which a square grid with a 10,000-meter spacing between reference lines is superimposed on a sectional map. The UTM grid is used between latitudes 80°N and 80°S (Fig. A–7). In the UTM system the Earth between latitudes 80°N and 80°S is divided into 60 sections, each 6° of longitude wide and each assigned a number of 1 through 60; sections 1, 2, and 3 are depicted in Figure A–7. Each 6° section, as shown in Figure A–7, is divided into twenty 6° by 8° quadrilaterals, with each quadrilateral assigned an index consisting of the number of its section and a letter from C through X (Fig. A–7).

The latitude range for the S-quadrilaterals runs from 32° to 40°N (Fig. A–7). The map of the Barnegat Inlet area in New Jersey (Fig. A–8) is in the quadrilateral that has an index of 18S in the UTM system. The Equator in the UTM system is assigned a northing of 0 meters for locations in the Northern Hemisphere and a northing of 10,000,000 meters for locations in the Southern Hemisphere. The UTM grid ticks in Figure A–8 are

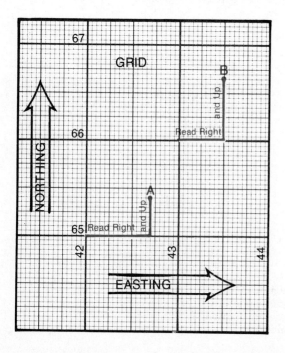

FIGURE 6–APPENDIX   This is a square grid. Numerical coordinates have been assigned to the grid's reference lines. If this grid is overlaid on a map, a point on the map—B, for example—can be located with respect to the coordinates of the grid. The grid reference for Point B is 436667.

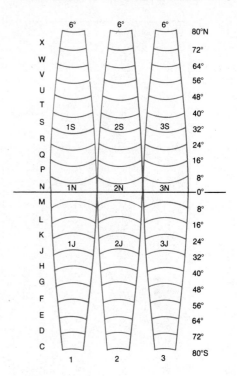

FIGURE 7–APPENDIX The UTM grid subdivides the Earth between 80°N and 80°S into 60 sections that are 6° of longitude wide, with each section assigned a number; sections 1, 2, and 3 are shown in this figure. Each 6° section is divided into 20 quadrilaterals, with each assigned a letter from *C* through *X*.

in color. The first northing UTM tick noted is 4,401,000 meters; the easting UTM ticks start with 575,000 meters. The UTM coordinates for Point A on the Barnegat map—between the lights on the northern rock jetty and southern rock jetty—are 578,000 meters E, 4,401,000 meters N.

STATE GRID

The United States has designed a grid system for each of the 50 states and Puerto Rico, called the *State Plane Coordinate System.* The basic grid square of the state coordinates is 10,000 feet on a side. Eastings and northings for the grid are listed in units of feet. State Coordinate ticks are in black in Figure A–8; Point B on the Barnegat map has coordinates of 2,160,000 feet E, 340,000 feet N based on the New Jersey system.

GEOLOGICAL SURVEY MAPS

The Geological Survey of the U.S. Department of Interior makes a series of topographic maps—the 7.5-minute topographic series—that covers the United States, Puerto Rico, Guam, American Samoa, and the Virgin Islands. The unit of survey is a quadrangle bounded by parallels of latitude and meridians of longitude. The portion of the Barnegat map shown in Figure A–8 is from this 7.5-minute series; it is a part of the Barnegat Light Quadrangle. The full Barnegat Light Quadrangle is bounded on the east by the meridan 74°W; its western border is at the meridian 74°7′30″W, which gives the quadrangle a 7.5′ width. The south-north span of the Barnegat Light Quadrangle is also 7.5′—from the parallel 39°45′N to 39°52′30″N.

The Plant City area of Florida, which is identified in Figure A–9, is bounded by the two parallels of latitude (28 and 29°N) and two meridians of longitude (82 and 84°W). The Red Level Quadrangle—located within the Plant City area—covers 7.5 minutes of latitude and longitude. The northern border of the Red Level Quadrangle is bounded

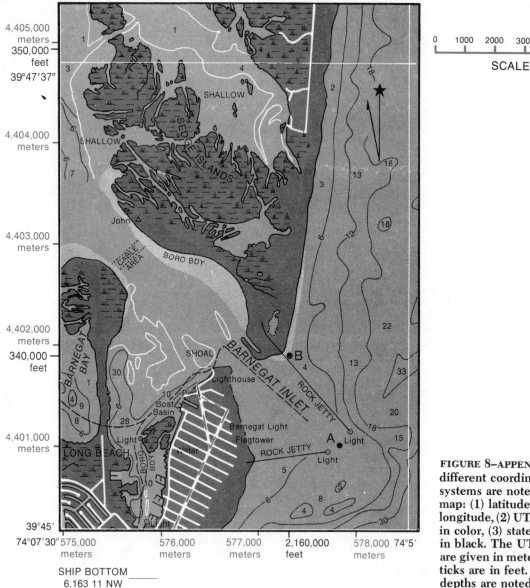

FIGURE 8–APPENDIX Three different coordinate systems are noted on this map: (1) latitude and longitude, (2) UTM grid ticks in color, (3) state coordinates in black. The UTM ticks are given in meters; the state ticks are in feet. Water depths are noted in feet.

by latitude 29°N; its southern border is bounded by latitude 28°52′30″N. The boundary of its eastern border is at longitude 82°37′30″W; its western border is at longitude 82°45′W.

Maps in the 7.5-minute series are published at the scale of 1:24,000 (1 inch = 2000 feet). Each 7.5-minute quadrangle is designated by the name of a city, town, or prominent natural feature within it.

SYMBOLISM

The section of the Barnegat Light Quadrangle shown in Figure A–8, like any map, is an assemblage of symbols. Even the line representing the coastline is a symbol because it attempts to indicate the mean sea level, which, in reality, does not exist in nature.

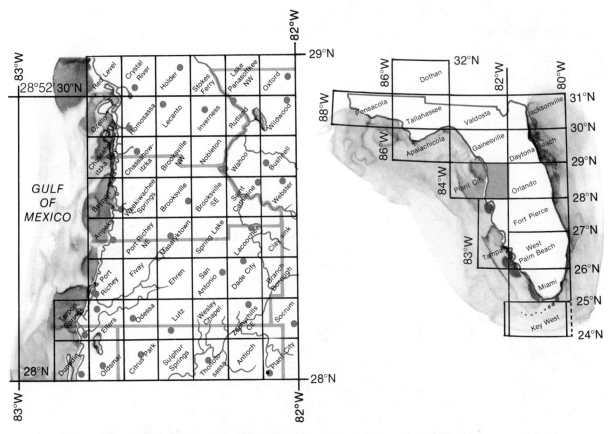

FIGURE 9–APPENDIX   The Red Level Quadrangle is located within the Plant City area in Florida. A 7.5 minute topographic map at a scale of 1:24,000 has been published for this quadrangle as well as the other 49 quadrangles in the Plant City area.

Most basic symbols have become standardized. For example, note the symbols appearing on the sedge islands to the north of the inlet on the Barnegat map; the sedge islands are marsh areas and the symbols used indicate this fact. Shoals, sand areas, roads, and houses are also shown by various symbols.

The roads depicted on the Barnegat map are very nearly to scale. On some maps, however, roads, if they are to show clearly, must be exaggerated in terms of width. Deliberate exaggeration is done only to call attention to features deemed to be important for a particular map; the South Carolina map (Figure A–10) showing its interstate highway system is a deliberate distortion to serve a particular purpose.

VALUE SYMBOLS

Each circle used on the map in Figure A–11 is an example of a value symbol. The location of a solid red circle tells us where electric power is being generated by nuclear means to serve South Carolina. But, in addition, the legend assigns a value to each circle: The larger circles indicate the station has a generating capacity of more than 500,000 kilowatts. The legend also gives us a way of differentiating between a nuclear energy facility and a station that generates its power by some other means.

A line connecting all points of equal value is called an *isopleth*. The lines of equal declination, isogonic lines, used in Figure A–4 are isopleths. An isopleth is a particular

FIGURE 10–APPENDIX The exaggeration in the width of the highways is deliberate and is used to serve a special purpose.

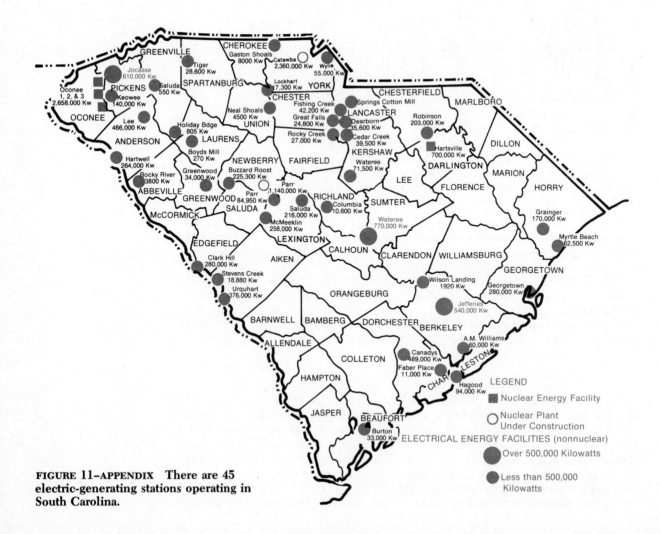

FIGURE 11–APPENDIX There are 45 electric-generating stations operating in South Carolina.

kind of value symbol that can be used in many ways on maps. A line connecting all points of equal temperature, for example, is an isopleth, which is generally referred to as an *isotherm*. The isothermal map in Figure A–12 depicts the average temperature pattern in South Carolina during the winter. An isopleth that connects all points of equal precipitation is referred to as an *isohyet*. And, in Figure A–3, isopleths—called *isobars*—are used to connect all points of equal barometric pressure. A most familiar isopleth, the contour line, which connects all points of equal elevation, is found on topographic maps.

MAP AND CHART PROJECTIONS

Any drawing on a flat surface that depicts a globe's network of meridians and parallels is called a *projection*. In fact, the term is derived from the technique of using a transparent globe with a light inside to project the lines and points from the globe's surface onto a flat surface (Fig. A–13). The lines and points projected by the light can be traced to make a flat map.

The framework of a map—the criss-cross network of parallels and meridians—is often referred to as the *graticule*. The graticule gives a general indication of directions on a map, but directional information is usually more accurately given by a north arrow.

PATTERNS OF DISTORTION

If we assume the Earth is a perfect sphere, the only way to represent the Earth so the representation is true in every detail is by a globe. When we attempt to transfer the Earth's curved surface onto a flat surface, the curved surface is twisted out of its natural, normal, and original shape. A curved surface cannot be transferred onto a flat surface without distortion and error; thus, there is no such thing as a perfect map and each map has an inherent defect.

In order to ascertain the distortions produced by the transfer of a curved surface to a flat surface, we must have a concept of what constitutes perfection—we must understand the basic geometric properties of the Earth and the coordinate system of longitude

FIGURE 12–APPENDIX   **The isopleths on this map are called *isotherms* because they join points of equal temperature. The 52°F isotherm, for example, runs through those areas in South Carolina that have 52°F as the average winter temperature.**

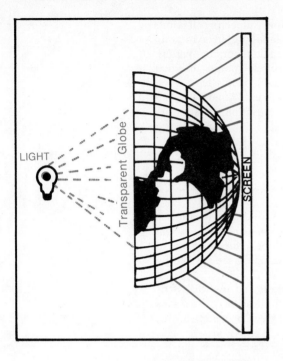

FIGURE 13–APPENDIX A light source inside a transparent globe can be used to project the lines and points from the globe's surface onto the flat surface of a screen.

and latitude. There are seven geometric properties that form the basis for any comparison: (1) scale on a perfect sphere is constant; (2) all latitude lines are parallel; (3) the length of latitude lines around the globe decreases from the Equator to the poles; (4) lines of longitude converge toward the poles and diverge toward the Equator; (5) meridians are equally spaced on parallels—distances along lines of longitude between any two latitude lines are equal; (6) all latitude lines meet and cross longitude lines at right angles; and (7) the area of the rectangle formed by two parallels and two meridians is the same everywhere between the same two parallels.

These seven geometric properties of the Earth and its coordinate system involve distance, direction, area, and shape. No flat map can be constructed to represent the globe and be true in all seven geometric properties. Distance, direction, area, and shape can be distorted in transferring a curved surface to a flat surface; but, depending on the method of projection, certain of these properties can be retained at the expense of others.

True map projections are produced by projecting the graticule onto a plane, a cone, or a cylinder. The matter of which projection to select relates very closely to the purpose of the map. Projections, for example, can be grouped according to which properties they represent best. Those projections that attempt to preserve true scale over the map are called *equidistant*. A projection that preserves direction is called *azimuthal* or *zenithal*. Equivalent projections are those that preserve areas in the correct proportions to one another. Those that attempt to preserve shape at the expense of accurate representation of area are called *conformal* or *orthomorphic* projections.

PROJECTIONS ONTO A CYLINDER

The cylindrical group of projections is developed as though a cylinder, or tube, were rolled around a globe (Fig. A–14). The cylinder makes contact with the globe along a great circle, called the *circle of tangency*. Cylindrical projections are usually designed so that the circle of tangency is the Equator or a meridian. With the Equator as the circle of tangency, the parallels and meridans are drawn as straight lines at right angles to each other. All the parallels are the correct distance apart, but they are represented as the

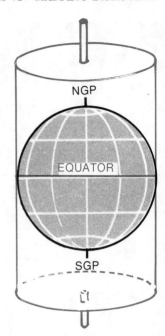

FIGURE 14–APPENDIX   The cylinder, which is wrapped around the globe, makes contact with the globe at the Equator.

same length as the Equator. The greatest distortions occur toward the poles because the parallels do not decrease in length as they do on the globe.

THE MERCATOR PROJECTION

From a navigator's point of view, the most significant invention in charting was the development of the Mercator projection by Gerhard Mercator in 1569. Using the standard Mercator projection with the Equator as the circle of tangency, a navigator can draw a straight line between two points and at any point on the line the same true compasss direction can be used.

The portion of the nautical chart shown in Figure A–15 is a Mercator projection; it represents an area in the Barnegat Bay of New Jersey between the Island Beach barrier bar on the right and the mainland on the left. A small boat is in the bay off Sunrise Beach Marsh; a line representing the boat's course runs from its present position to a point off Forked River. The line represents a course of 212° true; check the course against the true section of the compass rose.

A standard Mercator projection expands the meridians and parallels at the same rate with increasing latitude from the Equator. In other words, the Mercator projection is drawn so the distances between equally spaced parallels on the globe become greater and greater on the map as the parallels approach polar latitudes. This distortion makes up for the fact that the map distance between meridians remains the same from the Equator to the poles.

A standard Mercator projection makes areas in high latitudes seem much larger than they are. It is, however, an ideal projection for navigation charts because shapes of features such as islands, harbors, and bays are accurate and any compass course between two points can be shown as a straight line.

TRANSVERSE MERCATOR PROJECTION

The word transverse means to turn across. In a transverse Mercator projection, the cylinder is extended so that it is lying across the axis of the Earth. The circle of tangency is a meridian (Fig. A–16).

The transverse Mercator projection is a conformal projection in which shapes of small regions are represented accurately. Figure A–17 shows the parallels of a trans-

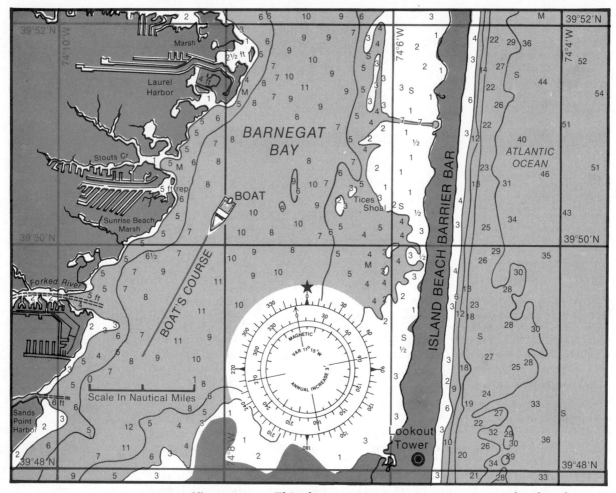

FIGURE 15–APPENDIX   This chart is a Mercator projection. It was developed as a projection onto a cylinder. The parallels and meridians are drawn as straight lines at right angles to each other. On a Mercator projection, a navigator can plot a course between two points as a straight line; at any point on the line the same true compass direction can be used.

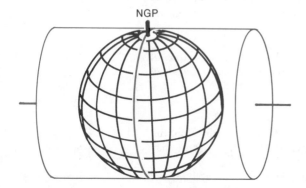

NGP

FIGURE 16–APPENDIX   The cylinder lies across the axis of the Earth in a transverse Mercator projection. The circle of tangency, which is shown in color, is a meridian.

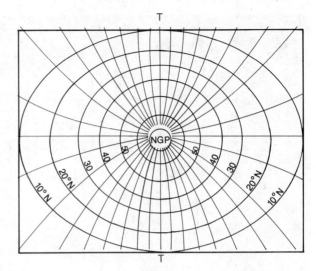

FIGURE 17–APPENDIX   The parallels of a transverse Mercator projection, which are shown in black, are unequally spaced ellipses; the color meridians are unequally spaced curve lines. The central vertical meridian, labeled *T*, is the meridian of the circle of tangency.

verse Mercator projection are unequally spaced ellipses and the meridians are un-equally spaced curved lines; thus, most straight lines on the map are not lines of constant azimuth on the Earth. Area distortion increases with increased distance from the central meridian. Scale distortion is constant along lines parallel to the central meridian.

The transverse Mercator projection is the basic map for the UTM grid system and for some State Plane Coordinate grids. The base maps for the UTM grid system are prepared using a transverse Mercator projection fitted to two arcs of tangency in each of the 60 separate sections between 80°N and 80°S latitude. Using this technique, no point on a sectional map is more than 150 miles (240 km) or so from a meridian of tangency; thus, distortion is correspondingly small.

PROJECTIONS ONTO A PLANE

In this type of projection, a flat surface—a plane—is placed next to the globe (Fig. A–18). The plane may be centered on a pole, the Equator, or any other point in an oblique arrangement. No matter what the point of tangency is, all these projections onto a flat surface are azimuthal since they tend to preserve direction. Azimuths measured around the point of tangency are mapped with no error.

The position of the light source, however, makes quite a difference in the way the grid system is projected (Fig. A–18). With a light source placed at infinity, the projection is called orthographic; a light source at the Earth's center produces a projection known as *gnomonic;* while light from the opposite point on the globe produces a flat-surface projection referred to as *stereographic.* Compare the projections of the line of latitude in Figure A–18. In each case, the plane is centered over the same point of tangency—one of the poles—but the light sources projecting the grid are placed at infinity, at the antipode or opposite point on the globe, and at the center of the Earth. The length of the projected line in Figure A–18 is different in each case; thus the spacing between lines of latitude within the grid system as a whole would also be different, depending on the position of the light source.

GNOMONIC PROJECTIONS

These are the best known of the azimuthal projections. The grid is established on a flat surface of the Earth from an eye point or light source that is at the center of the globe.

In a polar gnomonic projection—with a pole as the point of tangency—the meridians are straight lines outward from the pole; the parallels are concentric circles that are

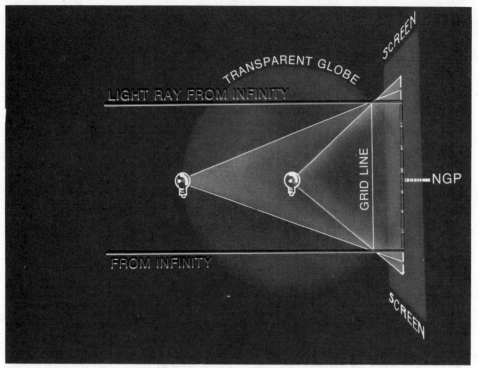

FIGURE 18–APPENDIX The point of tangency is a geographic pole. The projected length of this grid line depends on the position of the light source.

unequally spaced. Figure A–19 shows the spacing between the concentric circles in a gnomonic projection increases as you move outward from the poles. Distortion of shapes and area increases rapidly with increased distance from the pole.

The advantage of polar gnomonic, equatorial gnomonic, or oblique gnomonic maps is that a straight line between any two points represents a great circle route, that is, the shortest distance between two locations. Gnomonic projections are quite useful to navigators, especially in aerial navigation; with a gnomonic projection, a navigator can determine the latitude and longitude of places along the shortest route to a desired destination. When the coordinates are identified, they can be transferred to a Mercator chart of the same area; straight lines drawn between these positions on the Mercator chart give the true direction for each part of the trip. Using this technique the actual route of travel can be made to follow a great circle.

ORTHOGRAPHIC PROJECTIONS

The grid for these maps is established on a flat surface by projecting the curved surface of the Earth from an eye point or light source that is at infinity.

In a polar orthographic projection, the meridians are straight lines radiating outward from the pole just as they are in the polar gnomonic projection of Figure A–19. The parallels are concentric circles, too, but the space between the circles decreases rapidly as you move outward from the pole. Figure A–20 depicts the spacing of the parallels between two straight line meridians of a polar orthographic projection.

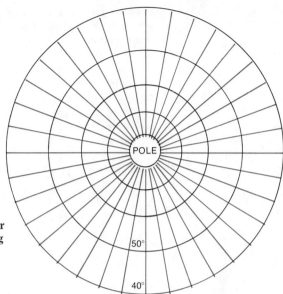

FIGURE 19–APPENDIX   The color meridians in this polar gnomonic projection of a grid are straight lines radiating outward from the pole. The black parallels are unequally spaced concentric circles. The spacing between the parallels increases as you move outward from the pole.

Directions from the point of tangency are correct. Distortion in shape and area increases moderately with increased distance from the point of tangency. Since a polar orthographic projection gives the Earth the appearance of being seen from deep space, it is used primarily for illustrations.

STEREOGRAPHIC PROJECTIONS

This projection makes use of an eye point or light source at the side of the globe opposite the point of tangency.

In a polar stereographic projection, the meridians are straight lines radiating outward from the pole. The parallels are concentric circles that are unequally spaced; the spacing

FIGURE 20–APPENDIX   The small wedge of a polar orthographic projection is shown in this figure. The wedge depicts only arcs of the concentric circles formed by the parallels. The two color meridians are straight lines radiating outward from the pole. The spacing between the black parallels decreases rapidly as you move outward from the poles.

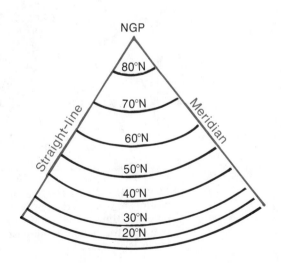

between parallels increases moderately as you move from polar to lower latitudes. This is the only azimuthal projection that is also conformal; in other words, shapes are represented well, but area distortion increases rapidly with increased distance from the point of tangency.

The surface weather map distributed by the National Oceanic and Atmospheric Administration is a polar stereographic projection. Figure A–21 depicts the polar stereographic base map used by NOAA. Note the way the spacing between parallels increases moderately as you move to lower latitudes. The shapes of states and provinces are reproduced with great accuracy, but can you detect the area distortion at lower latitudes?

PROJECTIONS ONTO A CONE

A cone laid over a globe contacts the globe along a circle of tangency. When the apex of the cone is above a pole, the circle of tangency coincides with a parallel of latitude (Fig.

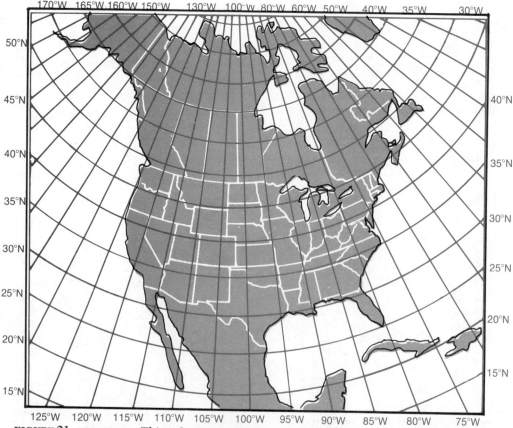

FIGURE 21–APPENDIX   This polar stereographic projection is used as a base map for the plotting of weather data by NOAA. The meridians are straight lines radiating outward from the pole and the parallels shown are portions or arcs of concentric circles. The spacing between the parallels increases moderately as you move outward from the pole to lower latitudes.

A–22). The parallel at which the cone makes contact with the globe is called the *standard parallel*. A cone can rest on the globe at any point. For example, a tall cone may touch the globe at the Equator, but a very short cone may touch the globe at 60°N.

SIMPLE CONIC PROJECTION

This is a projection in which the cone makes contact along one parallel (Fig. A–23). In flat map form, the meridians appear as radiating straight lines and the parallels, which are arcs of concentric circles, intersect with the meridians at right angles. In Figure A–23, the conical projection is based on the parallel of 45°N. The parallels of 30°N and 60°N are spaced along the meridians at the same distance from the standard parallel; beyond these parallels, however, the spacing increases.

A map produced by means of a simple conic projection has a scale that is exaggerated along the parallels, with the exception of the standard parallel. Although the simple conic projection is neither equal-area nor conformal, it is fairly accurate when used to depict a small region; the distortions are least nearest the standard parallel and, therefore, a simple conic projection is useful in mapping an area that extends along the east-west direction.

TWO-PARALLEL CONTACT

Alber's equal-area projection and Lambert's conformal projection are conic projections that are each built on two standard parallels. These projections are formed by imagining a cone that is too tight to fit over the globe, but the cone is so sharp that it slices into and out of the globe when it is forced over it (Fig. A–24). The map is made from the strip of the globe's surface cut off by the cone.

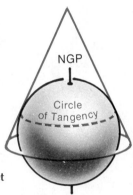

FIGURE 22–APPENDIX   **The color cone is tangent to the sphere. The points of contact between the cone and the sphere are referred to as a *circle of tangency*.**

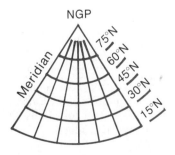

FIGURE 23–APPENDIX   **This simple conic projection is based on 45°N as the standard parallel. The meridians are radiating straight lines and the parallels are arcs of concentric circles. The parallels of 30° and 60°N are spaced at the same distance from the standard parallel; beyond these parallels the spacing increases.**

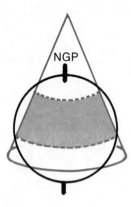

FIGURE 24–APPENDIX The cone intersects the sphere along the parallels. The broken color lines are both considered to be standard parallels. The map is made from the strip of the globe's surface (dark area) cut off by the cone.

The two standard parallels in Alber's conic projection are 29.5°N and 45.5°N for maps of the conterminous United States. The meridians are converging straight lines and the parallels are arcs of concentric circles. Alber's maps are equal-area projections. Distortion increases slowly with increased distance from the standard parallels. The scale along meridians is slightly too large between the standard parallels and somewhat too small beyond the parallels.

The projection most commonly used for aeronautical charts is the Lambert conic projection. Two standard parallels are used in the Lambert projection; for maps of the conterminous United States, the parallels are usually 33°N and 45°N. The parallels are arcs of concentric circles that are nearly equally spaced; the meridians are straight lines that converge toward the pole. This is a conformal projection; distortion increases slowly with increased distance from the standard parallels. The scale along the meridians is slightly too small between the standard parallels and somewhat too large beyond them.

POLYCONIC PROJECTIONS

A polyconic projection is commonly used for large-scale maps. The U.S. Geological Survey, for example, uses polyconic projections for its topographic maps because the projection is an excellent compromise between a conformal and an equal-area projection.

In the polyconic projection, every parallel is a standard parallel since the map is constructed as a projection of strips of the globe's surface on a series of touching cones. A map produced by a polyconic projection has a vertical straight line that marks the central meridian (Fig. A–25); all the other meridians are curves converging toward the poles. The parallels are arcs of nonconcentric circles that are equally spaced on the central meridian.

MAKING MAPS

Many people must work together to map a map. The information displayed as a map is the result of careful observations and measurements made of some portion of the Earth's surface. In general, the following steps are taken in producing a map: (1) observation and measurement, (2) organization and planning, and (3) drafting and reproduction.

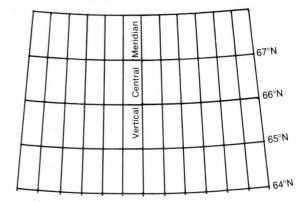

FIGURE 25–APPENDIX   The grid of a polyconic projection has a vertical central meridian. The other meridians are curves converging toward the poles. The parallels, which are equally spaced on the central meridians, are arcs of non-concentric circles.

OBSERVATION AND MEASUREMENT

Experts are constantly observing, recording, photographing, and describing surface features and other conditions at various locations on the Earth's surface. The accumulating information is most useful when it is related to the ground in such a way that its location, distance, and direction from a fixed point is known with great precision. Within the United States, ground control depends on a system of accurately located survey markers that are permanently implanted in the Earth. The exact latitude, longitude, and height above sea level have been determined for each of these markers through triangulation based on astronomical observations.

With ground-control markers as reference points, surveyors can relate features and conditions observed at one location on the Earth's surface to other locations. These markers and the data derived from them provide the basis for establishing the correct spacing between places and conditions on maps. Other specialists—such as the geologist, geographer, and meteorologist—using the established spatial relationships can collect additional data with every expectation that its location with reference to the ground-control markers can be accurately established.

Automatic cameras on airplanes and satellites have added to our ability to collect data rapidly and over wide-ranging areas. Photogammetry, the science of measuring from photographs, is one of the most useful tools available to the modern map-maker. Satellite photographs, for example, allow us to analyze surface features as well as atmospheric conditions. There are a number of sophisticated techniques available by which picture points, or recognizable surface features, can be related to the basic ground-control markers.

ORGANIZATION AND PLANNING

As new information accumulates, it is added to the facts already known about an area. The accumulating data must be organized, simplified, generalized, and then integrated with what is known. Eventually the data must be translated into the language of maps.

When volumes of specific information about an area are available, it becomes possible to select certain data to be presented in the form of a map. Once the purpose of a map is established, a cartographer can take the facts and plan the map. The objective of the cartographer is to develop the best visual display of the information within the limits of the accuracy required.

DRAFTING AND REPRODUCTION

The map or chart may be put on paper as an original drawing or series of drawings by a cartographic artist or draftsman. The drawing is usually made larger than the desired size of the final map. It is the job of the cartographic artist to bring together the facts and ideas of the scientists and cartographers and to make the map easy to read.

The printing process may follow any one of a number of methods when a map is to be printed in only one color. When color or background tints are to be used, the finished map drawing is rarely hand-colored. Instead, a succession of half-tone screens are developed, which are then used in conjunction with platemaking photography to get the final effect desired.

Airport Code Identification (1)

Airport Code	City Name
ABQ	Albuquerque, New Mexico
AMH	Amherst, Massachusetts
AMS	Ames, Iowa
ANN	Annette, Alaska
APL	Apalachicola, Florida
AST	Astoria, Oregon
ATL	Atlanta, Georgia
BET	Bethel, Alaska
BIS	Bismarck, North Dakota
BLU	Blue Hill, Massachusetts
BNA	Nashville, Tennessee
BOI	Boise, Idaho
BOS	Boston, Massachusetts
BOU	Boulder, Colorado
BRO	Brownsville, Texas
BRW	Barrow, Alaska
CAM	Cambridge, Massachusetts
CBU	Caribou, Maine
CHI	Chicago, Illinois
CHS	Charleston, South Carolina
CLE	Cleveland, Ohio
CMH	Columbus, Ohio
COU	Columbia, Missouri
COV	Corvallis, Oregon
DAV	Davis, California
DDC	Dodge City, Kansas
ELP	El Paso, Texas
ELY	Ely, Nevada
EWA	East Wareham, Massachusetts
FAI	Fairbanks, Alaska
FAT	Fresno, California
FGR	Flaming Gorge, Utah
FRD	Friday Harbor, Washington
FTW	Fort Worth, Texas
GEG	Spokane, Washington
GGW	Glasgow, Montana
GJT	Grand Junction, Colorado
GNV	Gainesville, Florida
GRF	Griffin, Georgia
GRL	Grand Lake, Colorado
GSO	Greensboro, North Carolina
GTF	Great Falls, Montana

Airport Code	City Name
HAT	Hatteras, North Carolina
HNL	Honolulu, Hawaii
IND	Indianapolis, Indiana
ITH	Ithaca, New York
IYK	Inyokern, California
JFK	New York, New York
LAR	Laramie, Wyoming
LAS	Las Vegas, Nevada
LAX	Los Angeles, California (WBO)
LAX	Los Angeles, California (WBAS)
LCH	Lake Charles, Louisiana
LEM	Lemont, Illinois
LEX	Lexington, Kentucky
LIT	Little Rock, Arkansas
LJO	LaJolla, California
LND	Lander, Wyoming
LNK	Lincoln, Nebraska
LYN	Lynn, Massachusetts
MAF	Midland, Texas
MAK	Matanuska, Alaska
MFR	Medford, Oregon
MHK	Manhattan, Kansas
MIA	Miami, Florida
MNL	Mauna Loa, Hawaii
MSN	Madison, Wisconsin
MSY	New Orleans, Louisiana
MWE	Mt. Weather, Virginia
NOM	North Omaha, Nebraska
NPT	Newport, Rhode Island
OAR	Oak Ridge, Tennessee
OKC	Oklahoma City, Oklahoma
PHX	Phoenix, Arizona
PIT	Pittsburgh, Pennsylvania
PHL	Pearl Harbor, Hawaii
PNB	Put-In Bay, Ohio
PRO	Prosser, Washington
PUW	Pullman, Washington
PWM	Portland, Maine
RAL	Riverside, California
RAP	Rapid City, South Dakota
RDU	Raleigh/Durham, North Carolina
SAT	San Antonio, Texas
SAY	Sayville, New York
SCE	State College, Pennsylvania
SCH	Schenectady, New York
SCL	St. Cloud, Minnesota
SDA	Soda Springs, California
SEA	Seattle/Tacoma, Washington
SEB	Seabrook, New Jersey
SHV	Shreveport, Louisiana
SJU	San Juan, Puerto Rico
SLC	Salt Lake City, Utah
STW	Stillwater, Oklahoma

Airport Code	City Name
SUM	Summit, Montana
TLH	Tallahasse, Florida
TPA	Tampa, Florida
TUS	Tucson, Arizona
TWF	Twin Falls, Idaho
UPT	Upton, New York
WAS	Washington, D.C.

Airport Code Identification (2)

Airport Code	City Name
ABQ	Albuquerque, New Mexico
ACV	Eureka/Arcata, California
ANC	Anchorage, Alaska
ATL	Atlanta, Georgia
AVL	Ashville, North Carolina
BIS	Bismarck, North Dakota
BNA	Nashville, Tennessee
BOI	Boise, Idaho
BOS	Boston, Massachusetts
BRW	Barrow, Alaska
BTV	Burlington, Vermont
CBU	Caribou, Maine
CHI	Chicago, Illinois
CHS	Charleston, South Carolina
CMH	Columbus, Ohio
CVG	Cincinnati, Ohio
DEN	Denver, Colorado
DET	Detroit, Michigan
DSM	Des Moines, Iowa
ELP	El Paso, Texas
FAI	Fairbanks, Alaska
FAT	Fresno, California
FLG	Flagstaff, Arizona
FTW	Fort Worth, Texas
GEG	Spokane, Washington
HNL	Honolulu, Hawaii
ICT	Wichita, Kansas
JFK	New York, New York
LAS	Las Vegas, Nevada
LAX	Los Angeles, California
LIT	Little Rock, Arkansas
LND	Lander, Wyoming
LNK	Lincoln, Nebraska
MAF	Midland/Odessa, Texas
MGM	Montgomery, Alabama
MIA	Miami, Florida
MLS	Miles City, Montana
MSP	Minneapolis/St. Paul, Minnesota
MSY	New Orleans, Louisiana
OKC	Oklahoma City, Oklahoma
PDX	Portland, Oregon
PHL	Philadelphia, Pennsylvania

Airport Code	City Name
RAP	Rapid City, South Dakota
RDU	Raleigh/Durham, North Carolina
SAT	San Antonio, Texas
SEA	Seattle, Washington
SFO	San Francisco, California
SHV	Shreveport, Louisiana
SJU	San Juan, Puerto Rico
SLC	Salt Lake City, Utah
STL	St. Louis, Missouri
TPA	Tampa, Florida
WAS	Washington, D.C.
WYS	West Yellowstone, Montana
YUM	Yuma, Arizona

APPENDIX IX

THE CANADIAN SYSTEM OF SOIL CLASSIFICATION

Canadian soil scientists have been profoundly influenced by the concept of soil as a natural body integrating the accumulative effects of climate and vegetation acting on surficial materials. The first Canadian taxonomic system of soil classification was outlined in 1955. The 1955 proposal, which is the basis for the Canadian system used today, had six categorical levels corresponding to order, great group, subgroup, family, series, and type. The seven taxa separated at the order level in the 1955 outline were: Chernozemic, Halomorphic, Podzolic, Forested Brown, Regosolic, Gleisolic, and Organic.

Progress in the Canadian system since 1955 has been toward more precise definitions of the taxa at all categorical levels. Some changes in taxa at the order, great group, and subgroup levels have been made in the system during the last 25 years; in 1968, for example, the former Podzolic order was divided into Luvisolic (clay translocation) and Podzolic (accumulation of aluminum and iron organic complexes). The Cryosolic order was added in 1973 to classify the soils with permafrost close to the surface. In 1980, the Canadian system of soil classification consisted of nine orders: Brunisolic, Chernozemic, Cryosolic, Gleysolic, Luvisolic, Organic, Podzolic, Regosolic, and Solonetzic. The major soil regions of Canada according to the Canadian system are shown in Figure A–26.

BRUNISOLIC ORDER

These soils develop under forest cover with generally brownish-colored B horizons. But the order—it should be noted—also includes soils of various colors. Most Brunisols are well to imperfectly drained. Some Brunisolic soils are calcareous to the surface and very slightly weathered; others are strongly acid and weathered.

The Brunisols are often found in association with Podzolic soils (Figure A–26). The Brunisols occur in a wide range of climatic environments throughout Canada. The vegetative environments of these soils is just as varied including boreal forest, mixed forest, shrubs, grass, heath, and tundra.

CHERNOZEMIC ORDER

The map in Figure A–26 shows the major area of Chernozemic soils in the cool, subarid to subhumid interior plains of western Canada. There are minor areas of Chernozemic soils, which are not shown on the map in Figure A–26, that occur in some valleys and mountain slopes in the Canadian cordillera; in some cases, these Chernozemic soils extend beyond the tree line.

The soils of the Chernozemic order occur in well to imperfectly drained areas. The surface horizons of these soils are darkened by the accumulation of organic matter from the decomposition of xerophytic or mesophytic grasses representative of grassland

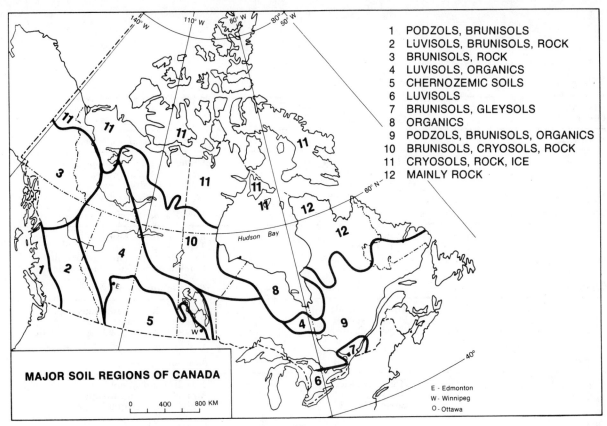

1 PODZOLS, BRUNISOLS
2 LUVISOLS, BRUNISOLS, ROCK
3 BRUNISOLS, ROCK
4 LUVISOLS, ORGANICS
5 CHERNOZEMIC SOILS
6 LUVISOLS
7 BRUNISOLS, GLEYSOLS
8 ORGANICS
9 PODZOLS, BRUNISOLS, ORGANICS
10 BRUNISOLS, CRYOSOLS, ROCK
11 CRYOSOLS, ROCK, ICE
12 MAINLY ROCK

MAJOR SOIL REGIONS OF CANADA

0   400   800 KM

E - Edmonton
W - Winnipeg
O - Ottawa

FIGURE 26–APPENDIX

communities. Most Chernozemic soils are frozen for a time during each winter. The mean annual temperature of these soils is generally higher than 32°F (0°C) and usually lower than 42°F (5.5°C).

CRYOSOLIC ORDER

Soils of the Cryosolic order occupy the northern third of Canada, which is, for the most part, above the tree line (Figure A–26). In this area of Canada, permafrost is always close to the surface of both mineral and organic deposits. Cryosolic soils are also common in the subarctic forest area and extend, to a limited extent, into boreal forest.

The soils of the Cryosolic order are formed in either mineral or organic materials that have permafrost within 3 to 6 feet (1 to 2 m) of the surface. The mean annual temperature of Cryosolic soils is below 32°F (0°C).

GLEYSOLIC ORDER

The principal area of Gleysolic soils shown on the map in Figure A–26 is in southeastern Canada. Generally these soils develop in shallow depressions and on lowlands that are saturated with water every spring. Gleysolic soils may also develop in an area with a high groundwater table.

TABLE A–1 CORRELATION BETWEEN THE CANADIAN AND U.S. SYSTEMS

CANADIAN SYSTEM Order	Great Group	CSCS OF U.S.
Brunisolic		*Inceptisol, some Psamments*
	Melanic Brunisol	Cryochrept, Eutrochrept, Hapludoll
	Eutric Brunisol	Cryochrept, Eutrochrept
	Sombric Brunisol	Umbric, Dystrochrept
	Dystric Brunisol	Dystrochrept, Cryochrept
Chernozemic		*Boroll, some Vertisols*
	Brown	Aridic Boroll subgroups
	Dark Brown	Typic Boroll subgroups
	Black	Udic Boroll subgroups, Rendoll
	Dark Gray	Boralfic Boroll subgroups, Alboll
Cryosolic		*Pergelic subgroups*
	Turbic Cryosol	Pergelic Ruptic subgroups
	Static Cryosol	Pergelic subgroups
	Organic Cryosol	Pergelic Histosol or Pergelic Histic subgroups of other orders
Gleysolic		*Aqu-suborders*
	Humic Gleysol	Aquoll, Humaquept
	Gleysol	Aquent, Fluvent, Aquept
	Luvic Gleysol	Argialboll, Argiaquoll, Aqualf
Luvisolic		*Boralf and Udalf*
	Gray Brown Luvisol	Hapludalf or Glossudalf
	Gray Luvisol	Boralf
Organic		*Histosol*
	Fibrisol	Fibrist
	Mesisol	Hemist
	Humisol	Saprist
	Folisol	Folist
Podzolic		*Spodosol, some Inceptisols*
	Humic Podzol	Cryaquod, Humod
	Ferro-Humic Podzol	Humic Cryorthod, Humic Haplorthod
	Humo-Ferric Podzol	Cryorthod, Haplorthod
Regosolic		*Entisol*
	Regosol	Entisol
	Humic Regosol	Entisol
Solonetzic		*Natric great groups, Mollisol & Alfisol*
	Solonetz	Natric great groups, Mollisol & Alfisol
	Solodized Solonetz	Natric great groups, Mollisol & Alfisol
	Solod	Glossic Natriboroll, Natralboll

The features of the Gleysolic order indicate periodic or prolonged saturation with water and reducing conditions. Thus, color is used as a diagnostic of these soils because color is an easily observable indicator of the oxidation–reduction status of a soil.

LUVISOLIC ORDER

The development of a Luvisolic soil involves the suspension of clay in the soil solution near the surface, that is, in the A horizon and then the downward movement and accumulation of the clay in the B horizons. The soils of this order can develop in well to imperfectly drained sites, in sandy loam to clay base-saturated parent materials, and in mild to very cold climates.

Refer to the map in Figure A–26 and you will find three major locations identified in which Luvisolic soils occur. There is a major concentration of these soils in southern Ontario. British Columbia also has a large area of Luvisols. The largest area of these soils is found in the central to northern Interior Plains where they develop under deciduous, mixed, and coniferous forests and extend to the zone of permafrost. Luvisolic soils—although not shown on the map—are also found in the Atlantic Provinces of Canada.

ORGANIC ORDER

The soils of this order are composed largely of organic materials and include most of the soils known as *peat, muck,* or *bog* soils. The Organic soils are generally found in poorly drained depressions. They are derived from the vegetation that grows on such sites. The Organic soils of Canada contain at least 30 percent organic matter by weight.

PODZOLIC ORDER

Podzolic soils have B horizons in which combinations of organic matter, aluminum, and iron have accumulated. The typical Podzolic soils of Canada occur in coarse- to medium-textured, acid parent materials. They may develop under forest or heath vegetation in cool to very cold humid climates. Most Podzolic soils have a reddish-brown to black B horizon with an abrupt upper boundary and a lower boundary that becomes progressively yellower in hue.

REGOSOLIC ORDER

The soils of this order are weakly developed without distinctive horizon development. They are found on recent alluvium and on slopes subject to mass wasting. Generally Regosolic soils are rapidly to imperfectly drained.

SOLONETZIC ORDER

The soils of this order have developed from parent materials that were salinized with salts high in sodium. Leaching of the salts by rainwater results in colloids being carried down and deposited in the B horizons. The B horizons of Solonetzic soils are very hard when dry and they swell to a sticky mass of low permeability when wet.

Most of the Solonetzic soils in Canada have a neutral to acidic A horizon. These soils are found on saline parent materials in some areas of the semiarid Interior Plains. They are often found in association with Chernozemic soils. The predominant vegetative cover for most Solonetzic soils is grass.

RELATIONSHIP OF CANADIAN AND U.S. SYSTEMS

The numerous national systems of soil taxonomy—those of the United States, Canada, Russia, and other European countries, for example—are an indication of the youthfulness of soil science. The Canadian system of soil taxonomy is more closely related to the United States system than to any other. Both the Canadian and United States systems are hierarchial and the taxa are based on measurable soil properties. The Canadian system, however, is designed to classify only soils that occur in Canada and is not a comprehensive system.

The CSCS of the United States has a suborder, which is a category not found in the Canadian system. But the main difference between the systems is that in the Canadian system all horizons to the surface may be diagnostic, whereas the CSCS of the United States emphasizes the horizons below the depth of plowing. This difference is probably the consequence of the fact that 90 percent of the area of Canada is not likely to be cultivated.

The correlation between the CSCS and the Canadian system is shown in Table A–1. Canadian soils differentiated at the order and great group level in the Canadian system have been assigned a corresponding classification in the CSCS.

Glossary

abrasion—Wearing away of sedimentary rock chiefly by currents of water and ice laden with sand and other rock debris.

absolute zero—The temperature of −459.69°F (−273.16°C) or 0 K, thought to be the temperature at which molecular motion dies and a body would have no heat energy.

abyssal plain—A flat, almost level area that occupies the deepest parts of many of the ocean basins.

acre—A unit of area, equal to 43,560 square feet, or to 4046.85 square meters.

acre-foot—The volume of water required to cover 1 acre to a depth of 1 foot, hence 43,560 cubic feet.

adhesive bond—The forces such as dipole bonds that attract adhesives and base materials to each other.

adiabatic—Refers to any change in which there is no gain or loss of heat.

adiabatic cooling—The process by which the temperature of a system is reduced without any heat being exchanged between the system and its surroundings.

advection—(1) The process of transport of an atmospheric property solely by the mass motion of the atmosphere. (2) The process of transport of water, or of an aqueous property, solely by the mass motion of the ocean.

advection fog—A type of fog that is caused by the horizontal movement of moist air over a cold surface and the consequent cooling of that air to below its dew point.

airglow—The quasi-steady radiant emission from the upper atmosphere over middle and low latitudes, as distinguished from the sporadic emission of auroras, which occur over high latitudes.

air mass—An extensive area of the atmosphere that approximates horizontal homogeneity in its characteristics, particularly with reference to temperature and moisture distribution.

air-mass analysis—The practice of synoptic surface-chart analysis by the so-called Norwegian methods, which involve the concepts of the polar front and of the broad-scale air masses that it separates.

air-mass climatology—The representation of the climate of a region by the frequency and characteristics of the air masses that pass over the area.

air pressure—The force per unit area that the air exerts on any surface in contact with it.

albedo—The reflection factor of a surface, that is, the fraction of the total light incident on a reflecting surface that is reflected back in all directions.

algae—The general name for the chlorophyll-bearing organisms in the plant subkingdom Thallobionta.

alpine—Any plant that is native to mountain peaks or boreal regions.

alpine glacier—A glacier that lies on or occupies a depression in mountainous terrain.

alpine tundra—Large, flat or gently sloping, treeless tracts of land above the timberline.

altocumulus cloud—A principal cloud type that is white or gray or both white and gray in color. It occurs as a layer or patch with a waved aspect, the elements of which appear as laminae, rounded masses, or rolls. Abbreviated Ac.

altostratus cloud—A principal cloud type that is in the form of a gray or bluish (never white) sheet or layer of striated, fibrous, or uniform appearance. Abbreviated As.

angiosperm—The common name for members of the plant division Magnoliophyta.

anticline—A fold in which the layered strata are inclined down and away from the axes.

anticyclone—A high-pressure atmospheric circulation whose relative direction of rotation is clockwise in the Northern Hemisphere and counterclockwise in the Southern Hemisphere. Also known as high; high-pressure area.

anticyclonic—Referred to a rotation about the local vertical that is clockwise in the Northern Hemisphere and counterclockwise in the Southern Hemisphere.

Appalachia—A borderland that existed along the southeastern side of North America, seaward of the Appalachian geosyncline during Paleozoic time.

Appalachian orogeny—A term that refers to Late Paleozoic diastrophism which began in the late Devonian and continued until the end of the Permian.

aquatic—Living or growing in, on, or near water.

aquiclude—A porous formation that absorbs water slowly but will not transmit it fast enough to furnish an appreciable supply for a well.

aquifer—A permeable body of rock capable of yielding quantities of groundwater to wells.

Archeozoic—The second era of geologic time when the lowest forms of life probably existed.

arctic air—An air mass whose characteristics are developed mostly in winter over northern surfaces of ice and snow.

arctic-alpine—Of or pertaining to areas above the timberline in mountainous regions.

Arctic Circle—The parallel of latitude 66°32′N (often taken as 66½°N).

arctic front—The semipermanent, semicontinuous front between the deep, cold arctic air and the shallower, basically warmer polar air of northern latitudes.

arctic high—A weak high that appears on charts of sea-level pressure over the Arctic Basin during late spring, summer, and early autumn.

Arctic Ocean—The north polar ocean lying between North America, Greenland, and Asia.

Arctic Zone—The area north of the Arctic Circle (66°32′N).

arcuate delta—A bowed or curved delta with its convex margin facing the body of water.

artesian well—A well in which the water rises, without the aid of a pump, above the top of the water-bearing bed.

asthenosphere—The portion of the upper mantle that is beneath the rigid lithosphere and which is plastic enough for rock flowage to occur.

A.39

asymmetrical fold—A fold in which one limb (side of a fold) dips more steeply than the other.

atoll—A ring-shaped coral reef that surrounds a lagoon in which there is no projecting land area.

aurora—The most intense of the several lights emitted by the Earth's upper atmosphere; the display is seen most often along the outer realms of the Arctic and Antarctic, where it is called the aurora borealis and aurora australis, respectively.

aurora australis—The aurora of southern latitudes. Also known as southern lights.

aurora borealis—The aurora of northern latitudes. Also known as northern lights.

axial plane—A plane that intersects a crest or trough in such a manner that the limbs or sides of the fold are more or less symmetrically arranged with reference to it.

axis of rotation—A straight line that passes through the stationary points of a rotating rigid body; the other points of the body move in circles about the axis.

Azoic—The first era of geologic time; it is that Precambrian time in which there is no trace of life.

bar—Any submerged or partially submerged ridge, bank, or mound of sand, gravel, or other unconsolidated sediment that is built up by waves or currents within stream channels, at estuary mouths, and along coasts.

barchan—A crescent-shaped dune of windblown sand, the arms of which point downwind.

baric wind law—See *Buys-Ballot's law*.

barrier bar—A ridge whose crest is parallel to the shore and which is made up of water-worn gravel put down by currents in shallow water at some distance from the shore.

barrier beach—A single, long, narrow ridge of sand that rises slightly above the level of high tide and lies parallel to the shore, from which it is separated by a lagoon.

barrier island—(1) An island that is similar to an offshore bar but differs from it in having multiple ridges, areas of vegetation, and swampy terraces extending toward the lagoon. (2) A detached portion of offshore bar between two inlets.

barrier reef—A coral reef that runs parallel to the coast of an island or continent and from which it is separated by a lagoon.

barycenter—The center of gravity of the Earth-moon system.

base-flow—The flow of water that enters a stream channel from groundwater sources.

batholith—A body of igneous rock, 40 square miles (100 km²) or more in area, emplaced in the Earth's crust.

bay head—A swampy region at the head of a bay.

bay head bar—A bar formed a short distance from the shore at the head of a bay.

bay head delta—A delta at the head of an estuary or a bay.

beach—The zone of unconsolidated material that extends landward from the low-water line to where there is marked change in material or to the line of permanent vegetation.

beach cusp. See cusp.

beach platform—See *wave-cut bench*.

beach ridge—A continuous mound of beach material behind the beach that was heaped up by waves.

beach scarp—A nearly vertical slope along the beach caused by wave erosion.

bight—A long, gradual bend or recess in the coastline that forms a large, open receding bay.

biogenic sediment—A deposit that results from the physiological activity of organisms.

biotic—Of or pertaining to life and living organisms.

biotic community—An aggregation of organisms characterized by a distinctive combination of both animal and plant species in a particular habitat.

biotic environment—That environment comprising living organisms, which interact with each other and their abiotic environment.

bird-foot delta—A delta formed by the outgrowth of fingers of natural levees at the mouth of river distributaries; the Mississippi Delta is an example.

block mountain—A mountain that was formed by the combined processes of uplifting, faulting, and tilting. Also known as fault-block mountain.

bog—A plant community that develops and grows in an area with permanently waterlogged peat substrates. Also known as a moor; *quagmire*.

boreal—Of or relating to the northern geographic regions.

bulk density—The mass of material per unit of volume.

Buys-Ballot's law—A law that describes the relationship between the horizontal wind direction in the atmosphere and the pressure distribution. If one stands with one's back to the wind, the pressure to the left is lower than to the right in the Northern Hemisphere; in the Southern Hemisphere the relation is reversed. Also known as *baric wind law*.

calendar—A system in which time is divided into days and longer periods, such as weeks, months, and years, and a definite order for these periods is established.

Canadian Shield—See *Laurentian Shield*.

cape—A prominent point of land jutting into a body of water. Also known as headland; point; promontory.

capillarity—The action by which the surface of a liquid where it contacts a solid is elevated or depressed because of the relative attraction of the molecules of the liquid for those of the solid.

capillary fringe—The lower subdivision of the zone of aeration that overlies the zone of saturation and in which the pressure of water in the interstices is lower than atmospheric.

capillary water—Soil water held by capillarity as a continuous film around solid particles and in interstices between particles above the phreatic line.

cave—A natural hollow or chamber beneath the Earth's surface, or in the side of a mountain or hill, with an opening to the surface.

channel—The deeper portion of a waterway carrying the main current of water.

chemical weathering—A weathering process whereby rocks and minerals are transformed into new, fairly stable chemical combinations by chemical reactions such as hydrolysis, oxidation, and solution.

chemosphere—The vaguely defined region of the upper atmosphere in which photochemical reactions take place.

chinook—The foehn on the eastern side of the Rocky Mountains.

circle of latitude—Also known as *parallel of latitude*, it is a meridian of the terrestrial sphere along which latitude is measured.

circle of longitude—See *parallel*.

cirrocumulus cloud—A principal cloud type that appears as a thin, white path of cloud without shadows, composed of very small elements in the form of grains, ripples, and so on. Abbreviated Cc.

cirrostratus cloud—A principal cloud type that appears as a whitish veil, usually fibrous but sometimes smooth, which may totally cover the sky and often produces halo phenomena. Abbreviated Cs.

cirrus cloud—A principal cloud type that is composed of detached cirriform elements in the form of white, delicate filaments, of white patches, or narrow bands. Abbreviated Ci.

clay—(1) A natural, fine-grained material that develops plasticity when mixed with a limited amount of water; composed primarily of silica, alumina, and water. (2) The fraction of an earthy material containing the smallest particles, that is, finer than 3 micrometers.

climate—The long-term manifestations of weather.

climatic diagram—A graphic presentation of climatic data that is generally limited to a plot of the simultaneous variations of two climatic elements. Also known as climagram; climagraph; climatograph; climogram; *climograph*.

climax—A mature, relatively stable community that will undergo no further change under the prevailing climate; represents the culmination of ecological succession.

climax plant formation—A mature, stable plant population in a climax community.

climograph—See *climatic diagram*.

cloud classification—(1) A scheme of distinguishing and grouping clouds according to their appearance and, where possible, to their process of formation. (2) A scheme of classifying clouds according to their altitudes: high, middle, or low clouds.

coast—The general region that extends from the sea inland to the first major change in terrain features.

coastal dune—A mobile mound of windblown sand found along a sea or lake shore.

cold front—A nonoccluded front that moves so that the colder air replaces the warmer air; the leading edge of a relatively cold air mass.

cold-front-like sea breeze—A sea breeze that moves slowly toward the land, and then moves inland quite suddenly. Also known as sea breeze of the second kind.

cold-front thunderstorm—A thunderstorm attending a cold front.

cold high—At a given level in the atmosphere, any high that is generally characterized by colder air near its center than around its periphery. Also known as cold anticyclone; cold-core high.

conifer—The common name for plants of the order Pinales.

coniferous forest—An area of wooded land dominated by conifers.

constant-level balloon—A balloon that is designed to float at a constant pressure level. Also known as constant-pressure balloon.

constant-pressure chart—The synoptic chart for a constant-pressure surface; it usually contains plotted data and analyses of the distribution of height of the surface, wind, temperature, and so on. Also known as isobaric chart; isobaric contour chart.

continental air—A type of air whose characteristics are developed over a large land area.

continental climate—A climate that is characteristic of the interior of a landmass of continental size; generally it is marked by annual, daily, and day-to-day temperature ranges, low relative humidity, and a moderate or small irregular rainfall.

continental high—A general area of high atmospheric pressure that overlies a continent during the winter. Also known as continental anticyclone.

continental polar air—Air from land areas of the Polar region. CP air generally has a low moisture content and a low surface temperature.

continentality—The degree to which a location on the Earth's surface is in all respects subject to the influence of a land mass.

continental rise—A transitional part of the continental margin; a gentle slope with a generally smooth surface, built up by the shedding of sediments from the continental block, and located between the continental slope and the abyssal plain.

continental shelf—The zone around a continent that extends from the shoreline to the continental slope. Also known as continental platform; shelf.

continental shield—Large areas of Precambrian rocks exposed as cratons within continents.

continental slope—The part of the continental margin from the edge of the continental shelf extending down to the continental rise.

continental tropical air—A type of tropical air produced over subtropical arid regions; it is hot and very dry.

convection current—(1) Mass movement of subcrustal or mantle material as a result of temperature variations. (2) Any current of air involved in convection; usually, the upward-moving portion of a convection circulation, such as a thermal or the updraft in cumulus clouds.

core—Center of the Earth.

Coriolis effect—Also known as Coriolis deflection. The deflection relative to the Earth's surface of any object moving above the Earth, caused by the Coriolis force; an object moving horizontally is deflected to the right in the Northern Hemisphere, to the left in the Southern.

craton—The large, relatively immobile portion of a continent that consists of both shield and platform areas.

creep—A slow, imperceptible downward movement of slope-forming rock or soil under sheer stress.

Cretaceous—The latest period of the Mesozoic Era.

crevasse—An open, nearly vertical fissure in a glacier or other mass of land ice or the Earth.

crust—The outermost solid layer of the Earth that extends no more than a few miles from the surface to the Mohorovicic discontinuity. Also known as Earth crust.

crystal—A homogeneous solid made up of an element, chemical compound, or isomorphous mixture throughout which the atoms or molecules are arranged in a regularly repeating pattern.

crystalline—Of, pertaining to, resembling, or composed of crystals.

crystalline porosity—Porosity in crystalline limestone and dolomite, making possible underground oil reservoirs.

crystalline rock—(1) Rock made up of minerals in a clearly crystalline state. (2) Igneous and metamorphic rock, as opposed to sedimentary rock.

cuesta—A gently sloping plain that terminates in a steep slope on one side.

cumuliform cloud—A fundamental cloud type that shows vertical development in the form of rising mounds, domes, or towers.

cumulonimbus cloud—A principal cloud type that occurs either as an isolated cloud or as a line or wall of clouds with separated upper portions.

cumulus cloud—A principal cloud type with detached elements that are generally dense and possess sharp, nonfibrous outlines; these elements develop vertically, appearing as rising mounds, domes, or towers, the upper parts of which often resemble a cauliflower.

current—The rate of flow of water across a surface per unit time.

cusp—One of a series of low, crescent-shaped mounds of beach material separated by smoothly curved, shallow troughs spaced at more or less regular intervals along and generally perpendicular to the beach face.

cuspate bar—A crescentic bar joining with the shore at each end.

day—One of various units of time equal to the period of rotation of the Earth.

daylight saving meridian—The meridian used for reckoning daylight saving time; generally 15° east of the zone of standard meridian.

daylight saving time—A variation of zone time, 1 hour more advanced than standard time.

deflation—The sweeping erosive action of the wind over the ground.

delta—An alluvial deposit, usually triangular in shape, at the mouth of a river, stream, or tidal inlet.

deltaic deposits—Sedimentary deposits in a delta.

delta plain—A plain formed by deposition of silt at the mouth of a stream or by overflow along the lower stream courses.

deposit—Consolidated or unconsolidated material that has accumulated by a natural process.

depositional sequence—A major but informal assemblage of formations, bounded by regionally extensive unconformities at both their base and top and extending over broad areas of continental cratons.

depression—(1) A hollow of any size on a plain surface having no natural outlet for surface drainage. (2) An area of low pressure; usually applied to a certain stage in the development of a tropical cycle.

desert—(1) A wide, open, comparatively barren tract of land with little rainfall. (2) Any waste, uninhabited tract, such as the vast expanse of ice in Greenland.

desert climate—A climate type that is characterized by insufficient moisture to support appreciable plant life; a climate of extreme aridity.

desertification—The creation of desiccated, barren, desertlike conditions due to natural changes in climate or possibly through mismanagement of the semiarid zone.

desert pavement—A mosaic of pebbles and large stones that accumulate as the finer dust and sand particles are blown away by the wind.

detritus—Any loose material removed directly from rocks and minerals by mechanical means.

diastrophism—The general process or combination of processes by which the Earth's crust is deformed.

dike—A tabular body of igneous rock that cuts across massive rock formations.

dike swarm—A large group of parallel, linear, or radially oriented dikes.

dip—The angle that a stratum or fault plane makes with the horizontal.

dipole—Any object or system that is oppositely charged at two points, or poles, such as a magnet or a polar molecule.

discharge—The flow rate of a fluid at a given instant expressed as volume per unit of time.

discharge head—Vertical distance between the intake level of a water pump and the level at which it discharges water freely to the atmosphere.

discharge hydrograph—A graph showing the discharge or flow of a stream with respect to time.

discordance—An unconformity characterized by lack of parallelism between strata that touch.

discordant pluton—An intrusive igneous body that cuts across the bedding or foliation of the intruded formations.

dissolve—To cause to pass into solution.

doldrums—A nautical term for the equatorial trough, with special reference to the light and variable nature of the winds. Also known as equatorial calms.

drift—Rock material picked up and transported by a glacier and deposited elsewhere.

drumlin—A hill of glacial drift or bedrock having a half-ellipsoidal form, with its long axis paralleling the direction of movement of the glacier that fashioned it.

dry season—In certain types of climate, an annually recurring period of one or more months during which precipitation is at a minimum for the region.

dune—A mound or ridge of unconsolidated granular material, usually of sand size and of durable composition (such as quartz), capable of movement by transfer of individual grains entrained by a moving fluid.

dystrophic—Pertaining to an environment that does not supply adequate nutrition.

earthquake—A series of suddenly generated waves in the Earth.

earthquake tremor—See *tremor*.

earthquake zone—An area of the Earth's crust in which movements, sometimes with associated volcanism, occur. Also known as seismic area.

earth resources technology satellite—One of a series of satellites designed primarily to measure the natural resources of the Earth; functions include mapping, cataloging water resources, surveying crops and forests, tracing sources of water and air pollution, identifying soil and rock formations, and acquiring oceanographic data. Abbreviated ERTS.

earth rotation—Motion about the Earth's axis that occurs 365.24 times over a year's period.

earth tide—The periodic movement of the Earth's crust caused by forces of the Moon and Sun.

eccentricity—The distance of the geometric center of a revolving body from the axis of rotation.

eccentric orbit—An orbit of a celestial body that deviates markedly from a circle.

ecliptic—The apparent annual path of the Sun among the stars; the intersection of the plane of the Earth's orbit with the celestial sphere.

eddy—A vortexlike motion of a fluid running contrary to the main current.

effluent—(1) Flowing outward or away from. (2) Liquid that flows away from a containing space or a main waterway.

electromagnetic spectrum—The total range of wavelengths or frequencies of electromagnetic radiation, extending from the longest radio waves to the shortest known cosmic rays.

ellipse—The locus of all points in the plane at which the sum of the distances from a fixed pair of points, the foci, is a given constant.

ellipsoid of revolution—An ellipsoid generated by rotation of an ellipse about one of its axes.

elliptical orbit—The path of a body moving along an ellipse.

embayed—Formed into a bay.

embayed coastal plain—A coastal plain that has been partly sunk beneath the sea, thereby forming a bay.

embayed mountain—A mountain that has been depressed enough for seawater to enter the bordering valleys.

energy flux—A vector quantity whose component perpendicular to any surface equals the energy transported across that surface by some medium per unit area per unit time.

environment—The sum of all external conditions and influences affecting the development and life of organisms.

environmental control—Modification and control of soil, water, and air environments of man and other living organisms.

eolation—Any action of wind on the land.

eolian—Pertaining to the action or the effect of the wind. Also spelled aeolian.

eolian dune—A dune resulting from entrainment of grains by the flow of moving air.

ephemeral stream—A stream channel that carries water only during and immediately after periods of rainfall or snowmelt.

epicenter—A point on the surface of the Earth that is directly above the seismic focus of an earthquake.

epoch—A major subdivision of a period of geologic time.

equinox—(1) Either of the two points of intersection of the ecliptic and the celestial equator, occupied by the Sun when its declination is 0°. (2) That instant when the Sun occupies one of the equinoctial points.

era—A division of geologic time of the highest order comprising one or more periods.

erosion—(1) The loosening and transporting of rock debris at the Earth's surface. (2) The wearing away of the land, chiefly by rain and running water.

escarpment—A cliff or steep slope generally separating two level or gently sloping areas, and produced by erosion or faulting. Also known as *scarp*.

estuarine deposit—A sediment deposited at the head and on the floor of an estuary.

estuarine environment—The physical conditions and influences of an estuary.

estuarine oceanography—The study of the chemical, physical, biological, and geological properties of estuaries.

estuary—A semienclosed coastal body of water that has a free connection with the open sea and within which seawater is measurably diluted with freshwater. Also known as branching bay; drowned river mouth; firth.

eutrophic—Pertaining to a lake containing a high concentration of dissolved nutrients; often with periods of oxygen deficiency.

eutrophication—Of bodies of water, the process of becoming better nourished either naturally by processes of maturation or artificially by fertilization.

evaporation—Conversion of a liquid to the vapor state by the addition of latent heat.

evaporation pan—A pan used in the measurement of the evaporation of water into the atmosphere.

evaporation tank—A tank used to measure the evaporation of water under controlled conditions.

evergreen—Pertaining to a perennially green plant.

exfoliation—Flaking away or peeling off in scales.

extrusion—Emission of magma or magmatic materials at the surface of the Earth.

extrusive rock—See *volcanic rock*.

fall line—(1) The zone or boundary between resistant rocks of older land and weaker strata of newer land. (2) The line indicated by the edge over which a waterway suddenly descends, as in waterfalls.

fault—A fracture in rock along which the adjacent surfaces are differentially displaced.

fault basin—A region depressed in relation to surrounding regions and separated from them by faults.

fault block mountain—See *block mountain*.

faulting—The fracturing and displacement processes that produce a fault.

fault line—Intersection of the fault surface with the surface of any other horizontal surface of reference.

fault line scarp—A cliff produced when a soft rock erodes against hard rock at a fault.

fault plane—A planar fault surface.

fault scarp—A steep cliff formed by movement along one side of a fault.

fault wall—The mass of rock on a particular side of a fault.

fell-field—A community of dwarfed, scattered plants or grasses above the timberline.

firn—Material transitional between snow and ice; it is formed from snow after existing through one summer melt season and becomes glacier ice when the permeability to liquid water drops to zero.

firn line—(1) The regional snow line on a glacier. (2) The line that divides the ablation area of a glacier from the accumulation area.

fjord—A narrow, deep inlet of the sea between high cliffs or steep slopes. Also spelled fiord.

fjord valley—A deep, narrow channel occupied by the sea that extends inland.

flood—The condition that occurs when water overflows the natural or artificial confines of a stream or accumulates by drainage over low-lying areas.

flood flow—Stream discharge during a flood.

floodplain—The relatively smooth valley floors adjacent to and formed by rivers that are subject to overflow.

flood stage—The stage, on a fixed river gage, at which overflow of the natural banks of the stream will occur.

flood wall—A levee or similar wall for the purpose of protecting the land from inundation by flood waters.

focus—The center of an earthquake and the origin of its elastic waves within the Earth.

foehn—A warm, dry wind on the lee side of a mountain range, the warmth and dryness being due to adiabatic compression as the air descends the mountain slopes.

foehn air—The warm dry air associated with foehn winds.

foehn wall—The steep leeward boundary of flat, cumuliform clouds formed on the peaks and upper windward sides of mountains during foehn conditions.

fog—Water droplets suspended in the air in sufficient concentration to reduce visibility appreciably.

fogbank—A fairly well-defined mass of fog observed in the distance, most commonly at sea.

fold—A bend in rock strata or other planar structure, usually produced by deformation; folds are recognized where layered rocks have been distorted into wavelike form.

folding—Compression of planar structure in the formation of fold structures.

fold system—A group of folds with common trends and characteristics.

foliation—A laminated structure formed by segregation of different minerals into layers that are parallel to the schistosity.

footwall—The mass of rock that lies beneath a fault.

foreland—An extensive area of land jutting out into the sea.

foreshore—The zone that lies between the ordinary high- and low-watermarks and is daily traversed by the rise and fall of the tide. Also known as beach face.

forest—An ecosystem consisting of plants and animals and their environment, with trees as the dominant form of vegetation.

forest ecology—The science that deals with the relationship of forest trees to their environment, to one another, and to other plants and to animals in the forest.

fringing reef—A coral reef attached directly to or bordering the shore of an island or continental landmass.

front—A sloping surface of discontinuity in the troposphere that separates air masses of different density or temperature.

frontal cyclone—Any cyclone associated with a front; often used synonymously with wave cyclone or with extratropical cyclone.

frontal fog—Fog associated with frontal zones and frontal passages.

frontal precipitation—Any precipitation attributable to the action of a front.

frontal profile—The outline of a front as seen on a vertical cross section.

frontal system—A system of fronts as they appear on a synoptic chart.

frontal thunderstorm—A thunderstorm associated with a front; limited to thunderstorms resulting from the convection induced by frontal lifting.

frontal zone—The three-dimensional zone or layer of large horizontal density gradient, bounded by frontal surfaces and surface front.

frontogenesis—The initial formation of a frontal zone or front.

frontolysis—The dissipation of a front or frontal zone.

frost point—The temperature to which atmospheric moisture must be cooled to reach the point of saturation with respect to ice.

fungi—Nucleated, usually filamentous, sporebearing organisms devoid of chlorophyll.

fungus—Singular of fungi.

geanticline—A broad land uplift; refers to the land mass from which sediments in a geosyncline are derived.

geographic latitude—A general term applying to either astronomical or geodetic latitude.

geographic longitude—A general term applying to either astronomical or geodetic latitude.

geographic position—The position of a point on the surface of the Earth expressed in terms of geographic coordinates either geodetic or astronomical.

geologic age—Any great time period in the Earth's history marked by special phases of physical conditions or organic development.

geostropic wind—That horizontal wind velocity for which the Coriolis force exactly balances the horizontal pressure force.

geosyncline—A part of the crust of the Earth that sank through time.

geyser—A natural spring or fountain that discharges a column of water or stream into the air at more or less regular intervals.

glacial deposit—A boulder moved to a point distant from its original site by a glacier.

glacial epoch—Any of the geologic epochs characterized by an ice age; thus, the Pleistocene epoch may be termed a glacial epoch.

glacial erosion—Movement of soil or rock from one point to another by the action of the moving ice of a glacier.

glacial till—See *till*.

glaciated terrain—A region that once was covered by great masses of glacial ice.

glaciation—(1) Alteration of any part of the Earth's surface by passage of a glacier, chiefly by glacial erosion or deposition. (2) The transformation of cloud particles from waterdrops to ice crystals, as in the upper portion of a cumulonimbus cloud.

glacier—A mass of land ice, formed by the further recrystallization of firn, flowing slowly from an accumulation area to an area of ablation.

glacier ice—Any ice that is or was once a part of a glacier.

glaze—A coating of ice, generally clear and smooth but usually containing some air pockets, formed on exposed objects by the freezing of a film of supercooled water deposited by rain, drizzle, or fog, or possibly condensed from supercooled water vapor.

Gondwanaland—The ancient continent that is supposed to have fragmented and drifted apart to eventually form the present continents.

graben—A block of the Earth's crust, generally with a length much greater than its width, that has dropped relative to the blocks on either side.

graded stream—A stream in which, over a period of years, slope is adjusted to yield the velocity required for transportation of the load supplied from the drainage basin.

gradient—The rate of descent or ascent (steepness of slope) of any topographic feature, such as streams or hillsides.

grass—The common name for all members of the family Gramineae; monocotyledonous plants having leaves that consist of a sheath that fits around the stem like a split tube, and a long narrow blade.

grassland—Any area of herbaceous terrestrial vegetation dominated by grasses and graminoid species.

gravity fault—See *normal fault*.

Great Basin high—A high-pressure system centered over the Great Basin of the western United States; it is a frequent feature of the surface chart in the winter season.

great circle—A circle, or near circle, described on the Earth's surface by a plane passing through the center of the Earth.

Greenwich mean time—Mean solar time at the meridian of Greenwich. Abbreviated GMT. Also known as Greenwich civil time; *universal time;* zulu time.

Greenwich meridian—The meridian passing through Greenwich, England, and serving as the reference for Greenwich time; it also serves as the origin of measurement of longitude.

Gregorian calendar—The calendar used for civil purposes throughout the world, replacing the Julian calendar and closely adjusted to the tropical year.

ground cover—(1) Prostrate or low plants that cover the ground instead of grass. (2) All forest plants except trees.

groundwater—All subsurface water, especially that part that is in the zone of saturation.

groundwater discharge—Water released from the zone of saturation.

groundwater flow—That portion of the precipitation that has been absorbed by the ground and has become part of the groundwater.

groundwater level—The level below which the rocks and subsoil are full of water.

gymnosperm—The common name for members of the division Pinophyta; seed plants having naked ovules at the time of pollination.

Gymnospermae—The equivalent name for Pinophyta.

hail—Precipitation in the form of balls or irregular lumps of ice, always produced by convective clouds, nearly always cumulonimbus.

hail stage—The thermodynamic process of freezing of suspended water drops in adiabatically rising air with temperature below the freezing point.

hailstone—A single unit of hail, ranging in size from that of a pea to that of a grapefruit.

halosere—The series of communities succeeding one another, from the pioneer stage to the climax, and commencing in salt water or on saline soil.

hanging glacier—A glacier lying above a cliff or steep mountainside.

hanging valley—A valley whose floor is higher than the level of the shore or other valley to which it leads.

hanging wall—The rock mass above a fault plane.

hardpan—A hard, impervious layer of soil clay cemented by insoluble materials.

hard water—Water that contains certain salts, such as those of calcium or magnesium, which form insoluble deposits in boilers and form precipitates with soap.

heat equator—(1) The line that circumscribes the Earth and connects all points of highest mean annual temperature for their longitudes. (2) The parallel of latitude of 10°N, which has the highest mean temperature of any latitude. Also known as *thermal equator.*

heat of vaporization—The quantity of energy required to evaporate 1 mole, or a unit mass, of a liquid, at constant pressure and temperature.

high—An area of high pressure, referring to a maximum of atmospheric pressure in two dimensions (closed isobars) in the synoptic chart, or a maximum height (closed contours) in the constant-pressure chart.

horizon—One of the layers, each of which is a few inches to a foot thick, that make up a soil.

horst—(1) A block of the Earth's crust uplifted along faults relative to the rocks on either side. (2) A mass of the Earth's crust limited by faults and standing in relief.

hot spring—A thermal spring whose water temperature is above 98°F (37°C).

hour—A unit of time equal to 3600 seconds. Abbreviated hr.

hour angle—Angular distance west of a celestial meridian or hour circle.

hour circle—An imaginary great circle passing through the celestial poles on the celestial sphere.

humidity—Atmospheric water vapor content, expressed in any of several measures, such as relative humidity.

humus—The amorphous, ordinarily dark-colored, colloidal matter in soil; a complex of organic matter of plant, animal, and microbial origin.

hydration—The incorporation of molecular water into a complex molecule with the molecules or units of another species; the complex may be held together by relatively weak forces or may exist as a definite compound.

hydraulic grade line—In a closed channel, a line joining the elevations that water would reach under atmospheric pressure.

hydraulic gradient—(1) With regard to an aquifer, the rate of change of pressure head per unit of distance of flow at a given point and in a given direction. (2) The slope of the hydraulic grade line of a stream.

hydrogeology—The science dealing with the occurrence of surface and groundwater, its utilization, and its functions in modifying the Earth, primarily by erosion and deposition.

hydrograph—A graphical representation of stage, flow, velocity, or other characteristics of water at a given point as a function of time.

hydrologic cycle—The complete cycle through which water passes, from the oceans, through the atmosphere, to the land, and back to the ocean. Also known as water cycle.

hydrology—The science that treats the occurrence, circulation, distribution, and properties of the waters of the Earth, and their reaction with the environment.

hydrophyte(1) A plant that grows in a moist habitat. (2) A plant requiring large amounts of water for growth. Also known as *hygrophyte*.

hydrosere—Community in which the pioneer plants invade open water, eventually forming some kind of soil such as peat or muck.

hygrophyte—See *hydrophyte*.

ice—(1) The dense substance formed by the freezing of water to the solid state; has a melting point of 32°F (0°C) and commonly occurs in the form of hexagonal crystals. (2) A layer or mass of frozen water.

ice age—A major interval of geologic time during which extensive ice sheets formed.

ice cap—(1) A perennial cover of ice and snow in the shape of a dome or plate on the summit area of a mountain through which the mountain peaks emerge. (2) A perennial cover of ice and snow on a flat land mass such as an Arctic island.

ice-cap climate—See perpetual frost climate.

ice desert—Any polar area permanently covered by ice and snow.

iced firn—A mixture of glacier ice and firn; firn permeated with meltwater and then refrozen.

ice field—A mass of land ice resting on a mountain region and covering all but the highest peaks.

ice island—A large tabular fragment of shelf ice found in the ocean.

ice-island iceberg—An iceberg having a conical or dome-shaped summit, often mistaken by mariners for ice-covered islands.

icelandic low—(1) The low-pressure center located near Iceland (mainly between Iceland and southern Greenland) on mean charts of sea-level pressure. (2) On a synoptic chart, any low centered near Iceland.

inselberg—A large, steep-sided residual hill, knob, or mountain, generally rocky and bare, rising abruptly from an extensive, nearly level lowland erosion surface in arid or semiarid regions. Also known as island mountain.

insolation—Solar energy received, often expressed as a rate of energy per unit horizontal surface.

interface—The boundary between any two phases: among the three phases (gas, liquid, and solid), there are five types of interfaces: gas-liquid, gas-solid, liquid-liquid, liquid-solid, and solid-solid.

intermittent spring—A spring that ceases flow after a long dry spell but flows again after heavy rains.

intermittent stream—A stream that carries water a considerable portion of the time, but which ceases to flow occasionally or seasonally because bed seepage and evaporation exceed the available water supply.

international date line—An arbitrary line, roughly equal to the 180° meridian, where a date change occurs; if the line is crossed from east to west a day is skipped; if from west to east the same day is repeated.

interstice—A space or volume between atoms of a lattice, or between groups of atoms or grains of a solid structure.

intertidal zone—The part of the littoral zone above low-tide mark.

intertropical convergence zone—The axis, or a portion thereof, of the broad trade-wind current of the tropics; this axis is the dividing line between the southeast trades and the northeast trades (of the Southern and Northern hemispheres, respectively). Also known as equatorial convergence zone; meteorological equator.

intertropical front—The interface or transition zone occurring within the equatorial trough between the Northern and Southern hemispheres. Also known as equatorial front; tropical front.

intrusion—The process of emplacement of magma in preexisting rock.

intrusive—Pertaining to material forced while still in a fluid state into cracks or between layers of rock.

ionosphere—That part of the Earth's upper atmosphere that is sufficiently ionized by solar ultraviolet radiation so that the concentration of free electrons affects the propagation of radio waves.

isobar—A line drawn through all points of equal atmospheric pressure along a given reference surface, such as a constant-height surface (notably mean sea level on surface charts).

isobaric map—A map depicting points in the atmosphere of equal barometric pressure.

isotherm—A line on a chart connecting all points having the same mean summer temperature.

jet—A strong, well-defined stream of compressible fluid issuing from an orifice or moving in a contracted duct.

jet stream—A relatively narrow, fast-moving wind current flanked by more slowly moving currents; observed principally in the zone of prevailing westerlies above the lower troposphere, and in most cases reaching maximum intensity with regard to speed and concentration near the troposphere.

Julian calendar—A calendar replaced by the Gregorian calendar; the Julian year was 365.25 days, the fraction allowed for the extra day every fourth year (leap year). There are 12 months, each 30 or 31 days except for February, which has 28 days or in leap year 29.

Julian day—The number of each day, as reckoned consecutively since the beginning of the present Julian period on January 1, 4713 B.C.; it is used primarily by astronomers to avoid confusion due to the use of different calendars at different times and places; the Julian day begins at noon, 12 hours later than the corresponding civil day.

Jura—See *Jurassic*.

Jurassic—The second period of the Mesozoic era of geologic time.

juvenile water—See *magmatic water*.

Kansas glaciation—The second glaciation of the Pleistocene epoch in North America.

kaolin—(1) Any of a group of clay minerals with a two-layer crystal in which silicon-oxygen and aluminum-hydroxyl sheets alternate; approximate composition is $Al_2O_3 \cdot 2SiO_2 \cdot 2H_2O$. (2) A soft, nonplastic white rock composed principally of kaolin minerals.

kaolinite—$Al_2Si_2O_5(OH)_4$. The principal mineral of the kaolin group of clay minerals; a white-yellowish, high-alumina mineral consisting of sheets of tetrahedrally coordinated silicon linked by an oxygen shared with octahedrally coordinated aluminum.

karst—A topography formed over limestone, dolomite, or gypsum and characterized by sinkholes, caves, and underground drainage.

karst plain—A plain on which karst features are developed.

kelvin—A unit of absolute temperature equal to 1/273 of the absolute temperature of the triple point of water. Symbolized K. Formerly known as degree Kelvin.

kerogen—The complex, fossilized organic material present in sedimentary rocks, especially in shales; converted to petroleum products by distillation.

kilo—A prefix representing 10^3 or 1000. Abbreviated k.

kilogram—The unit of mass in the meter-kilogram-second system.

kilometer—A unit of length equal to 1000 meters. Abbreviated km.

laccolith—A body of igneous rock that has been introduced into sedimentary rocks so that the overlying strata are notably lifted by the force of intrusion.

lake—An inland body of water, small to moderately large, with its surface water exposed to the atmosphere.

lake breeze—A wind, similar in origin to the sea breeze but generally weaker, blowing from the surface of a large lake onto the shores during the afternoon.

lake effect—Generally, the effect of any lake in modifying the weather about its shore and for some distance downwind.

lake effect storm—A severe snowstorm over a lake caused by the interaction between the warmer water and unstable air above it.

lake plain—One of the surfaces of the Earth that represent former lake bottoms; these featureless surfaces are formed by deposition of sediments carried into the lake by streams.

land breeze—A coastal breeze that blows from land to sea and is caused by a temperature difference when the sea surface is warmer than the adjacent land.

landform—All the physical, recognizable, naturally formed features of land, having a characteristic shape; includes major forms such as a plain, mountain, or plateau, and minor forms such as a hill, valley, or alluvial fan.

landscape—The distinct association of landforms that can be seen in a single view.

landslide—The perceptible downward sliding or falling of a relatively dry mass of rock under the influence of gravity.

langley—A unit of energy per unit area equal to 1 gram-calorie per square centimeter.

latent heat of vaporization—See heat of vaporization.

lateral moraine—Drift material that was deposited by a glacier in a valley as the glacier melted.

latitude—Angular distance from a primary great circle or plane, as on the celestial sphere or the Earth.

lattice—A regular periodic arrangement of points in three-dimensional space.

lattice constant—A parameter defining the unit cell of a crystal lattice, that is, the length of one of the edges of the cell or an angle between edges.

Laurasia—A continent that existed in the Northern Hemisphere; it broke up to form the present northern continents about the end of the Pennsylvania period.

Laurentian Shield—A Precambrian plateau extending over half of Canada from Labrador southwest along Hudson Bay and northwest to the Arctic Ocean. Also known as *Canadian Shield*; Laurentian Plateau.

Laurentide ice sheet—A major recurring glacier that at its maximum completely covered North America east of the Rockies from the Arctic Ocean to a line passing through the vicinity of New York, Cincinnati, St. Louis, Kansas City, and the Dakotas.

lava—(1) Molten extrusive material that reaches the Earth's surface through volcanic vents and fissures. (2) The rock mass formed by consolidation of molten rock issuing from volcanic vents and fissures.

lenticular—Of or pertaining to a lens. Having the shape of a double convex lens or a lentil.

levee—A low ridge deposited by a stream along its sides.

limb—The section of an anticline or syncline that is on either side of the axis.

lithification—Conversion of a newly deposited sediment into an indurated rock. Also known as *lithifaction*.

lithosere—A succession of plant communities that originate on rock.

lithosphere—Since the development of plate tectonics theory, a term referring to the rigid upper 100 kilometers of the crust and upper mantle, above the asthenosphere.

Little Ice Age—A period of expansion of mountain glaciers that began about 5500 years ago and extended to as late as A.D. 1550–1850 in some regions.

littoral current—A current produced by wave action, that runs parallel to the shore. Also known as *longshore current*.

loam—Soil mixture of sand, silt, clay, and humus.

loess—An unconsolidated, unstratified silt.

longitude—Angular distance, along the Equator, between the meridian passing through a position and, usually, the meridian of Greenwich.

longshore bar—A ridge of sand, gravel, or mud built by waves and currents, generally parallel to the shore.

longshore current—See *littoral current*.

longshore trough—A long, shallow depression of the sea floor that runs parallel to the shore.

lopolith—A large, floored, intrusive body that is sunken centrally into the shape of a basin.

low—See *depression*.

magma—The molten rock material from which igneous rock is formed.

magmatic water—Water derived from or existing in molten igneous rock or magma. Also known as juvenile water.

magnetopause—A boundary that marks the transition from the Earth's magnetosphere to interplanetary space.

magnetosphere—The region in which the geomagnetic field plays a dominant part in controlling the physical processes that take place.

mantle—The intermediate shell zone of the Earth below the crust and above the core.

maritime air—A type of air whose characteristics are developed over an extensive water surface.

maritime polar air—Polar air that initially possesses similar properties to that of *continental polar air,* but in passing over warmer water becomes unstable with a higher moisture content.

maritime tropical air—The principal type of tropical air, produced over the tropical and subtropical seas.

marsh—A transitional land-water area that is covered, at least part of the time, by estuarine or coastal waters; it is characterized by aquatic and grasslike vegetation.

mass wasting—Downslope transport of loose rock and soil material under the direct influence of gravity.

meander—The deviation in the flow pattern of a current.

meander belt—The zone along the floor of a valley across which a meandering stream periodically shifts its channel.

meandering stream—A stream that has a pattern of successive meanders.

meander plain—The flat area produced by the meandering process.

mediterranean—See *mesogeosyncline*.

mediterranean sea—A deep epicontinental sea that is connected with the ocean by a narrow channel.

Mediterranean Sea—A sea that lies between Europe and Africa and is completely landlocked except for the Strait of Gibraltar.

Mercalli scale—A scale for classifying the magnitude of an earthquake.

meridian—A great circle passing through the poles of the axis of rotation of a planet; a great circle through the geographical poles of the Earth.

mesic—(1) Of or pertaining to a habitat characterized by a moderate amount of water. (2) Of or pertaining to a mesophyte.

mesogeosyncline—A geosyncline between two continents, which is also known as a *mediterranean*.

mesopause—The top of the mesosphere; corresponds to a level of minimum temperature.

mesophyte—A plant that requires a moderate amount of moisture for optimum growth.

mesosere—A sere that originates in a mesic habitat and is characterized by mesophytes.

mesosphere—The atmospheric shell that extends from the top of the stratosphere to the mesopause.

mesotherm—A plant that grows successfully at moderate temperatures.

Mesozoic—The geologic era that falls between the Paleozoic and the Cenozoic; commonly referred to as the Age of Reptiles.

metamorphic rock—A rock formed from preexisting solid rocks by mineralogical, structural, and chemical changes produced in response to changes in temperature, pressure, and shearing stress.

meteoric water—Groundwater that originates in the atmosphere and reaches the zone of saturation by infiltration.

microorganism—A microscopic organism such as bacteria and protozoans.

microtherm—A plant that requires a mean annual temperature range of 0–14°C for optimum growth.

mid-oceanic ridge—A mountain range on the floor of the ocean, extending through the North and South Atlantic oceans, the Indian Ocean, and the South Pacific Ocean.

mid-ocean ridge—See *mid-oceanic ridge*.

millibar—A unit of pressure equal to one-thousandth of a bar. Abbreviated mb.

milliliter—A unit of volume equal to 10^{-3} liter or 10^{-6} cubic meter. Abbreviated ml.

millimeter—A unit of length equal to one-thousandth of a meter. Abbreviated mm.

mineral—A naturally occurring substance with a characteristic chemical composition.

miogeosyncline—The nonvolcanic portion of an orthogeosyncline.

Moho—See *Mohorovicic discontinuity*.

Mohorovicic discontinuity—A seismic discontinuity that separates the crust from the subjacent mantle.

monsoon—A large-scale wind system that strongly influences the climate of a large region and in which the direction of the wind flow reverses from winter to summer.

monsoon low—A seasonal low that is found over a continent in the summer and over the adjacent sea during the winter.

montane—The biogeographic zone composed of moist, cool slopes below the timberline; evergreen trees are the dominant lifeform in this zone.

moraine—An accumulation of glacial drift that is deposited by direct glacial action.

mountain and valley winds—A system of diurnal winds that blow uphill and upvalley by day and downhill and downvalley by night.

mudflow—A flowing mass of fine-grained material that has a high degree of fluidity during movement.

nacreous clouds—Clouds of unknown composition that show very strong irisation similar to that of mother-of-pearl; they occur at heights of about 20 or 30 kilometers.

nautical mile—A unit of distance considered the length of 1 minute of any great circle of the Earth.

nebula—Interstellar clouds of gas or small particles.

needleleaf—A slender-pointed leaf as of the firs and other evergreens.

neve—An accumulation of compacted, granular snow in transition from soft snow to ice.

noctilucent—A cloud of unknown composition that occurs at great heights; it resembles thin cirrus, but usually with a bluish or silverish color, although sometimes orange to red, standing out against a dark night sky.

noon—The instant at which the Sun is over the reference meridian.

normal fault—A fault, usually of 45–90°, in which the hanging wall appears to have shifted downward in relation to the footwall.

nutrient—Providing nourishment.

oasis—An isolated fertile area surrounded by desert and marked by vegetation and a water supply.

occluded cyclone—Any cyclone (or low) within which there has developed an occluded front.

occluded front—A composite of two fronts, formed as a cold front overtakes a warm front or quasi-stationary front. Also known as frontal occlusion; occlusion.

occlusion—See *occluded front.*

ocean basin—A great depression occupied by saltwater.

ocean current—A net transport of ocean water along a definable path.

ocean-floor spreading—See *sea-floor spreading.*

oceanic ridge—See *mid-oceanic ridge.*

oligotrophic—Of a lake, lacking plant nutrients and usually containing plentiful amounts of dissolved oxygen.

orogenic belt—A region that has undergone folding or other deformation during an orogenic cycle.

orogeny—The process or processes of mountain formation, especially the intense deformation of rocks by folding and faulting.

orographic precipitation—Precipitation that results from the lifting of moist air over an orographic barrier such as a mountain range.

orography—The branch of geography dealing with mountains.

orthogeosyncline—A linear geosynclinal belt lying between continental and oceanic cratons, and having internal volcanic belts (eugeosynclinal and external nonvolcanic belts (miogeosynclinal).

outgassing—The release of occluded gases or water vapor.

outwash plain—A broad, flat, or gently sloping alluvial deposit in front of or beyond the terminal moraine of a glacier.

oxbow—A closely looping, U-shaped stream meander.

oxbow lake—The crescent-shaped body of water located alongside a stream in an abandoned oxbow after a cutoff is formed and the ends of the original bends are silted up.

ozone—O_3, which is an unstable blue gas with pungent odor.

ozone layer—See *ozonosphere.*

ozonosphere—The general stratum of the upper atmosphere in which there is an appreciable ozone concentration.

Pacific high—The nearly permanent subtropical high of the North Pacific Ocean.

Pacific Ocean—The largest division of the hydrosphere.

Paleozoic—The era of geologic time from the end of the Precambrian to the beginning of the Mesozoic era.

Pangaea—Supercontinent supposedly composed of all the continental crust of the Earth, and later fragmented into Laurasia and Gondwana.

parallel—A circle on the surface of the Earth, parallel to the plane of the Equator and connecting all points of equal latitude. Also known as *circle of longitude;* parallel of latitude.

peat—A dark-brown residuum produced by the partial decomposition and disintegration of mosses, sedges, trees, and other plants that grow in marshes.

peat bog—A bog in which peat has formed.

perched water—Groundwater that is unconfined and separated from an underlying groundwater by an impermeable bed in the zone of aeration.

perched water table—The water table or upper surface of a body of perched water.

percolation—Gravity flow of groundwater through the pore spaces in rock or soil.

perennial stream—A stream that contains water at all times.

period—A unit of geologic time constituting a subdivision of an era.

permafrost—Perennially frozen ground.

permeability—The capacity of a porous rock, soil, or sediment for transmitting a fluid.

petrification—A fossilization process whereby inorganic matter dissolved in water replaces the original organic materials, converting them to a stony substance.

phacolith—A minor, concordant, lens-shaped intrusion into folded sedimentary strata.

phenocryst—A large, conspicuous crystal in a rock.

phreatic surface—See *water table.*

phreatic zone—See *zone of saturation.*

piedmont glacier—An ice sheet formed at the base of a mountain by the spreading out and coalescing of valley glaciers from higher elevations.

piezometric surface—See *potentiometric surface.*

pine—Any of the cone-bearing trees composing the genus Pinus; characterized by evergreen leaves (needles), usually in tight clusters of two to five.

pingo—A frost mound that resembles a volcano, it is a relatively large and conical mound of soil-covered ice, elevated by hydrostatic pressure of water within or below the permafrost of arctic regions.

Pinophyta—The gymnosperms, a division of seed plants characterized as vascular plants with roots, stems, and leaves, and with seeds that are not enclosed in an ovary but are borne on cone scales or exposed at the end of a stalk.

pipe—A vertical, cylindrical ore body. Also known as chimney; neck; ore chimney; ore pipe; *stock.*

plain—An extensive, broad tract of level or rolling land.

plateau basalt—A basaltic lava flow from fissure eruptions that accumulates to form a plateau. Also known as flood basalt.

plate tectonics—Global tectonics based on a model of the Earth characterized by a number of semirigid plates that float on a viscous underlayer in the mantle.

Pleistocene—An epoch of geologic time of the Quaternary period. Also known as Ice Age.

pluton—An igneous intrusion.

polar circle—A parallel of latitude whose distance from the pole is equal to the obliquity of the ecliptic.

polar climate—The climate of a geographical polar region, most commonly taken to be a climate that is too cold to support the growth of trees.

polar easterlies—The body of easterly winds located poleward of the subpolar low-pressure belt.

polar front—The semipermanent, semicontinuous front separating air masses of tropical and polar origin.

Polaris—A star in the constellation Ursa Minor; marks the north celestial pole.

Polestar—See *Polaris.*

polygonal ground—A ground surface that consists of polygonal arrangements of rock, soil, and vegetation formed on a level or gently sloping surface by frost action.

pond—A small natural body of standing freshwater filling a surface depression.

pore-size distribution—Variations in pore sizes in reservoir formations; each type of rock has its own typical pore size and related permeability.

porosity—(1) Property of a solid which contains many minute channels. (2) The fraction as a percent of the total volume occupied by these channels.

potentiometric surface—An imaginary surface that represents the static head of groundwater and is defined by the level to which water will rise.

prairie—An extensive level-to-rolling, treeless tract of land characterized by deep, fertile soil and a cover of coarse grass and herbaceous plants.

precipitation—Any or all of the forms of water particles, whether liquid or solid, that fall from the atmosphere and reach the ground.

pressure—A stress that is exerted uniformly in all directions; its measure is the force exerted per unit area.

pressure contour—A line connecting points of equal height of a given barometric pressure.

pressure gradient—The change in atmospheric pressure per unit horizontal distance, usually measured along a line perpendicular to the isobars.

pressure gradient-force—The force due to differences of pressure within the atmosphere.

pressure head—Also known as head. The pressure of water at a given point in a pipe arising from the pressure in it.

prevailing westerlies—The prevailing winds on the poleward sides of the subtropical high-pressure belts.

primary wave—The first seismic wave that reaches a station from an earthquake.

prisere—The ecological succession of vegetation that occurs in passing from barren rock or water to a climax community. Also known as primary succession.

Proterozoic—Third era of geologic time.

psammophyte—A plant that thrives on sandy soil.

pumice—A rock froth that is formed by the puffing up of liquid lava by expanding gases that are released from solution in the lava prior to solidification.

quagmire—See *bog.*

quaking bog—A peat bog floating or growing over water-saturated land which shakes or trembles when walked on.

quasi-stationary front—A front that is stationary or nearly so; conventionally, a front that is moving at a speed less than about 5 knots is generally considered to be quasi-stationary.

Quaternary—The second period of the Cenozoic.

quay—A solid embankment parallel to a waterway.

quicksand—A highly mobile mass of fine sand consisting of smooth, rounded grains that are usually thoroughly saturated with upward-flowing water; tends to yield under pressure and to readily swallow heavy objects.

radiation—The emission and propagation of waves transmitting energy through space or through some medium.

radiation fog—Produced over a land area when radiational cooling reduces the air temperature to or below its dew point.

radiosonde—Balloon-borne instruments for the measurement and transmission of meteorological data.

rain—Precipitation in the form of liquid water drops.

rainfall regime—The character of the seasonal distribution of rainfall at any place.

Rayleigh scattering—Scattering of electromagnetic radiation by independent particles that are much smaller than the wavelength of the radiation.

Recent—An epoch of geologic time (late Quaternary) following the Pleistocene; referred to as Holocene.

regional metamorphism—Geological metamorphism affecting an extensive area.

regolith—Unconsolidated rocky debris that overlies bedrock.

relative humidity—The ratio of the actual vapor pressure of the air to the saturation vapor pressure. Abbreviated RH.

relief—The configuration of a part of the Earth's surface, with reference to altitude and slope variations and to irregularities of the surface.

reverse fault—See *thrust fault*.

revolution—The motion of a body around a closed orbit.

Richter scale—A scale of numerical values of earthquake magnitude.

rift—A narrow opening in a rock caused by cracking horst block bounded by normal faults.

rift valley—A deep, central cleft with a mountainous flood in the crest of a mid-oceanic ridge.

rip current—The return flow of water that has been piled up on shore by incoming waves and wind.

rise—A long, broad elevation that rises gently from its surroundings, such as the sea floor.

river—A natural freshwater surface stream having a permanent or seasonal flow and moving toward a sea, lake, or another river in a definite channel.

rockfall—The free fall of newly detached bedrock segments from a cliff or other steep slope.

Rossby wave—A wave on a uniform current in a two-dimensional nondivergent fluid system.

rotation—The motion of a particle about a fixed point.

runoff—(1) Surface streams that appear after precipitation. (2) The flow of water in a stream, usually expressed in cubic feet per second.

saltation—Transport of a sediment in which the particles are moved forward in a series of short, intermittent bounces.

savanna—Any of a variety of environmentally similar vegetation types in tropical and extratropical regions; all contain grasses and one or more species of trees.

savanna-woodland—See *tropical woodland*.

scarp—See *escarpment*.

schist—A large group of coarse-grained metamorphic rocks that readily split into thin plates or slabs as a result of the alignment of lamellar minerals.

schistosity—A type of cleavage characteristic of metamorphic rocks, notably schists and phyllites, in which the rocks tend to split along parallel planes.

sclerophyllous—Characterized by thick, hard foliage.

sea breeze—A coastal, local wind that blows from sea to land, caused by the temperature difference when the sea surface is colder than the adjacent land.

sea-floor spreading—The ocean floor is spreading away from mid-oceanic ridges; the driving forces behind the spreading are convective cells in the Earth's mantle.

sea level—The level of the surface of the ocean.

sea-level pressure—The atmospheric pressure at mean sea level.

seamount—An elevation of the sea floor that is either flat-topped or peaked.

sea salt—The salt remaining after the evaporation of seawater.

secondary wave—See *S wave*.

sediment—A mass of organic or inorganic solid, fragmented material, or the solid fragment itself, that comes from weathering of rock and is carried by, suspended in, or dropped by air, water, or ice.

seismic sea wave—Any elastic sea wave, either body wave or surface wave, produced by earthquakes or by explosions. Also known as tsunami.

seismograph—An instrument that records vibrations in the Earth, especially earthquakes.

shallow-focus earthquake—An earthquake whose focus is located within 70 kilometers of the Earth's surface.

sheeting—The process by which thin sheets, slabs, scales, plates, or flakes of rock are successively broken loose or stripped from the outer surface of a large rock mass.

shelf edge—The demarcation between continental shelf and continental slope.

shingle—Pebbles, cobbles, and other beach material, coarser than ordinary gravel.

shingle beach—A narrow beach composed of shingle and commonly having a steep slope on both its landward and seaward sides.

shore—The narrow strip of land immediately bordering a body of water.

shoreline—The intersection of a specified plane of water, especially mean high water, with the shore; a limit that changes with the tide or water level.

shoreline emergence—A straight or gently curving shoreline formed by the dominant relative emergence of the floor of an ocean or a lake. Also known as emerged shoreline.

shoreline of submergence—A shoreline, characterized by bays, promontories, and other minor features, formed by the dominant relative submergence of a landmass.

sial—A petrologic term for the silica- and alumina-rich upper rock layers of the Earth's crust.

sidereal day—The time between two successive upper transits of the vernal equinox.

sidereal month—The time period of one revolution of the Moon about the Earth relative to the stars.

sidereal period—The length of time required for one revolution of a celestial body about its primary, with respect to the stars.

sidereal time—Time based on diurnal motion of stars.

sidereal year—The time period relative to the stars of one revolution of the Earth around the sun.

sill—(1) Submarine ridge in relatively shallow water that separates a partly closed basin from another basin or from an adjacent sea. (2) A tabular igneous intrusion that is oriented parallel to the planar structure of surrounding rock.

snow—The most common form of frozen precipitation, usually flakes of starlike crystals.

snowfield—A broad, level, relatively smooth and uniform snow cover on ground or ice at high altitudes or in mountainous regions above the snow line.

snowmelt—The water resulting from the melting of snow.

soft water—Water that is free of magnesium or calcium salts.

soil—(1) Unconsolidated rock material over bedrock. (2) Freely divided rock-derived material containing an admixture of organic matter and capable of supporting vegetation.

soil air—The air and other gases in spaces in the soil; specifically, that which is found within the zone of aeration. Also known as soil atmosphere.

soil creep—The slow, steady downhill movement of soil and loose rock on a slope. Also known as surficial creep.

soil erosion—The detachment and movement of topsoil by the action of wind and flowing water.

soil genesis—The mode by which soil originates, with particular reference to processes of soil-forming factors responsible for the development of true soil from unconsolidated parent material. Also known as pedogenesis; soil formation.

soil moisture—See *soil water*.

soil science—The study of the formation, properties, and classification of soil; includes mapping. Also known as pedology.

soil series—A family of soils having similar profiles, and developing from similar original materials under the influence of similar climate and vegetation.

soil structure—Arrangement of soil into various aggregates, each differing in the characteristics of its particles.

soil survey—The systematic examination of soils, their description and classification, mapping of soil types, and the assessment of soils for various agricultural and engineering uses.

soil water—Water in the belt of soil water.

solar constant—The rate at which energy from the Sun is received just outside the Earth's atmosphere on a surface normal to the incident radiation.

solar time—Time based on the rotation of the Earth relative to the Sun.

solifluction—A rapid soil creep, especially referring to downslope soil movement in periglacial areas.

solstice—The two instants during the year when the Earth is so located in its orbit that the inclination (about 23½°) of the polar axis is toward the Sun; the days are June 21 for the North Pole and December 22 for the South Pole.

specific humidity—In a system of moist air, the ratio of the mass of water vapor to the total mass of the system.

spheroid—See *ellipsoid of revolution*.

spit—A small point of land commonly consisting of sand or gravel and which terminates in open water.

spring—A general name for any discharge of deep-seated, hot or cold, pure or mineralized water.

stalactite—A conical or roughly cylindrical speleothem formed by dripping water and hanging from the roof of a cave.

stalagmite—A conical speleothem formed upward from the floor of a cave by the action of dripping water.

stationary front—See *quasi-stationary front*.

stock—See *pipe*.

stratocumulus—A principal cloud type predominantly stratiform, in the form of a gray or whitish layer of patch.

stratopause—The boundary or zone of transition separating the stratosphere and the mesosphere.

stratosphere—The atmospheric shell above the troposphere and below the mesosphere.

stratus—A principal cloud type in the form of a gray layer with a rather uniform base.

streamflow—A type of channel flow, applied to surface runoff moving in a stream.

stream gradient—The angle, measured in the direction of flow, between the water surface and the horizontal.

striation—One of a series of parallel scratches, small furrows, or lines on the surface of a rock or rock fragment.

sublimation—The process by which solids are transformed directly to the vapor state of vice versa without passing through the liquid phase.

subsequent—Referring to a geologic feature that followed in time the development of a consequent feature of which it is a part.

subsequent drainage—Drainage by a stream developed subsequent to the system of which it is a part; drainage follows belts of weak rocks.

subsequent stream—A stream that flows in the general direction of the strike of the underlying strata and is subsequent to the formation of the consequent stream of which it is a tributary.

subsidence—(1) A descending motion of air in the atmosphere. (2) A sinking down of a part of the Earth's crust.

subtropical forest—See *temperate rainforest*.

subtropical high—One of the semipermanent highs of the subtropical high-pressure belts.

succession—A gradual process brought about by the change in the number of individuals of each species of a community and by the establishment of new species populations that may gradually replace the original inhabitants.

summer—The period from the summer solstice about June 21, to the autumnal equinox, about September 22; popularly, and for most climatological purposes, summer is taken to include June, July, and August in the Northern Hemisphere and December, January, and February in the Southern Hemisphere.

summer solstice—(1) The Sun's position in the ecliptic (June 21); also known as the first point of Cancer. (2) The date (June 21) when the greatest northern declination of the Sun occurs.

surficial deposit—Unconsolidated alluvial, residual, or glacial deposits overlying bedrock or occurring on or near the surface of the Earth.

suspended load—The part of the stream load that is carried for a long time in suspension.

swamp—A waterlogged land supporting a natural vegetation of shrubs and trees.

S wave—A seismic body wave propagated in the crust or mantle of the Earth by a shearing motion of materials.

symbiont—A member of a symbiotic pair.

symbiosis—(1) An interrelationship between two different species. (2) An interrelationship between two different organisms in which the effects of that relationship are expressed as being harmful or beneficial.

symmetrical fold—A fold whose limbs have approximately the same angle of dip relative to the axial surface.

syncline—A fold having stratigraphically younger rock material in its core; it is concave upward.

synodic month—A month based on the Moon's phases.

tableland—A broad, elevated, nearly level, and extensive region of land that has been deeply cut at intervals by valleys or broken by escarpments. Also known as continental plateau.

tabular iceberg—An iceberg with clifflike sides and a flat top.

Taconian orogeny—A process of formation of mountains in the latter part of the Ordovician period, particularly in the northern Appalachians.

taiga—A zone of forest vegetation encircling the Northern Hemisphere between the arctic-subarctic tundras in the north and the steppes, hardwood forests, and prairies in the south.

talus—Coarse and angular rock fragment derived from and accumulated at the base of a cliff or steep, rocky slope.

talus slope—A steep, concave slope consisting of an accumulation of talus.

tarn—A landlocked pool or small lake that may occur in a marsh or swamp, or that may occupy a basin amid mountain ranges.

tectonic cycle—The orogenic cycle that relates larger crustal features, such as mountain belts, to a series of stages of development. Also known as geosynclinal cycle.

tectonic patterns—The arrangement of the large structural units of the Earth's crust, such as mountain systems, shields or stable areas, basins, arches, and volcanic archipelagoes.

tectonics—A branch of geology that deals with regional structural and deformational features of the Earth's crust.

temperate rain forest—A vegetation class in temperate areas of high and evenly distributed rainfall characterized by comparatively few species with large populations of each species; evergreens are somewhat short with small leaves, and there is an abundance of large tree ferns.

terminal moraine—An end moraine that extends as a ridge across a glacial valley.

terrace—(1) A narrow coastal strip sloping gently toward the water. (2) A benchlike structure bordering an underwater feature.

terrigenous sediment—Shallow, marine, sedimentary deposits composed of eroded terrestrial material.

Tethys—(1) A sea that existed for extensive periods of geologic time between the northern and southern continents of the Eastern Hemisphere. (2) A composite geosyncline from which many structures of the present Alpine-Himalayan orogenic belt were formed.

thermal equator—See *heat equator.*

thermal spring—A spring whose water temperature is higher than the local mean annual temperature of the atmosphere.

thermosphere—The atmospheric shell extending from the top of the mesosphere to outer space.

thornbush—A vegetation class that is dominated by tall suc-

culents and profusely branching, smooth-barked deciduous hardwoods which vary in density from the mesquite bush in the Caribbean to the open spurge thicket in Central Africa.

thorn forest—A type of forest formation, mostly tropical and subtropical, intermediate between desert and steppe; dominated by small trees and shrubs, many armed with thorns and spines.

thorn scrub—See *thornbush.*

thrust fault—A low-angle fault along which the hanging wall has moved up relative to the footwall.

thundercloud—A convenient and often used term for the cloud mass of a thunderstorm, that is, a cumulonimbus.

till—Unsorted and unstratified drift consisting of a heterogeneous mixture of clay, sand, gravel, and boulders which is deposited by and underneath a glacier. Also known as boulder clay; *glacial till;* ice-laid drift.

timberline—The elevation or latitudinal limits for arboreal growth. Also known as *tree line.*

topsoil—Surface soil, usually corresponding with the A horizon, as distinguished from subsoil.

trade wind—The wind system, occupying most of the tropics, that blows from the subtropical highs toward the equatorial trough.

trade wind desert—An area of very little rainfall and high temperature that occurs where the trade winds or their equivalent blow over land; the best examples are the Sahara and Kalahari deserts.

transcurrent fault—A strike-slip fault characterized by a steeply inclined surface. Also known as transverse thrust.

transpiration—The passage of a gas or liquid (in the form of vapor) through the skin, a membrane, or other tissue.

tree line—See *timberline.*

tremor—A minor earthquake. Also known as *earthquake tremor;* earth tremor.

trench—(1) A narrow, straight, elongate, U-shaped valley between two mountain ranges. (2) A narrow, stream-eroded canyon, gully, or depression with steep sides. (3) A long narrow, deep depression of the sea floor, with relatively steep sides. Also known as submarine trench.

tributary—A stream that feeds or flows into or joins a larger stream or a lake.

tropical rain forest—A vegetation class consisting of tall, close-growing trees, their columnar trunks more or less unbranched in the lower two-thirds, and forming a spreading and frequently flat crown; occurs in areas of high temperature and high rainfall.

tropical savanna—See *tropical woodland.*

tropical scrub—A class of vegetation composed of low woody plants (shrubs), sometimes growing quite close together, but more often separated by large patches of bare ground, with clumps of herbs scattered throughout.

tropical woodland—A vegetation class similar to a forest but with wider spacing between trees and sparse lower strata characterized by evergreen shrubs and seasonal graminoids.

tropical year—A unit of time equal to the period of one revolution of the Earth about the Sun measured between successive vernal equinoxes.

Tropic of Cancer—(1) A small circle on the celestial sphere connecting points with declination 23.45° north of the celestial equator, the northernmost declination of the Sun. (2) A parallel of latitude 23.45° north of the Equator, marking the northernmost latitude at which the Sun reaches its zenith.

Tropic of Capricorn—(1) A small circle on the celestial sphere connecting points with declination 23.45° south of the celestial Equator, the southernmost declination of the Sun. (2) A parallel of latitude 23.45° south of the Equator, marking the southernmost latitude at which the Sun reaches its zenith.

tropopause—The boundary between the troposphere and stratosphere.

troposphere—That portion of the atmosphere from the Earth's surface to the tropopause.

tundra—An area supporting some vegetation between the northern upper limit of trees and the lower limit of perennial snow on mountains.

tundra climate—The climate that produces tundra vegetation; it is too cold for the growth of trees but does not have a permanent snow-ice cover.

turbulent flow—Motion of fluids in which local velocities and pressures fluctuate irregularly, in a random manner. Also known as turbulence.

ultrabasic—Of igneous rock, having a low silica content, as opposed to the higher silica contents of acidic, basic, and intermediate rocks.

ultraviolet—Pertaining to ultraviolet radiation. Abbreviated UV.

umbra—That portion of a shadow that is screened from light rays emanating from any part of an extended source.

undercutting—Erosion of material at the base of a steep slope, cliff, or other exposed rock.

underfit stream—A misfit stream that appears to be too small to have eroded the valley in which it flows.

underground stream—A subsurface body of water flowing in a definite current in a distinct channel.

unfreezing—The upward movement of stones to the surface as a result of repeated freezing and thawing of the containing soil.

universal time—See *Greenwich mean time*.

upper air—The region of the atmosphere that is above the lower troposphere; although no distinct lower limit is set, the term is generally applied to levels above that at which the pressure is 850 millibars.

upper-air observation—A measurement of atmospheric conditions aloft, above the effective range of a surface weather observation. Also known as sounding; upper-air sounding.

urban heat island—Increased urban temperatures of 1–2°C higher for daily maxima and 1–9°C lower for daily minima compared to rural environs.

U-shaped valley—A type of valley with a broad floor and steep walls produced by glacial erosion.

U valley—See *U-shaped valley*.

vadose water—Water in the zone of aeration.

vadose zone—See *zone of aeration*.

valley breeze—A gentle wind blowing up a valley or mountain slope in the absence of cyclonic or anticyclonic winds, caused by the warming of the mountainside and valley floor by the Sun.

valley glacier—A glacier that flows down the walls of a mountain valley.

Van Allen radiation belt—One of the belts of intense ionizing radiation around the Earth formed by high-energy charged particles that are trapped by the geomagnetic field.

vapor pressure—The partial pressure of water vapor in the atmosphere.

volcanic ash—Fine pyroclastic material; particle diameter is less than 4 millimeters.

volcanic bombs—Pyroclastic ejecta; the lava fragments, liquid or plastic at the time of ejection, acquire rounded forms, markings, or internal structure during flight or upon landing.

volcanic foam—See *pumice*.

volcanic glass—Natural glass formed by the cooling of molten lava, or one of its liquid fractions, too rapidly to allow crystallization.

volcanic mud—Sediment containing large quantities of ash from a volcanic eruption, mixed with water.

volcanic rock—Finely crystalline or glassy igneous rock resulting from volcanic activity at or near the surface of the Earth. Also known as extrusive rock.

volcanic vent—Subcircular bodies composed of lava and fragmented volcanic rock; contacts are generally steep, and vents range in diameter from a few tens to thousands of feet.

volcanism—The movement of magma and its associated gases from the interior into the crust and to the surface of the Earth. Also known as volcanicity.

volcano—A vent in the surface of the Earth through which magma and associated gases and ash erupt; the structure produced by the ejected material is usually in the form of a cone.

wave-cut bench—A level or nearly level narrow platform that is produced by wave erosion; it extends outward from the base of a wave-cut cliff.

warm air mass—An air mass that is warmer than the surrounding air; an implication that the air mass is warmer than the surface over which it is moving.

water table—The planar surface between the zone of saturation and the zone of aeration.

weather—The state of the atmosphere, mainly with respect to its effects on life and human activities; as distinguished from climate, weather consists of the short-term (minutes to months) variations of the atmosphere.

weather forecast—A forecast of the future state of the atmosphere with specific reference to one or more associated weather elements.

weathering—Physical disintegration and chemical decomposition of earthy and rocky materials on exposure to atmospheric agents, producing an in-place mantle of waste.

weather map—A chart portraying the state of the atmospheric circulation and weather at a particular time over a wide area; it is derived from a careful analysis of simultaneous weather observations made at many observing points in the area.

wind—The motion of air relative to the Earth's surface; usually means horizontal air motion.

wind velocity—The speed and direction of wind.

windward—In the general direction from which the wind blows.

winter—The period from the winter solstice, about December 22, to the vernal equinox, about March 21; popularly and for most meteorological purposes, winter is taken to include December, January, and February in the Northern Hemisphere and June, July, and August in the Southern Hemisphere.

winter solstice—(1) The Sun's position on the ecliptic (about December 22). Also known as first point of Capricorn. (2) The date (December 22) when the greatest southern declination of the Sun occurs.

xeric—(1) Of or pertaining to a habitat having a low or inadequate supply of moisture. (2) Of or pertaining to an organism living in such an environment.

xerophyte—A plant adapted to life in areas where the water supply is limited.

xerosere—A temporary community in an ecological succession on dry, sterile ground such as rock, sand, or clay.

x-radiation—See *x-rays*.

x-rays—A penetrating electromagnetic radiation, usually generated by accelerating electrons to high velocity and suddenly stopping them by collision with a solid body.

Yarmouth interglacial—The second interglacial stage of the Pleistocene epoch in North America, following the Kansan glacial stage and before the Illinoian.

year—Any of several units of time based on the revolution of the Earth about the Sun; the tropical year to which the calendar is adjusted is the period required for the Sun's longitude to increase 360°; it is about 365.24220 mean solar days. Abbreviated yr.

yellow mud—Mud containing sediment having a characteristic yellow color, resulting from certain iron compounds.

Yellow Sea—An inlet of the Pacific Ocean between northeastern China and Korea.

Yucatan Current—A rapid northward-flowing current along the western side of the Yucatan Strait; generally loops to the north and exits as the Florida Current.

Zodiac—A band of the sky extending 8° on each side of the ecliptic, within which the Moon and principal planets remain.

zodiacal constellations—The constellations Aries, Taurus, Gemini, Cancer, Leo, Virgo, Libra, Scorpio, Sagittarius, Capricorn, Aquarius, and Pisces, which are assigned to 12 equal portions of the zodiac.

zonation—(1) Arrangement of organisms in biogeographic zones. (2) The condition of being arranged in zones.

zone meridian—The meridian used for reckoning zone time; this is generally the nearest meridian whose longitude is exactly divisible by 15°.

zone noon—Twelve o'clock zone time, or the instant the mean Sun is over the upper branch of the zone meridian.

zone of aeration—A subsurface zone containing water below atmospheric pressure and air or gases at atmospheric pressure.

zone of maximum precipitation—In a mountain region, the belt of elevation at which the annual precipitation is greatest.

zone of saturation—A subsurface zone in which water fills the interstices and is under pressure greater than atmospheric pressure. Also known as *phreatic zone*.

zone time—The local mean time of a reference or zone meridian whose time is kept throughout a designated zone.

Z time—See *Greenwich mean time*.

A Reference Guide
for Additional Readings

Abbotts, John: Radioactive Waste: A Technical Solution, *The Bulletin*, October 1979.

Abelson, Philip: Metallogenesis in Latin America, *Science*, March 14, 1980.

Acid From the Skies: Corrosive Rain Has Become an Insidious Menace, *Time*, March 17, 1980.

Adams, John: Contemporary Uplift and Erosion of the Southern Alps, New Zealand, *Geological Society of America Bulletin*, January 1980.

Adler, Cy: Change and Survival, *Science Digest*, Spring 1980 (Special).

Ahrends, Thomas: Dynamic Compression of Earth Materials, *Science*, March 7, 1980.

And Now It Is Pond Power: Salty Pools May Soon Provide Some Cheap Energy, *Time*, February 25, 1980.

Anderson, Alan; Mid-Atlantic Ridge: Diving to the Birth-Place of the Ocean, *Science Digest*, February 1975.

Andrews, James: Morphologic Evidence for Reorientation of Sea-Floor Spreading in the West Philippine Basin, *Geology*, March 1980.

Antaractic Sea Ice May Herald Ice Age, *Science News*, February 3, 1979.

Antarctica's Icy Assets, *Newsweek*, October 3, 1977.

Are U.S. Cities Ready for a Major Quake? *U.S. News and World Report*, March 3, 1980.

Atmospheric Mixing, *Environmental Science and Technology*, January 1, 1980.

Austin, Daniel: Florida's Tall Palms: The Beginning of the End? *Horticulture*, February 1979.

Baird, Thomas: March Snails: The Gentle Grazers, *Sea Frontier*, November 1979.

Barks, Edward: A Ray of Hope for Pine Barrens, *The New York Times*, May 14, 1978.

Barlow, Thomas: The Giveaway in the National Forests, *The Living Wilderness*, December 1979.

Barnett, Dick: Coal: The View from Down Below, *Mechanix Illustrated*, March 1980.

Barrow, John: The Structure of the Early Universe, *Scientific American*, April 1980.

Baxter, Patricia: Pine Barrens Under Pressure, *Parks and Recreation*, October 1977.

Beck, Eckardt, Environmental Protection in an Energy-Conscious Economy, *USA Today*, March 1980.

Beck, Robert: Production Flat; Demand Imports Off, *Oil and Gas Journal*, January 28, 1980.

Begley, Sharon: Guessing the Weather, *Newsweek*, December 10, 1979.

Berenbaum, May: The Cabbage White Butterfly: An American Success Story, Unfortunately, *Horticulture*, July 1979.

Berner, Robert: Dissolution of Pyroxenes and Amphiboles During Weathering, *Science*, March 14, 1980.

Blockwelder, Blake: Late Wisconsinan Sea Levels on the Southeast U.S. Atlantic Shelf Based on In-Place Shoreline Indications, *Science*, May 11, 1979.

Borgeson, Lillian: When the Heavens Go Berserk, *National Wildlife*, February/March 1980.

Brander, James: Seismic Risk in Britain, *New Scientist*, February 21, 1980.

Breeding Ground, *Reader's Digest*, February 1980.

Britton, Peter: Geothermal Goes East, *Popular Science*, February 1979.

Brower, Kenneth: Earth, *Omni*, December 1979.

Brower, Kenneth: Who'll Stop the Rain? *Omni*, March 1980.

Brown, Gwendolyn: North Africa Develops Scarce Water Resources, *Business America*, March 10, 1980.

Brown, Harriet: Pumping Uranium from Underground Wells, *Popular Science*, March 1980.

Bruce, J.: Somali Current: Recent Measurements during the Southwest Monsoon, *Science*, July 1, 1977.

Bryden, Harry, Heat Transport by Currents Across 25° N Latitude in the Atlantic Ocean, *Science*, February 22, 1980.

Bryson, Reed: Volcanic Activity and Climatic Change, *Science*, March 7, 1980.

Buckwaller, Len: Deadly Clash of the Winds, *Popular Mechanix*, March 1980.

Bull, William: Impact of Mining Gravel from Urban Stream Beds in the Southwestern United States, *Geology*, April 1974.

Cadet, D.: Low-Level Air Flow over the Western Indian Ocean as Seen from Meteostat, *Nature*, April 5, 1979.

Callahan, John: Florida's Borrowed Beaches, *Oceans*, March 1980.

Callis, Linwood: Ozone and Temperature Trends Associated with the 11-Year Solar Cycle, *Science*, June 1979.

Carr, Michael: Volcanic Activity and Great Earthquakes at Convergent Plate Margins, *Science*, August 12, 1979.

Carr, Robert: The Most Extravagant Bloom in 70 Years, *National Wildlife*, January 1980.

Carter, Luther: Carter and the Environment, *Science*, March 14, 1980.

Cerra, Frances: Suffolk Plan To Save Farms Is Still at Issue after 4 Years, *The New York Times*, April 2, 1980.

Challenger Rounds the Horn, *Science News*, March 15, 1980.

Changnon, Stanley: Rainfall Changes in Summer Caused by Stratospheric Lows, *Science*, July 27, 1979.

Chapin, Georganne: A Bitter Harvest, *The Progressive*, March 1980.

Chapman, Stephen: Gasaholics Agronomous, *The New Republic*, June 30, 1979.

Cheh, Albert: Nonvolatile Mutagens in Drinking Water: Production by Chlorination and Destruction by Sulfite, *Science*, January 4, 1980.

Chelminski, Rudolph: A Fungus Beats the Chestnut Blight at Its Own Game, *Audubon*, August 1979.

Chemical Solar Conversation, *Environmental Science and Technology*, March 3, 1980.

Choukroune, P.: In Situ Structural Observations along Transform Fault A in the Famous Area, Mid-Atlantic Ridge, *Geological Society of America Bulletin*, July 1978.

Christian, R.: Resistance of the Microbial Community within Salt Marsh Soils to Selected Perturbations, *Ecology*, Autumn 1978.

Clayton, Lee: Intersecting Minor Lineations on Lake Agassiz Plain, *Journal of Geology*, April 29, 1963.

Coal and the Coming Superinterglacial, *Science News*, June 14, 1977.

Coats, Robert: The Road to Erosion, *Environment*, February 1978.

Cogley, J.: Albedo Contrast and Glaciation Due to Continental Drift, *Nature*, June 21, 1979.

Cohen, Tedd: Millions for Boll Weevils? *Forbes*, March 3, 1980.

Collins, Peter: World Climate Conference Turns to the Weather, *Nature*, March 1, 1979.

Coombs, W.: Swimming Ability of Carnivorous Dinosaurs, *Science*, March 14, 1980.

Cornett, R.: Hypolimnetic Oxygen Deficits: Their Prediction and Interpretation, *Science*, August 10, 1979.

Cromie, William: Probing the Ocean's Metal Factories, *Sciquest*, December 1979.

Cromle, William: Which Is Riskier—Windmills or Reactors? *Sciquest*, March 1980.

Cronan, Christopher: Aluminum Leaching Response to Acid Precipitation: Effects on High-Elevation Watersheds in the Northeast, *Science*, April 20, 1979.

Cronin, Ron: Mosses, *New West*, March 24, 1980.

Cross, Robert: Smuggled in by the Wind, *The Conservationist*, April 1979.

Crossman, Elizabeth: Lobster Technology: A Grip on the Future, *The New York Times*, March 19, 1980.

Dale, Barrie: Toxicity in Resting Cysts of the Red-Tide Dinoflagellate Gonyaulax Excavata from Deeper Coastal Water Sediments, *Science*, September 29, 1978.

Davies, Teresa: Grasses More Sensitive to SO_2 Pollution in Conditions of Low Irradiance and Short Days, *Nature*, April 3, 1980.

Davies, Thomas: Core Drilling Through the Ross Ice Shelf (Antarctica), *Science*, March 28, 1980.

Davis, George: Geological Development of the Cordilleran Metamorphic Core Complexes, *Geology*, March 1979.

De Angeles, Dick: Hurricanes David and Frederic, *Weatherwise*, October 1979.

Deindorfer, Robert: Can We Make the Earth Green Again? *Parade*, January 13, 1980.

Devonshire, Brian: New Zealand's Forests, *American Forest*, August 1976.

Diaz, Henry: Three Extraordinary Winters, *Weatherwise*, February 1980.

Digging in to Turn the Tide, *National Wildlife*, May 1980.

Dolan, Robert: Barrier Islands, *American Scientist*, February 1980.

Dolan, Robert: Shoreline Erosion Rates along the Middle Atlantic Coast of the United States, *Geology*, December 1979.

Dowdell, R.: Losses of Nitrous Oxide Dissolved in Drainage Water from Agricultural Land, *Nature*, March 22, 1979.

Drummond, A.: Carbon Monoxide: The Hidden Threat, *Sciquest*, March 1980.

Duever, Linda: Dry Season, Wet Season: Annual Cycle of Life in Corkscrew Swamp, Florida, *Audubon*, November 1978.

Durham, Megan: Survey Shows Americans Support Wildlife Conservation, *National Parks and Conservation Magazine: The Environmental Journal*, March 1908.

Earth's Creeping Deserts, *National Geographic*, November 2, 1979.

The Earth's Evolving Atmosphere, *Chemistry*, February 1979.

Edelhart, Mike: New Sun, *Omni*, March 1980.

Edwards, Mike: Along the Great Divide, *National Geographic*, October 1979.

Einsele, Gerhard: Intrusion of Basaltic Silts into Highly Porous Sediments, and Resulting Hydrothermal Activity, *Nature*, January 31, 1980.

Elmore, R.: Black Shell Turbidite, Hatteras Abyssal Plain, Western Atlantic Ocean, *Geological Society of America Bulletin*, December 1979.

Elthon, Don: High Magnesia Liquids as the Parental Magma for Ocean Floor Basalts, *Nature*, April 2, 1979.

Endangered Fish Pose Problem in Developing Nevada's Deserts, *The New York Times*, March 2, 1980.

Eriksen, Mary: The Future of the Desert, *Sierra*, June 1978.

Erome, Lawrence: Diffusing the PCB Time Bomb, *Oceans*, May 1979.

Fabcan, P.: The August 1972 Solar Proton Event and the Atmospheric Ozone Layer, *Nature*, February 8, 1979.

Fighting the Spreading Sands, *Science Quest*, February 1980.

Five Million Urge U.S.-Canada Action on Acid Rain, *The Environmental Journal*, January 1980.

Forests, *National Wildlife*, March 1980.

Francillon, Gerard: Sumatra's Swampland Pioneers, *UNESCO Courier*, February 1978.

Franklin, Ben: A Rare Tidewater Blizzard Leaves Slush and Delays, *The New York Times*, March 5, 1980.

Friendly, David: Environmental Mediators, *Newsweek*, March 17, 1980.

Frisch, Joan: Wiring the Woods for Weather, *NOAA*, February 1980.

Galloping Glacier, *Omni*, March 1980.

Gambrell, Robert: Chemical Availability of Mercury, Lead, and Zinc in Mobile Bay Sediment Suspensions as Affected by pH and Oxidation-Reduction Conditions, *Environmental Science and Technology*, April 4, 1980.

Garmon, Linda: To Fish in Troubled Waters, *Science News*, March 8, 1980.

Gedicks, Al: Exxon, Copper, and the Sokaogon, *The Progressive*, February 1980.

Gentile, Arthur: Mining the Deep Sea, *Bioscience*, August 1979.

George, Uwe: An Ocean Is Born, *Reader's Digest*, September 1979.

Gilbert, Gil: A Range of Diversity: Arizona's Huachuca Mountains, *Sports Illustrated*, February 24, 1979.

Glass, Norman: Mounting Acid Rain, *EPA Journal*, November 1979.

Gleason, Patrick: Radiometric Evidence for Involvement of Floating Islands in the Formation of Florida Everglades Tree Islands, *Geology*, April 1980.

Gorove, Stephen: The Geostationary Orbit: Issues of Law and Policy, *American Journal of International Law*, July 1979.

Graham, Roberta: At Poland Spring, Sparkling Waters Run Fast, *Nation's Business*, March 1980.

Graham, S.: Evidence of 115 Kilometers of Right Slip on the San Gregorio–Hosgri Fault Trend, *Science*, January 13, 1978.

Grain Becomes a Weapon, *Newsweek*, February 18, 1980.

Graves, William: Earthquake, *National Geographic*, July 1964.

Gray, Robert: Piney Woods Workday, *Exploring*, February 1980.

Gresens, Randall: Deformation of the Wenatchee Formation and Its Bearing on the Tectonic History of the Chiwaukum Graben, Washington, During Cenozoic Time, *Geological Society of America Bulletin,* January 1980.

Grippi, J.: Submarine-Canyon Complex among Cretaceous Island–Arc Sediments, Western Jamaica, *Geological Society of America Bulletin,* March 1980.

Grover, Kathryn: Lake Weeds Can Be Controlled, *American City and County,* March 1980.

Groves, K.: Simultaneous Effects of CO_2 and Chlorofluoromethanes on Stratospheric Ozone, *Nature,* July 12, 1979.

The Growing Furor over Acid Rain, *U.S. News and World Report,* November 19, 1979.

Haberman, Clyde: Blasting of Rock at Niagara Falls Set for Summer, *The New York Times,* January 31, 1980.

Hallowell, Christopher, Trap or Die, *Science Digest,* Spring 1980 (Special).

Hamblin, Dora: Experiment in Tunisia Effects Pause in Spread of Sahara's Expanding Sands, *Smithosonian,* August 1979.

Hammond, Allen: Project Famous: Exploring the Mid Atlantic Ridge, *Science,* March 7, 1975.

Hance, William: The Sahara: Minerals and Energy, *Focus,* April 1979.

Hard Times Come to Environmentalists, *U.S. News and World Report,* March 10, 1980.

Harris, H.: Dynamic Chemical Equilibrium in a Polar Desert Pond: A Sensitive Index of Meteorological Cycles, *Science,* April 20, 1979.

Hartline, Beverly: Coastal Upwelling: Physical Factors Feed Fish, *Science,* April 4, 1980.

Hastings, David: Geological Structure and Evolution of the Keta Basins, West Africa, *Geological Society of America Bulletin,* September 1979.

Hathaway, John: U.S. Geological Survey Core Drilling on the Atlantic Shelf, *Science,* November 2, 1979.

Hayes, Denis: What's Ahead for Solar Energy? *U.S. News and World Report,* March 3, 1980.

Hays, J.: Variations in the Earth's Orbit: Pacemaker of the Ice Ages, *Science,* December 10, 1976.

Hebert, Paul: A Normal Year for Hurricanes, *Weatherwise,* February 1980.

Hedberg, Hollis: Ocean Floor Boundaries, *Science,* April 13, 1979.

Hejkal, Thomas: Survival of Poliovirus within Organic Solids During Chlorination, *Applied and Environmental Microbiology,* July 1979.

Hepper, F.: Plants that Live in Darwin's Islands: Flora of Galapagos, *Geographical Magazine,* March 1979.

Hershey, Robert: Hull's Fishing Industry Dying, *The New York Times,* March 13, 1980.

Hiels, Merrill: How Molecular Biology Is Spawning an Industry, *Newsweek,* March 17, 1980.

High, Colin: New England Returns to Wood, *Natural History,* February 1980.

Hoff: Taelings' End: Cleaner Days for Silver Bay, *Time,* March 31, 1980.

Holden, Constance: Egyptian Geologist Champions Desert Research, *Science,* September 28, 1979.

Holden, Constance: Rain Forests Vanishing, *Science,* April 25, 1980.

Hoyt, Douglas: Climatic Change and Solar Variability, *Weatherwise,* April 1980.

Hollie, Pamela: Seventy Days and Seventy Nights and Lake Still Rises, *The New York Times,* March 23, 1980.

Hollis, Ted: Rare Wetlands in a National Park, *Geographical,* February 1980.

Hough, Jack: Correlation of Glacial Lake Stages in the Huron-Erie and Michigan Basins, *Journal of Geology,* December 28, 1964.

Houston, Jourdan: The Plant Patrol, *Horticulture,* October 1979.

Howarth, Robert: Pyrite: Its Rapid Formation in a Salt Marsh and Its Importance in Ecosystem Metabolism, *Science,* January 5, 1979.

Hoyt, Douglas: An Empirical Determination of the Carbon Dioxide Greenhouse Effects, *Nature,* November 22, 1979.

Huebner, Albert: Forests: Lungs of the World, *The Humanist,* January 1980.

Huff, Warren: Mineralogy and Provenance of Pleistocene Lake Clay in the Alpine Region, *Geological Society of America Bulletin,* September 1974.

Hughes, Patrick: The Year without a Summer, *Weatherwise*, June 1979.

Idso, Sherwood: The Climatological Significance of a Doubling of Earth's Atmospheric Carbon Dioxide Concentration, *Science*, March 28, 1980.

Inflation Makes the Garden Grow, *U.S. News and World Report*, March 10, 1980.

Irvine, W.: Thermal History, Chemical Composition and Relationship of Comets to the Origin of Life, *Nature*, February 21, 1980.

Is the Sun Shrinking? *Science News*, September 20, 1979.

Janson, Donald: U.S. Legislation Aids Effort To Save Jersey's Pinelands, *The New York Times*, November 10, 1978.

Jeffrey, David: Preserving America's Last Great Wilderness, *National Geographic*, June 1975.

Jerome, Lawrence: Marsh Restoration, *Oceans*, January 1979.

Jordan, Thomas: The Deep Structure of the Continents, *Scientific American*, January 1979.

A Journey to the Eye of Hurricane Frederic, *U.S. News and World Report*, September 24, 1979.

Kerr, Richard: The Bits and Pieces of Plate Tectonics, *Science*, March 7, 1980.

Kerr, Richard: Climate Control: How Large a Role for Orbital Variations? *Science*, July 14, 1978.

Kerr, Richard: Concern Rising about the Next Big Quake, *Science*, February 24, 1980.

Kerr, Richard: Earthquake Prediction: Mexican Quake Shows One Way to Look for the Big Ones, *Science*, March 2, 1979.

Kerr, Richard: Plate Tectonics: Hot Spot Implicated in Ridge Formation, *Science*, November 3, 1978.

Kerr, Richard: Prospects for Earthquake Prediction Wane, *Science*, November 2, 1979.

Kerr, Richard: Weather Modification: A Call for Tougher Tests, *Science*, November 1979.

Kerr, Richard: When Disaster Rains Down from the Sky, *Science*, November 16, 1979.

Killingley, J.: Migrations of California Gray Whales Tracked by Oxygen-18 Variations in Their Epizoic Barnacles, *Science*, February 15, 1980.

King, Lester: Gondwanaland, Reunited, *Geology*, February 1980.

Ko, W.: Characteristics of Bacteriostases˙ in Natural Soils, *Journal of General Microbiology*, April 1977.

Krieger, James: Geothermal Energy Stirs Worldwide Action, *Chemical and Engineering*, June 9, 1975.

Labotka, Theodore: Stratigraphy, Structure, and Metamorphism in the Central Panamint Mountains, Death Valley Area, California, *Geological Society of American Bulletin*, March 1980.

The Lagging Cleanup of Great Lakes Pollution, *Business Week*, May 29, 1978.

Lansford, Henry: Tree Rings: Predictors of Drought? *Weatherwise*, October 1979.

Larsen, R.: Plutonium in Drinking Water: Effects of Chlorination on Its Maximum Permissible Concentration, *Science*, September 15, 1978.

The Last Straw: Battle at Storm King, *Readers Digest*, February 1980.

Laycock, George: Main Drag to Prudhoe Bay, *Audubon*, March 1979.

Lee, Douglas: Hokkaido: Japan's Last Frontier, *National Geographic*, January 1980.

Lehn, Waldemar: How a Foreshortened Arctic Led Norse Seamen to New World, *Science Digest*, April 1980.

Lewis, Harold: The Safety of Fission Reactors, *Scientific American*, March 1980.

Likens, Gene: Acid Rain, *Scientific American*, October 1979.

The Limits to the Weather's Predictability, *Mosaic*, December 1979.

Lindstedt, Stan: Desert Shrews, *National Wildlife*, February 1980.

Lindsley, E.: Heavy Crude, *Popular Science*, December 1979.

Loftness, Vivian: Climate and Architecture, *Weatherwise*, December 1978.

London Fights Off Disaster: A Giant Bulwark Rises on the Thames as a Flood Barrier, *Time*, February 11, 1980.

Ludlum, David: The Climate of Ireland, *Weatherwise*, April 1979.

MacDonald, Gordon: The Impact of Carbon Dioxide on Climate, *The Physics Teacher*, September 1979.

MAGSAT Takes Off, *Science World*, February 7, 1980.

McCann, William: Yakatata Gap, Alaska: Seismic History and Earthquake Potential, *Science*, March 21, 1980.

McCaslin, John: New Finds Heat Appalachian Basin Interest, *Oil and Gas Journal*, February 11, 1980

McCaslin, John: Oregon Hydrocarbons Hunt Broadens, *Oil and Gas Journal*, April 7, 1980.

McCaslin, John: Tests to Probe Undrilled South Dakota Land, *Oil and Gas Journal*, March 10, 1980.

McKenzie, Garry: Determination of Soil-Bank Erosion Rates in Ohio by Using Inter-bank Sediment Accumulations, *Geology*, August 1978.

McWethy, Jack: South Pole Diary, *U.S. News and World Report*, February 28, 1977.

McWhinnie, Mary: The High Importance of the Lowly Krill, *Natural History*, March 1980.

Maran, Stephen: Meteorites—How Science Reads Them for New Clues to the Birth of the Solar System, *Popular Science*, April 1980.

Marrero, Jose: Flash Floods, *Weatherwise*, February 1980.

Martin, Terry: How Can We Protect Southwestern National Parks? *National Parks & Conservation Magazine: The Environmental Journal*, March 1980.

Maugh, Thomas: Mining Could Increase Petroleum Reserves, *Science*, March 21, 1980.

Maugh, Thomas: Restoring Damaged Lakes, *Science*, February 2, 1979.

Maugh, Thomas: The Threat to Ozone Is Real, Increasing, *Science*, December 7, 1979.

Maugh, Thomas: Work on U.S. Oil Sands Heating Up, *Science*, March 14, 1980.

Meislin, Richard: Suffolk Will Be Warned That a Pesticide May Be Contaminating Water in Its Wells, *The New York Times*, March 4, 1980.

Menon, K.: Mystery of the Monsoon, *Oceans*, July 1976.

Mercer, James: Heat-Transport Processes in the Earth, *Geology*, March 1979.

Miller, Charles: Life Around a Lily Pad, *National Geographic*, January 1980.

Miller, James: Snow Drainage in the Black Hills, *Weatherwise*, June 1979.

Miller, Judith: As Eruption Nears Scientists Gather Round, *The New York Times*, January 29, 1980.

Miller, Mark: Washington's Yakima Valley, *National Geographic*, November 1978.

Miller, Maynard: Our Restless Earth, *National Geographic*, July 1964.

Mills, Hugh: Estimated Erosion Rates on Mount Rainier, Washington, *Geology*, July 1976.

Mitchell, J.: Jovian Cloud Structure, *Nature*, August 30, 1979.

Moffat, Anne: How Plants Make War, *Horticulture*, February 1978.

Moore, Gerald: Los Angeles Is Burning, *Reader's Digest*, August 1979.

Morrow, Lance: The Wonderful Art of Weathercasting, *Time*, March 17, 1980.

Morse, John: Magnesium Interaction with the Surface of Calcite in Seawater, *Science*, August 31, 1979.

Much of South Still Snowed Under. Freeze Hits Citrus Crops in Florida, *The New York Times*, March 4, 1980.

Myers, Norman: Tropical Rain Forests: Whose Hand Is on the Axe? *National Parks and Conservation Magazine*, November 1979.

National Academy of Science: The Sun, Weather and Climate, *Weatherwise*, March 11, 1980.

Nature Adds Acres to California, *Popular Mechanics*, April 1980.

New Theory Restructures Earth's Interior, *Science News*, December 1, 1979.

Newell, Reginald: Climate and the Ocean, *American Scientist*, August 1979.

Nicoles, C.: Environmental Fluctuation Effects on the Global Energy Balance, *Nature*, September 13, 1979.

Nightmare in Southern California: Torrential Rains Sweep in from Pacific, *Time*, March 3, 1980.

1980: A Good Year for Barrier Islands? *National Parks*, April 1980.

Norman, Geoffrey: The Firewood Obsession, *Esquire*, December 1979.

Novak, Daniel: Application of Laser Techniques in Rain Impact and Erosion Investigations, *Optical Engineering*, April 1979.

Nreagu, Jerome: Trace Metals in Humic and Fulvic Acids from Lake Ontario Sediments, *Environmental Science and Technology*, April 4, 1980.

Obstacles to Geothermal Development, *Environmental Science and Technology,* March 3, 1980.

Oliver, John: The Enigma of Climate, *USA Today,* March 1979.

Ortner, Everett: The Liberated House: It Rolls Anywhere and Lives Off Sun and Earth, *Popular Science,* April 1980.

Ozone, *Sciquest,* February 1980.

Ozone Depletion: Death to Shellfish? *Science News,* December 22, 1979.

Padgett, Joseph: Toxic Air, *EPA Journal,* August 1979.

Paerl, Hans: Nitrogen-Fixing Anabaena: Physiological Adaptations Instrumental in Maintaining Surface Blooms, *Science,* May 11, 1979.

Palmer, Tim: Don't Damn the Stanislaus, *National Parks and Conservation Magazine: The Environmental Journal,* March 1980.

Panofsky, Hans: Hello Ozone, Goodbye Smog, *Environment,* April 1978.

Parcells, Steven: How Long the Tallgrass . . ? *National Parks and Conservation,* April 1980.

Parisi, Anthony: Solar Power . . . Oil Crisis Boosts Grants for Growth, *Survival Rights— Journal of Environmental Debate,* March 1980.

Pearl, Richard: Earth's Active Crust, *Earth Science,* Fall 1979.

Pellenbarg, Robert: The Estuarine Surface Microlayer and Trace Metal Cycling in a Salt Marsh, *Science,* March 9, 1979.

Perfit, Michael: The Geology and Evolution of the Cayman Trench, *Geological Society of America Bulletin,* August 1978.

Peterson, Richard: An Australian Tornado, *Weatherwise,* October 1979.

Phillips, D.: A Record Cold Month in North America: February 1979, *Weatherwise,* February 1980.

Phung, H.: Soil Incorporation of Municipal Solid Wastes, *Public Works,* November 1977.

Pierce, Charles: Are Weather Forecasts Improving? *Weatherwise,* June 1976.

Pink Plankton: Ice Age Marker, *Science News,* September 1, 1979.

Pirone, P.: What Ails the London Plant? *Horticulture,* April 1978.

A Plague of Rain and Mud, *Newsweek,* March 3, 1980.

The Plains of Plenty, *Time,* January 21, 1980.

Posey, Carl: Heat Wave, *NOAA Magazine,* July 1979.

Post, Jonathan: Star Power for Supersocieties, *Omni,* March 1980.

Prospecting by Nodule, *Science News,* November 24, 1979.

Pryor, David: Beneath the Mississippi Delta, *Geographical,* January 1980.

Quirt, John: Socal Is Looking Homeward, *Fortune,* March 10, 1980.

Rampino, Michael: Can Rapid Climatic Change Cause Volcanic Eruptions? *Science,* November 16, 1979.

Ranwell, Derek: Washed by the Tides, *Oceans,* March 1980.

Ranwell, Derek: The Wildlife of the Wash, *Oceans,* March 1980.

Rea, David: Asymmetric Sea-Floor Spreading and Nontransform Axis Offset: The East Pacific Rise 20°S Survey Area, *Geological Society of America Bulletin,* July 1978.

Rearden, Jim: Caribou: Hardy Nomads of the North, *National Geographic,* December 1974.

Reardon, Susan, Cornbelt Conifers, *American Forest,* August 1977.

Reidel, Carl: The Forgotten Resource, *American Forest,* October 1979.

Renwick, William: Erosion Caused by Intense Rainfall in a Small Catchment in New York State, *Geology,* June 1977.

Reserve Mining Ends Dumping in Lake, *The New York Times,* March 18, 1980.

Reynolds, John: Icebergs Are a Frozen Asset, *Geographical,* December 1979.

Rhodes, Bob: The Pine Barrens: Can We Save the Ancient Forest? *The Environmental Journal,* November 1978.

Rich, Vera: Water Quality Threatens Polish Agriculture, *Nature,* July 26, 1979.

Roper, Margaret: Effect of Clay Particle Size on Clay Bacteriophage Interactions, *Journal of General Microbiology,* March 1978.

Rosen, Eric: Trees in Winter Wind and Ice: Miracle in Survival Methods, *Science Digest,* March 1980.

Rosen, Eric: White Continent, *Omni,* December 1979.

Rosenbaum, David: Coolant Spill Shuts Reactor in Florida, *The New York Times*, February 27, 1980.

Ross, David: The Red Sea: A New Ocean, *The Wilson Quarterly*, Winter 1980.

Roth, Louis: Cockroaches and Plants, *Horticulture*, August 1979.

Rubin, Hal: California Fights Toxic Wastes, *The Nation*, March 1, 1980.

Rummond, A. H.: Carbon Monoxide: The Hidden Threat, *Sciquest*, March 1980.

Sage, Bryan: Acid Drops from Fossil Fuels, *New Scientist*, March 6, 1980.

Sardar, Ziauddin: Pollution Threatens the Red Sea, *New Scientist*, March 6, 1980.

Sattar, Syed: Viral Pollution of Surface Waters Due to Chlorinated Primary Effluents, *Applied and Environmental Microbiology*, September 1978.

Savage, J.: Strain in Southern California: Measured Uniaxial North-South Regional Contraction, *Science*, November 24, 1978.

Schesgall, Oscar: Power Plants in the Sky, *Science Digest*, Spring 1980 (Special).

Schlee, John: Buried Continental Shelf Edge Beneath the Atlantic Continental Slope, *Oil and Gas Journal*, February 25, 1980.

Schneider, Eric: The Mexican Oil Gushes On and On, *The New York Times*, February 26, 1980.

Schumacher, Edward: Microwave Threat Stalls Trade Center TV Tower, *The New York Times*, February 26, 1980.

Science Plants an Oil Crop, *Newsweek*, March 24, 1980.

Sellers, A.: Clouds and the Long-Term Stability of the Earth's Atmosphere and Climate, *Nature*, June 28, 1979.

Selling the Power of Sun and Sand, *Nation's Business*, March 1980.

Settle, Dorothy: Lead in Albacore: Guide to Lead Pollution in Americans, *Science*, March 14, 1980.

Severo, Richard: Story of Safe Pesticide Ends as a Classic Case of Misuse: The Contamination of Long Island's Wells by Temik, *The New York Times*, March 4, 1980.

Shelly, Milton: Rainfall in Southern California, *Weatherwise*, June 1979.

Shelton, Marlyn: Drought to Flood: A Sudden Moisture Reversal in California, *Weatherwise*, March 11, 1980.

The Shifting Contour of the Strip Mine Law, *Business Week*, March 17, 1980.

Sigurdsson, Haraldur: An Active Submarine Volcano, *Natural History*, February 1980.

Silver Recovery Operations Become More Attractive in Soaring Price Market, *Graphic Arts Monthly*, February 1980.

Singh, Khushwant: Waiting for the Monsoon, *The New York Times*, August 26, 1973.

Singh, S.: The Oaxaca, Mexico, Earthquake of 29 November 1978: A Preliminary Report on Aftershocks, *Science*, March 14, 1980.

Smith, Jefferey: EPA Sets Rules on Hazardous Wastes, *Science*, March 14, 1980.

Smith, Robert: Yellowstone Park as a Window on the Earth's Interior, *Scientific American*, February 1980.

Snavely, Parke: Interpretation of Cenozoic Geologic History: Central Oregon Continental Margin, *Geological Society of America Bulletin*, March 1980.

Sobey, Ed: Ocean Ice, *Sea Frontiers*, March 1979.

Sofia, S.: Solar Constant: Constraints on Possible Variations from Solar Diameter Measurements, *Science*, June 22, 1979.

Solar Ultraviolet Radiation and Coral Reef Epifauna, *Science*, March 7, 1980.

South African Research Seeks to Lessen Impact of Seismic Events, *Engineering and Mining Journal*, January 1980.

Spaulding, James: Climate and Survival, *Soil Conservation*, July 1976.

Speth, Gus: The Sisyphus Syndrome: Acid Rain and the Public, *National Parks and Conservation*, February 1980.

Spurgeon, David: Agroforestry: New Hope for Subsistence Farmers, *Nature*, August 16, 1979.

Spurgeon, David: Full Steam Ahead to Save Kenya's Fuel, *Nature*, August 1979.

States Cooperate to Check Sea Pollution, *Nature*, February 28, 1980.

Steep, Clayton: What Is Happening to Our Weather, *Meteorology*, April 15, 1980.

Stepler, Richard: Solar Salts: New Chemical Systems Store the Sun's Heat, *Popular Science*, March 1980.

Stickney, Robert: Artificial Propagation of a Salt Marsh, *Sea Frontiers*, May 1979.

Stocker, Joseph: Birds in Transit, *The Elks Magazine,* March 1980.

Stokes, Henry: Japan Tries Using Ocean Waves as Power Source, *The New York Times,* March 4, 1980.

Stokes, Henry: Scaling Everest in a Search for a Legend, *The New York Times,* February 26, 1980.

Stone, Claudia: Hydrothermal Alteration of Basalts from Hawaii Geothermal Project Well-A, Kilauea, Hawaii, *Geology,* July 1978.

Stopping the Desert's March, *Nature,* May 31, 1979.

Strommen, Norton: Impact of Climate Variability on Global Grain Yields, *Environmental Data Information,* July 1978.

Strommen, Norton: NOAA—Yes-Maybe: Global Crop-Yield Modeling and Assessments, *Environmental Data and Information,* September 1978.

Stull, Robert: Effect of Off-Road Vehicles in Ballinger Canyon, California, *Geology,* January 1979.

Sugden, D.: A Case Against Deep Erosion of Shields by Ice Sheets, *Geology,* October 1976.

Sullivan, Walter: Earth's Ozone Layer Still Imperiled by Freon Gas, *The New York Times,* February 26, 1980.

Sullivan, Walter: Scientists Reviving Speculation on Climate and Slipping Antarctic Ice, *The New York Times,* March 9, 1980.

Sullivan, Walter: Stellar Explosions Now Believed Key to Formation of the Universe, *The New York Times,* February 17, 1980.

Sundquist, E.: The Homogeneous Buffer Factor: Carbon Dioxide in the Ocean Surface, *Science,* June 15, 1979.

Swanson, F.: Impact of Clear Cutting and Road Construction on Soil Erosion by Landslides in the Western Cascade Range, Oregon, *Geology,* July 1975.

Tanner, Henry: Urban Pollution Is Turning Glory That Was Rome to Dust, *The New York Times,* March 16, 1980.

Tehan, Rita: Greenwich Mean Time: Time the World Over, *Oceans,* September 1979.

Tension in a Vital Waterway: The Persian Gulf, *Life,* March 1980.

Thomas, Bill: The Last of the Bottomland Forests, *Living Wilderness,* July 1976.

Thomsen, Dietrick: Down among the Sheltering Pines, *Science News,* September 23, 1978.

Thousands Flee Waters in Flood in Palm Springs, *The New York Times,* February 22, 1980.

The Threat of Deforestation, *USA Today,* October 1979.

Three-Mile Island Casts a Pall on the Future, *U.S. News and World Report,* March 31, 1980.

Tjell, Jens: Atmospheric Lead Pollution of Grass Grown in a Background Area of Denmark, *Nature,* August 2, 1979.

Totemeier, Carl: Protecting Endangered Plant Species, *The New York Times,* March 16, 1980.

Toxic Wastes and Economic Incentives, *Resources,* January–April 1980.

Tucker, William: The Next American Dust Bowl and How To Avert It, *Atlantic Monthly,* April 1980.

Updating the Universe, *Sciquest,* January 1980.

Utilizing Species, *Salt Water Sportsman,* March 1980.

Valida, Ivan: The Nitrogen Budget of a Salt Marsh Ecosystem, *Nature,* August 23, 1979.

Verespy, Michael: Love Canal May Dump a New Role on Business, *Industry Week,* November 26, 1979.

Volcano Man, *Life,* March 1980.

Walker, Jearl, Stalking the Fossil Trilobite, Crinoid, and Seed Fern in Ohio, *Scientific American,* March 1980.

Walklet, Tracy: Industrial Pearls: Manganese Nodules, *Oceans,* November 1979.

Waring, R.: Evergreen Coniferous Forests of the Pacific Northwest, *Science,* June 29, 1979.

Water, *National Wildlife,* March 1980.

Weathermen for the Olympics, *Weatherwise,* January 1980.

Webster, Bayard: Plant Thermometer, *The New York Times,* March 11, 1980.

Webster, Bayard: U.S. Policy on Barrier Islands Often Hurts Them, *The New York Times*, March 18, 1980.

Wellemeyer, Marilyn: Rockhounds on the Prowl, *Fortune*, March 24, 1980.

Whipple, Fred: The Spin of Comets, *Scientific American*, March 1980.

White, Robert: Climate at the Millennium, *Environmental Data and Information*, May 1979.

Wilford, John: Promise of Energy Found in Subterranean Heat, *The New York Times*, March 25, 1980.

Wilford, John: Satellite to Map Earth's Magnetism, *The New York Times*, October 30, 1979.

Williams, Dennis: A Plague of Rain and Mud, *Newsweek*, March 3, 1980.

Wilson, A.: Geochemical Problems of the Antarctic Dry Areas, *Nature*, July 19, 1979.

Wood Famine in Developing Nations, *Science News*, February 24, 1979.

Woodwell, G.: The Flax Pond Ecosystem Study: Exchanges of Inorganic Nitrogen Between an Estuarine Marsh and Long Island Sound, *Ecology*, March 1979.

Wright, Penelope: Allison's Wind Engine, *Popular Science*, March 1980.

Young, G.: Middle and Late Proterozoic Evolution of the Northern Canadian Cordillera and Shield, *Geology*, March 1979.

Zahl, Paul: Portrait of a Fierce and Fragile Land, *National Geographic*, March 1972.

Zoback, Mark: Magnitude of Sheer Stress on the San Andreas Fault: Implications of a Stress Measurement Profile at Shallow Depth, *Science*, October 21, 1979.

Index

References to figures and photos are in italic type.
References to tables are followed by t.

X

Y

Z

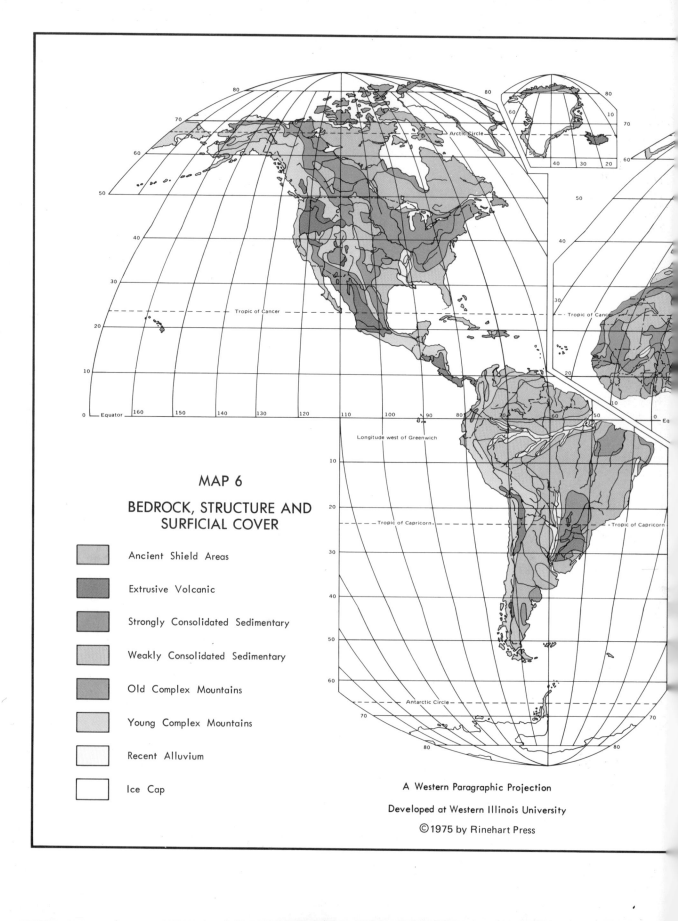

MAP 6

BEDROCK, STRUCTURE AND SURFICIAL COVER

Ancient Shield Areas

Extrusive Volcanic

Strongly Consolidated Sedimentary

Weakly Consolidated Sedimentary

Old Complex Mountains

Young Complex Mountains

Recent Alluvium

Ice Cap

A Western Paragraphic Projection

Developed at Western Illinois University

©1975 by Rinehart Press